EXOPLANETS: DETECTION, FORMATION & DYNAMICS

IAU SYMPOSIUM No. 249

COVER ILLUSTRATION: A two-planet system with a disk (Imagination)

A solar type proto-star was surrounded by its nascent gas disk and two protoplanets. The inner protoplanet whose mass is of 1 Jupiter mass has already open a clear gap and is undergoing slow type II migration. While, being affected by it, the outer protoplanet whose mass is 1 Saturn mass can not clear its horseshoe region and its migration is unsteady.

The orbit evolution of a multiple-planets-disk system is more complex than a single-planet-disk system. It is believed that protoplanets embedded in a disk will undergo migration induced by the gas disk and end up in mean motion resonace with other protoplanets.

This figure is an imagination comeing from the work of H. Zhang & J.-L. Zhou in this proceeding(p. 413). During their hydro-dynamics simulation the two planets first undergo convergent migration and locked in 5:3 MMR for a short while. Then the outer planet's migration reversed and it undergoes outward runaway migration.

IAU SYMPOSIUM PROCEEDINGS SERIES

2008 EDITORIAL BOARD

Chairman

I.F. CORBETT, IAU Assistant General Secretary
European Southern Observatory
Karel-Schwarzschild-Strasse 2
D-85748 Garching-bei-München
Germany
icorbett@eso.org

Advisers

K.A. VAN DER HUCHT, IAU General Secretary,
SRON Netherlands Institute for Space Research, Utrecht, the Netherlands
E.J. DE GEUS, *Dynamic Systems Intelligence B.V., Assen, the Netherlands*
U. GROTHKOPF, *European Southern Observatory, Germany*
M.C. STOREY, *Australia Telescope National Facility, Australia*

Members

IAUS251
SUN KWOK, *Faculty of Science, University of Hong Kong, Hong Kong, China*
IAUS252
LICAI DENG, *National Astronomical Observatories, Chinese Academy of Sciences, Beijing, China*
IAUS253
FREDERIC PONT, *Geneva Observatory, Sauverny, Switzerland*
IAUS254
JOHANNES ANDERSEN, *Astronomical Observatory, Niels Bohr Institute, Copenhagen University, Denmark*
IAUS255
LESLIE HUNT, *INAF - Istituto di Radioastronomia, Firenze, Italy*
IAUS256
JACOBUS Th. van LOON, *Astrophysics Group, Lennard-Jones Laboratories, Keele University, Staffordshire, UK*
IAUS257
NATCHIMUTHUK GOPALSWAMY, *Solar Systems Exploration Div., NASA Goddard Space Flight Center, MD, USA*
IAUS258
ERIC E. MAMAJEK, *Radio and Geoastronomy Division, Harvard Smithsonian CfA, Cambridge, MA, USA*
IAUS259
KLAUS G. STRASSMEIER, *Astrophysics Institute Potsdam, Potsdam, Germany*

INTERNATIONAL ASTRONOMICAL UNION
UNION ASTRONOMIQUE INTERNATIONALE

Exoplanets: Detection, Formation & Dynamics

PROCEEDINGS OF THE 249th SYMPOSIUM OF THE
INTERNATIONAL ASTRONOMICAL UNION
HELD IN SUZHOU, CHINA
OCTOBER 22–26, 2007

Edited by

Yi-Sui Sun
Nanjing University, Nanjing, China

Sylvio Ferraz-Mello
University of São Paulo, São Paulo, Brazil

Ji-Lin Zhou
Nanjing University, Nanjing, China

CAMBRIDGE UNIVERSITY PRESS
The Edinburgh Building, Cambridge CB2 2RU, United Kingdom
32 Avenue of the Americas, New York, NY 10013-2473, USA
477 Williamstown Road, Port Melbourne, VIC 3207, Australia
Ruiz de Alarcón 13, 28014 Madrid, Spain
Dock House, The Waterfront, Cape Town 8001, South Africa

© International Astronomical Union 2008

This book is in copyright. Subject to statutory exception
and to the provisions of relevant collective licensing agreements,
no reproduction of any part may take place without
the written permission of the International Astronomical Union.

First published 2008

Printed in the United Kingdom at the University Press, Cambridge

Typeset in System LaTeX 2_ε

A catalogue record for this book is available from the British Library

Library of Congress Cataloguing in Publication data

ISBN 9780521874717 hardback
ISSN 1743-9213

Table of Contents

Preface xi

Organizing committee xii

Conference photograph xiii

Conference participants xiv

Part 1. EXOPLANETS DETECTION

CoRoT: pioneer space mission for exoplanet transit search 3
    P. Barge, A. Baglin, M. Auvergne & CoRoT team

Finding Earth-size planets in the habitable zone: the *Kepler Mission* 17
    W. Borucki, D. Koch, G. Basri, N. Batalha, T. Brown, D. Caldwell, J.
    Christensen-Dalsgaard, W. Cochran, E. Dunham, T. N. Gautier, J. Geary,
    R. Gilliland, J. Jenkins, Y. Kondo, D. Latham, J. J. Lissauer & D. Monet

Extrasolar planet detections with gravitational microlensing 25
    S. D. Mao, E. Kerins & N. J. Rattenbury

Microlensing search for extrasolar planets: observational strategy, discoveries and
    implications 31
    A. Cassan, T. Sumi & D. Kubas

ARTEMiS (Automated robotic terrestrial exoplanet microlensing search) –
    Hunting for planets of Earth mass and below 35
    M. Dominik, K. Horne, A. Allan, N. J. Rattenbury, Y. Tsapras, C.
    Snodgrass, M. F. Bode, M. J. Burgdorf, S. N. Fraser, E. Kerins,
    C. J. Mottram, I. A. Steele, R. A. Street, P. J. Wheatley & Ł. Wyrzykowski

A HET search for planets around evolved stars 43
    A. Niedzielski & A. Wolszczan

The PSU/TCfA search for planets around evolved stars. Stellar parameters and
    activity indicators of targets 49
    A. Niedzielski, G. Nowak & P. Zieliński

A Korea-Japan planet search program: Current status and discovery of a brown
    dwarf candidate 53
    M. Omiya, H. Izumiura, B. Sato, M. Yoshida, E. Kambe, E. Toyota,
    S. Urakawa, S. Masuda, M. Takada-Hidai, I. Han, K.-M. Kim, B.-C. Lee &
    T. S. Yoon

Search for extrasolar planets with high-precision relative astrometry by ground-
    based and single-aperture observations 57
    T. Roell, A. Seifahrt & R. Neuhäuser

Preparing the exoplanet search with PRIMA: Searching for reference stars and
    target characterization 61
    R. Geisler, J. Setiawan, Th. Henning, D. Queloz, A. Quirrenbach,
    R. Launhardt, A. Müller, S. Reffert, P. Weise & ESPRI consortium

Finding new sub-stellar co-moving companion candidates - the case of CT Cha . . 65
  T. Schmidt & R. Neuhäuser

Closure phase studies toward direct detection of light from hot Jupiters 71
  M. Zhao, J. D. Monnier, T. T. Brummelaar, E. Pedretti & N. Thureau

First observation of planet-induced X-ray emission: The system HD 179949 79
  S. H. Saar, M. Cuntz, V. L. Kashyap & J. C. Hall

Photometric follow-up observation of some SuperWASP transiting planet candidates 83
  S. -H. Gu, A. C. Cameron, X. -B. Wang, L. -Y. Zhang, X. -S. Fang & X. -J. Li

Photometric observation of the transiting exoplanet WASP-1b 85
  X. -B. Wang, A. C. Cameron, S. -H. Gu & L. -Y. Zhang

Reconstruction of the transit signal in the presence of stellar variability 89
  A. Alapini & S. Aigrain

Observational window functions in planet transit searches 93
  K. von Braun & D. R. Ciardi

Several problems of exoplanetary orbits determination from radial velocity observations 101
  R. V. Baluev

Selection effects in Doppler velocity planet searches 111
  S. O'Toole, C. Tinney & H. Jones

Cadence optimisation and exoplanetary parameter sensitivity 115
  S. R. Kane, E. B. Ford & J. Ge

The astrometric data reduction software (ADRS) and error budget for PRIMA . 119
  N. M. Elias II, R. N. Tubbs, R. Köhler, S. Reffert, I. Stilz, R. Launhardt, J. de Jong, A. Quirrenbach, F. Delplancke, Th. Henning, D. Queloz & ESPRI Consortium

Unveiling exoplanet families 123
  S. Marchi, S. Ortolani

Part 2. PHYSICS OF ATMOSPHERES AND CLOSE-IN PLANETS

Diversity of close-in planets and the interactions with their host stars 131
  D. N. C. Lin & I. Dobbs-Dixon

Internal waves driven by stellar irradiation in a non-synchronized hot Jupiter . . 145
  P. -G. Gu & G. I. Ogilvie

The On/off nature of star-planet interactions in the HD 179949 and υ And systems 151
  E. Shkolnik, D. A. Bohlender, G. A. H. Walker & A. C. Cameron

Thermal evolution and magnetism of terrestrial planets 159
  C. Tachinami, H. Senshu & S. Ida

On uncertainty of Jupiter's core mass due to observational errors 163
    Y. Hori, T. Sano, M. Ikoma & S. Ida

Silicate, ruby, opal – Why gas giants keep their jewels in the atmosphere 167
    Ch. Helling

Comparison of cloud models for Brown Dwarfs. 173
    Ch. Helling, A. Ackerman, F. Allard, M. Dehn, P. Hauschildt, D. Homeier,
    K. Lodders, M. Marley, F. Rietmeijer, T. Tsuji & P. Woitke

Tidal friction in close-in planets 179
    A. Rodríguez, S. Ferraz-Mello & H. Hussmann

Tidal evolution of close-in extra-solar planets 187
    B. Jackson, R. Greenberg & R. Barnes

Modeling of evolution of the rotational axis of "hot Jupiter" planets under tidal
    perturbations 197
    I. Kitiashvili

Astrobiological effects of F, G, K and M main-sequence stars 203
    M. Cuntz, L. Gurdemir, E. F. Guinan & R. L. Kurucz

Part 3. PLANET FORMATION

Testing planet formation theories with Giant stars 209
    L. Pasquini, M. P. Döllinger, A. Hatzes, J. Setiawan, L. Girardi, L. da
    Silva, J. R. de Medeiros & A. Weiss

Orbital migration and mass-semimajor axis distributions of extrasolar planets . . 223
    S. Ida & D. N. C. Lin

Terrestrial planet formation in extra-solar planetary System 233
    S. N. Raymond

Planet formation in binary stars 251
    W. Kley

Observational tests of planet formation models 261
    A. Sozzetti, D. W. Latham, G. Torres, B. W. Carney, J. B. Laird,
    R. P. Stefanik, A. P. Boss, D. Charbonneau, F. T. ODonovan, M. J.
    Holman & J. N. Winn

Tidal barrier and the asymptotic mass of proto gas-giant planets 263
    I. Dobbs-Dixon, S. L. Li & D. N. C. Lin

Formation of heavy element rich giant planets by giant impacts 267
    H. Genda, M. Ikoma, T. Guillot & S. Ida

SPH simulations of star/planet formation triggered by cloud-cloud collisions . . . 271
    S. Kitsionas, A. P. Whitworth & R. S. Klessen

The formation of close-in planets by the slingshot model 279
    M. Nagasawa, S. Ida & T. Bessho

Migration and final location of hot super-Earths in the presence of gas giants .. 285
    J. -L. Zhou & D. N. C. Lin

Planet formation around intermediate mass stars............................ 293
    K. A. Kretke, D. N. C. Lin & N. J. Turner

Giant impact, planetary merger, and diversity of planetary-core mass......... 301
    S. -L. Li, C. Agnor & D. N. C. Lin

Formation of terrestrial planets from planetesimals around M dwarfs 305
    M. Ogihara & S. Ida

Retention of protoplanetary cores near the snowline 309
    X. J. Zhang, K. A. Kretke & D. N. C. Lin

Formation and detectability of Earth-like planets around Alpha-Centauri B.... 313
    E. Davis

Habitable planet formation in extreme planetary systems: systems with multiple stars and/or multiple planets 319
    N. Haghighipour

On the formation age of the first planetary system......................... 325
    T. Hara, S. Kunitomo, M. Shigeyasu & D. Kajiura

Part 4. PROTOPLANETARY DISKS AND MIGRATION

Planetary migration in gaseous protoplanetary disks 331
    F. S. Masset

On the solar system–debris disk connection............................... 347
    A. Moro-Martín

Disc signatures in a new population of low mass YSOs in ρ Ophiuchi 355
    C. A. de Oliveira & M. Casali

Searching for H_2 emission from protoplanetary disks using near- and mid-infrared high-resolution spectroscopy 359
    A. Carmona, M. E. van den Ancker, Th. Henning, Ya. Pavlyuchenkov, C. P. Dullemond, M. Goto, D. Fedele, B. Stecklum, W. F-. Thi, J. Bouwman & L. B. F. M. Waters

Astromineralogy of protoplanetary disks 369
    O. Schütz, G. Meeus, M. F. Sterzik & E. Peeters

Dust evolution in protoplanetary disks 375
    J. -F. Gonzalez, L. Fouchet, S. T. Maddison & G. Laibe

Origin of the dusty disks around white dwarfs 381
    R. B. Dong, Y. Wang, D. N. C. Lin & X. W. Liu

3D SPH simulations of grain growth in protoplanetary disks.................. 385
    G. Laibe, J. -F. Gonzalez, L. Fouchet & S. T. Maddison

Origin of debris disks and the supply of metals in DZ white dwarfs 389
  Y. Wang, R. B. Dong, D. N. C. Lin & X. W. Liu

Type I planetary migration in a self-gravitating disk 393
  C. Baruteau & F. S. Masset

On type I planetary migration in adiabatic disks 397
  C. Baruteau & F. S. Masset

The effect of poloidal magnetic field on type I planetary migration 401
  T. Muto, M. N. Machida & S. -I Inutsuka

Baroclinic generation of potential vorticity in an embedded planet-disk system . 407
  J. H. Ji, S. L. Ou & L. Liu

Runaway migration in a multiple-protoplanet system...................... 413
  H. Zhang & J. -L Zhou

Effects of dissipating gas drag on planetesimal accretion in binary systems..... 419
  J. -W. Xie & J. -L. Zhou

Part 5. DYNAMICS OF MULTIPLE EXOPLANET SYSTEMS

Orbital determination and dynamics of resonant extrasolar planetary systems .. 427
  C. Beaugé, S. Ferraz-Mello, T. A. Michtchenko & C. A. Giuppone

Dynamics and instabilities in exoplanetary systems 441
  E. B. Ford

Stability constraints in modeling of multi-planet extrasolar systems............ 447
  K. Goździewski, C. Migaszewski & A. Musieliński

On the dynamics of Trojan Planets in extrasolar planetary systems............ 461
  R. Dvorak, R. Schwarz & Ch. Lhotka

Extrasolar planet interactions... 469
  R. Barnes & R. Greenberg

Secular evolution of exoplanetary systems and close encounters 479
  M. Šidlichovský & E. Gerlach

Formation and transformation of the 3:1 mean-motion resonance in 55 Cancri System 485
  L. -Y. Zhou, S. Ferraz-Mello & Y. -S. Sun

Analysis of near-separatrix motion in planetary systems 491
  S. Wang & J. -L. Zhou

Habitable zones for Earth-mass planets in multiple planetary systems.......... 499
  J. H. Ji, L. Liu, H. Kinoshita & G. Y. Li

Habitability of super-Earths: Gliese 581c & 581d 503
  W. von Bloh, C. Bounama, M. Cuntz & S. Franck

Orbital stability of planets in binary systems: A new look at old results 507
   J. Eberle, M. Cuntz & Z. E. Musielak

Retrograde resonances in compact multi-planetary systems:
   a feasible stabilizing mechanism 511
   J. Gayon & E. Bois

Author index 517

Object index 521

Preface

This IAU Symposium No. 249, Exoplanets: Detection, Formation and Dynamics held in Suzhou, by invitation of the Nanjing University and Soochow University, was an initiative of the IAU Commission 7 (Celestial Mechanics and Dynamical Astronomy). It was centered on a simple idea. We have accumulated 10 years of observations of planets around MS stars and many disciplines evolved to search for the understanding of these new objects of the skies. It was time to meet and discuss the Physics and Dynamics of the exoplanets together with the new techniques and the most recent discoveries. In the last years, the subject has grown in many directions in both theory and observation, so much that it would be difficult to describe it in a few words. To assess it, the better is to look at the table of contents of this book with its 5 sections covering detection, orbits, dynamics, physics, formation, migration, evolution, disks, habitability, etc.

The success of the chosen program may be assessed from the fact that the program of the oral sessions was realized exactly as scheduled, without any absence or delay, and by the fact that the sessions room was full all the time, from the first to the last day. The international community of astronomers showed in this way how they were glad in meeting together and increasing their contacts.

We are glad that this Symposium was hosted by Chinese astronomers, a community with an increasing participation in the world scenario of Astronomy and Astrophysics. Researching results and planned projects presented in the Symposium by Chinese astronomers indicated that China is actively involved in the exoplanet field.

We would like to thank the support of the International Astronomical Union. The organization of IAU Symposium No. 249 was supported by many national agencies and institutions and special thanks are addressed to Nanjing University, Soochow University, National Natural Science Foundation of China and Chinese Astronomical Society.

This volume includes invited lectures and communications presented at IAUS 249. The oral and poster contributions were so uniformly good, that we do not make any distinction in these proceedings. More than 90 percent of the 77 contributions to the proceedings were reviewed despite of the limited editing time. We thank all lecturers and SOC members for the help that we received from them in the hard tasks of selecting referees. We thank all colleagues, participants and non-participants who acted as referees, for their fast response and for the many suggestions helping to improve the contents and the completeness of the papers.

Last but not least, we have to thank the members of the organizing committees: the members of the LOC for the large amount of work they have done in order to have this meeting organized properly and the members of the SOC for their active participation in the organization of the scientific program.

Sylvio Ferraz-Mello and Yi-Sui Sun, co-chairs SOC
Ji-Lin Zhou, chair LOC
Nanjing and São Paulo, January 31st, 2008

THE ORGANIZING COMMITTEE

Scientific

Yi-Sui Sun (co-chair, China)
F. Allard (France)
A.P. Boss (USA)
J.D. Hadjidemetriou (Greece)
R. Malhotra (USA)
M. Mayor (Switzerland)
K. Meech (USA)
R.P. Nelson (UK)

S. Ferraz-Mello (co-chair, Brasil)
P. Barge (France)
R. Dvorak (Austria)
D.N.C. Lin (USA)
G.W. Marcy (USA)
V.S. Meadows (USA)
A. Milani (Italy)
P. Sackett (Australia)

Local

Ji-Lin Zhou (chair)
Guo-Xuan Dong
Wen-Xin Qin
Hui Zhang

Yong-Luo Cao
Jian Li
Jian-Zhen Wang
Li-Yong Zhou

Acknowledgements

The proposal is supported by Division I: Fundamental Astronomy, Commission 7: Celestial Mechanics and Dynamical Astronomy, Commission 52: Exoplanets, Commission 16: Physical Study of Planets and Satellites, and Commission 51: Bioastronomy: Search for Extraterrestrial Life.

The Local Organizing Committee operated under the auspices of
Nanjing University and Soochow University.

Funding by the
International Astronomical Union,
National Natural Science Foundation of China,
is gratefully acknowledged.

IAU Symposium No. 249
中国苏州 2007.10.24

Participants

Aude **Alapini**, Astrophysics - School of Physics, University of Exeter, Exeter, UK — alapini@astro.ex.ac.uk
France **Allard**, Astrophysique, École Normale Supérieure de Lyon, Lyon, France — fallard@ens-lyon.fr
Catarina **Alves De Oliveira**, European Southern Observatory, Garching bei Muenchen, Germany — coliveir@eso.org
Alexandre **Andrei**, SYRTE, Observatoire de Paris, Paris, France — oat1@on.br
Roman **Baluev**, Saint Petersburg State University, Saint Petersburg, Russian Federation — roman@astro.spbu.ru
Pierre **Barge**, Laboratoire d'Astrophysique de Marseille, Marseille, France — pierre.barge@oamp.fr
Rory **Barnes**, Lunar and Planetary lab, University of Arizona, Tucson, AZ, USA — rory@lpl.arizona.edu
Clement **Baruteau**, DSM/DAPNIA/SAp, CEA Saclay, Gif-sur-Yvette, France — clement.baruteau@cea.fr
Cristian **Beauge**, Observatorio Astronomico, Universidad Nacional de Cordoba, Cordoba, Argentina — beauge@oac.uncor.edu
David **Bennett**, Physics Department, University of Notre Dame, Notre Dame, IN, USA — bennett@nd.edu
Carlo **Blanco**, Dipartimento di Fisica e Astronomia, Università di Catania, Catania, Italy — cblanco@oact.inaf.it
Eric **Bois**, Observatoire de la Côte d Azur, Nice, France — Eric.Bois@oca.eu
Xavier **Bonfils**, Centro de Astronomia e Astrofísica, University de Lisboa, Lisboa, Portugal — xavier.bonfils@oal.ul.pt
Herve **Bouy**, Astronomy Department, UC Berkeley, Berkeley, CA, USA — hbouy@astro.berkeley.edu
Andres **Carmona Gonzalez**, Max-Planck Institute for Astronomy, Heidelberg, Germany — carmona@mpia.de
Arnaud **CASSAN**, ARI, Center for Astronomy, Univ. of Heidelberg, Heidelberg, Germany — cassan@ari.uni-heidelberg.de
Carolina **Chavero**, Astronomy, Instituto de Astrofisica de Canarias, La Laguna - Tenerife, Spain — carolina@on.br
David **Ciardi**, Michelson Science Center, Caltech, Pasadena, CA, USA — ciardi@ipac.caltech.edu
Manfred **Cuntz**, Department of Physics, University of Texas at Arlington, Arlington, TX, USA — cuntz@uta.edu
Erica **Davis**, Department of Astronomy, University of California, Santa Cruz, CA, USA — ericadavis@gmail.com
Ian **Dobbs-Dixon**, Physics Department, McGill University, Montreal, Canada — iandd@physics.mcgill.ca
Ruobing **Dong**, Physics Department, Peking University, Beijing, China — rbdong@pku.edu.cn
Rudolf **Dvorak**, Institute of Astronomy, University of Vienna, Vienna, Austria — dvorak@astro.univie.ac.at
Sylvio **Ferraz-Mello**, IAG, University of São Paulo, São Paulo, Brasil — sylvio@astro.iag.usp.br
Scott **Fleming**, Department. of Astronomy, University of Florida, Gainesville, FL, USA — scfleming@astro.ufl.edu
Eric **Ford**, Department of Astronomy, University of Florida, Gainesville, FL, USA — eford@astro.ufl.edu
Julie **Gayon**, Observatoire de la Côte d Azur, Nice, France — gayon@obs-nice.fr
Jian **Ge**, Department of Astronomy, University of Florida, Gainesville, FL, USA — jge@astro.ufl.edu
Ronny **Geisler**, ZAH - Landessternwarte Koenigstuhl, Heidelberg, Germany — rgeisler@lsw.uni-heidelberg.de
Hidenori **Genda**, Earth and Planetary Sciences, Tokyo Institute of Technology, Tokyo, Japan — genda@geo.titech.ac.jp
Enrico **Gerlach**, Lohrmann-Observatory, Technical University Dresden, Dresden, Germany — enrico.gerlach@tu-dresden.de
Jean-François **Gonzalez**, Astrophysique, École Normale Supérieure de Lyon, Lyon, France — Jean-Francois.Gonzalez@ens-lyon.fr
Krzysztof **Gozdziewski**, Torun Centre, Nicolaus Copernicus Univ., Torun, Poland — k.gozdziewski@astri.uni.torun.pl
Carl **Grillmair**, Spitzer Science Center, Pasadena, CA, USA — carl@ipac.caltech.edu
Pin-Gao **Gu**, Institute of Astronomy & Astrophysics, Academia Sinica, Taipei, China — gu@asiaa.sinica.edu.tw
Sheng-hong **Gu**, Stellar physics, Yunnan Observatory, Kunming, China — shenghonggu@ynao.ac.cn
Nader **Haghighipour**, Institute for Astronomy, University of Hawaii, Honolulu, HI, USA — nader@ifa.hawaii.edu
Tetsuya **Hara**, Physics, Kyoto Sangyo University, Kyoto, Japan — hara@cc.kyoto-su.ac.jp
Leslie **Hebb**, School of Physics & Astronomy, Univ. of St Andrews, St Andrews, UK — lh61@st-andrews.ac.uk
Christiane **Helling**, School of Physics & Astronomy, Univ. St Andrews, St Andrews, UK — Christiane.Helling@st-and.ac.uk
Yasunori **Hori**, Earth and Planetary Sciences, Tokyo Institute of Technology, Tokyo, Japan — hori@geo.titech.ac.jp
Andrew **Howard**, Astronomy Department, University of California, Berkeley, CA, USA — andrew@alum.mit.edu
Bo **Hu**, High-Energy Astrophysics Group, Yunnan Observatory, Kunming, China — lijeter@126.com
Shigeru **Ida**, Earth & Planetary Science, Tokyo Institute of Technology, Tokyo, Japan — ida@geo.titech.ac.jp
Brian **Jackson**, Lunar and Planetary Laboratory, University of Arizona, Tucson, USA — bjackson@lpl.arizona.edu
Markus **Janson**, Planet and Star Formation, MPIA Heidelberg, Heidelberg, Germany — janson@mpia.de
Jianghui **Ji**, Division of Planetary Sciences, Purple Mountain Observatory, Nanjing, China — jijh@pmo.ac.cn
Liping **Jin**, Physics, Jilin University, Changchun, China — jinlp@jlu.edu.cn
Stephen **Kane**, Michelson Science Center, Caltech, Pasadena, CA, USA — skane@astro.ufl.edu
Mariko **Kato**, Earth and Planetary Sciences, Tokyo Institute of Technology, Tokyo, Japan — marikok@geo.titech.ac.jp
Irina **Kitiashvili**, Astronomy, Kazan State University, Kazan, Russia — Irina.Kitiashvili@ksu.ru
Spyridon **Kitsionas**, Astrophysikalisches Institut Potsdam, Potsdam, Germany — skitsionas@aip.de
Wilhelm **Kley**, Astronomy & Astrophysics, University of Tuebingen, Tuebingen, Germany — wilhelm.kley@uni-tuebingen.de
David **Koch**, NASA Ames Research Center, Moffett Field, CA, USA — d.koch@nasa.gov
Eiichiro **Kokubo**, Division of Theoretical Astron., National Astronomical Observatory, Tokyo, Japan — kokubo@th.nao.ac.jp
Katherine **Kretke**, Astronomy Department, University of California, Santa Cruz, CA, USA — kretke@ucolick.org
Daniel **Kubas**, European Southern Observatory, Santiago, Chile — dkubas@eso.org
Guillaume **Laibe**, Astrophysique, École Normale Supérieure de Lyon, Lyon, France — guillaume.laibe@ens-lyon.fr
Ralf **Launhardt**, Max Planck Intstitute for Astronomy, Heidelberg, Germany — rl@mpia.de
Man Hoi **Lee**, Department of Physics, University of California, Santa Barbara, CA, USA — mhlee@physics.ucsb.edu
Jian **Li**, Department of Astronomy, Nanjing University, Nanjing, China — ljian@nju.edu.cn
Shulin **Li**, Department of Astronomy, Peking University, Beijing, China — lisl.shulin@gmail.com
Xiang-Dong **Li**, Department of Astronomy, Nanjing University, Nanjing, China — lixd@nju.edu.cn
Douglas **Lin**, Kavli Institute of Astronomy & Astrophysics, Peking University, Beijing, China — lin@ucolick.org
Yujuan **Liu**, National Astronomical Observatories, Beijing, China — lyj@bao.ac.cn
Wladimir **Lyra**, Uppsala Astronomical Observatory, Uppsala University, Uppsala, Sweden — wlyra@astro.uu.se
Shude **Mao**, Jodrell Bank Observatory, University of Manchester, Macclesfiled, UK — shude.mao@manchester.ac.uk
Simone **Marchi**, Department of Astronomy, University of Padova, Padova, Italy — simone.marchi@unipd.it
Geoffrey **Marcy**, Astronomy, University of California, Berkeley, CA, USA — gmarcy@astro.berkeley.edu
Frederic **Masset**, Service d'Astrophysique - bat. 709, CE-Saclay, Gif/Yvette, France — fmasset@cea.fr
Michel **MAYOR**, Observatoire de Geneve, SAUVERNY, Switzerland — Michel.mayor@obs.unige.ch
John **Monnier**, Astronomy, University of Michigan, Ann Arbor, MI, USA — monnier@umich.edu
Ryan **Montgomery**, Astronomy and Astrophysics, University of California, Santa Cruz, CA, USA — rmontgom@ucolick.org
Amaya **Moro-Martin**, Astrophysical Sciences, Princeton University, Princeton NJ, USA — amaya@astro.princeton.edu
Takayuki **Muto**, Theoretical Astrophysics Group, Physics, Kyoto University, Kyoto, Japan — muto@tap.scphys.kyoto-u.ac.jp
Makiko **Nagasawa**, Global Edge Institute, Tokyo Institute of Technology, Tokyo, Japan — nagaswmk@geo.titech.ac.jp

Participants

Andrzej **Niedzielski**, Astronomy & Astrophysics, Nicolaus Copernicus University, Torun, Poland — aniedzi@astri.uni.torun.pl
Alberto **Noriega-Crespo**, Spitzer Science Center, Caltech, USA — alberto@ipacocaltech.edu
Robert **Noyes**, Solar, Stellar, and Planetary Division, Center for Astrophysics, Cambridge, MA, USA — noyes@cfa.harvard.edu
Masahiro **Ogihara**, Earth and Planetary Sciences, Tokyo Institute of Technology, Tokyo, Japan — ogihara@geo.titech.ac.jp
Masashi **Omiya**, Physics, Tokai University, Hiratuka,Kanagawa, Japan — ohmiya@peacock.rh.u-tokai.ac.jp
Simon **O'Toole**, Anglo-Australian Observatory, Epping, Australia — otoole@aao.gov.au
Luca **Pasquini**, ESO, Garching bei Muenchen, Germany — lpasquin@eso.org
Peter **Plavchan**, Michelson Science Center, Caltech, Pasadena, CA, USA — plavchan@ipac.caltech.edu
Dionyssia **Psychoyos**, Dept. of Physics, Aristotles Univ. of Thessaloniki, Thessaloniki, Greece — dpsyc@physics.auth.gr
Andreas **Quirrenbach**, Landessternwarte, Heidelberg University, Heidelberg, Germany — A.Quirrenbach@lsw.uni-heidelberg.de
Sohrab **Rahvar**, Physics Department, Sharif University of Technology, Tehran, Iran — rahvar@sharif.edu
Sean **Raymond**, CASA, University of Colorado, Boulder, USA — raymond@lasp.colorado.edu
Samuel **Regandell**, Department of Astronomy & Space physics, Uppsala University, Uppsala, Sweden — samreg@astro.uu.se
Jose **Robles**, Astronomy & Astrophysics, Australian National University, Weston Creek, Australia — josan@mso.anu.edu.au
Tristan **Roell**, Astrophysical Institute AIU, Friedrich-Schiller-Universität, Jena, Germany — troell@astro.uni-jena.de
Barbara **Rojas**, Astronomy Department, Cornell University, Ithaca, NY, USA — babs@astro.cornell.edu
Johannes **Sahlmann**, ESO, Garching bei Muenchen, Germany — jsahlman@eso.org
Kailash **Sahu**, Space Telescope Science Institute, Baltimore, MD, USA — ksahu@stsci.edu
Tobias **Schmidt**, Astrophysical Institute AIU, Friedrich-Schiller-Universität, Jena, Germany — tobi@astro.uni-jena.de
Oliver **Schuetz**, European Southern Observatory, Casilla, Chile — oschuetz@eso.org
Richard **Schwarz**, Astronomy, Eötvös Lorand, Pazmany Peter setany 1/A, Hungary — schwarz@astro.univie.ac.at
Damien **Segransan**, Observatoire de Genève, Université de Genèe, Versoix, Switzerland — Damien.Segransan@obs.unige.ch
Johny **Setiawan**, Planet and Star Formation, Max-Planck-Institut für Astronomie, Heidelberg, Germany — setiawan@mpia.de
Hongguang **Shan**, Yunnan Astronomical Observatory, CAS, Kunming, China — shg@ynao.ac.cn
Evgenya **Shkolnik**, Institute for Astronomy, University of Hawaii at Manoa,Honolulu, Hawaii, USA — shkolnik@ifa.hawaii.edu
Milos **Sidlichovsky**, Astronomical Institute, Academy of Sciences of the Czech Republic, Prague, Czech Republic — sidli@ig.cas.cz
Alessandro **Sozzetti**, Harvard-Smithsonian Center for Astrophysics, Cambridge, MA, USA — asozzetti@cfa.harvard.edu
Ning **Sui**, Department of Physics, Jilin University, Changchun, China — suining@email.jlu.edu.cn
Takahiro **Sumi**, Solar-Terrestrial Environment Laboratory, Nagoya University, Nagoya , Japan — sumi@stelab.nagoya-u.ac.jp
Yi-Sui **Sun**, Department of Astronomy, Nanjing University, Nanjing, China — sunys@nju.edu.cn
Chihiro **Tachinami**, Earth and Planetary Science, Tokyo Institute of Technology, Tokyo, Japan — ctchnm@geo.titech.ac.jp
Genya **Takeda**, Dept. of Physics and Astronomy, Northwestern Univ., Evanston, Illinois, USA — g-takeda@northwestern.edu
Stuart **Taylor**, 4059 Via Zorro B, Santa Barbara, USA — astrostuart@gmail.com
Robert **Tubbs**, Max Planck Institute for Astronomy, Heidelberg, Germany — tubbs@mpia-hd.mpg.de
Gerrit **Van Der Plas**, Office for Science, ESO, Garching bei Muenchen, Germany — gvanderp@eso.org
Charalampos **Varvoglis**, Depertment of Physics, University of Thessaloniki, Thessaloniki, Greece — varvogli@physics.auth.gr
Kaspar **Von Braun**, Michelson Science Center, California Institute of Technology, Pasadena, CA, USA — kaspar@ipac.caltech.edu
Xiaosheng **Wan**, Department of Astronomy, Nanjing University, Nanjing, China — xswan@nju.edu.cn
Liang **Wang**, National Astronomical Observatories, CAS, Beijing, China — wangliang@bao.ac.cn
Su **Wang**, Department of Astronomy, Nanjing University, Nanjing, China — suwang@nju.edu.cn
Tinggui **Wang**, Dept. of Astron. and Applied Physics, Univ. of Science and Technology of China, Hefei, China — twang@ustc.edu.cn
Xinming **Wang**, Department of Physics, Jilin University, Changchun, China — jluwxm@yahoo.com.cn
Yan **Wang**, Department of Astronomy, Peking University, Beijing, China — yanwang@pku.edu.cn
Zihua **Weng**, School of Physics and Mechanical & Electrical Engineering, Xiamen Univ., Xiamen, China — xmuwzh@xmu.edu.cn
Jason **Wright**, Department of Astronomy, University of California, Berkeley, CA, USA — jtwright@astro.berkeley.edu
Ji-Wei **Xie**, Department of Astronomy, Nanjing University, Nanjing, China — xjw0809@163.com
Weimin **Yuan**, Yunnan Astronomical Observatory, CAS, Kunming, China — wmy@ynao.ac.cn
Hui **Zhang**, Department of Astronomy, Nanjing University, Nanjing, China — huizhang@nju.edu.cn
Ke **Zhang**, Department of Astronomy, University of Maryland, College Park, USA — kzh@astro.umd.edu
Xiaojia **Zhang**, Department of Astronomy, Peking University, Beijing, China — xiaojia17@gmail.com
Ming **Zhao**, Astronomy Department, University of Michigan, Ann Arbor, MI, USA — mingzhao@umich.edu
Ji-Lin **Zhou**, Department of Astronomy, Nanjing University, Nanjing, China — zhoujl@nju.edu.cn
Li-Yong **Zhou**, Department of Astronomy, Nanjing University, Nanjing, China — zhouly@nju.edu.cn

Part 1

Exoplanet Detection

CoRoT: pioneer space mission for exoplanet transit search

Pierre Barge[1], Annie Baglin[2], Michel Auvergne[2] and the CoRoT team

[1]Laboratoire d'Astrophysique de Marseille
LAM/OAMP - BP8 - 13376 - Marseille - Cedex 12 - France
email: **pierre.barge@oamp.fr**

[2]Laboratoire d'Etudes Spatiales et d'Instrumentation en Astrophysique
LESIA/OPM - 5 pl. J. Janssen, F-92195 Meudon, France

Abstract. Led by the CNES space agency the CoRoT mission is born from a joint effort of France, Austria, Belgium, Brazil, Germany, Spain and ESA. In orbit around the Earth, CoRoT started its first observations in February 2007 and is, now, regularly producing ten thousand light-curves with a very high accuracy. Performances are better than expected and some Hot Jupiters have already been detected in the raw data. Once the fully corrected data will be delivered, much smaller transits should be detected giving access to the hot Neptunes and the big Terrestrial planet families. We briefly describes the status of the mission, the inflight performance and the ground based program follow up strategy. We also present some preliminary results issued from a first analysis of the data.

Keywords. instrumentation: photometers, planets and satellites: formation, stars: binaries: eclipsing, stars: variables: delta Scuti, stars: variables: other

1. Introduction

This paper presents an overview of the CoRoT mission (see also Baglin et al. 2006) and its status. A detailed description can be found in a pre-launch book (CoRoT 2006) and in a post-launch paper (Auvergne et al. 2008, in preparation) that describes how the instrument is actually working and its present in flight performances. CoRoT, which stands for COnvection ROtation and planetary Transits, is an experiment dedicated to stellar seismology and search for extrasolar planets. It was developed as a minisatellite CNES mission, i.e. a mission of intermediate size and low cost. It was preselected in 1994 for a phase A competitive study. However, political and financial difficulties have postponed the final decision, which was taken in October 2000. In between, the studies continued and a much richer scientific mission has been prepared with an extended exoplanet program and a much larger scientific community from the european countries.

CoRoT was launched December 27th from Baikonour and placed in orbit around the Earth at an altitude of 900km. Now, the satellite is collecting data that are downloaded to the Earth at a regular rate. A number of technical challenges have been taken up successfully : (i) a pointing stability of 0.5 arcsecond (the payload being in control loop with the platform), (ii) a very efficient protection against straylight thanks to a baffle whose rejection factor is better than 10^{-12}, (iii) a thermal stability better than 0.05C over 1 hour, (iv) the management of a cooperation with many international partners.

2. The scientific objectives

The CoRoT mission has two scientific programs, both requiring long and uninterrupted sequences of observations with a very high photometric accuracy. The first one is devoted to understanding the physics of stars and their evolution. The second is devoted to the discovery of other planetary systems and to the study of the way they form. During the observations, the telescope is pointing in fixed direction and the two scientific objectives are working simultaneously on adjacent parts of the sky. These two programs define the so called Core Program of the mission that splits into a central program organized around long observation runs of 150 days and an exploratory program based on short observation runs of 20-30 days. The exploratory program follows the requirements of the seismology program in order to get a reasonable coverage of the HR diagram. The short runs are also used by the exoplanet search program to enlarge the sample of Hot Jupiter planets.

2.1. The seismology program

Helioseismology has proved to be a powerful tool to probe Sun's interior, leading to identify many parameters of solar physics with an accuracy better than 0.1% down to the core, and even contributing to rethink the theory of the neutrinos. The observation from space of other stars than the Sun (with a precision and over time scales out of reach from the ground) was necessary to test different physical conditions and improve the modeling of the internal structure and evolution of stars. The light curve integrated over a stellar disk can bring accurate information on the modes and makes it possible to determine: the radius of the core, the deep limits of the convective external layers, the rotation profile inside the star, the angular momentum transport.

A good frequency resolution is necessary to discriminate a significant number of modes, to reveal the frequency splitting and rebuild the line profiles. The number of targets is of the order of 10 per run. The targets are chosen among F and G stars (supposed to have

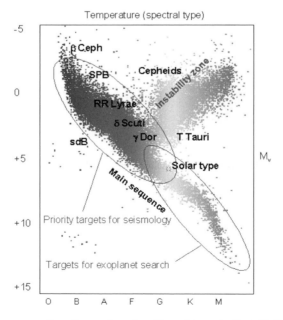

Figure 1. HR diagram reporting the various families of targets that CoRoT will observe (after Boisnard & Auvergne 2006)

solar like pulsations) but also among more classical pulsators as delta Scuti, gamma Dor and beta Ceph stars known as variables from the ground with only a small number of modes. During the long runs (150 days) the frequency resolution reaches 0.1 Hz in the Fourier space (central program). During the exploratory program the observation runs are shorter (20-30 days) and the aim is to widen the sample of target spectral types for a better coverage of the HR diagram (Fig. 1). This should be sufficient to produce statistical data about the excitation of the oscillating modes (see e.g. Michel *et al.* 2006).

2.2. *The exoplanet program*

The CoRoT planet finding program aims at detecting extrasolar planets when they transit in front of the disk of their parent stars. For the exoplanet channel the global design of the instrument is the same than for the seismology channel; the only difference is a smaller defocus and a longer integration time. Small flux variations down to 7.10^{-4} can be seen with CoRoT on a star of magnitude 15.5 in a one hour integration time. On the same duration, smaller variations are accessible to stars brighter than $m_V \simeq 12$, the magnitude at which saturation problems appear on the CCD.

The exoplanet program is performed during long runs of 150 days (180 days at maximum) but the data acquired during the short runs (20-30 days) are also used to search for short period planets. To detect a planet in complete confidence, the phase must stay coherent over three successive transits, so planets detected during a long run will have periods less than 50-60 days whereas those detected during a short run will have periods less than 10-15 days.

In order to partly overcome this limitation, a dispersion device has been implemented on the exoplanet channel, a few centimeters in front of the CCDs, providing a three-color signal that helps to discriminate planetary transits (achromatic events) from stellar "noise" (highly chromatic variations due to temperature variations at the surface of the star). Studies have shown that using this three-color information can improve transit detection for stars much more active than the Sun and also when the light curves contain less than three successive transits. Further, as the image of a star on the CCD surface is divided into three distinct regions, the three-color bands correspond to a spatial information that can be used to discriminate false candidates (due to background eclipsing binaries) from transit events on the target star.

During the nominal lifetime of the mission, CoRoT will observe some 60,000 stars during 5 long runs of 150 days and nearly some 60,000 stars during 5 short runs of 30 days. So, in total, some 120,000 light curves should be available to search for planet with periods less than 10 days and 60,000 light curves to search for planets with periods that range from 10 days to 50 days. With its photometric precision, mainly limited by the photon noise, CoRoT should detect big terrestrial planets 2-3 times bigger than the Earth (Barge *et al.* 2006)

The planet finding program also includes in-depth analyses of the data looking at: (i) transit timing variations due to an hypothetical non-transiting outer planet; (ii) peculiar transits due to planets in binary systems or to planets with rings or big moons; (iii) signatures of exoplanetary atmospheres. This program includes a complete characterization of the detected planets to get their orbital motion, masses, mean density and also to learn more about their composition and internal structure. The determination of planet radius from transit depth also depends on the precision the star radius is known. In a more general way, an optimal scientific return of the mission requires to characterize at best the planet and its host star. This is possible via the organization of a well co-ordinated program of follow-up and complementary observations like prepared by the CoRoT Exoplanet Science Team. CoRoT is expected to detect and characterize a large

large number of transiting planets that will help to constraint the models and to improve our understanding the way planets form.

2.3. Additional programs

Additional programs allows the astronomical community to propose observations devoted to specific target fields to address scientific cases different from the core program, as for example stellar activity (surface magnetism), binary systems, pulsating stars beyond the instability strip and possibly search for Kuiper belt objects (KBO), etc ... In the case of exoplanetary science, the search for the modulation of the star flux by the light reflected by an orbiting planet was selected as a specific additional program. The target is HD52265, a star known to host a planet from Doppler measurements, and that will benefit from the very high photometric accuracy available on the seismology CCDs. The additional programs have, however, a lower priority with respect to the seismology/exoplanet core program.

Figure 2. The CoRoT instrument on the Proteus platform

3. Organization and partners

CoRoT is a french national-lead program with a new type of collaboration with ESA. CNES is the prime contractor for the mission, the payload development is driven by a team with people from CNES and CNRS laboratories (LESIA in Meudon, LAM in Marseille and IAS in Orsay) and including European partners (Austria, Belgium, ESA and Germany). Alcatel Alenia space was responsible for the platform and the integration of the satellite. The ground segment is developed within the same frame of cooperation between CNES, CNRS laboratories (OMP in Toulouse) and international partners (Brasil and Spain). The satellite mission and control centers are located at the Toulouse Space center.

4. The satellite

The CoRoT spacecraft is based on a recurrent Proteus platform, developed by CNES and Alcatel Alenia Space. CoRoT is the third mission to use this platform in a low Earth orbit with its associated ground control segment. The payload is made up of a telescope, a wide field camera operating in the visible, an equipment bay hosting analogical and digital electronics and a flight software in charge of the aperture photometry processings and of the fine pointing mode (see Fig. 2).

The total mass at launch is 630 kg with a payload mass of 300 kg. The satellite is 4 meter long with a mean diameter of 2 meters. The pointing accuracy of the platform is 0.5 arcsec and the capacity of the telemetry is 1.5 Gbit/day. The nominal total duration of the mission is, at least, 2.5 years but a longer lifetime is expected. The payload is made up of three subsystems called CoRoTel, CoRoTcam and CoRoTcase that we briefly describe.

4.1. CoRoTel

CoRoTel is an afocal telescope composed of 2 parabolic mirrors with a pupil of 27cm and a two meter long cylindrical baffle. To reach the required photometric accuracy the baffle have to stop the straylight from the Earth with an efficiency that must be better than 10^{-12}. To obtain such a high efficiency was a real challenge, of a crucial importance for the success of the mission (see Fig. 3).

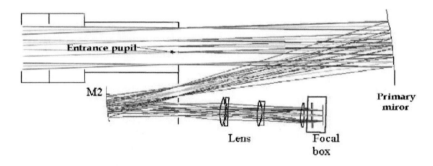

Figure 3. The optical design: the light is collected by an off-axis (afocal) parabolic system with 2 mirrors; the parallel beam at the output of the telescope is re-imaged on the focal plane by a dioptric objective.

4.2. CoRoTcam

This is a wide field camera composed of a dioptric objective (6 lenses) and a focal unit equipped with 4 frame transfer CCD 2048x4096. Two CCDs are devoted to the seismology program and the two others are devoted to the exoplanet program. The focusing is optimized as to spread at best the photon of the targets onto the CCDs; it is different for the two scientific objectives. In front of the 2 CCDs devoted to the exoplanets, a dispersive device (biprism) has been inserted to help discriminate planetary transits from stellar variability (see Fig. 4).

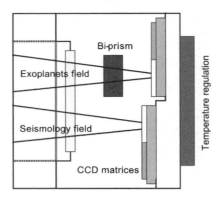

Figure 4. The two scientific channels inside the focal block

4.3. CoRoT case

The equipment bay host analogical and digital electronics and a flight software in charge of the aperture photometry processings and the delivery of angle error measurement data to the platform (fine pointing mode). The equipment bay supports the scientific data processing electronics and instrument housekeeping electronics.

5. The field of view and the photometry

In the focal plane of the instrument the four CCDs are mapping on the sky a field of view of $3.05° \times 2.8°$ and a single pixel represents 2.32 arcsec. On the two CCDs devoted to the seismology program the targets are bright stars whose magnitude ranges from 6 to 9. Each CCD contains a primary target with a magnitude of the order of 6 and a handful of stars brighter than magnitude 9. On the exoplanet CCDs the stars are fainter and much more numerous (see Fig. 5). Their magnitude ranges from 11 to 16 and their total number is limited to 12,000.

The stellar fluxes are measured as a function of time, every 32s in the seismology field and every 512s in the exoplanet field. For the exoplanet targets, the measure results from the piling up on board of 16 individual exposures of 32s obtained by aperture photometry. The apertures are in pixel unit and tailored from a reduced number (256 per CCD at maximum) of fixed patterns (or templates) that are uploaded at the beginning of a run of observation. The production of these templates results from a complex procedure that accounts for: photon noise from the targets and the background, readout noise from CCDs and electronics, jitter noise, thermoelastic breathing, variability of the background (Llebaria & Guterman 2006). This procedure is performed on ground on the full image of the CCDs that is downloaded at the very beginning of a run.

Due to the bi-prism implemented in the focal block between the dioptric objective and the plane of the detector, the image of each target star is a small spectrum whose area is about $300 arcsec^2$ on the sky. This low resolution spectrum is sufficient for a "colored photometry" obtained by splitting the aperture in three parts called "red", "green" and "blue" according to a flux criterion based on the fraction of energy in each color channel. The separation between the red, green and blue bands is adapted for each target star. The photometry is performed on board by piling up the pixel flux in each of the colored bands and the results are transmitted to the ground. In the case of faint stars, the colored photometry is very poor and reduces to standard one band ("white") photometry.

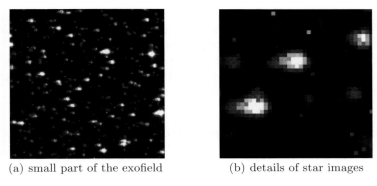

(a) small part of the exofield   (b) details of star images

Figure 5. Images of target stars on the exoplanet CCDs

So, in total and for each CCD, different types of windows are opened: 4500 "three-color" windows sampled at 512s; 1000 "white" windows sampled at 512s; 500 windows sampled at 32s (oversampling can be triggered from the ground); 20 "imagettes" i.e. windows of 10x15 pixels containing a single star image (all the pixels of the window are downloaded to the ground); 200 "black" windows for the correction of the background.

6. The mission constraints

As the PROTEUS platform only evolves on low Earth orbits, this puts constraints on the orientation of the satellite and on the possible observing regions. In order to observe the same direction of the sky for a long period of time (several months) without being blinded by the Sun or occulted by the Earth, the satellite must have a polar inertial orbit and a line of sight roughly perpendicular to the orbit plane (see Fig. 6).

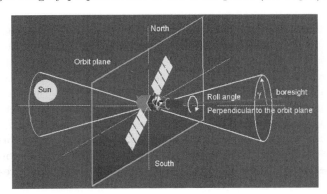

Figure 6. The observation cone and the pointing constraints.

Two reversal maneuvers are necessary once a year when the direction of the Sun is perpendicular to the line of sight. The available observing time, 180 days at maximum, is shared between long runs of 150 days and short runs of 20 to 30 days. The following scenario has been adopted: a short run of 20 days is inserted between two long runs of the central program (devoted to the seismology and the exoplanet core program). The short runs are equally shared by the exploratory program of seismology and the additional program. A dense region of the sky has been selected at the intersection between the galactic plane and the equatorial plane, defining two visibility zones (or "eyes") in the direction of the center and the anticenter of the Galaxy. Preparatory observations have permitted to find a compromise between the two scientific objectives; seismology requires a few number of well selected bright targets whereas planet search requires large number of stars with $11 < m_V < 16$ in regions of moderate star density.

So, during the nominal lifetime of the mission (at least 2.5 years), there will be a total of 5 long runs of 150 days and 5-10 short runs of 20-30 days, successively in the direction of the center or the anticenter of the Galaxy.

7. "Alarm mode" and follow up strategy

One specificity of the CoRoT exoplanet program is the "alarm mode", an operational loop between science team and command center on ground and the instrument in space (Quentin *et al.* 2006). Its goal is to optimize the science return of the mission by identifying, before the end of a run, transit candidates early in the process before the data be fully reduced. For these transit candidates, decision can be made to change the rate of the observations from one exposure in 512s to one exposure in 32s. The interest for this oversampling is two-fold: (i) to get a better coverage of the transit profile (in view to determine limb darkening and albedo); (ii) to reduce the level of noise by removing outliers from the corrupted exposures. Last but not least, another interest of this "on the stream" detection is to start follow-up operations as soon as possible.

Once transit signals are detected the goal is to secure at maximum the detections and to trigger the follow up observations necessary to completely characterize the discovered planets. Inside the CoRoT Exoplanet Science Team (CEST) the work necessary to reach this goal is shared into 5 different tasks performed in successive steps.

- *Task 1:* The first task, in direct relation with the detection procedures, is to make a detailed analysis of the light-curve and to remove the possible false alarms, for example by identifying indirect signatures of the eclipsing binaries (secondary transit, triangular shape of the primary transit, ellipsoidal modulations due to tidal effects).
- *Task 2:* It consists in using on/off photometry to test: (i) the level of variability of the background stars identified inside the target PSF, (ii) if one of the stars in the target PSF could mimic the transit observed on the target. The necessary information on the contaminating stars is found in the ExoDat preparatory database that contains all the necessary information on the field of view and the targets.
- *Task 3:* Once the transit candidate is secured, radial velocities are used to confirm the presence of a planet and to measure its mass. If accurate enough, radial velocity measurements can also be used to evidence the Rossiter-Mac-Laughlin effect that put constraints on the alignment of the star and planet rotation axes. Most of the CoRoT planet candidates will be within the reach of the SOPHIE and HARPS spectrographs.
- *Task 4:* It consists in obtaining precise stellar parameters using high resolution spectroscopy and to get a good estimate of the planet parameters.
- *Task 5:* Finally, the last task is to use space observations, like HST, FIRST, ...etc to go into deeper studies and to characterize at best all the planet parameters.

8. Summary of the in-flight performances

CoRoT is in orbit around the Earth since December 27th. The opening of the cover, on January 17th, was followed by a program of calibrations and in-orbit verifications. All the on board systems were checked, the optical performances were found to be excellent for the two scientific channels and the satellite successfully entered in fine pointing mode (the Proteus platform being fed by information from the payload). The scientific observations began on February 2nd, with a field of view in the direction of the anti-center of the Galaxy. This first run of observation lasted only 2 months and was used as a commissioning phase of the instrument.

- *Duty cycle:* Data losses occur either at the crossing of the South Atlantic Anomaly or due to random losses in high energy events. Before launch, the corresponding duty cycle was estimated of 90% . In reality losses due to random events are of the order of 2% and those due to SAA crossing are of the order of 6%; so, the resulting duty cycle is 92%, i.e. better than expected.

- *Hot or bright pixels:* The number of bright or "hot" pixels on the CCD matrices is found, however, 10 times greater than expected. This delayed the starting of the pipeline and slowed down the production of fully reduced data. Nine months after the end of the initial run data are ready and will be delivered to the Co-Is mid December 2007. A preliminary analysis of the data demonstrates the high quality of the instrument whose performances are better than expected and in very good agreement with the scientific specifications of the mission.

- *Pointing:* The actual performance in the pointing of the satellite is 0.12 pixels rms following the x axis and 0.15 pixels rms following the y axis. This corresponds on the sky to a mean value of 0.3 arcsec and is better than expected.

- *Photometric performances:* The noise level in the light curve of a star of magnitude $m_V = 15.4$ in a 1 hour integration time has been estimated to 700ppm; this value matches pretty well with the initial requirement for the instrument (700 ppm in 1 hour on a star of magnitude 15.5). In the case of a brighter star of magnitude $m_V \simeq 12.3$ the estimated noise level obtained in 1hour is 170ppm. Consequently, the noise level remains of the same order as the photon noise over all the range of magnitude.

- *Straylight:* The variations of the straylight all along the orbit of the satellite has been estimated to $0.4e^-$ (on average), a value that is significantly less than the expected value of $2-4e^-$. This is due to the very good performance of the baffle whose rejection factor is better than 10^{-12}.

9. Observations and preliminary results

For the time being CoRoT has observed four different fields of stars. The first run of observation (IRa01) was in the direction of the Galactic anticenter and lasted 55 days. The second and the third ones were in the direction of Galactic center: a short run of 25 days followed by a long run of 150 days. The fourth one (LRa01), in the direction of the Galactic anticenter, started October 24th and is still being observed.

In total, since the beginning of the observations, CoRoT monitored 30 bright stars for the seismology program and 34,560 stars for the exoplanet search program. The data, uncorrected from the jitter and still containing residuals of the orbital effects, were analyzed by the alarm mode software. A number of transit candidates were detected in the data and the sampling rate of the targets was changed correspondingly. Then, the resulting lists of planet candidates were examined following the procedure described in section 7, the best candidates being sent to to the follow-up group for confirmation and characterization.

- *Variable stars and eclipsing binaries*

From a quick inspection of the CoRoT light curves the most striking aspect is the large number of stars with a significant level of variability. These stars are identified as eclipsing binaries or periodic pulsators with periods that ranges from days to months. Fig. 7 gives an example of a δ Scuti star of magnitude 13.8 that shows variations of 3% with double modulation.

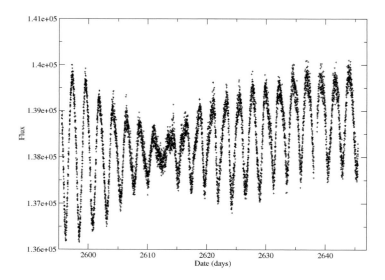

Figure 7. Example of variable stars in the CoRoT exofield

Eclipsing binaries are also frequently identified in the CoRoT field of view. They can be loose or tight systems with amplitudes that can be very large (several percents). A number of them shows clear evidence of ellipsoidal variations due to tidal interactions between the two components. An example of eclipsing binaries found by CoRoT is given in Fig. 8 with its succession of primary and secondary eclipses. This example also illustrates the possible confusion between eclipsing binaries and transiting planets when the amplitude of primary and secondary transits are nearly the same; indeed, noise may rub out the small differences between the two types of transit.

The high frequency of eclipsing binaries in the CoRoT field of view and the large extent of the CoRoT Point Spread Function (PSF) onto the sky (about 30 arcsec) are also favoring possible confusions between the periodic eclipses of a neighboring background star and the transits of a planet in front of the target star. A number of ambiguous situations can be removed with follow up observations: either on/off photometry of contaminating stars in the PSF or Doppler spectroscopy of the target.

- *The first planet detected by CoRoT*

Thanks to the alarm mode, the preliminary analysis of the data permitted to find two transit candidates during the observation of the initial run (IRa01). These two candidates were used as test cases to improve the organization and management of the necessary follow-up operations. One of them was rapidly identified as an eclipsing binary thanks to radial velocity measurements at the Tautenburg observatory. The other candidate was observed with the SOPHIE spectrograph at OHP; the first Doppler measurements gave us some positive clues about the planetary nature of the event but an additional effort was necessary to completely confirm the planet.

Thirty six transits were observed during the run which was ∼55-days-long. The duration of the transits is nearly 2.4 hours and the relative precision is 2.10^{-4} in 1 hour integration time. This planet was announced in a press release and named CoRoT-Exo-1b. A rough estimate of the main planet parameters was obtained using the folded light curve (see Fig. 9) : its orbital period is 1.5 days and its radius ranges ranges from 1.5 to 1.8 R_{Jup}. The large uncertainty in the radius comes from the bad determination of the

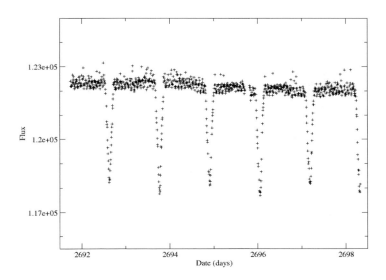

Figure 8. Example of eclipsing binaries in the CoRoT exofield

star's parameters (spectra with a good resolution and a high signal-to-noise were lacking at this moment). The best fitting of the folded light-curve is consistent with an orbital inclination of 84°.

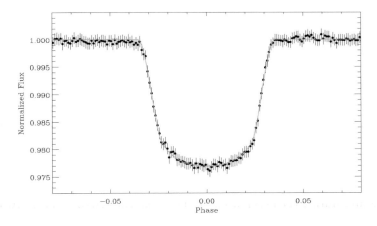

Figure 9. The transit of CoRoT-exo-1b folded at a period of 1.5 days. The bin size is 2.16 minutes and the precision $\sim 3.10^{-4}$.

With the high photometric accuracy achievable from space an interesting opportunity is to use the method of the transit timing variations to indirectly detect the presence of non-transiting outer planets. In the case of CoRoT-Exo1b we have 36 individual transits and a precision of 2.10^{-3} in 32s; so, time lags down to 20-30 sec should be accessible. The possibility to use transit timing variations to detect other planets around CoRoT-Exo-1 will be explored in a future CoRoT paper that will present the detailed analysis of the light-curve. This will be made with data fully corrected from the instrumental noises, once the last version of the pipeline will enter the production phase.

The estimation of the planet mass, based on the measurements of the SOPHIE spectrograph at Observatoire de Haute Provence, gives a value of $1.3 M_{Jup}$. Finally, with the present values of the parameters, the planet is found to orbit at some 0.04 AU from its parent star. More precise determinations of the radius and the mass of the planet will be given in a forthcoming CoRoT paper (Barge *et al.* 2008, *in preparation*).

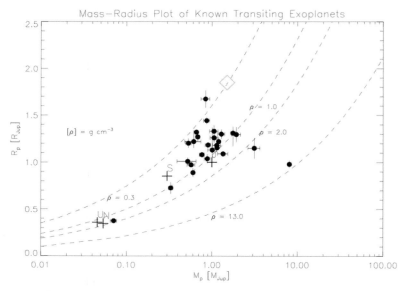

Figure 10. The mass-radius diagram: dots represent known transiting planets; crosses represent Solar system planet; the diamond corresponds to CoRoT-Exo-1b.

The new planet discovered by CoRoT has been reported on a mass-radius diagram simultaneously with all the other transiting planets discovered so far (see Fig. 10). It is located at the extreme border of the plot and detaches from all the other points. This new object seems to challenge the recent models on the physics of the hot giant planets. However, we must avoid hasty conclusions since the radius of the star is still poorly known. With an increasing number of transiting planets CoRoT will fill up this diagram with an homogeneous sample of data that will help to better constraint the models. Of course, the most interesting parts of the diagram that CoRoT should supply are, at the bottom-left, the domain of the icy giants and the domain of the big rocky terrestrials.

• *Other planet candidates*

Early transit detection with the alarm mode also permitted to identify during a single long run (LRc01) 20 planet candidates that are reported below in a list ordered following their orbital period:
— 8 candidates with $P < 5$ days
— 4 candidates with $5 < P < 10$ days
— 5 candidates with $10 < P < 27$ days
— 3 single transit events.

The variability of each target star was first checked in the database of the BEST preparatory survey. Then, we used the follow up procedure described in section 7 and triggered the following operations: (i) on/off photometry with telescopes at IAC and OHP, but also with EULER and WISE; (ii) radial velocity measurements with the spectrographs SOPHIE, FLAMES, HARPS and Tautenburg. As a result, eclipsing binaries were identified in five cases.

10. Conclusions

CoRoT is the first space survey devoted to a photometric search for extrasolar planets. The satellite, launched in December 2006, is in a low Earth orbit where the instrument is working well and all systems are nominal or better. Since February 2007, CoRoT is regularly providing 12,000 light curves with a very high photometric precision. On the total duration of the mission, CoRoT will observe in ten different directions of the sky and will produce some 120,000 light curves. The level of noise in the data is nearly equal to the photon noise, in accordance with the initial requirements of the mission. So, according to pre-launch simulations (Moutou et al. 2005), the detection of big and hot terrestrial planets should be within the reach of CoRoT. The number of hot pixels on the CCD matrices, which is higher than expected, slowed down the starting of the pipeline. When the release of the fully corrected data will be available (a few weeks), the detection of short-period planets smaller than Uranus will become effective.

For the time being, the preliminary analysis of the raw data performed during the alarm mode permitted to identify many variable stars and eclipsing binaries, but also a number of transit candidates. Follow up operations were triggered to secure the detected candidates and to characterize the discovered planets. One planet was found in the initial run of observation and 20 other candidates in the first long run, in the direction of the Galactic center. The first planet, CoRoT-Exo-1b, is now completely confirmed (Barge et al., in preparation) and 15 candidates are still in the Follow Up process.

With the ground based discoveries of a number of transiting systems, the characterization of extrasolar planets just passed an important step and is leading the way to transit surveys like CoRoT and Kepler. Indeed, for such systems, mass and radius measurements can reach a precision that is only limited by the stellar parameters. This is of a crucial importance to determine the mean density of the exoplanets and to better understand their internal structure. So, the most direct impact of CoRoT will be the filling up of the mass/radius diagram with families of planets in the domains of the icy giants and the big terrestrials.

The high photometric accuracy available with CoRoT can also give access to: (i) the method of the transit timing variations (to identify non-transiting outer planets); (ii) the identification of dark spots on the surface of the stars (in possible connection with presence of a planet); (iii) the detection of tiny secondary transits due to the planet reflected light (indicative of the orbit eccentricity).

Once a transit is detected it is also interesting to search for the Rossiter-McLaughlin effect using spectroscopic measurements during the transits. This is important because this effect provides an insight on the rotation of the star and on the orientation of its spin axis with respect to the planet orbital plane.

So, CoRoT will help to learn more about the short-period planets (orbital distribution, density and structure), their possible interaction with their parent stars (planet induced stellar activity), their dynamical evolution (through tides and magnetic fields) and their formation.

To learn more on CoRoT, see the web site: http://corot.oamp.fr.

References

Baglin, A., et al. 2006, 36th COSPAR Scientific Assembly, 36, 3749

Barge, P. et al. 2006 *The CoRoT Mission: pre-launch status* ESA-SP 1306, eds M. Fridlund, A. Baglin, L. Conroy & J. Lochard, p. 419.

Boisnard, L. & Auvergne, M., 2006 *The CoRoT Mission: pre-launch status* ESA-SP 1306, eds M. Fridlund, A. Baglin, L. Conroy & J. Lochard, p. 19.

CoRoT 2006, *The CoRoT Mission: pre-launch status* ESA-SP 1306, eds M. Fridlund, A. Baglin, L. Conroy & J. Lochard

Llebaria, A. & Guterman, P. 2006 page *The CoRoT Mission: pre-launch status* ESA-SP 1306, eds M. Fridlund, A. Baglin, L. Conroy & J. Lochard, p. 293

Michel, E. *et al.* 2006 *The CoRoT Mission: pre-launch status* ESA-SP 1306, eds M. Fridlund, A. Baglin, L. Conroy & J. Lochard, p. 39.

Moutou, C., Pont, F. & Barge, P. and 13 co-authors, 2005, Astron. & Astrophys, 437, 355

Quentin, C. *et al.* 2006 *The CoRoT Mission: pre-launch status* ESA-SP 1306, eds M. Fridlund, A. Baglin, L. Conroy & J. Lochard, p. 409.

Finding Earth-size planets in the habitable zone: the *Kepler Mission*

William Borucki[1], David Koch[1], Gibor Basri[2], Natalie Batalha[3], Timothy Brown[4], Douglas Caldwell[5], Jørgen Christensen-Dalsgaard[6], William Cochran[7], Edward Dunham[8], Thomas N. Gautier[9], John Geary[10], Ronald Gilliland[11], Jon Jenkins[5], Yoji Kondo[12], David Latham[10], Jack J. Lissauer[1] and David Monet[13]

[1] NASA Ames Research Center, Moffett Field, CA 94035, USA
[2] University of California-Berkeley, Berkeley, CA, 94720, USA
[3] San Jose State University, San Jose, CA, 95192, USA
[4] Las Cumbres Observatory Global Telescope, Golenta, CA 93117, USA
[5] SETI Institute, Mountain View, CA, 94043, USA
[6] University of Aarhus, Denmark
[7] University of Texas at Austin, Austin, TX, 78712, USA
[8] Lowell Observatory, Flagstaff, AZ, 86001, USA
[9] Jet Propulsion Laboratory, Pasadena, CA, 91109, USA
[10] Smithsonian Astrophysical Observatory, Cambridge, MA, 02138, USA
[11] Space Telescope Science Institute, Baltimore, MD, 21218, USA
[12] NASA Goddard Space Flight Center, Greenbelt, MD, 20771, USA
[13] United States Naval Observatory, Flagstaff, AZ, 86002, USA

Abstract. The *Kepler Mission* is a space-based mission whose primary goal is to detect Earth-size and smaller planets in the habitable zone of solar-like stars. The mission will monitor more than 100,000 stars for transits with a differential photometric precision of 20 ppm at V=12 for a 6.5 hour transit. It will also provide asteroseismic results on several thousand dwarf stars. It is specifically designed to continuously observe a single field of view of greater than 100 square degrees for 3.5 or more years.

This overview describes the mission design, its goals and capabilities, the measured performance for those photometer components that have now been tested, the Kepler Input Catalog, an overview of the analysis pipeline, the plans for the Follow-up Observing Program to validate the detections and characterize the parent stars, and finally, the plans for the Guest Observer and Astrophysical Data Program.

Keywords. Planet detection, exoplanets, differential photometry, space-based telescope

1. Introduction

Over 250 exoplanets have been detected as of the time of this symposium (Marcy, report in IAUS 249, 2007). Most of these are gas giants. A few approaching super-Earths in short period orbits have also been found (Rivera *et al.*, 2005). However, the holy grail is to find habitable planets, that is, those in the habitable zone (HZ) (Kasting *et al.*, 1993), where liquid water can exist on their surfaces, and with a size and density such that they can have a life-sustaining atmosphere, that is, from about 0.8 to 2.2 $R_\oplus$ or, if one assumes an Earth-like density, from about 0.5 to 10 $M_\oplus$. Finding planets more than three hundred times less massive than Jupiter is not trivial. Even finding extra-solar Jupiters took until 1995 (Mayor & Queloz, 1995). However, the transit method proposed by Borucki and Summers (1984) can detect Earth-size planets in the HZ. Thus, it was necessary to show that: 1) the variability of the Sun and presumably most stars similar to the Sun on the time scale of a transit (on the order of 10-12 hours) is substantially smaller in amplitude than that of a Sun-Earth transit analog (84 ppm) (Jenkins 2002) and 2) to demonstrate that a space-based photometer with all the known forms of realistic noise has a combined differential photometric precision ≤20 ppm, one-sigma on the time scale of a transit (Koch, et al. 2000). The concept was proposed as a Discovery mission four times before finally being selected in December 2001 as NASA's tenth Discovery mission.

2. Mission Overview

Various aspects of the mission have been described in a number of papers (Borucki, et al. 2005, Koch, et al. 2004, Koch, et al. 2006). The top level scientific goals are to:

1. Determine the frequency of terrestrial and larger planets in or near the habitable zone of a wide variety of spectral types of stars;
2. Determine the distributions of sizes and orbital semi-major axes of these planets;
3. Estimate the frequency of planets and orbital distribution of planets in multiple-stellar systems;
4. Determine the distributions of semi-major axis, albedo, size, mass and density of short-period giant planets;
5. Identify additional members of each photometrically discovered planetary system using complementary techniques; and,
6. Determine the properties of those stars that harbor planetary systems.

To achieve these goals, three fundamental design requirements were established: the photometric precision, the mission life time and the number of stars observed. The photometric requirement is to detect individual Earth-size (R=1.0 $R_\oplus$) transits of 6.5 hrs (half of a central transit duration for a planet at 1 AU) of a twelfth magnitude solar-like star with an SNR of greater than or equal to four, when all sources of noise (stellar variability, shot and instrument) are included. Since a periodic sequence of at least three transits is required, the mission must last three or more years to detect planets in the HZ. Finally, the photometer is required to observe enough solar-like stars to produce a statistically meaningful result. Hence, the aperture size, field of view and location on the sky have been chosen to provide at least 100,000 dwarf stars.

The photometer design is based upon a classical Schmidt telescope with a 95 cm diameter aperture and more than one-hundred square degree field of view (FOV). The FOV is equivalent to about six Palomar Schmidt plates. The completed flight focal plane is shown in Figure 1.

The position of the FOV on the sky is in the Cygnus-Lyra region centered on RA=19^h 22^m 40^s, Declination=44° 30', just above the galactic plane and looking down the Orion arm of the Galaxy. This provides a sample of stars similar to our local neighborhood and is a rich star field that is continuously viewable throughout the year as the spacecraft drifts away from the Earth as it orbits the Sun. The typical stellar distances for most of the usable stars are from a few hundred parsecs to about 1 kpc. Shot noise limits the magnitude of usable stars to about V=15-16 for F-, G- and K-dwarfs and about V=16.5 for M-dwarfs.

The single FOV will be viewed for the entire mission including any extended mission life. But to keep the fixed-solar array pointed toward the Sun and the focal-plane radiator pointed to deep space, the photometer-spacecraft must be rotated 90° about the optical axis every 93 days. The orientations of the CCD modules have been chosen so that the focal plane is four-fold symmetric (Figure 1). Therefore, after a 90° rotation all of the selected target stars remain on active pixels and the orientation of the columns remains the same on the sky except for the central module which is only two-fold symmetric.

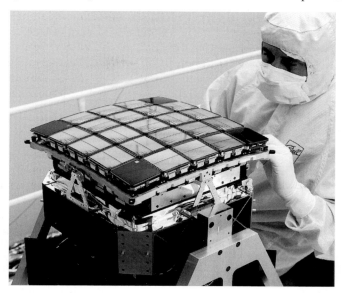

Figure 1. Completed flight focal plane for *Kepler* with 42 science CCDs and 4 fine guidance sensor CCDs in the corners. The science CCDs are thinned, back-illuminated, and AR-coated with 1024 by 2200 pixels each with two readouts per CCD for a total of 84 science outputs. The pixels are 27 micrometers with a plate scale of 3.98 arcsec/pixel. The FOV is 13.9° side to side.

3. Measured Photometric Performance

The ability to detect transits depends on the photometric precision of the entire system on the time scale of transits, that is, from a few hours to about a half of a day. Absolute photometric accuracy is not necessary, since one is only looking for temporal variations. Common-mode noise can be removed by performing ensemble normalization within each of the 84 data channels, thereby taking out effects, e.g., gain variations and DC offsets.

The total noise we describe as the Combined Differential Photometric Precision (CDPP) in units of parts per million (ppm). The CDPP includes variability of the source, the photon shot noise and the measurement noise. Stellar variability is inherent in the source and uncontrollable. From the ACRIM-SMM (Willson *et al.* 1991) and DIARAD-SOHO (Froelich, 1997) measurements show that on the time scale of a transit, the solar variability is $\leq$10 ppm during solar-max. Note that solar variability is red noise and does not scale with $1/t^{1/2}$. Shot noise is determined by the brightness and spectral type of the star and the photometer design (aperture, obscuration, transmissions, reflectivity, bandpass filtering, quantum efficiency, etc.) To minimize the shot noise, the photometer has a single broad bandpass. Figure 2 shows the measured spectral response for the *Kepler* photometer. When convolved with the spectrum of a G2V (solar-like) star with V=12, the photometer will measure 4.7×10^9 photoelectrons in 6.5 hrs (91% duty cycle). This results in a shot noise contribution of 14.6 ppm.

The final contribution to CDPP is the measurement noise. This noise includes not only that from the CCD, but also electronic and optical cross talk, pointing jitter and any other noise introduced by the methodology used to make the measurements and reduce the data. At least fourteen individual terms are tracked and measured.

The Kepler Technology Demonstration (Koch, *et al.* 2000) was used once again in 2007 to test a single string engineering model of the flight detector system design. These tests demonstrated that the instrument noise requirement was met (after making modifications

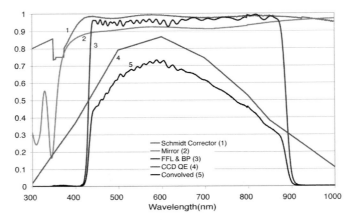

Figure 2. Spectral response curves. The FFL & BP is the average for the field flattener lenses on the CCD modules, which also have the bandpass filters. The CCD QE is the average from all 42 CCDs. The convolved also includes a 3-4.5% loss for contamination. The red cutoff is to avoid fringing in the CCDs and to minimize the potential for false positives produced by faint reddened background eclipsing binaries. The blue cutoff is designed to avoid the UV and Ca II H&K lines. For the Sun, 60% of the irradiance variation is at wavelengths less than 400 nm, but it only accounts for 12% of the total flux (Krivova, et al., 2006).

in the design to eliminate under/overshoot due to an analog bandwidth issue) and that Earth-size transit signals in individual stars could be detected. For bloomed star images, it was again shown that photometric precision was preserved and transits were detected. A comprehensive test program of the components and full up focal plane has been performed at operating temperatures to measure and characterize such parameters as pixel response non-uniformity (PRNU), 2-D biases, under/overshoot, clocking cross talk (due to the fine guidance sensors which are read out at 10 Hz) and smear (due to reading out the CCD without having a shutter). All of this characterization information is being incorporated into the data reduction pipeline. The net result of the measurement noise contribution is 7.7 ppm. Figure 3 is a histogram of the individual CDPP values. The design requirement is a CDPP$\leq$20 ppm for a V=12 G2V star and 6.5 hour integration.

4. Target Selection

The primary goal of *Kepler* is to detect terrestrial planets around solar-like stars, that is, late-dwarf (F, G, K and M) stars. Since *Kepler* does not send down data from every pixel in the focal plane, but rather only approximately 3-5% of the pixels of interest that are used for each star of interest, one has to preselect the stars to observe. The magnitude range is roughly V=9 to 15. Since there did not exist a catalog to this depth near the galactic plane with the necessary information on which to base the selection, the project undertook the process of creating the *Kepler Input Catalog* (KIC) led by a team from SAO. The KIC is based on new multi-band photometric observations using the same g-r-i-z filter sequence as in the Sloan survey (Abazajian et al. 2003) plus an additional filter for the Mg b lines at 516.7, 517.3 and 518.4 nm. The net result is a catalog with classification information for the two million stars in the *Kepler* FOV to K<14. Using model fitting and a newly expanded Kurucz library of spectra (Castelli and Kurucz, 2003), the catalog contains the effective temperature, log(g), metallicity [Fe/H], reddening-extinction, mass and radius. The catalog was federated at the USNO Flagstaff Station with other catalogs, including 2MASS and USNO-B for cross reference and contains about 15 million objects.

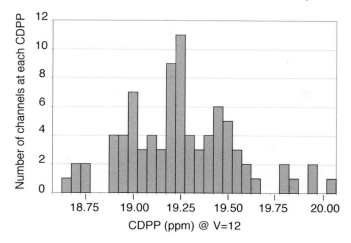

Figure 3. Combined Differential Photometric Precision (CDPP) for each of the 84 data channels for a V=12 solar-like star taking into account the measured noise from each channel, the effects of the predicted pointing stability, shot noise and the canonical 10 ppm for stellar variability.

Prior to launch, each star will be ranked for its potential for terrestrial planet detection using a merit function to determine the minimum detectable planet size in or near the HZ of each star. This process will be reapplied post-launch incorporating the measured CDPP for each star to re-rank the stars on the target list as part of a necessary on-going down-selection process.

5. Data Processing

On-board the integration time for the CCDs may be set from 2.5 to 8.0 sec. The longer integration time helps to improve the CDPP for the fainter objects, but also results in saturation of more of the brighter stars. With an integration time of 5 sec, stars brighter than about V=12 saturate. However, we have demonstrated in laboratory testing that precision photometry can still be achieved with bloomed pixels provided the rail voltages are properly set on the CCD and that the full scale on the analog-to-digital converter is greater than full well. The CCDs are read out in half a second without a shutter. Individual integrations are co-added on-board for thirty minutes, although for a subset of 512 target stars, one-minute co-additions are preserved. Once a thirty-minute co-add is accumulated in computer memory for all the 95 megapixels in the focal plane, the pixels of interest for the target stars are read out. There are additional collateral pixels from over-clocking and from masked regions. These are used for removing the bias, smear and determining the dark level. The values are re-quantized to account for the larger value of noise on the high end of the scale. The data are then Huffman encoded (compressed) and stored on the solid-state recorder for later transmission to the ground.

Smear is a result of clocking out the data without a shutter. Every pixel in a column passes under every piece of sky in a column during a read out. This produces a column-unique constant offset. And it also produces an optical fat-zero that helps to keep the traps full.

On the ground, the raw data are unpacked, decompressed and archived. First a 2-D bias correction is made. Then the correction for undershoot is made, followed by the gain and non-linearity corrections. Cosmic ray hits are identified and removed. This is followed by the smear, dark current and flat-field corrections. Then two parallel paths are followed to obtain stellar flux time series, one using optimal aperture photometry and the other using difference image analysis. Ensemble normalization is applied to the raw flux time series to remove common-mode noise at each cadence in time. These data are then used to produce de-trended relative flux time series for each target. These light curves are then archived and will be made available for others to use as well, for example, for asteroseismic analysis and eclipsing binary modeling. To each time series a wavelet transform is applied and conditioned with a whitening matched filter. The time series are then folded modulo all possible orbital periods and searched for a multiple-event-detection statistic above a threshold of seven sigma for planetary transits, yielding an 84% detection rate for an eight sigma folded transit signal. A similar process using a Fourier transform rather than a wavelet transform is used to conduct the reflected light search for short period non-transiting giant planets (Jenkins 2004).

All threshold crossing events then go through a validation process before they are considered viable candidates. This includes ensuring that each individual transit detected is consistent within the statistical limits to the character of the folded event. The individual pixels that make up each star are also examined to ensure that they do not exhibit any anomalous behavior. Finally, the results from the difference image analysis (DIA) method are examined to determine if there was any centroid shift during the event, which would be indicative of either a background eclipsing binary or a background transiting giant planet that is slightly off center from the target star and within the unresolved image.

6. Follow-up Observing Program

First, any transiting planet candidate with sufficient SNR to warrant it will be switched from the long cadence (thirty-minute) target list to the short cadence (one-minute) target list, to provide better resolution of the transit time, duration and shape.

Considerable ground-based observing resources will be utilized to eliminate false positives and to characterize both the host star and the planet itself where possible. This will be a step wise process, which will begin with using moderate precision radial velocity measurements to eliminate binary systems, such as with the Tillinghast Reflector Echelle Spectrograph on Mt. Hopkins and using high spatial resolution images from WIYN at Kitt Peak to eliminate any background object not detected with DIA. Once the viability of the candidate has been established, then more precious resources will be utilized such as the HET at McDonald Observatory, Keck and ultimately HARPS-North. The latter will have the capability of confirming the detection of planets as small as 2 Earth-masses out to 0.04 AU with 20% uncertainty with 50 hrs of observing of a V=12 G2V star (Lovis, et al. 2007). All of these spectra will also be used to improve our understanding of the parent star and to search for any companion non-transiting planets in the system.

7. Additional Planned Analyses

In addition to using the data to search for sequences of transits and reflected light from (not necessarily transiting) close-in giant planets, the data will also be used for: obtaining parallax measurements, thereby obtaining the stellar size from the luminosity and distance, which is essential to knowing the size of the transiting planet; determining stellar rotation rates; and performing asteroseismic analysis (Christensen-Dalsgaard et al., 2007). The latter analysis will be performed by the Kepler Asteroseismic Science Consortium (KASC) to measure p-mode oscillations (Brown and Gilliland, 1994) of stars in the FOV brighter than about V=11.5, which have been observed with a one-minute cadence. This will yield the mass, radius, density and age of those stars.

8. Community Participation and Data Access

The community can participate in the *Kepler Mission* in several ways: a participating scientist program (PSP), a guest observer (GO) program, and an Astrophysical Data Analysis Program (ADP), all of which will be competed by NASA Headquarters through the annual Research Opportunities in Space and Earth Sciences (ROSES). Although the program is open to scientists worldwide, only US proposals may receive funding.

8.1. *Participating Scientist Program*

The PSP is conceived to solicit proposals from scientists in the community to provide for potential direct enhancements to achieving the primary goals of the *Kepler Mission*. As described in the NASA ROSES-2007 section D.10, this includes such things as, detection of non-transiting planets using timing variations, improvements in determining the size of stars, and performing ground-based observing to aid in elimination of false-positives. Eight proposals have been selected for addition to the *Kepler* science team.

8.2. *Guest Observer Program*

Within the GO program, scientist may propose to view objects within the *Kepler* FOV which are not already on the planet detection target list, whether galactic or extragalactic. The *Kepler Target Catalog* (KTC) is expected to become publicly available shortly before launch. In general one may assume that all F-, G-, and K-dwarf stars to V=14–15 and M-dwarf stars to V=16.5 are already on the list and any other object is not. Proposed objects will typically be observed for a minimum of three months and to as long as the mission duration, based on justification. Capacity for three thousand objects has been set aside for this program. The data will be processed using the standard *Kepler* pipeline to produce de-trended light curves. Solicitation for the GO is expected prior to launch with observing to begin shortly after commissioning.

8.3. *Astrophysical Data Analysis Program*

There are potentially a host of other astrophysical uses for the *Kepler* data (Granados & Borucki, 1994), given the uniqueness in precision, completeness, duration and number of stars in the archived data. Potential uses include such things as; analysis of white light flaring and stellar activity, which can yield star spot cycles, especially if the mission is extended beyond about half of a solar cycle; the frequency of Maunder minimums for solar-like stars, which has implications for paleoclimatology and perhaps the future of our Earth's climate; cataclysmic variables, providing pre-outburst activity and mass transfer rates; and active galactic nuclei, providing a measure of the "engine" size in BL Lacs, quasars and blazers.

9. Status and Summary

Kepler is NASA's first mission capable of detecting Earth-size and smaller exoplanets in the HZ. The hardware and software are progressing through development toward a launch in February of 2009. The photometer assembly, integration and test will be completed in early 2008 and delivered for spacecraft integration. The mission is designed to be capable of detecting hundreds of terrestrial planets, if they are common, or provide a significant null result if they are not. Either result would be profound.

Acknowledgements

Funding for this mission is provided by NASA's Discovery Program Office.

References

Abazajian, Kevork *et al.* 2003, The First Data Release Of The Sloan Digital Sky Survey, *ApJ*,126, 2081-2086

Borucki, W. J. & Summers, A. L. 1984, *Icarus* 58, 121

Borucki, W. J., Koch, D., Basri, G., Brown, T., Caldwell, D., DeVore, E., Dunham, E., Gautier, T., Geary, J., Gilliland, R., Gould, A., Howell, S., Jenkins, J. & Latham, D. 2005, *Kepler Mission*: Design, Expected Science Results, Opportunities to Participate, in (ed. M. Livio),*A Decade of Extrasolar Planets around Normal Stars* (Cambridge: Cambridge University Press) in preparation

Brown, Timothy, M. & Gilliland, Ronald L. 1994, Asterioseismology,*ARAA*, 32, 37–82

Castelli, F. & Kurucz, R. L. 2003, New grids of ATLAS9 model atmospheres, in (eds. N.E. Piskunov, W. W. Weiss. and D. F. Gray) *IAU Symposium 210, Modelling of Stellar Atmospheres*

Christensen-Dalsgaard, J., Arentoft, T., Brown, T. M., Gilliland, R. L., Kjeldsen, H., Borucki, W. J. & Koch, D., 2007, The *Kepler* Asteroseismic Investigation. in (eds L. Gizon & M. Roth) *Proc. HELAS II International Conference: Helioseismology, Asteroseismology and the MHD Connections, (Göttingen, , J. Phys.: Conf. Ser.)*, in the press.

Frohlich, C. 1987, *JGR* 92, 796

Granados, A. F. & Borucki, W. J., (eds.) 1994 *Astrophysical Science with a Spaceborne Photometric Telescope NASA Conf.Publ. 10148* (Mt View)

Jenkins, Jon M, 2002, The Impact of Solar-like Variability on the Detectability of Transiting Terrestrial Planets, *ApJ* 575, 493

Jenkins, Jon M.& Doyle, Laurance R. ,2003, Detecting Reflected Light from Close-in Extrasolar Giant Planets with the Kepler Photometer, *ApJ*, 595, 429

Kasting, J. F., Whitmire, D. P., & Reynolds, R. T. 1993, *Icarus* 101, 108

Koch, D. G., Borucki, W., Dunham, E., Jenkins, J., Webster, L., & Witteborn, F. 2000, CCD Photometry Tests for a Mission to Detect Earth-Size Planets in the Extended Solar Neighborhood, in *SPIE Conference 4013, UV, Optical and IR Space Telescopes and Instruments*, (Munich, Germany)

Koch, D., Borucki, W., Dunham, E., Geary, J., Gilliland, R.,, Jenkins, J., Latham, D., Bachtell, E., Berry, D., Deininger, W., Duren, R., Gautier, T. N., Gillis, L., Mayer, D., Miller, C., Shafer, D., Sobeck, C., Stewart, C., & Weiss, M. 2004, Overview and status of the *Kepler Mission*, in *SPIE Conf 5487, Optical, Infrared, and Millimeter Space Telescopes*, (Glasgow, Scotland)

Koch, D. G., Borucki, W., Basri, G., Brown, T., Caldwell, D., Christensen-Dalsgaard, J., Cochran, W., Dunham, E., Gautier, T., Geary, J., Gilliland, R., Jenkins, J., Kondo, Y., Latham, D., Lissauer, J., & Monet., D. 2006, The *Kepler Mission*: Astrophysics and Eclipsing Binaries, *ApSS* 304, 389

Krivova, N. A., Solanki, S. K., & Floyd, L. 2006, *A&A*, 452, 631

Lovis. C. *et al.*, 2007, in (eds. Debra Fischer, Fred Rasio, Steve Thorsett & Alex Wolszczan)*ASP Conference Series, Proceedings of the Santorini conference on Extreme Solar Systems*, in preparation.

Mayor, M. & Queloz, D. 1995, A Jupiter-Mass Companion to a Solar-Type Star, *Nature*, 378, 355

Rivera, Eugenio J., Lissauer, Jack J., Butler, R. Paul, Marcy, Geoffrey W., Vogt, Steven S., Fischer, Debra A., Brown, Timothy M., Laughlin, Gregory, & Henry, Gregory W. 2005, A 7.5 $M_\oplus$ Planet Orbiting the Nearby Star, GJ 876, *ApJ*, 634, 625

Sahu, K. C. & Gilliland, R. L. 2003, *ApJ*, 584, 1042

Willson, R. C., Gulkis, S., Janssen, M., Hudson, H. S. & Chapman, G. A. 1981, *Science* 211, 700

Extrasolar planet detections with gravitational microlensing

Shude Mao, Eamonn Kerins and Nicholas J. Rattenbury

Jodrell Bank Centre for Astrophysics, University of Manchester, Manchester M13 9PL, UK
email: (shude.mao, eamonn.kerins, nicholas.rattenbury)@manchester.ac.uk

Abstract. Microlensing light curves due to single stars are symmetric and typically last for a month. So far about 4000 microlensing events have been discovered in real-time, the vast majority toward the Galactic centre. The presence of planets around the primary lenses induces deviations in the usual light curve which lasts from hours (for an Earth-mass [$M_\oplus$] planet) to days (for a Jupiter-mass [M_J] planet). Currently the survey teams, OGLE and MOA, discover and announce microlensing events in real-time, and follow-up teams (together with the survey teams) monitor selected events intensively (usually with high magnification) in order to identify anomalies caused by planets. So far four extrasolar planets have been discovered using the microlensing technique, with half a dozen new planet candidates identified in 2007 (yet to be published). Future possibilities include a network of wide-field 2m-class telescopes from the ground (which can combine survey and follow-up in the same setup) and a 1m-class survey telescope from space.

Keywords. gravitational lensing, Galaxy: centre, planetary systems

1. Introduction

Gravitational microlensing refers to the fact that when an intervening object (the lens) is well-aligned with a background star, then the background star will be amplified. Due to relative motions, the source appears brighter as the lens moves closer to the line of sight, and falls back to the baseline as the lens moves away. The resulting light curve is symmetric and follows a characteristic shape. Since microlensing does not depend on whether the lens is luminous or dark, it was first proposed as a way to detect massive compact dark matter object in the Milky Way (Paczynski 1986).

For a source in the Local Group, the chance for a star to be microlensed is small, of the order of 10^{-6}. To have a realistic yield of microlensing events, very dense fields have to be monitored. The targets so far include the Galactic centre, the Magellanic clouds, and M31. Hundreds of millions of stars are routinely monitored, and thousands of microlensing events have been discovered, with diverse applications (see section 2 and the review by Paczynski 1996).

It was first realised that microlensing can be used to discover extrasolar planets more than 15 years ago (Mao & Paczynski 1991; Loeb & Gould 1992). The importance of observing high-magnification events was realised by Griest & Safizadeh (1998) while the sensitivity of microlensing down to Earth masses was highlighted by Bennett & Rhie (1996).

It was optimistically written in Mao & Paczynski (1991) that "A massive search for microlensing of the Galactic bulge stars may lead to a discovery of the first extrasolar planetary systems". In fact, the first discovery by microlensing did not come until 2004 (Bond *et al.* 2004), while other methods (particularly the radial velocity method, see the contributions by Mayor and Marcy for details) made rapid progress. In the last few years, the pace of microlensing discoveries clearly quickened; we expect many more exciting discoveries yet to come.

Two excellent reviews on microlensing detection of extrasolar planets appeared very recently (Rattenbury 2006; Gaudi 2007); the readers are referred to those papers for more in-depth discussions. Future possibilities from the ground and space are discussed in two recent white papers (Gould *et al.* 2007; Bennett *et al.* 2007b).

2. Status of microlensing surveys

After the proposal by Paczynski (1986), several collaborations started to survey the Galactic centre and the Magellanic clouds (MACHO, Alcock *et al.* 1993; EROS, Aubourg *et al.* 1993; OGLE, Udalski *et al.* 1992; MOA, Muraki *et al.* 1999). Several groups (e.g. Calchi Novati *et al.* 2005; de Jong *et al.* 2006; Kerins *et al.* 2006) are also observing the M31 where the stellar population is unresolved (Crotts 1992; Gould 1996).

So far, about 4000 events have been discovered real-time, the vast majority towards the Galactic centre; an equal number is likely yet to be identified in the archive. The current discovery rate of microlensing events in real-time by the OGLE† and MOA‡ collaborations is close to 1000 events per year (with some overlapping events). The experiment toward M31 can also now issue real-time alerts of microlensing events (Darnley *et al.* 2007), which may be particularly relevant for the detection of extrasolar planets in M31 (Chung *et al.* 2006).

The Einstein radius crossing time for a typical event toward the Galactic centre is about 20 days, but can be as short as one day and as long as four years (OGLE-1999-BUL-32, Mao *et al.* 2002). A typical peak magnification is a factor of a few, but events (for bright stars) with peak magnification a few percent above unity have been identified; the highest peak magnification is about 3000 (Dong *et al.* 2006).

The microlensing surveys have not only detected thousands of microlensing events, they also accumulated photometric and astrometric data for hundreds of millions of stars for over a decade. Unfortunately, much of the database has yet to be fully explored. For example, the data can be used to provide strong constraints on the Galactic structure using the maps of optical depths (Kiraga & Paczynski 1994), proper motions (e.g. Kozlowski *et al.* 2006; Rattenbury *et al.* 2007a), stellar populations (e.g., Stanek *et al.* 1997; Rattenbury *et al.* 2007b). However, this potential has yet to be fulfilled partly due to the lack of man-power in the field.

Undoubtedly the most exciting highlight of microlensing in the last few years is the discovery of extrasolar planets from analysis of individual microlensing events, a topic we will turn to next.

3. Principles of extrasolar planet detection

A single point lens always creates two images of a background source, one magnified and one de-magnified. One special case is when the source, lens and observer are are all perfectly aligned, then due to axis-symmetry, a ring of images (called Einstein ring) will form (see Fig. 1 for an illustration).

The presence of a planet can perturb the two existing images (and create an extra faint image close to itself). In addition, in some cases, the planet can create an extra pair of bright images when the source is located inside the so-called caustics (see Fig. 1). Thus for caustic-crossing trajectories, much more dramatic deviations in the light curve can be seen. Both types of light curves have been observed in planetary microlensing events.

† see www.astrouw.edu.pl/~ogle/ogle3/ews/ews.html
‡ www.phys.canterbury.ac.nz/moa/microlensing_alerts.html

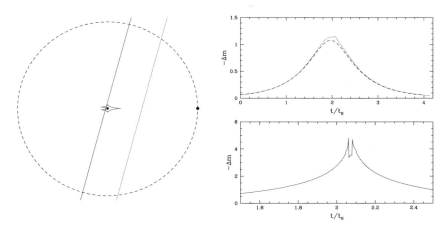

Figure 1. Geometry and light curves of planetary lensing. The primary star is located at the centre of the Einstein ring (dashed curve). A planet of 0.1% of the mass of the primary star is located to the right on the Einstein ring (indicated by a dot). The central cuspy feature around the primary star is the (central) caustic within which an extra pair of images are created. Two source trajectories are shown, with the corresponding light curves (magnitudes vs. time in units of the Einstein radius crossing time, t_0) shown on the right. The bottom light curve corresponds to the light curve for the trajectory on the left which intercepts the caustic four times. The top light curve corresponds to the trajectory on the right; the dashed curve shows the best fit from the single lens model. The source size is assumed to be 1.67×10^{-3} Einstein radius.

Currently the survey teams discover the microlensing events and announce these in real-time. Follow-up teams, together with the survey teams, then observe selected events (typically with high magnification) to identify deviations from the standard single light curve, through either visual inspection (e.g., for μFun¶) or automated algorithms (Dominik et al. 2007). The latter may allow a better evaluation of the detection efficiency (Cassan 2008). Once an anomaly is detected, a much more intensive observing campaign is launched, involving multiple sites and large international teams. Detailed modelling can then indicate the presence of an extrasolar planet. So far, four extrasolar planets have been discovered (and published) using the gravitational microlensing technique, with a probable six new discoveries made in 2007 (yet to be published).

The rate and duration of the planetary perturbations roughly scale as $q^{1/2}$, where q is the mass ratio between the planet and its host star. Typically, planetary deviations last from days for $1 M_{\rm J}$ planets to hours for one $M_\oplus$ planet. It is worth emphasizing that the deviation amplitude can be high even for a $1 M_\oplus$ planet, especially when the source crosses the caustics (see Fig. 1). Microlensing is most sensitive to low-mass planets in the so-called 'lensing zone', between 0.6-1.6 Einstein radii (corresponding to roughly 0.5-2.4 AU in physical units).

Microlensing is also sensitive to free-floating planets, which manifest themselves as short single events lasting hours to days. Furthermore, the method is sensitive to multiple planets (see the contribution from Bennett in these proceedings for an example).

From fitting of a binary light curve, the two most important quantities one can learn about the planet is the mass ratio, q, and the (projected) separation in units of the Einstein radius, b. If $q \lesssim 10^{-2}$, then the binary companion is likely a planet since typical primary lensing stars have $M \sim 0.3 M_\odot$. It is well known that for sparsely sampled data

¶ http://www.astronomy.ohio-state.edu/~microfun/

(e.g., Di Stefano & Mao 1996, Dominik 1999, Gaudi & Gould 1997) that different solutions (with distinct mass ratios and separations) can fit the same light curve. However, for a densely sampled light curve, there is often only a unique solution (in terms of q and b), except for the wide and close binary degeneracy, Dominik 1999). Densely monitored light curves frequently allow one to derive extra physical constraints, for example the angular Einstein radius; in some cases, the lens light can also be seen (e.g., Alcock et al. 2001; Kozlowski et al. 2007). In a few cases cases, the planet mass (not only q) can be inferred directly (for an illustration, see Bennett's contribution to these proceedings).

4. What have we learned from microlensing extrasolar planets?

So far four microlensing extrasolar planets have been published (Bond et al. 2004; Udalski et al. 2005; Beaulieu et al. 2006; Gould et al. 2006). The duration of the planetary deviations lasting from 7 days (for the first case, Bond et al. 2004) to about one day, for the lowest mass ($\sim 5.5 M_\oplus$) extrasolar planet (Beaulieu et al. 2006).

Microlensing clearly probes a different part of the parameter space in the plane of separation vs. planet mass (see Fig. 2 in Bennett et al. 2007b). Compared with other methods (such as the radial velocity method), microlensing can probe lower-mass planets at larger radii, including analogues of the Solar system.

While the statistics of extrasolar planets are still limited, one important conclusion can already be drawn: super-Earths are common (Gould et al. 2006). Considering the modest resources devoted to this method, this is a very important and cost-effective discovery, which is fully consistent with the conclusions from the radial velocity method (see Mayor's contribution in these proceedings) and the planet formation theory (Ida & Lin 2004).

The lack of detections in high signal-to-noise ratio light curves also allow us to put constraints on the frequency of planets around other stars. Gaudi et al. (2002) studied 43 high-quality light curves (with no planetary signals) and concluded that less than 1/3 of $\sim 0.3 M_\odot$ stars have Jupiter-mass companions between 1.5-4 AU (see also Snodgrass et al. 2004), consistent with the planet frequencies around solar-type stars from the radial velocity method.

5. Future

The current discovery rate of extrasolar planets from microlensing still lags behind the radial velocity method. To improve the efficiency, one must consider more ambitious strategies, including possibilities from both the ground and space. Theoretically, faster algorithms for searching the best planetary models are also desirable.

Gould et al. (2007) proposed to establish a network of four 2m telescopes (or eight 1.3m telescopes for the same photon gathering power) strategically located across the globe, with the primary goal of searching for Earth-mass planets. Ideally, each telescope would have 4 square degrees of view, with a cadence of about 10 minutes. Such a network would monitor 4 fields with a total of 16 square degrees continuously. The main advantage of such a network is that it combines the survey and follow-up teams in the same setup, thus avoiding the division between survey and follow-up networks. The upgraded MOA telescope (MOA-II) has a field of view of about 2 square degrees. This large field of view already combines survey and follow-up to some degree, which enabled the collaboration to make several important discoveries in 2007 (see the contribution by Sumi in these proceedings). Gould et al. (2007) showed that if each host star has two identical mass planets between 0.4-20 AU (uniformly distributed in per decade of separation) an aggressive wide-field survey can detect half a dozen $1 M_\oplus$ planets each year. The MOA-II and the soon-to-be-upgraded OGLE (OGLE-IV) can already be regarded as part of such a network. However, to complete the whole network, more funding is still needed. Efforts are currently under way to secure funding from various agencies.

Bennett et al. (2007b) proposed the Microlensing Planet Finder satellite mission to search for extrasolar planets. This will put a 1.1m telescope in space, equipped with a CCD camera with 0.7 sq. degree of field of view. The advantage is that there are no atmospheric effects and the spatial resolution is much higher, allowing fainter stars to be probed. Since the lower limit of the extrasolar mass we can detect is determined by the finite source size (which smooths and reduces the magnification), a space satellite can detect lower mass planets. Furthermore, the lens light can be routinely detected from space Bennett et al. (2007a), which, combined with other information, allows us to not only detect but also characterize extrasolar planets from $0.1 M_\oplus$ upwards in mass and ranging from 0.5 AU to essentially infinity in separation. The cost of the mission is not cheap, about $390 million, but the reward is also substantially higher. Its capability very nicely complements the Kepler mission in the discovery space.

Binary/planetary light curves are diverse. For an observed light curve, it is non-trivial to find the best-fit model and explore the parameter space fully. There are two difficult issues that make the exercise non-trivial; both are related to the presence of caustics. When a point source sits on a caustic, in principle its magnification is singular (infinite under geometric optics). However, stars have finite sizes and thus their microlensed magnification is always finite. To obtain the precise value, one must integrate over the singularity. Several methods (Gould & Gaucherel 1997; Dominik 2007; including ray-shooting techniques, see Bennett & Rhie 1996, Rattenbury et al. 2002) have been proposed, but it remains to be seen whether there are better ways of doing this. The second difficulty is that the χ^2 surface is not smooth, particularly for caustic-crossing events. Most optimization routines cannot easily jump over the sharp features to find the best solution. The problems are tractable for a single planet system, but become much more challenging for multiple planets. A completely satisfactory solution has yet to be found.

6. Summary

The progress in the last few years clearly demonstrates that the microlensing method can be used to discover extrasolar planets. The method is based on General Relativity and simple geometry. So far, about ten extrasolar planets have been discovered. These extrasolar planets are further away from the host stars than those discovered by the radial velocity method, and so it probes a different part of parameter space. The method will provide important statistical information about extrasolar planets, e.g., the microlensing planets already allow us to draw the remarkable conclusion that super-Earths are quite common around stars (Gould et al. 2006). A drawback is that microlensing is a one-time observation and most of the detected planets are at a distance of several kpc and thus too far away for direct imaging follow-up.

With more funding, microlensing has the possibility to make ground-breaking discoveries about Earth-mass planets from both the ground and space. The method complements other methods quite well in the discovery space.

Acknowledgements

SM thanks the local organisers, in particular Prof. Ji-lin Zhou, for travel support and a stimulating meeting at the ancient Chinese city of Suzhou.

References

Alcock, C., et al. 1993, *Nature*, 365, 621

Alcock, C., et al. 2001, Nature, 414, 617
Aubourg, E., et al. 1993, Nature, 365, 623
Beaulieu, J. -P., et al. 2006, Nature, 439, 437
Bennett, A. & Rhie, S. H. 1992, ApJ, 472, 660
Bennett, D. P., Anderson, J., & Gaudi, B. S. 2007a, ApJ, 660, 781
Bennett, D. P., et al. 2007b, arXiv:0704.0454
Bond, I. A., et al. 2004, ApJ, 606, L155
Calchi Novati, S. et al. 2005, A&A, 443, 911
Cassan, A., Sumi, T., & Kubas, D. 2008, in: Y.-S. Sun, S. Ferraz-Mello & J.-L. Zhou, (eds.), *Exoplanets: Detection, Formation and Dynamics*, Proc. IAU Symposium No. 249 (Suzhou,China), p. 31
Chung, S. -J., Kim D., Darnley, M. J., Duke, J. P., Gould, A., Han, C., Jeon, Y. -B., Kerins, E., Newsam, A., & Park, B. -G. 2006, ApJ, 650, 432
Crotts, A. P. S. 1992, ApJ, 399, 43
Darnley, M. J., Kerins, E., Newsam, A., Duke, J. P., Goud, A., Han, C., Ibrahimov, M. A., Im, M., Jeon, Y. -B., Karimov, R. G., Lee, C. -U., & Park, B. -G. 2007, ApJ, 661, L45
Di Stefano, R., Mao, S. 1996, ApJ, 457, 93
Dominik, M. 1999, A&A, 349, 108
Dominik, M. 2007, MNRAS, 377, 1679
Dominik, M., Rattenbury, N. J., Allan, A., Mao, S. et al., 2007 MNRAS, 380, 792
Dong, S., DePoy, D. L., Gaudi, B. S., Gould, A. et al. 2006, ApJ, 642, 842
Gaudi, B. S. & Gould, A. 1997, ApJ, 486, 85
Gaudi, B. S., et al. 2002, ApJ, 566, 463
Gaudi, B. S. 2007, in: D. Fischer, F. Rasio, S. Thorsett and A. Wolszczan (eds.), *Extreme Solar Systems* (ASP Conference Series), arXiv:0711.1614
Gould, A. 1996, ApJ, 470, 201
Gould, A. & Loeb, A. 1992, ApJ, 396, 104
Gould A. & Gaucherel C. 1997, ApJ, 477, 580
Gould, A., et al. 2006, ApJ, 644, L37
Gould, A., Gaudi, B. S., & Bennett, D. P 2007, arXiv:0704.0767
Griest, K. & Safizadeh, N. 1998, ApJ, 500, 37
Ida, S. & Lin, D. N. C. 2004, ApJ, 616, 567
de Jong, J. et al. 2006, A&A, 446, 855
Kerins, E. 2006, MNRAS, 365, 1099
Kiraga, M. & Paczynski, B. 1994, ApJ, 430, L101
Kozlowski, S., Wozniak, P. R., Mao, S., Smith, M. C., Sumi, T., Vestrand, W. T., & Wyrzykowski, L. 2006, MNRAS, 370, 435
Kozlowski, S., Wozniak, P. R., Mao, S., & Wood, A. 2007, ApJ, in press
Mao, S., Smith, M. C., Wozniak, P., Udalski, A., Szymanski, M., Kubiak, M., Pietrzyski, G., Soszyski, I., & Zebrun, K. 2002, MNRAS, 329, 349
Mao, S. & Paczynski, B. 1991, ApJ, 374, L37
Muraki, Y., Sumi, T., Abe, F., Bond, I. et al. 1999, *Progress of Theoretical Physics*, 133, 233
Paczynski, B. 1986 ApJ, 304, 1
Paczynski, B. 1996 ARAA, 34, 419
Rattenbury, N. J., Bond, I. A., Skuljan, J., & Yock, P. C. M. 2002, MNRAS, 335, 159
Rattenbury, N. J. 2006, *Modern Physics Letters A*, 21, 919
Rattenbury, N. J., Mao, S. Debattista, V. P., Sumi, T., Gerhard, O., & de Lorenzi, F. 2007, MNRAS, 378, 1165
Rattenbury, N. J., Mao, S., Sumi, T., & Smith, M. C. 2007 MNRAS, 378, 1064
Snodgrass, C., Horne, K., & Tsapras, Y. 2004, MNRAS, 351, 967
Stanek, K. Z., Udalski, A., Szymanski, M., Kaluzny, J., Kubiak, M., Mateo, M., & Krzeminski, W. 1997, ApJ, 477, 163
Udalski, A., Szymański, M., Kałużny, J., Kubiak, M., & Mateo, M. 1992, *Acta Astronomica*, 42, 253
Udalski, A., et al. 2005, ApJ, 628, L109

Microlensing search for extrasolar planets: observational strategy, discoveries and implications

Arnaud Cassan[1,4]† and Takahiro Sumi[2,5]‡ and Daniel Kubas[3,4]

[1] ARI/ZAH Heidelberg University, Germany
[2] Nagoya University, Japan
[3] ESO, Chile
[4] PLANET/RoboNET Collaborations
[5] MOA Collaboration

Abstract. Microlensing has proven to be a valuable tool to search for extrasolar planets of Jovian- to Super-Earth-mass planets at orbits of a few AU. Since planetary signals are of very short duration, an intense and continuous monitoring is required. This is achieved by ground-based networks of telescopes (PLANET/RoboNET, μFUN) following up targets, which are identified as microlensing events by single dedicated telescopes (OGLE, MOA). Microlensing has led to four already published detections of extrasolar planets, one of them being OGLE 2005-BLG-390Lb, a planet of only $\sim 5.5\,M_\oplus$ orbiting its M-dwarf host star at ~ 2.6 AU. Very recent observations (May–September 2007) provided four more planetary candidates, still under study, that will double the number of detections. For non-planetary microlensing events observed from 1995 to 2006 we compute detection efficiency diagrams, which can then be used to derive an estimate of the Galactic abundance of cool planets in the mass regime from Jupiters to Sub-Neptunes.

Keywords. Extrasolar planets, Gravitational microlensing

1. Introduction

A Galactic microlensing event occurs when a massive compact intervening object (the lens) deflects the light coming from a more distant background star (the source). It leads to an apparent flux brightening (or magnification) of the source star. In a typical scenario, the source belongs to the Galactic bulge, while the lens can be part either of the bulge (2/3 of the events) or the disk (1/3 of the events) population. Mao & Paczyński (1991) were the first to suggest that microlensing could provide a powerful tool to search for extrasolar planets at distances of a few kpc, provided a continuous monitoring of bulge stars. Since the detection of planets by microlensing does not rely on their light but on their mass, the planetary host mass function basically follows the stellar mass function of the Galaxy, implying that planet hosts are preferably low-mass K to M dwarfs.

The relative motion between source, lens and observer ($\sim 15\,\mu$as/day) induces a variation of the magnification factor with time, with a typical time scale of $t_\mathrm{E} \simeq 40 \times \sqrt{M_*/M_\odot}$ days, where M_* is the lens mass and assuming a source and lens distance of respectively 8.5 and 6.5 kpc. The duration of the planetary light curve signal then scales as $t_\mathrm{p} \approx 2\sqrt{q} \times t_\mathrm{E}$, where q is the planet-to-star mass ratio, which means few days for a giant planet to only few hours for Neptune- to Earth-mass planets.

† Based on the talk "Microlensing search for extrasolar planets".
‡ Based on the talk "Exoplanet Candidates from the MOA-II Microlensing Survey in 2007".

The microlensing method is remarkable in the sense that it probes a domain in the planet mass-orbit diagram that is mainly out of reach of other techniques, for it is mostly sensitive to Jovian- down to Earth-mass planets (e.g. Bennett & Rhie 2002) with orbits of $\sim 1 - 10$ AU, at several kpc.

2. Alerts and follow-up of microlensing events

Microlensing surveys (OGLE & MOA) currently monitor more than $\sim 10^8$ Galactic bulge stars by 1-2 m class telescopes on a daily basis to find and alert microlensing events. The second phase of MOA, MOA-II, carries out survey observations toward the Galactic bulge to find exoplanets via microlensing using a 1.8m telescope in New Zealand. We observe our target fields (~ 50 deg^2) very frequently (10 to 50 times/night) and analyze data in real-time to issue microlensing alerts. This high cadence is specifically designed to find the short duration time signatures characteristic of planets orbiting the lens star. In 2007, real-time event monitoring started in order to search for planetary signatures in ongoing microlensing events. Each new data point on the light curves is available within 5 minutes after image exposure. During 2007, MOA has detected around 500 microlensing events.

While single alert telescopes are able to identify and follow microlensing planetary candidates, they somehow suffer from gaps in the data coverage. Network of telescopes, as operated by the PLANET/RoboNET or μFUN collaborations perform a "round-the-clock" monitoring of a reduced number of selected targets, which significantly increases the planet detection efficiency. For example, with currently five 1m-class telescopes located in Chile, two in South Africa, Australia and Tasmania, as well as using three robotic telescopes, PLANET/RoboNET currently has unequaled capability for covering microlensing events, by minimizing data gaps in which planetary signatures could hide.

Recent efficient and interactive communication between alert and network collaborations has played a major role in improving the ability to quickly focus on suspected planetary signal in many events.

3. Results

3.1. Detections and new planetary candidates

As of now four extrasolar planets detected by microlensing are reported in the literature (see eg. Rattenbury 2006 for a more detailed review). Among them there are two giants of few Jupiter masses, MOA 2003-BLG-53Lb (Bond et al. 2004) and OGLE 2005-BLG-071Lb (Udalski et al. 2005), as well as two Super-Earth-mass planets, OGLE 2005-BLG-169Lb (Gould et al. 2006) and OGLE 2005-BLG-390Lb (Beaulieu et al. 2006), the latter being one the lightest ever discovered planets, with only $\sim 5.5\,M_\oplus$ and a wide orbit of ~ 2.6 AU. This first detection of a cool rocky/icy sub-Neptune mass planet has thus opened a new observing window for the exoplanet field.

Very recent observations (May–September 2007) have revealed four more planetary candidates which are in the process of final analysis and are expected to be published in 2008. These new detections will double the number of detections and should provide a better understanding of the statistical properties of the microlensing planet population and also help to optimize he detection efficiency of the current observational set up (see 3.3).

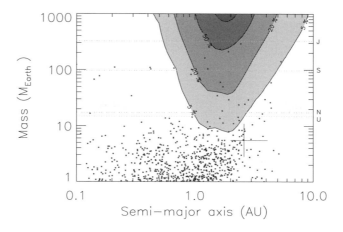

Figure 1. Detection efficiency for an additional planet orbiting OGLE-2005-BLG-390L as a function of its orbital separation and mass. Efficiency contours of 5%, 20%, 50% and 70% are shown. The cross marks the median values for the properties of OGLE 2005-BLG- 390Lb along with 1 σ confidence intervals and the dashed horizontal lines mark the masses of Jupiter (J), Saturn (S), Neptune (N) and Uranus (U) for comparison. The blue dots represent the predicted final distribution of a seed of 2×10^4 planetary cores around an M-dwarf of 0.2 $M_\odot$ resulting from a core-accretion model assuming inefficient migration (taken from Fig. 9b of Ida & Lin (2005)).

3.2. Limits on the multiplicity of planetary systems

Microlensing also allows the direct detection of a multi-planetary system (see the contribution from D. Bennett, this volume), as well as of circum-binary planets.

When only a single planet is detected, it is still possible to put limits on the presence of further planets in the microlensing event, such as in the case of OGLE 2005-BLG-390 (Kubas *et al.* 2007), presented in Fig. 1. Although the detection efficiency depends strongly on the reached peak magnification, a good data coverage can result in significant detection sensitivities even for low peak magnification events. In this particular case one finds that more than 50% of potential planets with a mass in excess of 1 M_J between 1.1 and 2.3 AU around OGLE 2005-BLG-390L would have revealed their existence, which was however not observed. For gas giant planets above 3 M_J in orbits between 1.5 and 2.2 AU, the detection efficiency exceeds 70%. Furthermore we find a detection probability for an OGLE-2005-BLG-390Lb-like planet, given an idealization of the microlensing technique, to be around 1 − 3 %. In agreement with current planet formation theories this quantitatively supports the prediction that sub-Neptune mass planets are common around low mass stars.

3.3. Detection efficiencies

Apart from the detection of planets a main goal of microlensing observations is to estimate the planet detection efficiency in order to put constraints the Galactic planet abundance. From 42 densely monitored events between 1997 and 1999, PLANET was able to provide the first significant upper abundancy limit of Jupiter- and Saturn-mass planets around M-dwarfs, namely that less than 1/3 of the lens stars have Jupiter-mass companions at orbital radii between 1.5 and 4 AU, and less than 2/3 have Saturn-mass companions for the same range of orbital radii, assuming circular orbits (Gaudi *et al.* 2002).

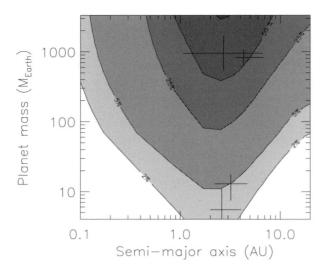

Figure 2. PLANET detection efficiency from the 2004 season (preliminary diagram), as a function of planet mass and orbital separation. The crosses are the detected planets with their parameter error bars.

By using an adequate Galactic model for the distribution of lens masses and velocities (Dominik 2006), we aim to pursue and improve the study, moreover taking into account more than ten years of observations (Cassan *et al.* 2008). The Fig. 2 shows a preliminary planet detection efficiency diagram, computed from well-covered events of the 2004 season.

4. Summary and prospects

Microlensing has proven to be a robust method to search for extrasolar planets at large separations from their parent stars ($\sim 1 - 10$ AU). It is sensitive to masses down to the mass of the Earth using ground based telescopes and even capable to detect planets of a few fractions of Earth masses when considering space-based telescope scenarios.

Microlensing is also very well-suited for statistical studies on planet abundance in the Galaxy. In fact, the method is by essence not limited to our close solar neighborhood or to a particular type of host stars.

References

Mao & Paczyński, 1991, ApJ, 374, L37
Bennett & Rhie 2002, ApJ, 574, 985
Bond *et al.* 2004, ApJ, 606, L155
Udalski *et al.* 2005, ApJ, 628, L109
Beaulieu *et al.* 2006, Nature, 439, 437
Gould *et al.* 2006, ApJ, 644, L37
Rattenbury 2006, Modern Physics Letter A, 21, 919
Kubas *et al.* 2007, to appear in A&A (astro-ph/0710.5306)
Gaudi *et al.* 2002, ApJ, 566, 463
Dominik 2006, MNRAS, 367, 669
Cassan *et al.* 2008, in preparation

ARTEMiS (Automated Robotic Terrestrial Exoplanet Microlensing Search) – Hunting for planets of Earth mass and below

Martin Dominik[1]†, Keith Horne[1], Alasdair Allan[2], Nicholas J. Rattenbury[3], Yiannis Tsapras[4], Colin Snodgrass[5], Michael F. Bode[4], Martin J. Burgdorf[4], Stephen N. Fraser[4], Eamonn Kerins[3], Christopher J. Mottram[4], Iain A. Steele[4], Rachel A. Street[6], Peter J. Wheatley[7] and Łukasz Wyrzykowski[8,9]

[1] SUPA, University of St Andrews, School of Physics & Astronomy,
North Haugh, St Andrews, KY16 9SS, United Kingdom
email: (md35,kdh1)@st-andrews.ac.uk

[2] School of Physics, University of Exeter,
Stocker Road, Exeter EX4 4QL, United Kingdom
email: aa@astro.ex.ac.uk

[3] Jodrell Bank Centre for Astrophysics, The University of Manchester,
Manchester, M13 9PL, United Kingdom
email: (Nicholas.Rattenbury,Eamonn.Kerins)@manchester.ac.uk

[4] Astrophysics Research Institute, Liverpool John Moores University,
Twelve Quays House, Egerton Wharf, Birkenhead, CH41 1LD, United Kingdom
email: (yt,mfb,mjb,snf,cjm,ias)@astro.livjm.ac.uk

[5] European Southern Observatory (ESO),
Casilla 19001, Santiago de Chile, Chile
email: csnodgra@eso.org

[6] Las Cumbres Observatory Global Telescopes Network,
6740B Cortona Dr, Goleta, CA 93117, United States of America
email: rstreet@lcogt.net

[7] Department of Physics, University of Warwick,
Coventry, CV4 7AL, United Kingdom
email: P.J.Wheatley@warwick.ac.uk

[8] Institute of Astronomy, University of Cambridge,
Madingley Road, Cambridge CB3 0HA, United Kingdom
email: wyrzykow@ast.cam.ac.uk

[9] Warsaw University Astronomical Observatory,
Al. Ujazdowskie 4, 00-478 Warszawa, Poland

Abstract. Gravitational microlensing observations will lead to a census of planets that orbit stars of different populations. From 2008, ARTEMiS will provide an expert system that allows to adopt a three-step strategy of survey, follow-up and anomaly monitoring of gravitational microlensing events that is capable of detecting planets of Earth mass and below. The SIGNALMEN anomaly detector, an integral part, has already demonstrated its performance during a pilot season. Embedded into eSTAR, ARTEMiS serves as an open platform that links with existing microlensing campaigns. Real-time visualization of ongoing events along with an interpretation moreover allows to communicate "Science live to your home" to the general public.

Keywords. gravitational lensing, planetary systems

---

† Royal Society University Research Fellow

1. Introduction

While the overwhelming majority of all exoplanets known so far are gas giants, and most of these are in close orbits around their respective host star, the last few years have seen the detection of the first rocky or icy planets, with subsequently more Earth-like ones. The increasing sensitivity of several techniques will soon lead to exploring the abundance of planets that could harbour life. However, an understanding of their origin will only arise once models of planet formation and orbital migration can be made to match observations over an embracing wide region of parameter space, thereby providing a test of sufficient power to distinguish between alternatives.

The technique of gravitational microlensing is currently unique in its capability to provide a sample of terrestrial exoplanets at galactic distances rather than just in the Solar neighbourhood, and moreover can probe planets in wider orbits, whose orbital periods prevent them being detected by other indirect techniques. In particular, a planetary census for different stellar populations, namely the Galactic disk and bulge, can be obtained, and even the abundance of planets around stars in neighbouring galaxies, such as M31, can be determined (Covone et al. 2000; Chung et al. 2006).

With several planets having been detected, the most spectacular one of about 5 Earth masses (Beaulieu et al. 2006; Dominik et al. 2006), microlensing has become well-established. As shown here, with ground-based observations, it competes for the first detection of an Earth-mass planet against space missions such as COROT or Kepler, and it could even go further below in mass.

2. Revealing exoplanets by microlensing

Gravitational microlensing is understood as the transient brightening of an observed star due to the bending of light caused by the gravitational field of a foreground star. A substantial effect requires the source star at distance D_S and the lens star with mass M at distance D_L to be located within an angle $\theta \lesssim \theta_E$ on the sky, where

$$\theta_E = \sqrt{\frac{4GM}{c^2}\left(D_L^{-1} - D_S^{-1}\right)} \quad (2.1)$$

is the *angular Einstein radius* (Einstein 1936). For source stars in the Galactic bulge, gravitational microlensing events last about a month, while source and lens star carry out their transverse motion according to the kinematics of the Milky Way.

A planet orbiting the lens star can reveal its existence by causing a short deviation to the observed light curve that is otherwise symmetric with respect to a peak, preferably if it is separated from its host star by an angle $\delta \sim \theta_E$ (Mao & Paczyński 1991). Depending on the mass of the planet, the relative proper motion between the lens and source star, and the angular size of the source star, such planetary anomalies last between several hours and two weeks. Therefore, they constitute a snapshot of the planet, irrespective of its orbital period. Given that typically $\theta_E \sim 350~\mu$as and $D_L \sim 6.5$ kpc, microlensing is most sensitive to the detection of planets at an orbital separation of 1 to 10 AU.

With only about one in a million monitored stars being significantly brightened by the gravitational field of a foreground star at a given time (Kiraga & Paczyński 1994), the OGLE (Optical Gravitational Lensing Experiment) and MOA (Microlensing Observations in Astrophysics) surveys monitor more than 100 million stars on a daily basis, which results in 700-1000 microlensing events per year being alerted on-line while they are in progress (Udalski 2003; Bond et al. 2001). While their sampling is however insufficient for detecting planets with masses significantly below that of Jupiter (Snodgrass *et al.* 2004), PLANET (Probing Lensing Anomalies NETwork) established the first microlensing follow-up network that was both capable of hourly sampling and a round-the-clock coverage on selected events (Dominik *et al.* 2002). While this network of 1m-class telescopes relies on human observers and dedicated observing time, the demand of not only an immediate response, but also a flexible scheduling makes robotic telescopes ideally suited to carry out such an observing programme. Since 2004 – and since 2005 in cooperation with PLANET –, microlensing observations have been carried out with the RoboNet-1.0 network of UK-built 2m robotic telescopes (Burgdorf *et al.* 2007), the largest of their kind. In contrast to PLANET/RoboNet, the MicroFUN team concentrate on a few quite promising events, with a network only being activated on target-of-opportunity basis.

The success of discovering a planet by microlensing critically depends on the availability of a telescope in a suitable location at the right time, which can mean within 10 min. To encourage follow-up observations, microlensing campaigns are therefore releasing photometric data in real time. For ongoing planetary anomalies, world-wide efforts are being undertaken to make sure that sufficient data are obtained, since there is no second chance.

3. An anomaly detector to hunt for (sub-) Earth-mass planets

In 2005, microlensing observations led to the discovery of OGLE-2005-BLG-390Lb, the first cool rocky/icy exoplanet ever found (Beaulieu *et al.* 2006; Dominik *et al.* 2006), with a mass of about 5 times that of Earth the least massive and most Earth-like exoplanet known at the time of its discovery. The light curve of the respective microlensing event OGLE 2005-BLG-390 showed a $\sim 15\,\%$ deviation over about a day due to the planet orbiting its lens star.

As Fig. 1 shows, an Earth-mass planet in the same spot would still have caused a signal amplitude of $\sim 3\,\%$ and a duration of the planetary anomaly of ~ 12 h. While the standard follow-up sampling of 2 h would have been insufficient for characterizing the observed deviation and thereby claiming the presence of a planet, this would have been possible with high-cadence (10–15 min) anomaly monitoring, triggered upon the first suspicion of a departure from the symmetric ordinary light curve. Since 2007, the SIGNALMEN anomaly detector (Dominik *et al.* 2007) scans incoming microlensing data in real time for ongoing anomalies, where further observations are requested successively until an anomaly can be confirmed or rejected with the required significance.

By means of robust-fitting techniques, SIGNALMEN eliminates the effect of outliers, and triggers on residuals whose absolute value is among the upper $5\,\%$ of all residuals for the respective site. Thereby, the fact that reported photometric error bars frequently do not properly represent the true uncertainties and in general do not follow a Gaussian distribution is addressed.

The source star in event OGLE-2005-BLG-390 was identified as a clump giant of spectral type G4 III with a radius $R_\star \sim 9.6\ R_\odot$. As compared to a main-sequence star, this led to a larger probability to detect a planetary signal and increased its duration, but reduced its amplitude. Provided that the exposure times are chosen long enough for achieving a photometric accuracy of 1–2 %, SIGNALMEN also allows to reveal an Earth-mass planet from a 5 % deviation for a main-sequence source star ($R_\star \sim 1.2\ R_\odot$). While the microlensing searches face a fair chance of detecting an exoplanet of Earth mass within two years, the detection of planets with masses as small as $0.1\ M_\oplus$ is challenging both by means of the short signal duration and the tiny probability for signals of appropriate amplitudes to occur, but nevertheless possible (Dominik *et al.* 2007).

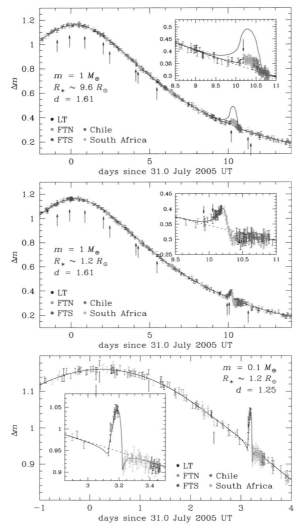

Figure 1. Simulations showing the possible detection of planets of Earth mass or below with the three robotic 2m-telescopes that constitute the RoboNet-1.0 network, namely the Liverpool Telescope (LT), the Faulkes Telescope North (FTN), and the Faulkes Telescope South (FTS), supplemented by two hypothetical further similar telescopes located in Chile and South Africa. For three different configurations, arrows indicate the epochs at which the SIGNALMEN anomaly detector requested further observations. (top) 1-$M_\oplus$ planet in the same spot as OGLE-2005-BLG-390Lb ($d = \delta/\theta_E = 1.61$), with the original model light curve also plotted; (middle) 1-$M_\oplus$ planet in the same spot, but with a main-sequence source star ($R_\star \sim 1.2\,R_\odot$ instead of $R_\star \sim 9.6\,R_\odot$); (bottom) 0.1-$M_\oplus$ planet at a closer distance ($d = 1.25$) to the lens star, and a main-sequence source star. For the middle and bottom panels, the orientation angle of the source trajectory has been adjusted for creating significant deviations. Light curves that would be observed in absence of any planets are shown as dashed lines.

4. SIGNALMEN performance during the 2007 pilot season

From April to October 2007, SIGNALMEN performed its pilot season, marking "phase zero" of ARTEMiS (Automated Robotic Terrestrial Exoplanets Microlensing Search), on observed data as these were released. SIGNALMEN provided 34 anomaly alerts on OGLE events that are clearly anomalous, and a further 3 on cases that are unclear due to the lack of follow-up data. Most notably, several candidate planetary signals have received crucial coverage as a result of SIGNALMEN flagging the respective event and the news being spread to the observing campaigns.

The art in designing a detection algorithm is in finding a suitable trade-off between the aims to maximize the number of correct detections, and to keep the number of false positives at a minimum. In particular, if very few false positives arise, the algorithm is almost certainly not well-balanced, but likely to be rather inefficient. After the elimination of obvious failures that can be grouped into 4 types that are fixed now by slight adjustments or further filtering, there were 3 false positives. While two of these refer to suggested signals at magnification $A \sim 1$, a required revision of the adopted model of an ordinary light curve has been mistaken for an anomaly in the remaining third case, which is a known and not unwelcome side-effect, because a dense sampling is required in order to sufficiently constrain the model parameters for keeping the opportunity for detecting anomalies (Dominik et al. 2007). For comparison, PLANET publicly circulated 35 anomaly alerts on OGLE events from 2004 to 2006, of which 4 were false positives. While SIGNALMEN detected a much larger number of anomalies per season, it provided a comparable fraction of false positives of 10–15 %.

SIGNALMEN did not fail to alert on any ongoing anomaly apparent by visual inspection that it should clearly have alerted on, while it considered insufficient evidence for an anomaly in 2 marginal cases. Coincidentially, it alerted on one real ongoing anomaly based on the false assumption of a significant deviation due to a shortcoming on finding an appropriate model, which is now fixed.

However, anomalies are missed under certain circumstances that deprive SIGNALMEN of any chance to detect an ongoing anomaly. In particular, SIGNALMEN cannot properly judge about anomalies that are already in progress with the first release of data on the respective event. Moreover, SIGNALMEN is designed to detect anomalies in progress in order to trigger the collection of further data that allow a proper characterization. If data are released in blocks rather than point-by-point, the anomaly may already be found over at the time of assessment, so that no ongoing anomaly can be reported. Finally, opportunities are missed when SIGNALMEN decides that an anomaly *may* be in progress, but an insufficient amount of data are subsequently collected for providing evidence before the potential anomaly is over. Given the fact that the microlensing surveys monitor substantially more events than current follow-up campaigns, the former harbour a potential for detecting anomalies, including such caused by planets, that so far is not efficiently exploited.

5. The ARTEMiS concept

In order to succeed with the detection of planets of Earth mass and below, several crucial requirements need to be met. Planetary anomalies in progress need to be sampled frequently, i.e. every 10 min or less, which also means that a telescope needs to be available in a suitable location. Moreover, the small probability for an Earth-mass planet to produce a deviation in a given event implies the need for many events to be monitored regularly. A brute-force approach would be a network of wide-field telescopes (e.g. Gould et al. 2007), carrying out a survey with a 10-min sampling, which comes with a substantial price tag for fully-dedicated facilities, and will not be realized in near time. However, a cooperative effort involving resources that already exist or are about to be commissioned, using a well-coordinated three-step strategy of survey, follow-up, and anomaly monitoring, is a serious alternative.

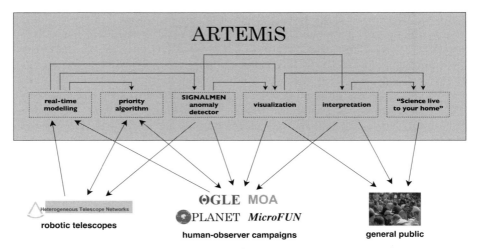

Figure 2. The components of ARTEMiS (Automated Robotic Terrestrial Exoplanet Microlensing Search) and how they interact with the outside world.

Even for an Earth-mass planet, a follow-up sampling interval of 1–1.5 h is sufficient for properly characterizing an anomaly from more frequent observations taken after a first trigger by SIGNALMEN. However, the current PLANET network of 1m-class telescopes is only able to monitor about 20 % of all events found by the OGLE and MOA survey, and thereby falls short of exploiting their full potential for detecting planets. In contrast, a network of either 2m or 4 × 1m telescopes at each site, of which RoboNet-1.0 is a prototype, is well-suited to make full use of the capabilities of current surveys. Nevertheless, the participation of further smaller telescopes can critically decide about whether data can be obtained at crucial stages of an ongoing anomaly or not.

As shown in Fig. 2, ARTEMiS, whose first phase will go active in April 2008, bundles several components into an expert system that can coordinate the efficient detection of planets of Earth mass and below between microlensing observing campaigns using either robotic or staffed telescopes, and moreover provides outreach and live information to the general public.

In order to be able to react within a few minutes, ARTEMiS relies on real-time observational data being provided, which are then used to derive model parameters for the light curve of the respective event. Based on these models, the SIGNALMEN anomaly detector will decide on a suspected or ongoing anomaly and suggest further observations at specific sites, or circulate a public anomaly alert, respectively. The same real-time models are also used to prioritize the selection of ongoing events for follow-up monitoring by each observing site, which can define its specific capabilities and commitment, in order to maximize the expected number of planet detections. Moreover, ARTEMiS guides the human observers with the visualization of the collected data and model light curves as well as the interpretation of anomalous events. The selection of targets for follow-up and anomaly monitoring by means of a fully-deterministic automated procedure allows to predict the detection efficiency by carrying out simulations.

The ARTEMiS concept shares most of its paradigm with the eSTAR (e-Science Telescopes for Astronomical Research) project (Steele *et al.* 2002), which builds a meta-network between existing proprietary robotic-telescope networks by providing a uniform interface based upon a multi-agent contract model (Allan *et al.* 2006). By embedding ARTEMiS into eSTAR and using the VOEvent (Virtual Observatory Event) protocol, a communication standard adopted by the IVOA (International Virtual Observatory Alliance), a direct interaction with the telescopes that are part of the HTN (Heterogeneous Telescope Networks) consortium becomes possible. Beyond that, ARTEMiS is also equipped with means of communication with human observers and the general public.

The real-time availability of both the data and their (preliminary) analysis allows to bring forefront science as a live event to the interested general public, who not only can watch a planet being discovered, but as amateur astronomers even can contribute to the discovery themselves.

References

Allan, A., Naylor, T., & Saunders, E.S. 2006, *Astronomische Nachrichten* 327, 767

Beaulieu, J.-P., Bennett, D. P., Fouqué, P., Williams, A., Dominik, M., Jørgensen, U. G., Kubas, D., Cassan, A., Coutures, C., Greenhill, J., Hill, K., Menzies, J.,. Sackett, P. D., M. Albrow, Brillant, S., Caldwell, J. A. R., Calitz, J. J., Cook, K. H., Corrales, E., Desort, M., Dieters, S., Dominis, D., Donatowicz, J., Hoffman, M., Kane, S., Marquette, J.-B., Martin, R. Meintjes, P., Pollard, K., Sahu, K., Vinter, C., Wambsganss, J., Woller, K., Horne, K., Steele, I., Bramich, D. M., Burgdorf, M., Snodgrass, C., Bode, M., Udalski, A., Szymański, M. K., Kubiak, M., Wieckowski, T., Pietrzyński, G., Soszyński, I., Szewczyk, O., Wyrzykowski, Ł., Paczyński, B., Abe, F., Bond, I. A., Britton, T. R., Gilmore, A. C., Hearnshaw, J. B., Itow, Y., Kamiya, K., Kilmartin, P. M., Korpela, A. V., Masuda, K., Matsubara, Y., Motomura, M., Muraki, Y., Nakamura, S., Okada, C., Ohnishi, K., Rattenbury, N. J., Sako, T., Sato, S., Sasaki, M., Sekiguchi, T., Sullivan, D. J., Tristram, P. J., Yock, P. C. M., & Yoshioka, T. 2006, *Nature*, 439, 437

Chung, S.-J., Kim, D., Darnley, M. J., Duke, J. P., Gould, A., Han, C., Jeon, Y.-B., Kerins, E., Newsam, A., & Park, B.-G. 2006, *ApJ*, 650, 432

Bond, I. A., Abe, F., Dodd, R. J., Hearnshaw, J. B., Honda, M., Jugaku, J., Kilmartin, P. M., Marles, A., Masuda, K., Matsubara, Y., Muraki, Y., Nakamura, T., Nankivell, G., Noda, S., Noguchi, C., Ohnishi, K., Rattenbury, N. J., Reid, M., Saito, T., Sato, H., Sekiguchi, M., Skuljan, J., Sullivan, D. J., Sumi, T., Takeuti, M., Watase, Y., Wilkinson, S., Yamada, R., Yanagisawa, T., & Yock, P. C. M. 2001, *MNRAS*, 327, 868

Burgdorf, M.J., Bramich, D.M., Dominik, M., Bode, M. F., Horne, K. D., Steele, I. A., Rattenbury, N., & Tsapras, Y. 2007, *Planetary & Space Science*, 55, 582

Covone, G., De Ritis, R., Dominik, M., & Marino, A. A. 2000, *A&A*, 357, 816

Dominik, M., Albrow, M. D., Beaulieu, J.-P., Caldwell, J. A. R., DePoy, D. L. , Gaudi, B. S., Gould, A., Greenhill, J., Hill, K., Kane, S., Martin, R., Menzies, J., Naber, R. M., Pel, J.-W., Pogge, R. W., Pollard, K. R., Sackett, P. D., Sahu, K. C., Vermaak, P., Watson, R., & Williams, A. 2002, *Planetary & Space Science*, 50, 299

Dominik, M., Horne, K., & Bode, M. F. 2006, *Astronomy & Geophysics*, 47, 3.25

Dominik, M., Rattenbury, N. J., Allan, A., Mao, S., Bramich, D. M.. Burgdorf, M. J.. Kerins, E., Tsapras, Y., & Wyrzykowski, L. 2007, *MNRAS*, 380, 792

Einstein, A. 1936, *Science*, 84, 506

Gould, A., Gaudi, B. S., & Bennett, D. P. 2007, *Ground-based Microlensing Surveys*, White paper submitted to the NASA/NSF Exoplanet Task Force, arXiv e-print 0704.0767

Kiraga, M., & Paczyński, B. 1994, *ApJ*, 430, L 110

Mao, S., & Paczyński, B. 1991, *ApJ*, 374, L37

Snodgrass, C., Tsapras, Y., & Horne, K. 2004, *MNRAS*, 351, 967

Steele, I. A., Naylor, T., Allan, A., Etherton, J., & Mottram, C. J. 2002, in: R. I. Kilbrick (ed.), *Advanced Global Communications Technologies for Astronomy II*, Proc. SPIE, Vol. 4845, p. 13

Udalski, A. 2003, *AcA* 53, 291

A HET search for planets around evolved stars

Andrzej Niedzielski[1,2] and Alex Wolszczan[2,1]

[1]Toruń Centre for Astronomy, Nicolaus Copernicus University
Gagarina 11, 87-100 Toruń, Poland

[2]Dept. of Astronomy & Astrophysics, The Pennsylvania State University,
525 Davey Laboratory, University Park, 16802 PA, USA

Abstract. We present our ongoing survey of ∼1000 GK-giants with the 9.2-m Hobby-Eberly Telescope in search for planets around evolved stars. The stars selected for this survey are brighter than 11 mag and are located in the section of the HR-diagram, which is approximately delimited by the main sequence, the instability strip, and the coronal dividing line. We use the High Resolution Spectrograph to obtain stellar spectra for radial velocity measurements with a 4-6 m s^{-1} precision. So far, the survey has discovered a planetary-mass companion to the K0-giant HD 17092, and it has produced a number of plausible planet candidates around other stars. Together with other similar efforts, our program provides information on planet formation around intermediate mass main sequence-progenitors and it will create the experimental basis with which to study dynamics of planetary systems around evolving stars.

Keywords. Extrasolar planets, red giants, radial velocity

1. Introduction

Precision radial velocity (RV) studies have established more than a decade ago that GK-giant stars exhibit RV variations ranging from days to many hundreds of days (e.g. Walker *et al.* 1989, Hatzes & Cochran 1993, Hatzes & Cochran 1994). Enough observational evidence has been accumulated to identify three distinct sources of this variability, namely stellar pulsations, surface activity and a presence of substellar companions. A possibility to discover planets around post-MS giants, in numbers comparable to the current statistics of planets around MS-dwarfs (e.g. Butler *et al.* 2006), offers a very attractive way to provide the much needed information on planet formation around intermediate mass MS-progenitors ($\geqslant 1.5 M_\odot$) and to create a foundation for studies of the dynamics of planetary systems orbiting evolving stars (e.g. Duncan & Lissauer 1998).

Fourteen planet discoveries around GK-giants have been reported so far (Niedzielski *et al.* 2007, and references therein, Lovis & Mayor 2007, Johnson *et al.* 2007). Locations of giants with planets in the HR diagram are shown in Fig. 1. All detections have been made using the Doppler velocity technique with the RV precision ranging from ∼5 to ∼25 m s^{-1}, exploiting the availability of many narrow absorption features generated in the cool atmospheres of evolved stars. These developments demonstrate that sufficiently large surveys of post-MS giants should soon furnish enough planet detections to meaningfully address the above problems.

Initial analyses based on the currently available statistics (Lovis & Mayor 2007, Johnson *et al.* 2007) suggest that the frequency of massive planets is correlated with stellar mass. Because more massive stars probably have more massive disks, these results appear to support the core accretion scenarios of planet formation (Pollack *et al.* 1996). Furthermore, Pasquini *et al.* 2007 have used the apparent lack of correlation between the frequency of planets around giants and stellar metallicity to argue that this effect may imply a pollution origin of the observed planet frequency - metallicity correlation for main sequence stars (Fischer & Valenti 2005). Finally, for planets around giants, the absence of planets on tight orbits can be explained as the effect of post-MS evolution of their parent stars, but, as discussed by Johnson *et al.* (2007), other scenarios must also be considered. For example, the observed paucity of small orbital radii can be the result of faster depletion of disks around more massive stars, as suggested by simulations carried out by Burkert & Ida (2007) .

In this paper, we describe our contribution to searches for planets around post-MS stars with a survey of ~1000 GK-giants with the 9.2-m Hobby-Eberly Telescope. Our program has already discovered a number of interesting planet candidates, first of which has been recently published by Niedzielski *et al.* (2007).

2. The survey

Our long-term project to search for planets around evolved stars with the 9.2-m Hobby-Eberly Telescope (Ramsey *et al.* 1998) and its High Resolution Spectrograph (Tull 1998) has been established in early 2004. The sample of stars we have been monitoring is composed of two groups, approximately equal in numbers. The first one falls in the "clump giant" region of the HR-diagram (Jimenez *et al.* 1998), which contains stars of various masses over a range of evolutionary stages. The second group comprises stars,

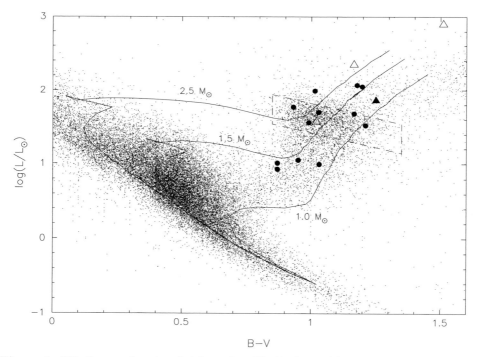

Figure 1. HR diagram for stars brighter than V=10 observable with the HET. The solid lines are evolutionary tracks from Girardi *et al.* (1996) for Z=0.02. The dashed box delineates the red giant clump region (Jimenez *et al.* 1998). Symbols mark the stars with published planet detections and planets from this survey discussed in the text. *(Filled triangle)* Niedzielski *et al.* (2007); *(Open triangles)* unpublished detections from this survey; *(Filled circles)* other published detections. The diagram is based on data from the *Hipparcos* catalogue.

which have recently left the MS and are located ∼1.5 mag above it. Generally, as shown in Fig. 1, all our targets, a total of ∼1000 GK-giants brighter than ∼11 mag, occupy the area in the HR-diagram, which is approximately defined by the MS, the instability strip, and the coronal dividing line (a narrow strip in the HR-diagram marking the transition between stars with steady hot coronae and those with cool chromospheric winds Linsky & Haisch 1979).

The HET observations and data analysis for this survey have been described by Niedzielski et al. (2007). Briefly, we observe with the HET in its queue-scheduling mode and use the HRS at the R=60,000 resolution with the gas cell (I_2) inserted in the optical path. In our target selection, we avoid bright objects, which are accessible to smaller telescopes. Consequently, more than 66% of our target stars are fainter than V=8 mag. The observing scheme follows the standard practices implemented in precision radial velocity measurements with the iodine cell (Marcy & Butler 1992). The spectral data used for RV measurements are extracted from the 17 echelle orders, which cover the 505 to 592 nm range of the I_2 cell spectrum. The observing strategy consists of the initial set of measurements of a target star (2-3 exposures, typically 3-6 months apart), to check for any RV variability exceeding a 30-50 m s^{-1} threshold, followed by more frequent observations, if a significant variability is detected. If the RV variability is confirmed, the star becomes part of the high priority list.

Radial velocities are measured by means of the commonly used I_2 cell calibration technique (Butler et al. 2006). A template spectrum is constructed from a high-resolution Fourier Transform Spectrometer (FTS) I_2 spectrum and a high signal-to-noise stellar spectrum measured without the I_2 cell. Doppler shifts are derived from the least-squares fits of template spectra to stellar spectra with the imprinted I_2 absorption lines. The average radial velocity for each epoch is calculated as a mean value of the independent determinations from the 17 usable echelle orders. The corresponding uncertainties of these measurements are estimated assuming that errors obey the Student's t-distribution. Typically, they fall in the 4-5 m s^{-1} range at 1σ-level. Radial velocities are referred to the Solar System barycenter using the Stumpff (1980) algorithm, which is accurate enough given the RV precision limitations that are intrinsic to the evolved stars.

As the intrinsic variability may contribute to the observed RV variations (e.g. Gray 2005), stellar line profiles are studied in detail in search for any signatures of a rotation induced spot activity. Also the existing photometry databases like Hipparcos, Tycho or Northern Variability Sky Survey (Woźniak et al. 2004) are used to study possible integrated light variations that might be interpreted as a result of pulsations. These analyses are reviewed elsewhere (Niedzielski et al. this vol.).

3. Results

In almost four years of observations, we have obtained more than one RV measurement for >600 GK-giant stars with a 4-6 m s^{-1} precision. Adopting a working definition of RV scatter ⩽40 m s^{-1} for a stable (single) red giant, we find that 55 % of stars in that sample are single, 20 % are new binaries and 25% stars possibly have low-mass companions.

We have been currently monitoring more than 30 planetary candidate companion stars and have obtained preliminary orbital solutions for most of them. Each one obviously requires a thorough examination of stellar activity, which includes bisector analysis and a study of H$_\alpha$ variations. The RV curves and the corresponding best-fit orbital models for our first published planet around the K0 giant, HD 17092 (Niedzielski et al. 2007), and for three other examples of the detections that are being prepared for publication, are shown in Fig. 2. The observed RV curves are highly repeatable and their periods are not reproduced in the measured line bisector and photometric variations. Provisional stellar mass estimates using Girardi *et al.*(2000) evolutionary tracks indicate a planetary nature of the companions. It is quite clear that star 162 has a third, long-period companion, whose nature will be established in the course of further observations. The star 37 planet has the most compact orbit among the existing detections (a = 0.6 AU), whereas the planet around star 18 may be orbiting the most massive star in the existing sample (5.5 $M_\odot$) and has an exceptionally high intrinsic RV noise.

A steadily increased number of stars observed in this survey makes it possible to carry out statistical studies of RV noise properties of GK-giants. Our preliminary results confirm the intrinsic RV jitter of red giants with the maximum of its distribution at about 20 m s^{-1}. Furthermore, the RV scatter increases with B-V, easily reaching 100 m s^{-1} for stars later than K5. Clearly, more observations are needed to understand the nature of the scatter, part of which may be contributed by short-period pulsations, which remain unresolved by the sparse sampling of our survey.

Searches for planets around evolved stars are still in their infancy compared to similar programs for solar-type stars, which have been steadily furnishing new planet detections to bring the count up to over 250 at the time of this writing. However, it is the former searches that are now needed to obtain new information on stellar mass and time-dependent aspects of planet formation and evolution that is not accessible through the latter ones. A continuation of the survey described in this proposal, together with other similar programs, is already creating a base of planet detections around GK-giants, which

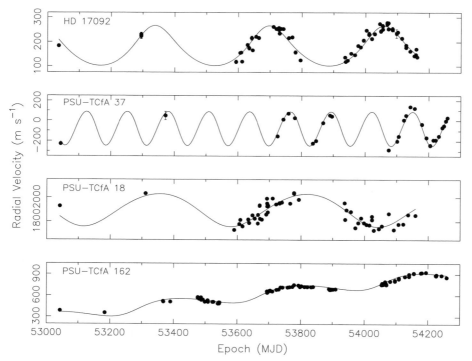

Figure 2. Radial velocity measurements (filled circles) and the best-fit orbital models (solid lines) for a sample of four K-giant stars monitored by the HET survey. For most data points, the circle size is larger than the sizes of error bars.

will soon become sufficient to fully address the questions of stellar mass and chemical composition dependence of planet formation for masses $> 1 M_\odot$ and of the possible fates of planetary systems under the influence of an evolving parent star. This knowledge will improve our understanding of the astrophysics of planetary systems, it will provide an experimental base for theories of the far future of the Solar System and it will broaden our knowledge of the astrophysical aspects of long-term survival of life on Earth and elsewhere, including a possibility of the emergence of life on planets in the expanded habitable zones of red giants (Lopez et al. 2005).

Acknowledgements

AN and AW were supported in part by the Polish Ministry of Science and Higher Education grant 1P03D 007 30. AW also acknowledges a partial support from the NASA Astrobiology Program. The Hobby-Eberly Telescope (HET) is a joint project of the University of Texas at Austin, the Pennsylvania State University, Stanford University, Ludwig-Maximilians-Universität München, and Georg-August-Universität Göttingen. The HET is named in honor of its principal benefactors, William P. Hobby and Robert E. Eberly.

References

Burkert, A., & Ida, S. 2007, *Astrophysical Journal*, 660, 845
Butler, R. P., et al. 2006, *Astrophysical Journal*, 646, 505
Butler, R. P., Marcy, G. W., Williams, E., McCarthy, C., & Dosanjh, P. 1996, *Publications of the Astronomical Society of the Pacific*, 108, 500
Duncan, M. J., & Lissauer, J. J. 1998, *Icarus*, 134, 303
Fischer, D. A., & Valenti, J. 2005, *Astrophysical Journal*, 622, 1102
Girardi, L., Bressan, A., Chiosi, C., Bertelli, G., & Nasi, E. 1996, *Astronomy & Astrphysics Supplement Series*, 117, 113
Girardi, L., Bressan, A., Bertelli, G., & Chiosi, C. 2000, *Astronomy & Astrphysics Supplement Series*, 141, 371
Gray, D. F. 2005, *Publications of the Astronomical Society of the Pacific*, 117, 711
Hatzes, A. P., Cochran, W. D. 1993, *Astrophysical Journal*, 413, 339
Hatzes, A. P., & Cochran, W. D. 1994, *Astrophysical Journal*, 422, 366
Jimenez, R., Flynn, C. & Kotoneva, E. 1998, *Monthly Notices of the Royal Astronomical Society*, 299, 515
Johnson, J. A., et al. 2007, *Astrophysical Journal*, 665, 785
Linsky, J. L., & Haisch, B. M. 1979, *Astrophysical Journal Letters*, 229, L27
Lopez, B., Schneider, J., & Danchi, W. C. 2005, *Astrophysical Journal*, 627, 974
Lovis, C., & Mayor, M. 2007, *Astronomy & Astrphysics*, 472, 657
Marcy, G. W. & Butler, R. P. 1992, *Publications of the Astronomical Society of the Pacific*, 104, 270
Niedzielski, A., et al. 2007, *Astrophysical Journal*, 669, 1354
Pasquini, L., Döllinger, M. P., Weiss, A., Girardi, L., Chavero, C., Hatzes, A. P., da Silva, L., & Setiawan, J. 2007, *Astronomy & Astrphysics*, 473, 979
Pollack, J. B., Hubickyj, O., Bodenheimer, P., Lissauer, J. J., Podolak, M., & Greenzweig, Y. 1996, *Icarus*, 124, 62
Ramsey, L. W., et al. 1998, *Proceedings of the SPIE*, 3352, 34
Stumpff, P. 1980, *Astronomy & Astrphysics Supplement Series*, 41, 1
Tull, R. G. 1998, *Proceedings of the SPIE*, 3355, 387
Walker, G. A. H., Yang, S., Campbell, Bruce, I., Alan W. 1989 *Astrophysical Journal*, 343, 21
Woźniak, P. R., et al. 2004, *Astronomical Journal*, 127, 2436

The PSU/TCfA search for planets around evolved stars. Stellar parameters and activity indicators of targets.

Andrzej Niedzielski[1,2], Grzegorz Nowak[1] and Paweł Zieliński[1]

[1]Toruń Centre for Astronomy, Nicolaus Copernicus University,
ul. Gagarina 11, 87-100 Toruń, Poland,
email: aniedzi, grzenow, pawziel @astri.uni.torun.pl

[2]Department of Astronomy and Astrophysics Pennsylvania State University,
525 Davey Laboratory, University Park, PA 16802

Abstract. The main objective of the Penn State/Toruń Centre for Astronomy search for planets around evolved stars is the detection of planetary systems around massive, evolved stars. We are also interested in the evolution of these systems on stellar evolution timescales. In this paper we present our approach to determine the basic physical parameters of our targets GK-giants. We also discuss the stellar activity indicators used in our survey: line bisector and curvature, and Hα variability.

Keywords. stars: fundamental parameters, stars: activity, (stars:) planetary systems

1. Introduction

Proper interpretation of the results from precision RV studies of GK-giants requires a detailed knowledge of their physical parameters. Effective temperatures and gravitational accelerations are needed to obtain luminosities, and, with the additional knowledge of metallicities, estimates of stellar masses and ages can be derived by means of the isochrone fitting. Together with estimates of stellar radii and rotation periods, these data allow one to address the influence of stellar surface inhomogeneities (spots) on the observed RV variations.

All alternative sources of RV variations in GK-giants have to be ruled out before substellar companion interpretation becomes acceptable. Unfortunately, the long period variations, if present, cannot usually be studied using data other than the existing photometry (usually of moderate quality), or the data collected from the RV survey. Therefore, the detailed activity discussion is usually based on the indicators defined on the basis of the same spectra as those used for the RV measurements. In this paper, we illustrate our methodology of the determination of stellar parameters and activity analysis based on the case of the K-giant PSU-TCfA 18, a potential planet hosting star.

2. Observations

Observations were made with the Hobby-Eberly Telescope (HET) and the High Resolution Spectrograph (HRS). The HRS was used in the R=60,000 resolution mode with a gas cell (I_2) inserted into the optical path, and it was fed with a 2 arcsec fiber. Typically, the signal-to-noise ratio per resolution element (at 594 nm) was $\sim$200 for the stellar spectra taken with the gas cell, and $\geqslant$250 for the templates.

3. Basic stellar parameters

The atmospheric parameters of the program stars were obtained with the spectroscopic method (Takeda et al. 2005a, b), which is based on analysis of Fe I and Fe II lines and relies on conditions resulting from the assumption of the LTE. Typically, over 200 FeI and about 25 FeII lines were measured for every star.

We tested a reliability of our determinations with the Takeda et al. (2005a) TGVIT code by applying it to 8 stars, for which the parameters have been published by Butler et al. (2006). A comparison of the results shows that T_{eff} values agree to within 49 K, log g to within 0.11 dex and $[Fe/H]$ to within 0.11 dex.

Stellar masses were derived by comparing the positions of stars in the HR diagram with the theoretical evolutionary tracks of Girardi et al. (2000) and Salasnich et al. (2000) for a given metallicity. For stars for which the parallax determinations are precise enough, the metallicity may introduce a significant uncertainty in mass because of the choice of an evolutionary track. We assume that for an average red giant with known parallax and good photometry the mass may be estimated to within $0.3 M_\odot$. The precision depends however on the actual star position in HRD. We also note that for stars in the red giant clump, which are in the fast evolution phase with mass-loss, the derived masses are probably the upper limits.

Stellar radii were determined using the calibration given in Alonso et al. (2000). Stellar ages are usually estimated with the aid of the theoretical stellar isochrones.

In the case of the PSU-TCfA 18 star, we have measured equivalent widths of up to 195 Fe I and 11 Fe II lines for further analysis. For this star, the basic physical parameters are $T_{eff} = 4246 K$, log $g = 2.43$, $v_t = 1.52 kms^{-1}$, $[Fe/H] = 0.11$, $M = 5.5 M_\odot$, and $R = 27.4 R_\odot$. Intrinsic uncertainties of our determinations are $\sigma T_{eff} = 29 K$, $\sigma \log g = 0.09$, $\sigma v_t = 0.13 kms^{-1}$ and $\sigma [Fe/H] = 0.06$.

4. $V \sin i$ measurements

Rotation periods represent a parameter of particular importance in searches for planetary companions to red giants. As the rotation periods of these stars are very similar to expected orbital periods, their knowledge is critical for an unambiguous interpretation of observations. Any correlation of stellar activity indicators variations with the rotation period make a substellar companion hypothesis unlikely. To estimate rotation periods from our spectra we have used the cross-correlation technique, as described in Benz & Mayor (1984).

We have computed the CCFs by cross correlating the high S/N blue spectra with a numerical mask. To measure $V \sin i$ we have worked out a $V \sin i$ calibration for the HET/HRS. To determine the σ_0 vs. $(B-V)$ relationship, we have used 16 slow rotators with accurately known projected rotational velocities, preferably from Gray (1989), Fekel (1997) and de Medeiros & Mayor (1999). For these stars we have determined σ_0 using the formula from Benz & Mayor (1984) ($V \sin i = A \sqrt{\sigma_{obs}^2 - \sigma_0^2}$) and assuming the constant $A = 1.9$ following the Queloz et al. (1998) and Melo et al. (2001). We have carried out a least-squares fit to the data of the analytical function $\sigma_0 = a_2 (B-V)^2 + a_1 (B-V) + a_0$, which yields the following calibration: $\sigma_0 = 15.592(B-V)^2 - 26.753(B-V) + 14.559$.

Using this calibration we have obtained $V \sin i = (3 \pm 1) kms^{-1}$ for our star. Adopting the radius for this star as determined above, we have estimated its rotational period to be 220 - 950 days. The large uncertainty in the rotation period is caused by uncertainties related to the determination of the radius and $V \sin i$.

5. Stellar activity indicators

One of possible sources of the observed RV variations in GK-giants is due to their pulsations. Therefore, photometric data that span long periods of time are needed for the interpretation of the results of our survey. Because we do not conduct our own parallel photometric observations, we must rely on the existing photometric databases like *Hipparcos* or *NSVS* (Woźniak *et al.* 2004). These moderate quality data provide time-series, which are long enough to be useful in searches for long-term periodicities. However, these measurements were usually performed many years prior to our RV survey. In the particular case of the PSU-TCFA 18 star, no detectable variability is present in the existing photometric data.

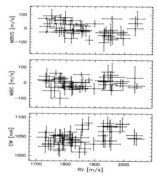

Figure 1. MBVS, MBC and EW of the $H\alpha$ for PSU-TCfA 18 as a function of radial velocity.

5.1. Line bisectors

The basic tool to study the origin of RV variations derived from the stellar spectra is the analysis of the shapes of spectral lines via line bisectors (Gray 1983).

We have computed line bisectors for 5 strong, unblended spectral features of a moderate intensity, which were located close to the center of echelle orders: Cr I 663.003 nm, Ni I 664.638 nm, Ca I 671.77 nm, Fe I 675.02 nm, and Ni I 676.784 nm. All these lines show well defined bisectors.

The changes in the spectral line bisectors were quantified using the bisector velocity span (BVS) parameter, which is simply the velocity difference between the upper and the lower points of the line bisector ($BVS = v_3 - v_2$), and the bisector curvature (BC), which is the difference of the velocity span of the upper half of the bisector and its lower half ($BC = (v_3 - v_2) - (v_2 - v_1)$). It is important to examine both BVS and BC, because it is possible for a star to show variations in one of these parameters only. In choosing the span points, it is important to avoid the wings and cores of the spectral line, where the errors in the bisector measurements are large. For our span measurements we chose $v_1 = 0.29$, $v_2 = 0.57$, and $v_3 = 0.79$ in terms of the line depth at the line core. Using the bisector measurements of all 5 spectral lines we have computed the average velocity span and curvature after subtracting the mean value for each spectral line.

In Figure 1, we present the mean bisector velocity span ($MBVS$) and the mean bisector curvature (MBC) for our star, as a function of RV. Uncertainties in the derived values of $MBVS$ and MBC were estimated as standard deviations of the mean. The correlation coefficients were found to be $r = -0.32 \pm 0.10$ for $MBVS$ and $r = 0.07 \pm 0.02$ for MBC. It is clear that they are not correlated with radial velocities which supports the planetary mass companion hypothesis. The available RV measurements are not uniformly distributed over the estimated period, which is visible in Fig. 1 as a scatter varying with RV. More observations are needed to confirm the apparent lack of correlation.

5.2. Hα variations

Since our spectra start at 407.6 nm we cannot investigate the variation of the Ca II K emission line (393.4 nm). Also the infrared CaII triplet lines 849.8-854.2 are outside the range of our spectra. Therefore, we use Hα line (656.28 nm) as a chromospheric activity indicator. The EW measurements of the $H\alpha$ line can be made in our spectra with a typical precision of a few percent. To minimize contamination from telluric lines we measured the EW of the central part of the line profile defined by $I/I_c \leqslant 0.775$. In the case of our star, the measurements give a mean value of $1049 \pm 20 m\text{Å}$. The rms value of $20 m\text{Å}$ corresponds to 2 % variation in the EW. In Figure 1 we present EW measurements for $H\alpha$ as a function of RV. The correlation coefficient of $r = 0.44 \pm 0.10$ shows marginal relationship.

6. Conclusions

A detailed knowledge of stellar parametres of red giants is very important for interpretation of their RV variations. To rule out stellar activity as the source of such variations, one needs precise rotation periods and several other indicators to be measured at many epochs.

Acknowledgements

AN & GN acknowledge the financial support from the MNiSW through grant 1P03D 007 30. GN is a recipient of a graduate stipend of the Chairman of the Polish Academy of Sciences. PZ was supported by MNiSW grant SPB 104E-3376. The Hobby-Eberly Telescope (HET) is a joint project of the University of Texas at Austin, the Pennsylvania State University, Stanford University, Ludwig-Maximilians-Universität München, and Georg-August-Universität Göttingen. The HET is named in honor of its principal benefactors, William P. Hobby and Robert E. Eberly.

References

Alonso, A., Salaris, M., Arribas, S., Martínez-Roger, C. & Asensio Ramos, A. 2000, *A&A*, 355, 1060
Benz, W. & Mayor, M. 1984, *A&A*, 138, 183
Butler, R. P., Wright, J. T., Marcy, G. W., Fischer, D. A., et al. 2006, *ApJ*, 646, 505
Fekel, F. C. 1997, *PASP*, 109, 514
Fischer, D. A., & Valenti, J. 2005, *ApJ*, 622, 1102
Frink, S., Mitchell, D. S., Quirrenbach, A., Fischer, D. A., Marcy, G. W., & Butler, R. P. 2002, *ApJ*, 576, 478
Girardi, L., Bressan, A., Bertelli, G., & Chiosi, C. 2000, *A&AS*, 141,371
Gray, D. F. 1983, *PASP*, 95, 252
Gray, D. F. 1989, *ApJ*, 347, 1021
Hatzes, A. P., & Cochran, W. D. 1993, *ApJ*, 413, 339
Hatzes, A. P., & Cochran, W. D. 1994, *ApJ*, 422, 366
de Medeiros, J. R., & Mayor, M. 1999, *A&AS*, 139, 433
Melo, C. F. H., Pasquini, L., & de Medeiros, J. R. 2001, *A&A*, 375, 851
Queloz, et al. 1998, *A&A*, 335, 183
Salasnich, B., Girardi, L., Weiss, A., Chiosi, C. 2000, *A&A*, 361, 1023
Sato, B., Izumiura, H., Toyota, E., Kambe, E., Takeda, Y., et al. 2007, *ApJ*, 661, 527
Takeda, Y., Ohkubo, M., Sato, B., Kambe, E., & Sadakane, K. 2005, *PASJ*, 57, 27
Takeda, Y., Sato, B., Kambe, E., Izumiura, H, Masuda, S. & Ando, H. 2005, *PASJ*, 57, 109
Walker, G.A.H., Yang, S., Campbell, Bruce, I., & Alan W. 1989, *ApJ*, 343, 21
Woźniak, P. R., et al. 2004, *AJ*, 127, 2436

A Korea-Japan planet search program: Current status and discovery of a brown dwarf candidate

Masashi Omiya[1], Hideyuki Izumiura[2], Bun'ei Sato[3], Michitoshi Yoshida[2], Eiji Kambe[2], Eri Toyota[4], Seitaro Urakawa[5], Seiji Masuda[6], Masahide Takada-Hidai[7], Inwoo Han[8], Kang-Min Kim[8], Byeong-Cheol Lee[8,9] and Tae Seog Yoon[9]

[1] Department of Physics, Tokai University, Hiratsuka, Kanagawa 259-1292, Japan
email: ohmiya@peacock.rh.u-tokai.ac.jp

[2] Okayama Astrophysical Observatory, National Astronomical Observatory of Japan, Asakuchi, Okayama 719-0232, Japan

[3] Global Edge Institute, Tokyo Institute of Technology, Meguro-ku, Tokyo 152-8550, Japan

[4] Graduate School of Science, Kobe University, Kobe, Hyogo 657-8501, Japan

[5] Japan Spaceguard Association, Shibuya-ku, Tokyo 151-0073, Japan

[6] Tokushima Science Museum, Asutamu Land Tokushima, Itano-gun, Tokushima 779-0111, Japan

[7] Liberal Arts Education Center, Tokai University, Hiratsuka, Kanagawa 259-1292, Japan

[8] Korea Astronomy and Space Science Institute, Daejeon 305-348, South Korea

[9] Department of Astronomy and Atmospheric Sciences, Kyungpook National University, Daegu 702-701, South Korea

Abstract. Since 2005, we have been carrying out a precise radial velocity survey of about 190 intermediate-mass (1.5-5 $M_\odot$) G and K giants at Bohyunsan Optical Astronomy Observatory (BOAO) in Korea and Okayama Astrophysical Observatory (OAO) in Japan, which aims to reveal statistical properties of planetary systems around intermediate-mass stars. We have finished the first screening of 120 stars so far and have identified 5 candidates with large periodic radial velocity variations. One of the candidates turned out to be orbited by a brown dwarf mass companion with minimum mass of 37.6 M_{Jup} and semimajor axis of 1.71 AU. The primary star has a mass of 3.9 $M_\odot$, which ranks among the most massive stars with substellar companions. Our discovery may support the current view obtained from results of planet searches around intermediate-mass stars that massive substellar companions tend to form around massive stars.

Keywords. stars: low-mass, brown dwarfs, techniques: radial velocities

1. Introduction

To date, more than 200 exoplanets have been detected around solar-type stars by precise Doppler surveys. These surveys have unveiled various statistical characteristics of the planets. On the other hand, only 10 planets have been discovered around intermediate-mass stars (1.5-5 $M_\odot$) and their statistical properties have not been yet clear. Planetary systems around such massive stars are particularly important for constructing planet formation theories because they can constrain key processes of the planet formation such as dependence on host star's mass (probably related with mass of proto-planetary disk), role of radiation of central stars, and timescale of planet formation. However, early-type stars (B-A dwarfs), intermediate-mass stars on the main sequence, are not suitable for precise radial velocity survey due to fewer metallic absorption lines in their spectra and higher surface activity. Late G and early K type giants, intermediate-mass stars in the evolved stages, are promising targets for this purpose because they have many sharp absorption lines in their spectra and their surface activity is relatively low compared to early-type stars and more evolved late K and M type giants.

So far, radial velocity surveys of intermediate-mass stars (e.g. Johnson *et al.* 2006, Sato *et al.* 2007, Niedzielski *et al.* 2007) have revealed some properties of substellar companions around them. For example, frequency of massive substellar companions is higher than that of solar-type stars (Lovis & Mayor 2007, Johnson *et al.* 2007), and many massive host stars have lower metallicity than typical one of solar-type host stars (e.g. Sato *et al.* 2003). These properties seem to be not similar to those of solar-type stars and such different properties should be explored by further studies. In the case of solar-type stars, a "brown dwarf desert" is widely known as a paucity of substellar companions falling in the brown dwarf mass region (13-80 M_{Jup}) (Marcy & Butler 2000, Halbwachs *et al.* 2000, Grether & Lineweaver 2006), which suggests distinct formation mechanisms between planets and stellar companions. To uncover whether a brown dwarf desert also exist around intermediate-mass stars will give us a clue to understand formation mechanism of planets around them.

2. Korea-Japan Planet Search Program

We are monitoring our targets using the 1.8 m telescope at BOAO with Bohyunsan Optical Echelle Spectrograph (BOES, $R=\lambda/\Delta\lambda=50000$, 3500Å-10500Å), which is a fiber-fed high resolution echelle spectrograph (Kim *et al.* 2007) and the 1.88 m telescope at OAO with HIgh Dispersion Echelle Spectrograph (HIDES, $R=65000$, 5000Å-6100Å) installed at the coude focus of the telescope (Izumiura 1999). For precise radial velocity measurements, an iodine absorption cell (I2 cell) is installed at the optical path in front of the fiber or slit of each spectrograph (Kim *et al.* 2002, Kambe *et al.* 2002). We also take stellar spectra without I2 cell for abundance analysis. Radial velocities are derived by using Sato *et al.* (2002)'s code which models an I2-superposed stellar spectrum with high resolution stellar and iodine templates and reproduced instrumental profiles of the spectrographs. We have achieved long term radial velocity precisions of $\sim$ 11 m/s and $\sim$

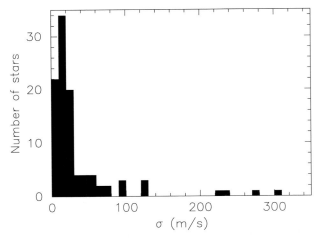

Figure 1. Velocity RMS of 116 BOAO & OAO targets observed more than three times for 2.5 years.

6-7 m/s with BOES and HIDES, respectively, over a time span of 2.5 years during the project. A Korea-Japan planet search program has been conducting a precise Doppler survey of about 190 G and K giants at BOAO in Korea and OAO in Japan since 2005. This joint planet search program is an extended version of on-going OAO planet search program (Sato et al. 2005) and a part of the international collaborations between Korea, China and Japan (East-Asian Planet Search Network, EAPSNet, Izumiura 2005) aiming to obtain the properties of planetary systems around intermediate-mass stars by surveying a total of "1000" giant stars for planets.

3. Current status and Orbital solution of a brown dwarf companion

Up to now, we observed about 120 targets more than three times. Seventeen of the targets exhibited large radial velocity variations with amplitudes of 50 m/s - 350 m/s and five of them showed probable periodic variations (See Figure 1).

One of such targets clearly shows large periodic radial velocity variations as seen in Figure 2. Low surface activity of this star favors an orbital motion as the cause of the observed variability. On the basis of Keplerian orbital fit, we obtained a velocity semi-amplitude of 413.5 ± 2.6 m/s, period of 410.5 ± 0.6 days and eccentricity of 0.082 ± 0.007. Adopting a host star's mass of 3.9 ± 0.4 $M_\odot$, which was estimated from evolutionary track and fundamental stellar parameters of $L = 251 \pm 95$ $L_\odot$, $T_{eff} = 5083 \pm 103$ K, and [Fe/H] = 0.04 ± 0.18, we obtained a minimum mass for the companion $m_p \sin i = 37.6 \pm 2.6$ M_{Jup} and a semimajor-axis of $a = 1.71 \pm 0.06$ AU. If we assume the orbit is randomly oriented, there is a 12% chance that the true mass exceeds 80 M_{Jup} ($i \geqslant 28°$).

4. Summary

We have been carrying out a precise redial velocity survey of G and K giants at BOAO and OAO and discovered a brown dwarf mass companion together with some planetary candidates so far. This brown dwarf companion is the forth one among those discovered around intermediate-mass giants. The host star has a mass of 3.9 $M_\odot$, which is one of the most massive stars harboring substellar companions. Three stars with masses of larger

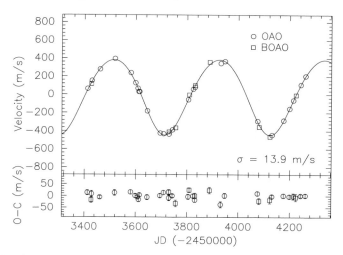

Figure 2. Radial velocities and orbital fit of a star with a brown dwarf candidate from BOAO (open squares) and OAO (open circles). O-C indicates difference between each observational point and orbital curve.

than 3 $M_\odot$ are currently known to harbor substellar companions (Hatzes et al. 2005, Lovis & Mayor 2007, this work) and all the companions have brown dwarf mass. The result supports the current view that more massive substellar companions tend to form around more massive stars (Lovis & Mayor 2007, Johnson et al. 2007).

So far, we have detected two brown dwarf mass companions (Liu et al. 2008 and this work) among a total of about 500 targets from OAO and BOAO-OAO surveys. The current detection rate of brown dwarf companion in our sample is thus estimated to be 0.4 %, which seems to be comparable to that of less than 1% for solar-type stars (Marcy & Butler 2000, Grether & Lineweaver 2006). It is still unclear, however, whether a brown dwarf desert also exists among intermediate-mass stars. Since less massive planets are normally difficult to detect around evolved giants due to their larger intrinsic radial velocity variability (up to 10 - 20 m/s), a mass distribution of planetary mass companions around them has not been well established yet. Further investigation of a mass distribution of substellar companions is required to elucidate formation mechanism of planets and brown dwarfs around massive stars.

References

Grether, D., & Lineweaver, C. H. 2006, *ApJ*, 640, 1051
Halbwachs, J. L., Arenou, F., Mayor, M., Udry, S., & Queloz, D. 2000, *A&A*, 355, 581
Hatzes, A. P., Guenther, E. W., Endl, M., Cochran, W. D., Döllinger, M. P., & Bedalov, A. 2005, *A&A*, 437, 743
Izumiura, H. 1999, *in Proc. 4th East Asian Meeting on Astronomy*, ed. P.S. Chen (Kunming: Yunnan Observatory), 77
Izumiura, H. 2005, *Journal of Korean Astronomical Society*, 38, 81
Johnson, J. A., Marcy, G. W., Fischer, D. A., Henry, G. W., Wright, J. T., Isaacson, H., & McCarthy, C. 2006, *ApJ*, 652, 1724
Johnson, J. A., Butler, R. P., Marcy, G. W., Fischer, D. A., Vogt, S. S., Wright, J. T., & Peek, K. M. G. 2007, *ApJ*, 670, 833
Kambe, E., Sato, B., Takeda, Y., Ando, H., Noguchi, K., Aoki, W., Izumiura, H., Wada, S., Masuda, S., Okada, N., Shimizu, Y., Watanabe, E., Yoshida, M., Honda, S., Kawanomoto, S. 2002, *PASJ*, 54, 865
Kim, K.-M., Jang, B.-H., Han, I., Jang, J. G., Sung, H. C., Chun, M.-Y., Hyung, S., Yoon, T.-S., Vogt, S. S. 2002, *Journal of Korean Astronomical Society*, 35, 221
Kim, K. M., Han, I., Valyavin, G., Plachinda, S., Jang, J. G., Jang, B. -H., Seong, H. C., Kang, D. I., Park, B. G., Yoon, T. S. and Vogt, S., 2007, *PASP*, 119, 1052.
Liu, Y. J., Sato, B., Zhao, G., Noguchi, K., Wang, H., Kambe, E., Ando, H., Izumiura, H., Chen, Y. Q.; Okada, N., Toyota, E., Omiya, M., Masuda, S., Takeda, Y., Murata, D., Itoh, Y., Yoshida, M., Kokubo, E., & Ida, S., 2008, *ApJ*, 672, 553
Marcy, G. W., & Butler, R. P. 2000, *PASP*, 112, 137
Niedzielski, A., Konacki, M., Wolszczan, A., Nowak, G., Maciejewski, G., Gelino, C. R., Shao, M., Shetrone, M., & Ramsey, L. W. 2007, *ApJ*, 669, 1354
Sato, B., Kambe, E., Takeda, Y., Izumiura, H., & Ando, H. 2002, *PASJ*, 54, 873
Sato, B., Ando, H., Kambe, E., Takeda, Y., Izumiura, H., Masuda, S., Watanabe, E., Noguchi, K., Wada, S., Okada, N., Koyano, H,. Maehara, H., Norimoto, Y., Okada, T., Shimizu, Y., Uraguchi, F., Yanagisawa, K., & Yoshida, M. 2003, *ApJ*, 597, L157
Sato, B., Kambe, E., Takeda, Y., Izumiura, H., Masuda, S., & Ando, H. 2005, *PASJ*, 57, 97
Sato, B., Izumiura, H., Toyota, E., Kambe, E., Takeda, Y., Masuda, S., Omiya, M., Murata, D., Itoh, Y., Ando, H., Yoshida, M., Ikoma, M., Kokubo, E., & Ida, S. 2007, *ApJ*, 661, 527

Search for extrasolar planets with high-precision relative astrometry by ground-based and single-aperture observations

Tristan Roell[1] and Andreas Seifahrt[1,2] and Ralph Neuhäuser[1]

[1] Astrophysikalisches Institut und Universitäts-Sternwarte Jena,
email: troell@astro.uni-jena.de
email: rne@astro.uni-jena.de

[2] Institut für Astrophysik, Göttingen
email: seifahrt@astro.physik.uni-goettingen.de

Abstract. We present our search program for substellar companions using high-precision relative astronomy. Due to its orbital motion around the star, an unseen substellar companion would produce a periodic "wobble" of the host star, which is the astrometric signal of the unseen companion. By measuring the separation between the components of stellar double and triple systems, we want to measure this astrometric signal of a possible unseen companion indirectly as a relative and periodic change of these separations. Using a new observation mode (the "cube-mode") where the frames were directly saved in cubes with nearly no loss of time during the readout, an adaptive optics system to correct for atmospheric noise and an infrared narrow band filter in the near infrared to suppress differential chromatic refraction (DCR) effects we achieve for our first target (the double star HD 19994) a relative precision for the separation measurements of about $100\ldots150\,\mu as$ per epoch. To reach a precision in the μas-regime, we use a statistical approach. We take several thousand frames per target and epoch and after a verification of a Gaussian distribution the measurement precision can be calculated as the standard deviation of our measurements divided by the square root of the number of Gaussian distributed measurements. Our first observed target is the stellar binary HD 19994 A & B, where the A component has a known radial velocity planet candidate.

Keywords. extrasolar planets, relative astrometry, double stars, triple stars, 47 Tuc, cube-mode

1. Introduction

Up to now, most of the extrasolar planets have been detected with the radial velocity (RV) technique. But due to the unknown inclination angle i this technique just yields the lower mass limit $M \sin i$ and not the true mass of the substellar companion. Therefore, all radial velocity planets should be regarded as planet candidates, until their true mass is determined. In contrast to the radial velocity technique, astrometry yields the inclination angle by measuring the astrometric signal of the substellar companion and hence its true mass. Until now, three radial velocity planet candidates could be confirmed with absolute astrometry using the Fine Guiding Sensor (FGS) of the Hubble Space Telescope (HST), GJ 876 b by Benedict *et al.* (2002), 55 Cancri d by McArthur *et al.* (2004) and Epsilon Eridani b by Benedict *et al.* (2006). Recently, Bean *et al.* (2007) measured the astrometric signal of the radial velocity planet candidate HD 33636 b ($M \sin i = 9.3\,M_J$) and obtained a value for the true mass of the companion of $M = 142 \pm 11\,M_J$, thus it is a low mass star. This clarifies the importance of astrometric follow up observations to determine the true mass of radial velocity planet candidates. The mass is one of the most important stellar and substellar parameters and plays a key role in our understanding of the distribution, forming and evolution of substellar objects. Besides all other methods to determine the mass, which are using theoretical predictions (like evolutionary models), astrometry is a method, which is independent from theoretical assumptions (hence also from theoretical uncertainties) by measuring the dynamical mass of the objects.

Besides the confirmation of RV planet candidates, we also want to search for unknown substellar companions in stellar double and triple systems. This kind of search program will provide new insights into the formation and evolution of extrasolar planets in multiple systems, which is important due to the fact that more than 50% of all stars are members of a multiple system.

2. Observation method

To reach a precision comparable to the HST observation, we observe double and triple stars and measure the separation between all stellar components, thus using relative astrometry. In the case of an unseen substellar companion, we would measure the astrometric signal indirectly as a relative and periodic change in the separations. The quest of measuring the astrometric signal of a substellar companion needs a carefully handling of all noise sources, which are among others atmospheric noise, photon noise, background noise, readout-noise and DCR. Our observations on the southern hemisphere are done with the 8.2 meter telescope UT4 of the ESO Very Large Telescope (VLT) and the NACO S13 (NAOS-CONICA) infrared camera. Using the adaptive optics system NAOS (Nasmyth Adaptive Optics System) we correct for atmospheric turbulences and by using

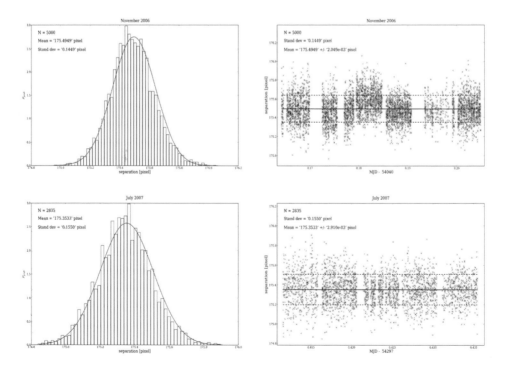

Figure 1. Separation measurements (right) and the Gaussian distribution of the measurements (left) of the stellar binary HD 19994 from 2006 (top) and 2007 (bottom)

a narrow band filter centered in the near infrared ($\lambda_{cen} = 2.17\,\mu m$) we suppress DCR effects. Due to the use of the double-correlated readout mode, we suppress readout noise and by choosing a suitable exposure time (to reach a high signal to noise ratio) we can neglect photon and background noise. The pixel scale of the used detector (NACO S13) is about $13.25\,mas$, which means a Field of View (FoV) of about $14'' \times 14''$. The guide star for the AO system is always one of our stellar components. The separation of our observed multiple systems is typically four arcseconds. Hence, the angular separation is (with normal seeing conditions) always smaller than the isoplanatic angle.

Furthermore (besides the use of relative astrometry), we use a new observation mode, called "cube-mode". This mode saves frames directly in a cube and thus has nearly no loss of time during the readout. With the minimal exposure time of 0.35 seconds using the double-correlated readout mode it is possible to obtain 2500 frames within 15 minutes.

The following statistical principle is similar to the method of measuring the radial velocity with hundreds of spectral lines to reach a higher precision, which is used in the radial velocity technique. In our astrometric case we measure the separation between all stellar components in every frame and obtain several thousand measurements of the same separation. After a verification of Gaussian distributed measurements (done with a Kolmogorov-Smirnov-Test) and a two sigma clipping (to eject frames with bad quality, due to the non-constant performance of the AO system and the dynamical seeing behaviour), the measurement precision (Δ_{meas}) can be calculate as the standard deviation of the measurements (σ_{meas}) divided by the square root of the number of Gaussian distributed measurements (N), $\Delta_{meas} = \dfrac{\sigma_{meas}}{\sqrt{N}}$ (see Fig. 1).

But we have to keep in mind that the above value is just the measurement precision and includes only the statistical distributed error sources. To determine the systematic error sources, which affect in the case of relative astrometry the pixel scale and the position angle, we need a calibration system (see next section).

3. Calibration

Because we are dealing with relative astrometry we do not need an absolute astrometric calibration of our data, but we have to monitor the stability of our pixel scale to correct our measurements for possible variations of the pixel scale. The "normal way" of relative astrometric calibration is to use a Hipparcos binary system. This method results in the case of the NACO S13 camera in a pixel scale of typically $13.25 \pm 0.05\,mas$ per pixel (see Neuhäuser et al. (2005)). We do not need the absolute value of the pixel scale, but we have to determine changes very precisely. Hence, what we need is a calibration system with a high and known intrinsic stability (higher than a Hipparcos binary). We choose the old globular cluster 47 Tuc as our calibration cluster for targets on the southern hemisphere. The reasons are a high number of "calibration stars" in the FoV of the S13 NACO camera and a known intrinsic stability. McLaughlin et al. (2006) determined the transversal velocity dispersion of the 47 Tuc cluster members and obtained a value of about $630\,mas$ per year. To monitor the pixel scale we take hundreds of frames of 47 Tuc per epoch, measure the separation from each star to each star and compute the mean of all these separation measurements on every single frame. This mean of the separations represents the relative alignment of all observed cluster members and should have, within the errors (intrinsic instability and measurement errors), the same value in every epoch. Using a Monte-Carlo-Simulation with our observed cluster members and a Gaussian distributed transversal velocity dispersion of $630\,mas$ per year, we obtain an intrinsic stability of our used calibration cluster of about 3/100000 per pixel and year, which results for a given pixel scale of 13.25 mas for the S13 camera an intrinsic stability of about $0.4\,\mu as$ per pixel and year. This means, with 47 Tuc we are able to determine a change in the pixel scale down to $0.5\ldots1\,\mu as$ per pixel and year (including the typical measurement errors of the 47 Tuc cluster members). Due to the fact, that we are using a fixed and given "reference pixel scale" of 13.25 mas in our first epoch, we can not determine the absolute value of the pixel scale, but we are able to detect changes between the epochs very precisely.

Furthermore, we have to consider that the observed and measured change in the separation is a product of the orbital movement of the unseen companion and the stellar system (double or triple system). Regarding the timescale of both influences (hence the orbital periods of the substellar unseen companion and the seen stellar companion) we see a large difference. The orbital period of stellar binaries with a separation of about three arcseconds is typically more then thousand years. Astrometric search programs are most sensitive for the intermediate semi-major-axis regime, which means a period of about one year up to about three years. Thus, the influence of the stellar binary during our search program is nearly linear or just slightly curved. The advantage of a large difference in the orbital periods are that we are not depend on the exact orbital elements of the stellar binary and we are able to distinguish both influences due to the different timescale. There are two possibilities to determine the influence of the stellar binary. Either to add the stellar influence as an additional free parameter in the final fit or to determine the influence directly by measuring the separation of the stellar binary at two (linear influence) or three (sligthly curved influence) epochs where the substellar companion has the same orbital position. The second possibility can be applied for RV planet candidates (the orbital period of the candidate is well known) but to search for unknown companions, we have to use the first method thus to consider the nearly linear stellar influence in the final fit.

Our first observed stellar binary (HD 19994) has a separation of about 2.5 arcsec and an orbital period of about 1500 years. Mayor *et al.* (2004) discovered a RV planet candidate around the A component with $M\sin i = 1.68\,M_J$, $a = 1.4$ AU and a period of about 535 days. For HD 19994 we achive a total relative precision of $100\ldots150\,\mu as$ per epoch. The expected change in the separation (due to the astrometric signal of the planet candidate) depends on the orientation of both orbits (especially from the difference in the inclination angle between the planetary and the double star orbit) and is about $450\,\mu as$ for $\Delta i \approx 50°$ ($M_{true} \approx 5\,M_J$, $2\,\sigma$ detection). In the case of an astrometric non-detection we are able to exclude differences in the inclination angle of more than $\Delta i \approx 65°$ ($3.5\,\sigma$ detection) and thus to exclude true masses of more than $9\,M_J$.

References

Bean, J. L., McArthur, B. E., Benedict, G. F., Harrison, T. E. et al. 2007, *ApJ*, 134, 749
Benedict, G. F., McArthur, B. E., Forveille, T., Delfosse, X. et al. 2002, *ApJ*, 581, 115
Benedict, G. F., McArthur, B. E., Gatewood, G., Nelan, E. et al. 2006, *ApJ*, 132, 2206
Gatewood, G. & Eichhorn, H. 1973, *AJ*, 78, 769
Lippincott, S. L. 1978, *Space Sci. Revs*, 22, 153
Mayor, M., Udry, S., Naef, D. et al. 2004, *A&A*, 74, 238
McArthur, B. E., Endl, M., Cochran, W. D., Benedict, G. F. et al. 2004, *ApJ*, 614, 81
McLaughlin, D. E. & Anderson, J. & Meylan, G. et al. 2006, *ApJS*, 166, 249
Neuhäuser, R., Guenther, E. W., Wuchterl, G. et al. 2005, *A&A*, 435, 13
Pravdo, S. H. & Shaklan, S. B. 1996, *ApJ*, 456, 264
van de Kamp, P. 1969, *AJ*, 74, 238

Preparing the exoplanet search with PRIMA: searching for reference stars and target characterization

R. Geisler[1,2,*], J. Setiawan[2], Th. Henning[2], D. Queloz[3], A. Quirrenbach[1], R. Launhardt[2], A. Müller[2], S. Reffert[1], P. Weise[2] and ESPRI consortium[1,2,3,4]

[1]ZAH - Landessternwarte, Königstuhl 12, D-69117 Heidelberg, Germany

[2]Max-Planck-Institut für Astronomie, Königstuhl 17, D-69117 Heidelberg, Germany

[3]Observatoire Astronomique de l'Université de Genève, CH-1290 Sauverny, Switzerland

[4]European Southern Observatory (ESO), D-85748 Garching, Germany

[*] Fellow of the IMPRS for Astronomy and Cosmic Physics at the University of Heidelberg, email: `rgeisler@lsw.uni-heidelberg.de`

Abstract. The PRIMA (Phase-Referenced Imaging and Micro-arcsecond Astrometry) facility at ESO VLTI (Paranal observatory) is expected to be commissioned in mid 2008. The ESPRI (Exoplanet Search with PRIMA) consortium is currently preparing an astrometric survey to search for extrasolar planets. To achieve the scientific goal of this survey, a careful selection of target and reference stars is necessary. Apart from catalog search and modelling, extensive and dedicated preparatory observations are indispensable. Here we present two aspects of the preparatory observation programs: A high dynamic range near infrared (NIR) imaging survey to search for astrometric reference stars around the preselected target stars and characterization of the target stars by using high-resolution spectroscopy.

Keywords. astrometry, stars: planetary systems, techniques: interferometric, photometric, spectroscopic, radial velocities

1. Introduction

Relative astrometry with an accuracy of 10 μarcsec with PRIMA by using interferometric delays relies on phase reference stars at small angular separation (within 10 to 15 arcsec) from the target stars. Therefore, to prepare our planet search program ESPRI (Launhardt *et al.* 2008a/b), a dedicated search for reference stars around our potential planet-search targets is mandatory. In addition, accurate determination of fundamental stellar parameters and characterization of the stellar activity are useful to select appropriate target stars for ESPRI.

For this reason the usefulness of available catalogs in various archives was tested. The limiting magnitude in K-band with PRIMA and the Auxiliary Telescopes (ATs) for the bright target star will be K$\approx$12 mag and for the fainter reference star K$\approx$14-16 mag. Due to saturations and artifacts of the bright star, it is not possible to identify faint potential reference stars close to the bright star with the required brightness contrast in the available catalogs (e.g. 2MASS: Figure 1). This makes the majority of available catalogs useless for this task and an additional high dynamic range NIR imaging survey is necessary. In these proceedings we present the preliminary results of our NIR imaging survey and target characterization by high-resolution spectroscopy.

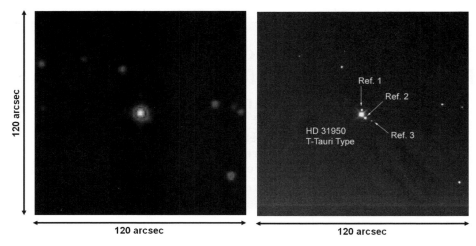

Figure 1. K-band image of HD 31950 obtained with SOFI (right panel). This star of T-Tauri type is part of the young star group. Three phase reference stars are found within 10 arcsec of the target star. In the contrast enhanced 2MASS K-band image no close reference stars are visible (left panel).

2. Observations and Preliminary Results

For our NIR imaging observations we used SOFI at the NTT (ESO, La Silla) and OMEGA Cass at the 3.5 m telescope (Calar Alto observatory). To achieve the required limiting magnitude for the astrometric reference stars we searched down to a detection limit of K≈17 mag. To characterize the reference stars (e.g., spectral type, photometric distance) all target stars with identified reference stars are additionally observed in J-band. The spectroscopic observations were done by using FEROS ($R = 48000$) at the 2.2 m MPG/ESO telescope in La Silla.

The ESPRI target list consists currently of nearby stars (main-sequence stars within 15 pc), young stars (within 100 pc and ages 5...300 Myr) and stars with known radial velocity planets, and contains 865 stars altogether. Almost half of of them have been observed already with SOFI and OMEGA Cass. The analysis of the images has been completed. We found at least one reference star within 10 arcsec and with K⩽16 mag in 22 % of all observed targets (Table 1). The rate predicted by modelling is 20 % (Figure

Star group	Number of stars	Observed		Reference stars found (K≤16mag)		2MASS
		SOFI	OMEGA Cass	2"- 10"	2"- 20"	2"- 10"
Nearby Stars	336					0.6%
Young Stars	381	163	259	91	148	11.8%
Radial Velocity	148					8.8%
Total:	865	49%		22%	35%	7.0%

Table 1. 422 of 865 stars (49 %) from our target list have been observed with SOFI and OMEGA Cass already. Out of these we found 91 stars (22 %) with at least one reference star between 2 and 10 arcsec with K⩽16 mag. Reference stars with separations between 10 and 20 arcsec are still useful for lower anticipated accuracy and thus would still allow the detection of a giant planet.

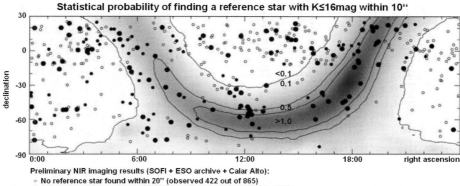

Figure 2. Sky map showing the statistical probability of finding at least one reference star with K⩽16 mag within 10 arcsec around any given star (grey scale: less then 10 % to 100 %). The map was constructed from COBE DIRBE NIR all sky maps and calibrated with Besançon synthetic stellar population models. Plotted as dots are the results from our NIR observations. The actual detection rates agree with the predicted ones to within 2 %, thus proving the applicability and usefulness of the model.

2) and very close to the actual detection rate. Only 7 % of the stars have a reference star with the required properties in 2MASS.

From the spectroscopic observations we determined the fundamental parameters of the target stars. In addition, we also monitor radial velocity variations of the targets to identify possible stellar/substellar companions or starspots. In case of stellar companions and starspots, the targets should be examined carefully whether they are still suitable for the anticipated astrometric observations. In case of substellar companions (planets/brown dwarfs), those targets with available reference stars are of high-priority to be observed with PRIMA. Both results, from NIR imaging and spectroscopy, will be included in a dedicated target database (ASTRID).

3. ASTRID Database (http://www.mpia-hd.mpg.de/ASTRID/)

ASTRID (AStrometric Target and Reference stars Interactive Database) was developed to collect relevant parameters for the target and reference star pairs. The database includes archival data (basically from Hipparcos and 2MASS) on the stars as well as the measurements obtained from the preparatory programs (NIR imaging, spectroscopy and astrometry). Parts of the database will be publicly accessible. When ESPRI observations commence with the VLTI, astrometric results will also be added to the ASTRID database.

References

Launhardt, R., A., Henning, Queloz, D., Quirrenbach, T., Elias, N., Pepe, F., Reffert, S., Segransan, D., Setiawan, J., & ESPRI Consortium 2008a, The ESPRI project: Astrometric Exoplanet Search with PRIMA, *These Proceedings*.

Launhardt, R., A., Henning, Queloz, D., Quirrenbach, T., Elias, N., Pepe, F., Reffert, S., Segransan, D., Setiawan, J., & ESPRI Consortium 2008b, The ESPRI project: Narrow-angle astrometry with VLTI-PRIMA, *Proc. IAUS, 248*, in press.

Finding new sub-stellar co-moving companion candidates - the case of CT Cha

Tobias Schmidt[1] and Ralph Neuhäuser[1]

[1] Astrophysikalisches Institut und Universitäts-Sternwarte, Universität Jena,
Schillergäßchen 2-3, 07745 Jena, Germany
email: tobi@astro.uni-jena.de

Abstract. We have searched for close and faint companions around T Tauri stars in the Chamaeleon star forming region. Two epochs of direct imaging data were taken with the VLT Adaptive Optics instrument NaCo in February 2006 and March 2007 in Ks band for the classical T Tauri star CT Cha together with a Hipparcos binary for astrometric calibration. Moreover a J band image was taken in March 2007 to get color information. We found CT Cha to have a very faint companion ($Ks_0 = 14.6$ mag) of ~ 2.67" separation corresponding to ~ 440 AU. We show that CT Cha A and the faint object form a common proper motion pair and that the companion is not a non-moving background object (with 4σ significance).

Keywords. stars: planetary systems, pre-main-sequence, imaging, individual: CT Cha

1. Introduction

CT Cha introduced in the 65th Name-List of Variable stars by Kholopov et al. (1981), was originally found by Henize & Mendoza (1973), often called HM 9, an emission-line star in Chamaeleon exhibiting variations in its Hα line from plate to plate and partial veiling (Rydgren 1980).

While the star was first classified as T Tauri star by Whittet et al. (1987) it was later on found to be a classical T Tauri star by Weintraub (1990) and Gauvin & Strom (1992) from IRAS data, before Natta et al. (2000) could find evidence for a silicate feature disk with $L_{sil} = 10^{-2} L_\odot$ and $L_{sil}/L_* = 0.014$ using ISO data.

The variations of the Hα line could later be interpreted when Hartmann et al. (1998) measured a mass accretion rate of $\log \dot{M} = -8.28 M_\odot/yr$. Additional variations of infrared (Glass 1979) and optical photometry can possibly be explained by surface features on CT Cha and a possible rotation period of 9.86 days found by Batalha et al. (1998).

All additional properties of the K7 (Gregorio Hetem et al. 1988) star CT Cha, such as e.g. its age estimated to be 3 Myr by Feigelson et al. (1993) or 0.9 Myr by Natta et al. (2000) as well as its equivalent width of the lithium absorption line of $W_\lambda(\mathrm{Li}) = 0.40 \pm 0.05$ Å (Guenther et al. 2007) and its radial velocity of 15.1 ± 0.1 km/s (Joergens 2006) and proper motion (table 1) are consistent with a very young member of the Cha I star-forming cloud with an age of 2 ± 2 Myr and a medium radial velocity of 14.9 ± 1.7 km/s (Melo et al. 2003).

2. Direct observations of a wide companion

2.1. AO imaging detection

We observed CT Cha in two epochs in February 2006 and in March 2007 with the European Southern Observatory (ESO) Very Large Telescope (VLT) at Cerro Paranal in the programmes with IDs 076.C-0292(A) & 078.C-0535(A). All observations were done using the Adaptive Optics (AO) instrument Naos-Conica (NaCo, Lenzen et al. 2003 & Rousset et al. 2003). In all cases the S13 camera (∼13 mas/pixel) and the double-correlated read-out mode were used.

For the raw data reduction we subtracted a mean dark from all science frames and the flatfield frames, then divided by the normalized dark-subtracted flatfield and subtracted the mean background by using ESO *eclipse/jitter*.

In all three images a companion candidate is found 2.67" northwest of CT Cha (Fig. 1) corresponding to ∼ 440 AU at a distance of 165 ± 30 pc estimated from the combination of data by Bertout et al. (1999) and Whittet et al. (1997) for Cha I members.

2.2. Astrometry

To check for common proper motion of the tentative companion of CT Cha we used the proper motion (here PM) of the star published in the literature (table 1). We use the weighted mean proper motion for checking, whether the two objects show common proper motion below.

We calibrated the NaCo data using the wide binary star HIP 73357 for our two measurements in 2006 and 2007. The astrometric calibration for our images was done relative, hence the error bars of separation and position angle given in table 2 include possible orbital motion of the binary inbetween our observations as well as their measurement uncertainties.

To determine the positions of both components we constructed a reference PSF from both objects. Thus, we obtained an appropriate reference PSF for each single image. By using IDL/starfinder we scaled and shifted the reference PSF simultaneously to both components in each of our individual images by minimizing the residuals. As a result we obtained positions (see table 2) and relative photometry (see table 3) including realistic error estimates for each object by averaging the results of all single images taken within each epoch.

From separation in figure 2, we can exclude by $3.4\,\sigma$ that CT Cha b is a non-moving background object. Due to the location of the companion (northwest of A) and the proper motion (towards northwest), the position angle information does not give much additional significance ($1.0\,\sigma$ deviation from the background hypothesis), hence resulting in a combined significance of $3.7\,\sigma$. For a K7 star and a sub-stellar companion (see below) and the given separation (at ∼ 165 pc), the orbital period is ∼ 11000 yrs, so that the

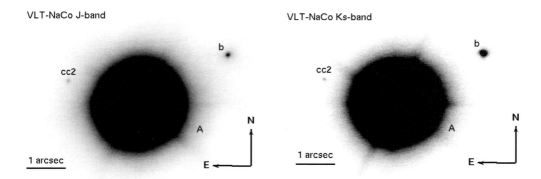

Figure 1. VLT-NaCo J-band and Ks-band images of CT Cha and its 6.3 mag fainter companion candidate (in Ks-band) CT Cha b 2.670 ± 0.036" northwest from March 1st and 2nd 2007. The object marked as 'cc2' was found to be a background object.

Table 1. Proper motions of CT Cha

Reference	$\mu_\alpha \cos\delta$ [mas/yr]	μ_δ [mas/yr]
UCAC2 (Zacharias et al. 2004)	-22.2 ± 5.2	7 ± 5.2
ICRF ext. (Camargo et al. 2003)	-18 ± 10	4 ± 9
weighted mean	-21.3 ± 4.6	6.3 ± 4.5

maximum change in separation due to orbital motion (for circular edge-on orbit) is ~ 1 mas/yr or ~ 0.07 pix/yr ($\sim 0.03\,°$/yr in PA for pole-on orbit). Neither in separation nor in position angle (measured from north over east to south), being $\sim 300\,°$, signs of orbital motion could be detected given the short epoch difference (~ 1 year).

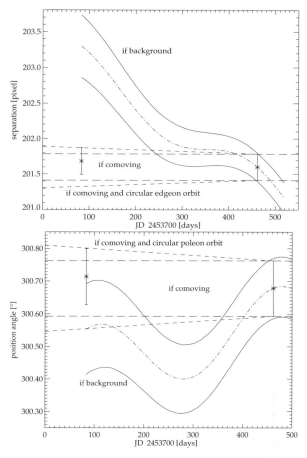

Figure 2. Top: Observed separation between CT Cha A and its companion. Our two measurements from 2006 and 2007 are shown. The long dashed lines enclose the area for constant separation. The dash-dotted line is the change expected if the companion is a non-moving background star. The opening cone enclosed by the dash-double-dotted lines are its estimated errors. The waves of this cone show the differential parallactic motion which has to be taken into account if the other component is a non-moving background star with negligible parallax. The opening short dashed cone is for the combination of co-motion and the maximum possible orbital motion. Bottom: Same for the position angle (measured from north over east to south) of the companion.

Table 2. Relative astrometric results for CT Cha

Epoch difference [days]	Target	Change in separation [pixel]	Change in PA[1] [deg]
378.08299	CT Cha Ab	-0.086 ± 0.268	-0.036 ± 0.121

Notes:
All Ks band images. [1]PA is measured from N over E to S. PA is given relative to the first epoch (absolutely measured it is 300.71 ± 1.24°) in Feb 2006.

Table 3. Apparent magnitudes of the companion CT Cha b

Epoch	J-band	Ks-band
17 Feb 2006	- ± -	14.951 ± 0.302
1/2 Mar 2007	16.607 ± 0.302	14.891 ± 0.301

While the negligible differences seen in PA and separation between the different observations are consistent with common proper motion, a possible difference in proper motion between both objects of up to a few mas/yr cannot be excluded from the data; such a difference in proper motion would be typical for the velocity dispersion in star forming regions like Cha I (Ducourant et al. 2005), so that we cannot yet exclude that both objects are independent members of Cha I, but not orbiting each other. Even if this would be the case, age and distance would be the same as assumed below for both objects, hence also the spectral type and its mass.

2.3. Photometry

As described in the last section we obtained from the PSF fitting of both components also the relative brightnesses. Using the photometry of CT Cha A from the Two Micron All Sky Survey (2MASS) catalogue of $J = 9.715 \pm 0.024$ mag and $K = 8.661 \pm 0.021$ mag and adding 0.3 mag variability of A estimated from data by Batalha et al. (1998), Ghez et al. (1997) and Lawson et al. (1996) to the error of b, we obtain the photometry of the companion using our measured brightness difference, see table 3.

The Ks band magnitudes agree within the 1σ errors of the two epochs, giving no indication of photometric variability. From the extinction corrected $J_0 - Ks_0 = 1.240 \pm 0.490$ mag (from A_V and extinction law by Rieke & Lebofsky et al. (1985)) we estimate a spectral type of M5 – L5 for the faint companion CT Cha b using (Golimowski et al. 2004).

2.4. Conclusions

We have found a new co-moving companion candidate of CT Cha and estimated its spectral type from photometry to be M5 – L5. After this result we have taken integral field spectra of the object with the VLT instrument SINFONI and could determine temperature and surface gravity of the object by comparison with synthetic spectra by Brott & Hauschildt (2005). From comparison with the eclipsing binary brown dwarf 2M0535 (Stassun et al. 2006) we conclude that the object has less mass than 36 M_{Jup}, similar to the GQ Lup companion found by Neuhäuser et al. (2005). Our best mass estimate from the spectral modelling gives a few Jupiter masses, so that the companion can be a planet imaged directly, hence has to be seen as planetary mass companion candidate, hence called b (small letter for planets and planet candidates). A detailed discussion of the spectra and further informations can be found in Schmidt et al. (submitted to A&A).

References

Batalha, C. C., Quast, G. R., Torres, C. A. O. et al. 1998, *A&AS*, 128, 561
Bertout, C., Robichon, N., Arenou, F. 1999, *A&A*, 352, 574
Brott, I. & Hauschildt, P. H. 2005, *ESASP*, 576, 565
Camargo, J. I. B., Ducourant, C., Teixeira, R. et al. 2003, *A&A*, 409, 361
Ducourant, C., Teixeira, R., Périé, J. P. et al. 2005, *A&A*, 438, 769
Feigelson, E. D., Casanova, S., Montmerle, T. et al. 1993, *ApJ*, 416, 623
Gauvin, L. S., Strom, K. M. 1992, *ApJ*, 385, 217
Ghez, A. M., McCarthy, D. W., Patience, J. L., & Beck, T. L. 1997, *ApJ*, 481, 378
Glass, I. S. 1979, *MNRAS*, 187, 305
Golimowski, D. A., Leggett, S. K., Marley, M. S. et al. 2004, *AJ*, 127, 3516
Gregorio Hetem, J. C., Sanzovo, G. C., & Lepine, J. R. D. 1988, *A&AS*, 76, 347
Guenther, E. W., Esposito, M., & Mundt, R. 2007, *A&A*, 467, 1147
Hartmann, L., Calvet, N., Gullbring, E. et al. 1998, *ApJ*, 495, 385
Henize, K. G. & Mendoza V., E. E. 1973, *ApJ*, 180, 115
Joergens, V. 2006, *A&A*, 448, 655
Kholopov, P. N., Samus, N. N., Kukarkina, N. P. et al. 1981, *IBVS*, 1921, 1
Lawson, W. A., Feigenlson, E. D., & Huenemoerder, D. P. 1996, *MNRAS*, 280, 1071
Lenzen, R., Hartung, M., Brandner, W. et al. 2003, *SPIE*, 4841, 944
Melo, C. H. F. 2003, *A&A*, 410, 269
Natta, A., Meyer, M. R., & Beckwith, S. V. W. 2000, *ApJ*, 534, 838
Neuhäuser, R., Guenther, E. W., & Wuchterl, G. 2005, *A&A*, 435, L13
Rieke, G. H., & Lebofsky, M. J. 1985, *ApJ*, 288, 618
Rousset, G., Lacombe, F., Puget, P. et al. 2003, *SPIE*, 4839, 140
Rydgren, A. E. 1980, *AJ*, 85, 444
Schmidt, T. O. B., Neuhäuser, R., Seifahrt, A. et al. 2007/8, *submitted to A&A*
Stassun, K. G., Mathieu, R. D., & Valenti, J. A. 2006, *Nature*, 440, 311
Weintraub, D. A. 1990, *ApJS*, 74, 575
Whittet, D. C. B., Kirrane, T. M., Kilkenny, D. et al. 1987, *MNRAS*, 224, 497
Whittet, D. C. B., Prusti, T., Franco, G. A. P. et al. 1997, *A&A*, 327, 1194
Zacharias, N., Urban, S. E., Zacharias, M. I. et al. 2004, *AJ*, 127, 3043

Closure phase studies toward direct detection of light from hot Jupiters

Ming Zhao[1], John D. Monnier[1], Theo ten Brummelaar[2], Ettore Pedretti[3], Nathalie Thureau[4]

[1] Department of Astronomy, University of Michigan,
Ann Arbor, MI 48109, USA
email: mingzhao@umich.edu

[2] The CHARA Array, Georgia State University

[3] University of St. Andrews, Scottland, UK

[4] University of Cambridge, UK

Abstract. Precision closure phase measurements obtained with ground-based long baseline optical interferometers is a promising way to directly detect light from nearby hot Jupiters. Here we present our closure phase simulations for the CHARA array for several bright hot Jupiters, v And b, 51 Peg b, and τ Boo b. The maximum possible closure phase signals from these hot Jupiters are very small, for example, only ~ 0.17 degrees for v And b, requiring very high precision and stable closure phase measurements. We present preliminary results of a closure phase study on test object β Tau and hot Jupiter system v And, both obtained with the MIRC instrument at the CHARA array. We demonstrate that directly detecting the light from hot Jupiters is feasible using high precision closure phase measurements obtained by CHARA-MIRC along with its sub-milli-arcsecond resolution, although challenges remain.

Keywords. techniques: interferometry, techniques: high angular resolution, exoplanets: hot Jupiters: v And b, 51 Peg b, τ Boo b

1. Introduction

Detecting the light from extrasolar planets is the most direct way to characterize and understand their structures and atmospheres. It is, however, a very challenging task because of the planets' close angular distances and extremely high brightness contrast to their host stars. Among the more than 200 detected extrasolar planets so far, a large population of them are "hot Jupiters", i.e., close-in (< 0.1AU) extrasolar giant planets with high temperatures ($T_{eff} > 1000K$) and large masses ($\sim 1 M_j$). These "hot Jupiters" are best suited for direct detection and characterization because of their relatively high brightness ratios to their host stars at convenient bands (e.g., J, H, K) and also because of their well-studied atmospheric models. Theoretical models have predicted many features in the atmospheres of hot Jupiters, e.g., molecular bands, dusty clouds, day/night flux variation, etc. (e.g., Burrows et al. 1997; Barman et al. 2005). These features, if confirmed by observations, can provide valuable information on the physical structure, chemical composition and energy distribution of the atmospheres of hot Jupiters (e.g., Barman et al. 2005). To date, four "hot Jupiters" have been directly detected by the Spitzer Space Telescope: HD 209458b, TrES-1, HD 189733b, and most recently, v And b (Deming et al. 2005, 2006; Charbonneau et al. 2005; Harrington et al. 2006). These measurements provided strong constraints to the current planetary atmospheric models and have opened a new era of direct study of these extrasolar planets. However, the number of current measurements is still too few to break model degeneracies and provide

Table 1. Hot Jupiter candidates for CHARA-MIRC

Star Name	Dist. pc	H mag	K mag	Period day	e	Semimajor axis AU	(mas)	T_0 JD	R_* mas
υ And	13.5	2.957	2.859	4.6170	0.034	0.059	(4.42)	2450088.64	0.569
τ Boo	15.6	3.546	3.507	3.3128	0.018	0.049	(3.13)	2451653.968	0.45
51 Peg	15.4	4.234	3.911	4.2310	0.01	0.051	(3.31)	2450203.947	0.35

References: Marcy, J., *et al.* 1997; Drake, S., *et al.* 1998; Butler, P. *et al.* 1999; Henry, G., *et al.* 2000; Burrows, A., *et al.* 2000

detailed conclusions (see e.g., Burrows *et al.* 2005), and they only provide information in mid-IR. To better understand the physical and chemical characters of these planets, more direct detections are necessary and are undoubtly the best way to proceed.

Among the methods used to detect light from hot Jupiters, a promising way is to use ground-based long baseline optical/IR interferometers along with precision closure phase and/or differential phase measurements. These two methods, especially closure phase, are well studied and have been applied to modern optical/IR interferometers (see, e.g., Segransan *et al.* 2000; Lopez *et al.* 2000; Joergens & Quirrenbach 2004; Monnier 2007). These methods are applied in the near infrared so any detections will be very valuable in addition to the Spitzer results. For instance, planetary T_{eff} and atmospheric characters can be more accurately determined by combining detections across near and mid-IR bands. In this proceeding we only give a brief introduction to these methods.

Closure Phase. Interferometers obtain information of distant objects through fringe contrast and phase. The presence of a planet causes a small phase shift in the stellar fringe. However, the phase information is always corrupted because the atmospheric turbulence always induces random and extra optical paths to the light. Nevertheless, these extra phases can be canceled if we sum the phases from 3 telescopes in a closed triangle (i.e., $\Delta\Phi_{AB} + \Delta\Phi_{BC} + \Delta\Phi_{CA}=0$, where A, B, and C stand for telescopes, see Monnier 1999 for details). This phase closure, or closure phase, is immune to any phase shifts induced by the atmosphere as well as other systematic errors hence is a good observable (Monnier 2007). Closure phase is very sensitive to asymmetry and can sense the structure of star-planet system as analog of high-contrast binaries, which allows us to extract their full orbital parameters (most importantly, the inclination angle) as well as the planet-to-star flux ratio.

Differential Phase. Hot Jupiters and their host stars' brightness contrast differ at different wavelengths (e.g., Sudarsky *et al.* 2003, Burrows *et al.* 2007). The difference in the brightness contrast between two wavelengths can cause a slight shift to the system's photo-center, therefore inducing a phase shift to the observed fringe visibility (e.g., Vasisht *et al.* 2004). This phase shift, also called differential phase, is also immune to the atmospheric turbulence in that the turbulence changes the phases in the same way for different wavelengths, which therefore can be eliminated by the differential measurement. Differential phase can yield the flux ratio and orbital parameters of hot Jupiter systems and, in addition, it also provides valuable spectral information about the planet atmospheres.

In this proceeding we perform simulation studies using both techniques for several hot Jupiters, and show our test observations. Mostly we concentrate on the closure phase method.

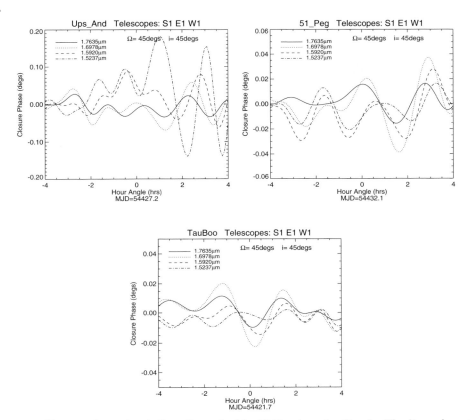

Figure 1. Closure phase simulations for υ And b, 51 Peg b and τ Boo b. The lines show the results for 4 different wavelength channels (out of 8 in total) of MIRC.

2. Simulations for nearby hot Jupiters

Among our list of hot Jupiter candidates chosen for CHARA-MIRC, 3 of them are currently most favorable because of the high brightness of their host stars (see Table 1). Because hot Jupiter-hosting stars and their closest Jupiters are similar to high contrast close binaries, we simulate closure phase signals using binary models (Zhao et al. 2007) for these three systems. The simulated closure phases are obtained using the CHARA interferometer array (ten Brummelaar et al. 2005) along with its MIRC instrument (Monnier et al. 2004) at H band. The orbital properties of these systems are well studied and are listed in Tab. 1. The diameter of the star υ And is newly measured to be 1.138±0.058 mas using the FLOUR instrument (Mérand et al. 2006) at CHARA (Mérand 2007, private communication). The infrared flux ratios of the planets are adopted from the models of Sudarsky et al.(2003). The inclination i and the position angle of the ascending node Ω are unknown for these systems. We assume 45° inclination and $\Omega=45^o$ for our first set of models. Figure 1. shows the resulting closure phase simulations for the longest telescope triangle, S1-E1-W1. For υ And b (the first panel), we see large variations between channels and the signal level is higher than the other two candidates because its host star is resolved by CHARA, which lowers the fringe visibility of the star and boosts the closure phase. The closure phase of the shortest wavelength channel peaks at $\sim 0.17^o$ for υ And b and the difference between the first and the last channel is even bigger. For 51 Peg b,

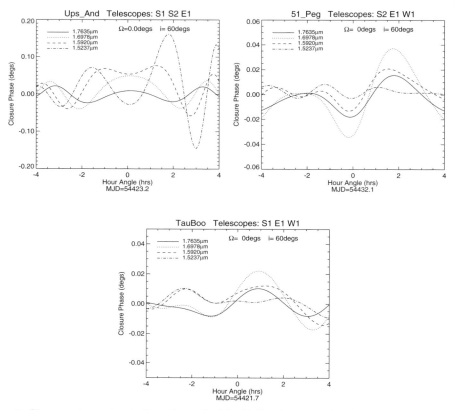

Figure 2. Closure phase simulations for υ And b, 51 Peg b and τ Boo b, similar to Fig. 1. but for different i & Ω, i.e., $i = 60°$ and $\Omega = 0°$

as shown in the second panel, closure phase peaks at ∼0.04 degs. Because the star 51 Peg is not resolved by CHARA, the closure phase is significantly lower than υ And b even though its flux contrast is higher at H band. For τ Boo b, although its host star is slightly bigger than 51 Peg, its flux ratio is much lower thus the closure phase is weaker and only peaks at ∼$0.02°$.

Although i and Ω are arbitrarily chosen in Fig. 1, Figure 2 indicates that for a different set of i and Ω we can always get similar signal levels for the three objects using different telescope triangles and/or at a different date. These plots also suggest that in order to detect such small signals, our precision on closure phases needs to be better than ∼ $0.17°$ for υ And b and much better than that for 51 Peg b and τ Boo b.

In addition to closure phases, we also simulate differential phases for the three candidates. Specifically, we first simulate complex visibilities for the 8 individual channels, and use the two channels at the edge (i.e., at $1.52\mu m$ & $1.76\mu m$) to interpolate the phases for the middle channels to eliminate the OPD drift at different wavelengths due to atmospheric pistons. The differential phase for a certain channel is then calculated by subtracting the interpolated phase. Figure 3 shows the results at 1.66 μm (the central channel) using CHARA's longest baseline (S1-E1, 331m). The signals are weaker than closure phases, therefore the required precision for detection is also higher, for instance, need to be better than $0.1°$ for υ And b.

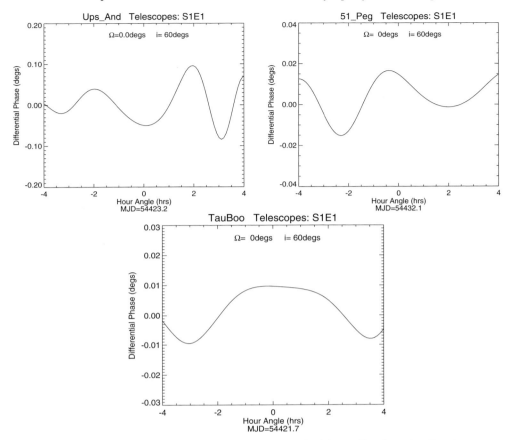

Figure 3. Simulated differential phases for υ And b, 51 Peg b and τ Boo b.

3. Test Observations

In order to verify the feasibility of detecting weak signals using CHARA-MIRC as simulated above, we need test observations to study the stability and precision of our measurements. We conducted our first test observation on a bright star β Tau (H=1.9) on Sep. 15, 2005. The top panel of Fig. 4 shows the closure phase for each data file averaged over all wavelength channels in H band. The middle panel shows the result of averaging 10 files, and the bottom averages 20 files. The error reduces as $\sqrt{N}$, suggesting the measurement is stable and immune from effects due to changes in the seeing. Encouragingly, the error for $\sim$1.5 hour integration is <0.1 degs, small enough for the required precision, 0.17^o, for υ And b.

Observations on our first candidate υ And was performed in August, 2007. Figure 5 shows the results in 3 panels similar to Fig. 4. We notice that the error is 0.25^o for only 0.6 hours of integration. Although it is larger than the required precision, the error can in fact reach comparable precision as that of β Tau when taking actual integration time and brightness of the star into account. We also notice that the intrinsic errors of the data points are much lower than the scatter and the measurements drift a little with time, which may relate to unknown calibration issues.

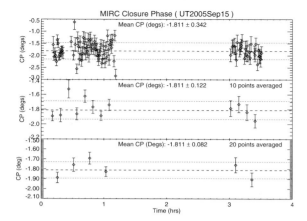

Figure 4. Test observation on bright star β Tau. Each point in the top panel corresponds to 45 seconds of integration. The middle panel shows the average of 10 points and the bottom shows the average of 20 points.

4. Conclusions

Our simulations show that in order to detect the signal from a hot Jupiter like υ And b, the errors of the closure phase and differential phase need to be below 0.1degs and much lower for other hot Jupiters. Our test observations show that for a bright star with long integration, the precision of our closure phase measurement is enough for the requirement. For a real hot Jupiter, this precision is feasible but requires much longer integration as well as higher sensitivity from the facilities to shorten the integration time. There are indications of some calibration issues in our latest observations. We are investigating these issues to improve the calibration and precision. In the near future, a fringe tracker, CHAMP (Berger *et al.* 2006), will be commissioned to increase the sensitivity of MIRC, and some optics of the array will be improved with better coating and focusing. With the improvements on both precision and sensitivity, the goal of detecting light from hot Jupiters will be feasible for CHARA-MIRC.

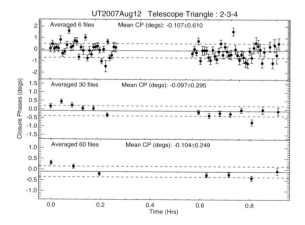

Figure 5. Test observation on hot Jupiter system υ And b. Each point in the top panel is an average of 6 files, which corresponds to 32 seconds of integration. Points in the middle panel are averages of 30 files and in the bottom are average of 60 files.

Acknowledgements

MZ thanks the support from the Michelson Fellowship. We also thank the support from a NASA-TPF Foundation research grant.

References

Barman, T. S., Hauschildt, P. H., & Allard, F. 2005, *ApJ*, 632, 1132
Berger, D. H., Monnier, J. D., Millan-Gabet, R., ten Brummelaar, T. A., Muirhead, P., Pedretti, E., & Thureau, N. 2006, in *Proceedings of the SPIE*, Volume 6268, pp. 62683K (2006).
Burrows, A., Marley, M., Hubbard, W. B., Lunine, J. I., Guillot, T., Saumon, D., Freedman, R., Sudarsky, D., & Sharp, C. 1997, *ApJ*, 491, 856
Burrows, A., Hubeny, I., & Sudarsky, D. 2005, *ApJL*, 625, L135
Burrows, A., Guillot, T., Hubbard, W. B., Marley, M. S., Saumon, D., Lunine, J. I., & Sudarsky, D. 2000, *ApJL*, 534, L97
Burrows, A., Hubeny, I., Budaj, J., Knutson, H. A., & Charbonneau, D. 2007, *ApJL*, 668, L171
Butler, R. P., Marcy, G. W., Fischer, D. A., Brown, T. M., Contos, A. R., Korzennik, S. G., Nisenson, P., & Noyes, R. W. 1999, *ApJ*, 526, 916
Charbonneau, D., Allen, L. E., Megeath, S. T., Torres, G., Alonso, R., Brown, T. M., Gilliland, R. L., Latham, D. W., Mandushev, G., O'Donovan, F. T., & Sozzetti, A. 2005, *ApJ*, 626, 523
Deming, D., Harrington, J., Seager, S., & Richardson, L. J. 2006, *ApJ*, 644, 560
Deming, D., Seager, S., Richardson, L. J., & Harrington, J. 2005, *Nature*, 434, 740
Drake, S. A., Pravdo, S. H., Angelini, L., & Stern, R. A. 1998, *AJ*, 115, 2122
Harrington, J., Hansen, B. M., Luszcz, S. H., Seager, S., Deming, D., Menou, K., Cho, J. Y.-K., & Richardson, L. J. 2006, *Science*, 314, 623
Henry, G. W., Baliunas, S. L., Donahue, R. A., Fekel, F. C., & Soon, W. 2000, *ApJ*, 531, 415
Lopez, B., Petrov, R. G., & Vannier, M. 2000, *Proceedings of SPIE*, 4006, 407
Marcy, G. W., Butler, R. P., Williams, E., Bildsten, L., Graham, J. R., Ghez, A. M., & Jernigan, J. G. 1997, *ApJ*, 481, 926
Mérand, A., Coudé du Foresto, V., Kellerer, A., ten Brummelaar, T., Reess, J.-M., & Ziegler, D. 2006, in *Proceedings of the SPIE*, Volume 6268, pp. 62681F (2006).
Monnier, J. D. 1999, in Principles of Long Baseline Interferometry, Edited by Peter R. Lawson, 203–226
Monnier, J. D. 2007, *New Astronomy Review*, 51, 604
Monnier, J. D., Berger, J.-P., Millan-Gabet, R., & Ten Brummelaar, T. A. 2004, in *Proceedings of SPIE* Volume 5491., 2004., p.1370, ed. W. A. Traub, 1370–+
Joergens, V., & Quirrenbach, A. 2004, *Proceedings of SPIE*, 5491, 551
Segransan, D., Beuzit, J.-L., Delfosse, X., Forveille, T., Mayor, M., Perrier-Bellet, C., & Allard, F. 2000, *Proceedings of SPIE*, 4006, 269
Sudarsky, D., Burrows, A., & Hubeny, I. 2003, *ApJ*, 588, 1121
ten Brummelaar, T. A., McAlister, H. A., Ridgway, S. T., Bagnuolo, Jr., W. G., Turner, N. H., Sturmann, L., Sturmann, J., Berger, D. H., Ogden, C. E., Cadman, R., Hartkopf, W. I., Hopper, C. H., & Shure, M. A. 2005, *ApJ*, 628, 453
Vasisht, G. & Colavita, M. M. 2004, in *Proceedings of SPIE*, Volume 5491., 2004., p.567, ed. W. A. Traub, 567–+
Zhao, M., Monnier, J. D., Torres, G., Boden, A. F., Claret, A., Millan-Gabet, R., Pedretti, E., Berger, J.-P., Traub, W. A., Schloerb, F. P., Carleton, N. P., Kern, P., Lacasse, M. G., Malbet, F., Perraut, K. 2007, *ApJ*, 659, 626

First observation of planet-induced X-ray emission: The system HD 179949

S. H. Saar[1], M. Cuntz[2], V. L. Kashyap[1] and J. C. Hall[3]

[1]Harvard-Smithsonian Center for Astrophysics,
Cambridge, MA 02138, USA
email: saar@cfa.harvard.edu, kashyap@cfa.harvard.edu

[2]Department of Physics, University of Texas at Arlington,
Arlington, TX 76019-0059, USA
email: cuntz@uta.edu

[3]Lowell Observatory, Flagstaff, AZ 86001, USA
email: jch@lowell.edu

Abstract. We present the first observation of planet-induced stellar X-ray activity, identified for the HD 179949 system, using Chandra / ACIS-S. The HD 179949 system consists of a close-in giant planet orbiting an F9 V star. Previous ground-based observations already showed enhancements in Ca II K in phase with the planetary orbit. We find an $\sim$30% increase in the X-ray flux over quiescent levels coincident with the phase of the Ca II enhancements. There is also a trend for the emission to be hotter at increased fluxes, confirmed by modeling, showing the enhancement at $\sim$1 keV compared to $\sim$0.4 keV for the background star.

Keywords. Planetary systems, stars: activity, stars: coronae, stars: individual (HD 179949), stars: late-type, stars: magnetic fields

1. Introduction

Planets have been discovered around a large number of stars, mostly by the cycle Doppler shift of their photospheric lines. Most of these planets have been found around F, G, and K-type main-sequence stars. Moreover, about 20% of these planets are at an orbital distance of 0.1 AU or less (e.g., Butler *et al.* 2006), commonly referred to as close-in extrasolar giant planets (CEGPs). An interesting question is whether the CEGPs have any effect on the atmosphere of their parent star. Using observed star-planet systems as a basis, Cuntz *et al.* (2000) were first to propose that CEPGs can increase chromospheric and coronal activity. This effect was thereafter identified through high-quality data, obtained by Ca II K observations by Shkolnik *et al.* (2003) for five stars (i.e., HD 179949, HD 209458, τ Boo, 51 Peg, υ And). The observations indicated unambigious star-planet activity enhancement in the HD 179949 system. Subsequent results were given by Shkolnik *et al.* (2005).

The enhancement was found to be phased with the planet orbital period ($P_{\rm orb} \simeq$ 3.09 d), and not the (poorly known) stellar rotation period, $P_{\rm rot}$, estimated to lie between 7 and 11 d. This clearly implies that the Ca II emission enhancement is caused by some form of star-planet interaction. The peak excess amounts to $\sim$0.7% in the continuum-normalized K line core strength (which translates to a $\sim$12% increase in the [basal-subtracted] chromospheric flux; see Saar *et al.* 2008), and is shifted in phase by $\Delta \phi_{\rm orb} \sim$ 0.18 from the planet's inferior conjunction. The presence of only one emission peak per $P_{\rm orb}$ seems to rule out a tidal interaction that would result in a period of $P_{\rm orb}/2$; i.e., one peak per tidal bulge. A variety of models have been proposed to explain the observations, which indicate that planet-induced stellar activity enhancements can be an important

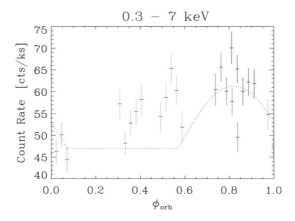

Figure 1. Background subtracted ACIS-S count rate (0.3 − 7 keV; 6 ksec bins) as a function of the planet orbital phase $\phi_{\rm orb}$ for various $P_{\rm rot}$. Observations were taken in 30 ksec segments separated by 1, 8, 9, and 10 × $P_{\rm orb}$ (orange, green, light & dark blue, respectively) from the first (red). We also show the best fitting Ca II H+K model following Shkolnik et al. (2003) (dotted).

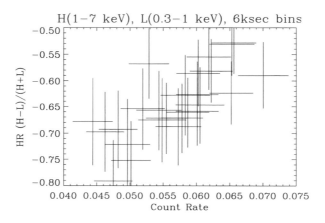

Figure 2. Hardness ratio between high (1 − 6 keV) and low (0.3 − 0.6 keV) energies as a function of the total count rate.

probe of (1) the close-in stellar magnetic field structure, (2) stellar wind properties, and/or (3) the planetary magnetosphere (Saar et al. 2004, Grießmeier et al. 2004, Preusse et al. 2005, McIvor et al. 2006, Zarka 2007, Cranmer & Saar 2008). In the following, we report the first observation of planet-induced X-ray emission, found in the exoplanetary system HD 179949.

2. Observations and Interpretation

To reduce complications due to varying stellar activity, the observations were taken at nearly the same stellar rotational phase $\phi_{\rm rot}$ and within 3 $P_{\rm rot}$, minimizing possible spatial and temporal changes, respectively; see Saar et al. (2008) for details. For 7.5 d $\leqslant P_{\rm rot} \leqslant 10$ d, the phase span $\Delta\phi_{\rm rot} \leqslant 0.3$. Plots versus $\phi_{\rm rot}$ (not given here) show either scatter ($P_{\rm rot} = 8 - 9$ d) or sharp changes over multiple orbits difficult to explain with rotational modulation. A Lomb-Scargle periodogram analysis yields $P = 3.289$ d (false alarm probability $= 2 \times 10^{-11}$) with no other P consistent with possible $P_{\rm rot}$ values; thus a planet-related origin for any variation is preferred.

The data phased to $\phi_{\rm orb}$ show a minimum around $\phi \sim 0.0 - 0.1$ and a gradual rise to maximum around $\phi \sim 0.7 - 0.9$ (Fig. 1). Nearly all the variation is at high energies (> 1 keV); additionally, the hardness ratio correlates well with total count rate (Fig. 2). The variation at high energies is also not smooth. Thus the "planet effect" produces hot, fluctuating (flare-like?) variability. Contemporaneous Ca II H+K data from the Lowell Observatory Solar-Stellar Spectrograph is consistent (within the large errors) with variations seen in Shkolnik *et al.* (2005). The best fit to a scaled version of their H+K model, however (fixing the minimum to the average flux in $\phi \sim 0.0 - 0.1$ and the emission peak phase shift $\Delta\phi_{\rm orb} = 0.18$), shows significant unmodeled excess flux in the range $\phi \sim 0.4 - 0.6$ (Fig. 1).

3. Summary

We have detected, in the HD 179949 exoplanet system, the X-ray counterpart to the excess Ca II H+K emission previously obtained that is phased to the planet's orbit. The following results are forwarded by the observations:

(*a*) Peak X-ray enhancement (0.3 − 7 keV) over the background is ~ 6 times that seen in Ca II H+K and shows a similar phase shift from inferior conjunction, which is $\Delta\phi_{\rm orb} \sim 0.18$.

(*b*) Thermal plasma models indicate the background has $T \sim 0.4$ keV, consistent with the corona of a modestly active star. The component responsible for the variability, and associated with the planet, is hotter with $T \sim 1$ keV and $\rm EM_{hot}/EM_{cool} \sim 0.3$.

(*c*) There is significant additional excess flux around $\phi \sim 0.4 - 0.6$, which is also hot ($T \sim 1$ keV), but does not follow the variation seen in Ca II H+K. The source of this emission is unclear; it may come from interaction with a second loop (possibly rooted at high stellar latitude) yielding a larger $\Delta\phi$, or it may be emission from the planet's magnetosphere itself (most visible near $\phi = 0.5$). These different scenarios will be explored in future studies (Saar *et al.* 2008).

Additional observations and modeling of this phenomenon are needed to obtain further insight into the dominant physical processes.

References

Butler, R. P., *et al.* 2006, *ApJ*, 646, 505
Cranmer, S. R., & Saar, S. H. 2008, astro-ph/0702530
Cuntz, M., Saar, S. H., & Musielak, Z. E. 2000, *ApJ* (Letters), 533, L151
Grießmeier, J. -M., *et al.* 2004, *A&A*, 425, 753
McIvor, T., Jardine, M., & Holzwarth, V. 2006, *MNRAS*, 367, L1
Preusse, S., Kopp, A., Büchner, J., & Motschmann, U. 2005, *A&A*, 434, 1191
Saar, S. H., Cuntz, M., & Shkolnik, E. 2004, in: A. K. Dupree & A. O. Benz (eds.), *Stars as Suns: Activity, Evolution and Planets*, IAU Symp. 219 (San Francisco: ASP), p. 355
Saar, S. H., Kashyap, V. L., Cuntz, M., Shkolnik, E., & Hall, J. C. 2008, *ApJ*, in preparation
Shkolnik, E., Walker, G. A. H., & Bohlender, D. A. 2003, *ApJ*, 597, 1092; Erratum 609, 1197 [2004]
Shkolnik, E., Walker, G. A. H., Bohlender, D. A., Gu, P. -G., & Kürster, M. 2005, *ApJ*, 622, 1075
Zarka, P. 2007, *Planetary and Space Science*, 55 (5), 598

Photometric follow-up observation of some SuperWASP transiting planet candidates

Sheng-hong Gu[1], Andrew Collier Cameron[2], Xiao-bin Wang[1], Li-yun Zhang[1], Xiang-song Fang[1] and Xue-jing Li[1]

[1]National Astronomical Observatories/Yunnan Observatory, CAS,
Kunming 650011, China
email: shenghonggu@ynao.ac.cn

[2]School of Physics and Astronomy, University of St Andrews,
Fife KY16 9SS, UK
email: acc4@st-and.ac.uk

Abstract. Three SuperWASP transiting planet candidates were observed through R or I filters using the 1-meter telescope and CCD camera of Yunnan Observatory from 2006 to 2007. The relative photometric data were corrected for the systematic errors by means of Tamuz et al. (2005) and Collier Cameron et al. (2006)'s algorithms. The resulting light curves demonstrate that one of three targets is a potential exoplanet candidate, which is worthy to perform further follow-up observation to clarify.

Keywords. exoplanet, transit, photometry

1. Introduction

The SuperWASP project (Pollacco et al. 2006) is an UK wide-field survey for exoplanet transits. Since this project was started in 2004, many candidates have been discovered by the WASP consortium (Christian et al. 2006, Clarkson et al. 2007, Collier Cameron et al. 2007, Lister et al. 2007, Street et al. 2007). Due to the small apertures of the SuperWASP telescopes, it is very important to make follow-up photometry for these candidates by using telescopes with larger apertures. Thus, we began a follow-up observation project at the 1.0m telescope of Yunnan Observatory in 2006. Up to now, we have finished photometric observations for 3 SuperWASP targets: One of them was already classified to be a SB1 binary (Collier Cameron et al. 2007); for the second one, we could not find the light variation; we found a transit-like light curve for the third candidate with a magnitude of 9.37 in V band. Here, we present the results of candidate 3.

2. Observation

We performed the photometric observation for candidate 3 of our SuperWASP targets using the 1.0m telescope at Yunnan Observatory on September 18, 2007. The equipment used in the observation is a 1k×1k CCD camera with the I filter. The field of view is about 6'.5×6'.5. The exposure time is between 20 and 40 seconds depending on the observing condition.

3. Data reduction

The observed images are reduced with the IRAF package, including the bias subtraction, flat-fielding and cosmic ray removal. The target star and 5 comparisons in the

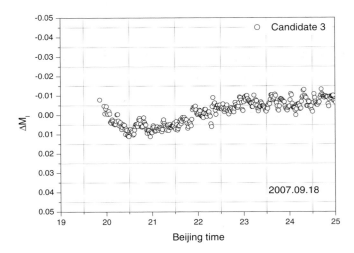

Figure 1. The reduced light curve of candidate 3.

same field of view are measured by using the APPHOT sub-package of IRAF through an aperture with radius of 12.5". The result light curve consists of 500 data points.

It is well-known that the transit depth of exoplanets is very shallow. In order to hunt the transit signal, we need to use some new methods to improve the quality of the observed light curve. Here, we have utilized the algorithms of Tamuz *et al.* (2005) and Collier Cameron *et al.* (2006) to remove the systematic effects in photometric data so as to obtain the final light curve with high quality.

4. Preliminary result

Fig. 1 shows the final light curve for candidate 3 on Sept. 18, 2007 after removing the systematic effects. This is potentially a transit-like light curve, it is worthy to do further follow-up observations to confirm. From this light curve, we can find that the transit center time deviates from the ephemeris formulae provided by the SuperWASP consortium. In our observation, the transit center time is coming earlier by several ten minutes than the prediction, which suggests that the orbital period derived before is a little longer.

We shall carry out further photometric and spectroscopic observations in this winter to clarify whether the candidate 3 of our samples is an exoplanet.

Acknowledgements

This work is partly supported by NSFC under grant No. 10673027.

References

Christian, D. J. *et al.* 2006, *MNRAS*, 372, 1117
Clarkson, W. I. *et al.* 2007, *MNRAS*, 381, 851
Collier Cameron, A. *et al.* 2006, *MNRAS*, 373, 799
Collier Cameron, A. *et al.* 2007, *MNRAS*, 380, 1230
Lister, T. A. *et al.* 2007, *MNRAS*, 379, 647
Pollacco, D. L. *et al.* 2006, *PASP*, 118, 1407
Street, R. A. *et al.* 2007, *MNRAS*, 379, 816
Tamuz, O., Mazeh, T. & Zucker, S. 2005, *MNRAS*, 356, 1466

Photometric observation of the transiting exoplanet WASP-1b

Xiao-bin Wang[1], Andrew Collier Cameron[2], Sheng-hong Gu[1] and Li-yun Zhang[1]

[1] National Astronomical Observatories/Yunnan Observatory, CAS,
Kunming 650011, China
email: wangxb@ynao.ac.cn

[2] School of Physics and Astronomy, University of St Andrews,
Fife KY16 9SS, UK
email: acc4@st-and.ac.uk

Abstract. The extra-solar planet WASP-1b, discovered by SuperWASP consortium in 2006, was observed through R filter using the 1-meter telescope with CCD camera at Yunnan Observatory on November 11, 2006. The systematic errors in photometric data were reduced by means of Tamuz *et al.* (2005) and Collier Cameron *et al.* (2006)'s algorithms. In order to estimate the parameters of the system, the MCMC (Markov Chain Monte Carlo) analysis is applied to fit the observed light curve. The following parameters of the system are derived: $R_p=1.44R_J$, $M_p=0.88M_J$, $R_*=1.52R_\odot$, $a=0.0396$AU. The new parameters of the planet imply its low density, which agrees to the previous results.

Keywords. exoplanet, transit, photometry

1. Introduction

Photometric observation of transiting exoplanet can provide us the size and mass of the host star and the planet. Furthermore, we can know the density of the planet, which is very important for understanding the formation of stellar system with planets. The SuperWASP project (Pollacco *et al.* 2006) is UK wide-field survey for exoplanet transits. The extra-solar planet WASP-1b was discovered by the WASP consortium (Collier Cameron *et al.* 2007a) in 2006. Later, the follow-up photometric and spectroscopic observations (Charbonneau *et al.* 2007, Shporer *et al.* 2007, Stempels *et al.* 2007) for the system were made, and the basic parameters of the system were derived.

2. Observation

We made the photometric observation for WASP-1b during its transit period using the 1-meter telescope at Yunnan Observatory on November 11, 2006. The observational data were gathered by using a 1024×1024 pixel CCD detector through the R filter. The scale of one pixel is 0.38", the field of view of the camera is about 6.5'×6.5'. The observation for this transit event was done at air mass less than 1.2 with an exposure time of 30 seconds.

3. Data reduction

The observed CCD images are reduced with the IRAF package, including the bias subtraction, flat-fielding and cosmic ray removal. The instrumental magnitude of the

WASP-1 system and 5 comparison stars in the field of view are measured by using APPHOT sub-package of IRAF through an aperture with radius of 6.5". Finally, the observed light curve contains 312 data points.

For the amplitude of light variation due to transit of a planet is small, the high-quality photometric data are necessary for searching new exoplanets and re-determining the parameters of known systems. Here, we have applied Tamuz et al. (2005) and Collier Cameron et al. (2006)'s algorithms to minimize the systematic effects in photometric data so as to enhance the signal to noise ratio of the light curve. Fig. 1 shows the light curve on Nov. 11, 2006 after removing the systematic effects.

4. Photometric parameters

Assuming the planet's orbit is circular, we model the flux of the transiting system with the parameters $\{T_c, p, \Delta F, t_T, b, M_*\}$ considering the 4-coefficient limb-darkening law of Claret (2000). The limb darkening coefficients for the R-band are interpolated from Claret (2000)'s table according to the effective temperature of the host star. The basic parameters of the host star are adopted from the recent results of Stempels et al. (2007). All observed data points are involved in MCMC (Markov Chain Monte Carlo) analysis to search the optimal parameters $\{T_c, p, \Delta F, t_T, b, M_*\}$ under the condition of minimizing the quantity χ^2 (Collier Cameron et al. 2007b). Here, the chi-squared $\chi^2 = \sum_j (m_j - \mu_j)^2 / \sigma_j^2$, m_j and μ_j are observed and model flux, respectively. For the span of our observational data is short, only five parameters $\{T_c, \Delta F, t_T, b, M_*\}$ are analyzed in practice.

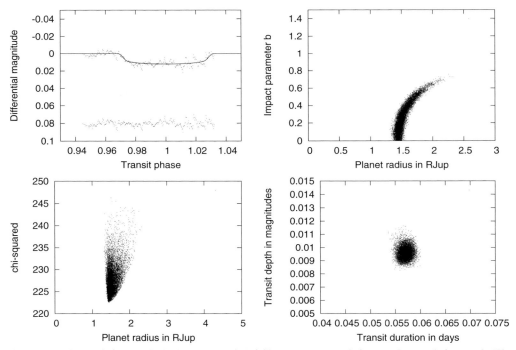

Figure 1. Upper-left panel: Light curve of WASP-1 system and fitting. Lower-left panel: The radius of planet (in unit of radius of Jupiter) vs chi-squared. Upper-right panel: The relation between radius of planet and impact parameter. Lower-right panel: The relation between transit duration and transit depth.

Table 1. The optimal parameters derived using the MCMC method.

Transit epoch T_c [HJD]	$2454051.1087\,^{+0.0013}_{-0.0006}$
Transit depth ΔF	$0.0094\,^{+0.0006}_{-0.0002}$ mag
Transit width t_T	$0.1436\,^{+0.0020}_{-0.0021}$ days
Impact parameter b	$0.012\,^{+0.452}_{--0.095}$ R_*
Orbital separation a	$0.0396\,^{+0.0003}_{-0.0003}$ AU
Orbital inclination i	$89.87\,^{+-0.98}_{-5.25}$ degrees
Stellar radius R_*	$1.52\,^{+0.19}_{--0.006}$ $R_\odot$
Planet radius R_p	$1.44\,^{+0.22}_{-0.002}$ R_J
Stellar mass M_*	$1.30\,^{+0.03}_{-0.03}$ $M_\odot$
Planet mass M_p	$0.879\,^{+0.014}_{-0.013}$ M_J
Orbital period p	2.51996 days

With the stellar mass of $1.30 M_\odot$ (Stempels et al. 2007) and orbital period of 2.51996days (Shporer et al. 2007), the optimal values of other parameters are derived (see Fig. 1). Table 1 lists the optimal parameters of this system. From our new simulated parameters, WASP-1b is a low density planet among known transiting exoplanets, which is similar to the results derived by other authors (Charbonneau et al. 2007, Shporer et al. 2007).

Acknowledgements

This work is partly supported by NSFC under grant No. 10673027.

References

Charbonneau, D., Joshua, N., Mark, E. et al. 2007, *ApJ*, 658, 1322
Claret, A. 2000, *A&A*, 363, 1081
Collier Cameron, A. et al. 2006, *MNRAS*, 373, 799
Collier Cameron, A., Bouchy, F., Hebrard, G. et al. 2007a, *MNRAS*, 375, 951
Collier Cameron, A., Wilson, D. M., West, R. G. et al. 2007b, *MNRAS*, 380, 1230
Pollacco, D. L. et al. 2006, *PASP*, 118, 1407
Shporer, A., Tamuz, O., Zucher, S. & Mazeh, T. 2007, *MNRAS*, 376, 1296
Stempels, H. C., Collier Cameron, A., Hebb, L. et al. 2007, *MNRAS*, 379, 773
Tamuz, O., Mazeh, T. & Zucker, S. 2005, *MNRAS*, 356, 1466

Reconstruction of the transit signal in the presence of stellar variability

Aude Alapini[1] and Suzanne Aigrain[1]

[1] Astrophysics Group, School of Physics, University of Exeter,
Stocker Road, Exeter EX4 4QL, United Kingdom
email: alapini@astro.ex.ac.uk, suz@astro.ex.ac.uk

Abstract. Intrinsic stellar variability can hinder the detection of shallow transits, particularly in space-based data. Therefore, this variability has to be filtered out before running the transit search. Unfortunately, filtering out the low frequency signal of the stellar variability also modifies the transit shape. This results in errors in the measured transit depth and duration used to derive the planet radius, and orbital inclination. We present an evaluation of the magnitude of this effect based on 20 simulated light curves from the CoRoT blind exercise 2 (BT2). We then present an iterative filter which uses the strictly periodic nature of the transits to separate them from other forms of variability, so as to recover the original transit shape before deriving the planet parameters. On average with this filter, we improve the estimation of the transit depth and duration by 15% and 10% respectively.

Keywords. exoplanet, transit, stellar variability

---

1. Introduction

1.1. Planet parameters from observations and associated errors

The radius (R_p, eq. 1.1) and the mass (M_p, eq. 1.2) of an exoplanet can be fully solved when measuring both the flux and the radial velocity variations of the parent star due to its orbiting planetary companion.

$$R_p = R_\star \sqrt{\frac{\Delta F}{F}} \quad (1.1)$$

$$M_p = M_\star^{\frac{2}{3}} \frac{K}{\sin i} \left(\frac{P}{4\pi G}\right)^{\frac{1}{3}} \quad (1.2)$$

where, $R_\star$ and $M_\star$ are the radius and the mass of the parent star, $\frac{\Delta F}{F}$ the flux variation due to the planet transiting the disc of its parent star, K the amplitude of the radial velocity variation of the parent star due the gravitational influence of its orbiting planet, i and P the orbital inclination and period of the planet, and G the gravitational constant. $\frac{\Delta F}{F}$, P, and i can be measured from the light curve. A common way to measure $R_\star$ and $M_\star$ is by comparing the stellar spectrum to stellar atmosphere models, allowing to derive the stellar parameters (T_{eff}, $\log g$, [Fe/H]) used to obtain the stellar mass and radius. $R_\star$ can also be measured more precisely, and without the use of models, with interferometry, or with transit fitting (in the case of high precision light curves).

1.2. Planet parameters and planet evolution and formation models

Improving the precision on observational planet masses and radii is important for both planet structure and planet formation models. The internal structure of a planet can be studied by comparing its mass and radius to model predictions of planets with different composition. Determining planet structure is important to derive observational statistics on planet types, which can then be compared to the predictions of planet formation models. Seager *et al.* (2007) show that to determinate the composition of sub-Uranus planets, error bars of 2% on the planet parameters are required. The current uncertainties on planet masses and radii are of the order of 10%. Improving these measurements is thus vital to help confirm the models.

1.3. *Sources of uncertainties on planet parameters*

The uncertainties on the planet mass, radius, and inclination depend on the uncertainties on the host star mass and radius, on the uncertainties on the transit parameters ($\frac{\Delta F}{F}$, total transit duration), and on the uncertainties on the radial velocity measurements. For large planets (>Jupiter), the uncertainties on the planet mass and radius are mainly due to the uncertainties on the stellar parameters. For sub-Uranus planets around active stars, the uncertainties on the planet mass and radius can be dominated by the uncertainties on the transit parameters.

2. Side effects of pre-detection stellar variability filters

2.1. *Pre-detection stellar variability filter*

Active stars show intrinsic flux variations due to temporal evolution and rotational modulation of structures on their surface (stellar spots, plages, granulation). Intrinsic stellar variations can have amplitudes much greater than transits, and thus, can hinder transit detection. These variations occur at a lower frequency (longer time length) than transit events (Aigrain, S. (2005), chap. 3, fig. 3.2). Pre-detection stellar variability filters use this difference to separate variations typical of stellar variability, from those on the time scale of transits (minutes to hours).

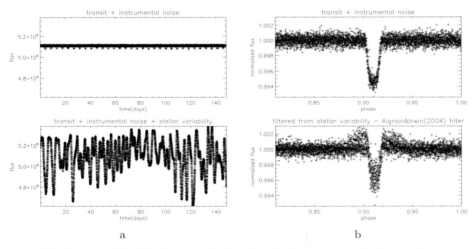

Figure 1. Stellar variability hinders transit detection (*column a*): the *bottom plot* is an example of a CoRoT BT2 light curve with Jupiter like transits on an active host star. This light curve is composed of a simulated transit signal (*top plot*), the instrumental noise expected in CoRoT data, and simulated stellar variability. Pre-detection filters can deform the shape of transits (*column b*): the *top plot* of *column b* is the phase fold of the *top plot* of *column a*, centered on the transit. The *bottom plot* of *column b* is the phase fold of the *bottom plot* of *column a*, filtered from stellar variability using Aigrain & Irwin (2004) filter.

The work presented in this paper is based on the pre-detection stellar variability filter described in Aigrain & Irwin (2004). This filter is a combination of median filtering (to reduce the level of random noise), and of boxcar averaging/smoothing (to extract the long term variations: the stellar variability).

2.2. Deformation of the transit signal

We tested the effect of Aigrain & Irwin (2004) pre-detection filter on the transit signal of 20 simulated light-curves, produced for the CoRoT blind test exercise 2 (BT2, Moutou et al. (2005)). The three main components in these light-curves are the stellar variability, the transit signal, and the instrumental noise. Figure 1a (bottom panel) shows a BT2 light curve of an active star with Jupiter-like transits.

We noticed that Aigrain & Irwin (2004) filter deforms the transit shape (Figure 1b, bottom panel). This effect is stronger for shallow transits on active stars. On average, for filtered light curves, the transit depth is under-estimated by 20% and its duration by 15% (compared to the value measured from the original transit with instrumental noise only). Miss-estimating the transit depth and duration leads to a miss-estimation of the planet radius, and orbital inclination (used to derive the planet mass).

3. Iterative filtering

We designed an iterative process to filter stellar variability and extract the original transit shape, based on Aigrain & Irwin (2004) filter, and using the additional post-detection knowledge on the planet orbital period.

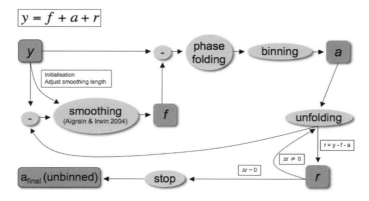

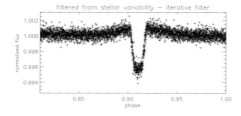

Figure 2. *Top:* Chart of the iterative filter process. The light curve (y) is decomposed into three components: the transit signal (a), the stellar variability (f), and the residual noise (r). The final estimation of the transit signal (a_{final}) is obtained after iterating on the pre-detection filter (step called "smoothing"). *Bottom:* Transit (a_{final}) resulting from the iterative filter applied to the light curve of Figure 1a *bottom plot*.

3.1. *Implementation of the post-detection filter*

Our post-detection filter is based on Kovács et al. (2005) idea of decomposing the light curve into three components (transit signal, correlated noise, and white noise), to filter out systematics due to atmospheric fluctuations. In our case, we filter out stellar intrinsic variability extracted using an iterative process (top chart of Figure 2).

We estimate the stellar variability by applying Aigrain & Irwin (2004) pre-detection filter. We subtract the estimation from the original light curve, phase-fold the result, and bin it to reduce the random noise. We then remove the resulting transit signal from the original light curve, and start a new iteration re-evaluating the stellar variability in the light curve freed from the transit signal. We stop iterating when the residuals (original light curve minus last estimations of the transits and stellar variability) stop evolving. The best estimation of the transit signal is the original light curve minus the last estimation of the stellar variability (after 3-4 iterations).

3.2. *Evaluation of performance*

We have applied the iterative stellar variability filter described in section 3.1, on 20 CoRoT BT2 light-curves with transits. Iterating on the estimation of the stellar variability, and using the additional knowledge on the orbital period, appears to successfully better recover the original transit shape (compare bottom of Figure 2 with *bottom plot* of Figure 1b). On average, light curves filtered with the iterative filter gives a better estimation of the transit depth by 15% and of the transit duration by 10%, than when filtered with Aigrain & Irwin (2004) pre-detection filter.

The iterative filter has also been found efficient in reconstructing shallow transits that had become un-detectable after pre-detection stellar variability filtering. Another use for the iterative filter could thus be to confirm borderline transit detections in active stars.

4. Summary and future work

Based on 20 simulated CoRoT light curves from the BT2 light curve sample, we have shown that Aigrain & Irwin (2004) pre-detection filter, used to remove stellar variability prior to transit detection, deforms the transit depth by 20% and the transit duration by 15%. To circumvent this, we have adapted Kovács et al. (2005) iterative filtering method to the case of filtering stellar variability present in space-based light curves. The resulting post-detection iterative filter improves the estimation of the transit depth and duration by 15% and 10% respectively.

The two areas where we plan to focus future efforts are *a)* to further automate the filtering process (some user interaction is currently needed to initialize the filter smoothing length), and *b)* to evaluate the improvement in the planet parameters uncertainties resulting from the improvement in the transit parameters from this work. As *b)* depends on the particulars of each system, we plan to derive these uncertainties using Monte Carlo simulations for known planets. Further tests will include using other pre-detection filters, including ones which can be applied to data with significant temporal gaps.

References

Aigrain, S. 2005, *PhD Thesis, Cambridge, UK*, http://www.astro.ex.ac.uk/people/suz/docs/thesis/
Aigrain, S. & Irwin, M. 2004, *MNRAS*, 350, 331
Kovács, G., Bakos, G., & Noyes, R. W. 2005, *MNRAS*, 356, 557
Moutou, C., Pont, F., Barge, P. *et al.* 2005, *A&A*, 437, 355
Seager, S., Kuchner, M., Hier-Majumder, C. A., & Militzer, B. 2007, *ApJ*, 669, 1279

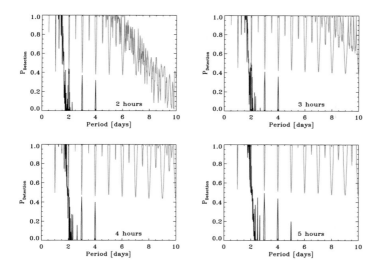

Figure 6. Transit duration in hours (shown in each of the four panels), in the absence (grey line) and presence (black line) of red noise ($\sigma_r = 3$ mmag). Other input parameters are 50 consecutive nights of observing, a 10-min observing cadence, $\sigma_w = 5$ mmag, $\mathrm{SNR}_{threshold} = 10$, and a transit depth of 0.01 mag. The extent to which the detection probability is affected by transit duration is higher for white noise dominated data.

4.4. Transit Duration

The duration of a transit is dependent upon the combination of stellar and planetary radii, the orbital period, and the inclination angle of the system, as described in, e.g., Mandel & Agol (2002) and Seager & Mallén-Ornelas (2003). To simulate the effect of a non-central transit, Fig. 6 shows how the detection probability changes as the duration of transit varies between 2 and 5 hours, typical for planets with periods of around a few days. The impact of shorter transit durations is noticeably higher in the white noise dominated regime (grey line). This appears to indicate that, as long as the system actually transits, the dependence of the detection efficiency upon inclination angle is relatively weak.

5. Summary and Conclusions

In this presentation, we illustrate the influence of several parameters on the probability that an existing planetary transit is detectable in a data set with given noise properties. Red noise dominates in the regime in which planets are typically found (the brightest stars in one's sample). In order to beat down red noise effects and improve detection efficiency, detrending is a vital instrument.

Transit survey strategy can be employed to maximize the projected yield of a given survey. We examine how much the detection efficiency for different orbital periods would suffer when changing one's observing strategy, e.g., to increase the number of monitored stars. We show examples involving the number of consecutive observing nights, typical night length, as well as the observing cadence, and we find that sacrificing full nights or parts of nights can significantly lower transit detection probability. Finally, we show the effects of non-central transits and the associated change of transit duration upon detection efficiency. Our results indicate that, for $\sigma_r = 3$ mmag (typically found in some of the major transit surveys before detrending is applied), small changes in inclination angle and associated transit duration do not greatly affect detection efficiency.

In all simulations presented here, the presence of red noise (for $\sigma_r = 3$ mmag) dominates all other effects. However, the variation of some of the parameters examined in this work (in particular the observing cadence and the transit duration) appears to have a bigger effect on white noise dominated (e.g., detrended) data.

References

Aigrain, S. & Pont, F. 2007, *MNRAS*, 378, 741
Kane, S. R., *et al.* 2007, ArXiv e-prints, 711, arXiv:0711.2581
Kovács, G., Bakos, G., & Noyes, R. W. 2005, *MNRAS*, 356, 557
Mandel, K., & Agol, E. 2002, *ApJL*, 580, L171
Mallén-Ornelas, G., Seager, S., Yee, H. K. C., Minniti, D., Gladders, M. D., Mallén-Fullerton, G. M., & Brown, T. M. 2003, *ApJ*, 582, 1123
Pepper, J., & Gaudi, B. S. 2005, *ApJ*, 631, 581
Pont, F. 2006, in "Tenth Anniversary of 51 Peg-b: Status of and prospects for hot Jupiter studies" (eds. Arnold, L., Bouchy, F., & Moutou, C.), p.153 (astro-ph/0612540)
Pont, F., Zucker, S., & Queloz, D. 2006, *MNRAS*, 373, 231
Seager, S., & Mallén-Ornelas, G. 2003, *ApJ*, 585, 1038
Tamuz, O., Mazeh, T., & Zucker, S. 2005, *MNRAS*, 356, 1466
von Braun, K., Lee, B. L., Seager, S., Yee, H. K. C., Mallén-Ornelas, G., & Gladders, M. D. 2005, *PASP*, 117, 141

Several problems of exoplanetary orbits determination from radial velocity observations

Roman V. Baluev

Sobolev Astronomical Institute, Saint Petersburg State University,
Universitetskij prospekt 28, Petrodvorets, Saint Petersburg 198504, Russia.
email: roman@astro.spbu.ru

Abstract. Existing algorithms of analysis of radial velocity time series are improved for the purposes of extrasolar planets detection and characterizing. Three important effects are considered: the poorly known radial velocity jitter, periodic systematic errors, and statistical bias due to non-linearity of models. Mathematical tools to account for these effects are developed and applied to a number of real planetary systems. In particular, it is shown that two outer planets of HD37124 are likely trapped in the 2/1 resonance. The dwarf star GJ876 may host an extra, Neptune-mass, planet which is in resonance with two giant planets in this system.

Keywords. methods: data analysis, techniques: radial velocities, stars: planetary systems

1. Introduction

For now, more than 200 exoplanets orbiting main-sequence stars are known (Schneider 2007). Among the known planetary systems more than 20 contain two planets at least. Significant fraction of multiple planet systems possesses an interesting dynamics like mean-motion resonances and apsidal corotations. Still, the majority of planetary systems was discovered by radial velocity (RV) techniques.

Unfortunately, orbital parameters (and masses) in many planetary systems are still poorly determined, especially in multiple ones. The best-fitting orbital solutions for such systems may appear dynamically unstable in short timescales. In accordance with Goździewski et al. (2008), the stability constraint may be accounted for during data analysis, but sometimes this approach simply imposes a preferred result. Often, when recovering a chronology of orbital fits for a given planetary system we can note that parameters for subsequent fits may differ surprisingly largely.

The above reasoning makes us to assume that something is not accounted for during analysis of RV data. The main aim of this work is to discuss three important effects which could be this 'something'. They are (i) the poorly constrained RV jitter, (ii) periodic systematic errors, and (iii) statistical bias due to non-linearity of models. They are described in more details in Section 2. Section 3 describes briefly mathematical algorithms I used to account for these effects. Section 4 discusses preliminary results obtained for several planetary systems. Section 5 ends this contribution by conclusions.

2. Overview

2.1. Radial velocity jitter

During analysis of RV data from planet search surveys, it is routinely assumed that the total variance σ_i^2 of an $i^{\rm th}$ RV measurement is a sum of the instrumental error variance $\sigma_{{\rm mes},i}^2$ and of the constant 'jitter' $\sigma_\star^2 \equiv p$. In the astrophysical part, this jitter is inspired by various processes on the star leading to an apparent instability of its radial velocity. It also depends on the instrument, on the way of observation and its reduction to final radial velocity measurement. For example, the spectrum exposition as long as $20-30$ min averages out apparent RV variations inspired by stellar oscillations (which have periods $\sim 5-10$ min for solar-like stars). This decreases the astrophysical part of the RV jitter. However, the astrophysical jitter does not represent the only source of RV variations beyond the expected noise level. Many other sources like extra systematic RV errors lie in the instrument (but they may also depend on stellar properties).

It is very difficult to split the extra RV variability in parts originating from different physical sources. Moreover, this problem seems to be rather well separated from the problem of determination of planetary parameters. Due to this reasoning, I will use in this paper the generalized notion 'RV jitter' (or 'full RV jitter') as any short-term RV variability beyond the currently expected instrumental noise, for the current model of the RV curve. The full RV jitter consists of the astrophysical jitter (originating in the star), of the instrumental RV errors not accounted for in the current RV uncertainties (originating in the instrumentation and its nearest environment), and of the extra RV variations not accounted for in the current RV model. In Section 4 we will see that full RV jitters may be quite different for different observatories, as well their zero velocities are. Hence, when we perform a joint analysis of RV time series from different observatories, we should account for possible differences in RV jitters. Any undetected planets would also increase the effective jitter, but this increasing does not depend on an instrument.

Usually, the jitter is determined using empirical models based on stellar characteristics (Wright 2005). Unfortunately, this way of jitter estimation allows accuracies of only $\sim 50\%$ or sometimes worse. Jitter of a few m/s remains almost unconstrained $a\ priori$ (in comparison with magnitudes of instrumental errors) and represents, in fact, an extra unknown parameter. Note, that statistical algorithms which are used usually for obtaining orbital fits require an input sequence of weights of measurements ($w_i \propto \sigma_i^2$). Therefore, they assume multiplicative model $\sigma_i^2 = \kappa/w_i$ and estimate implicitly the unspecified factor κ. There is no principle obstacles to modify these algorithms for the case of the square-additive model $\sigma_i^2 = \sigma_{{\rm mes},i}^2 + \sigma_\star^2$ with the last term being considered as a free parameter. Such extension may be done by transition from χ^2-minimization to likelihood function maximization. Related mathematical questions are discussed in Section 3.

2.2. Periodic instrumental errors

Instrumental errors always contain a systematic part, along with a random one. Periodic or quasiperiodic systematic errors in radial velocities can significantly affect parameter estimations of also periodic planetary RV signals. In an extreme case, these systematic errors may be interpreted as Doppler signature of a planet. Fortunately, periods of instrumental systematic errors often may be identified $a\ priori$. For ground-based observations, they usually coincide with well-known periods like one year, one day (and one sidereal day), one month (either sidereal or synodic). Note that the timings for a given RV time series are usually well concentrated near some fixed sidereal time (in order to observe a given star near a fixed position on the sky). This concentration (usually within a one third part of the sidereal day) makes any diurnal periodicity approximately equivalent to an annual one. Hence, we are left with three prior periods only: 1 year ($\approx$ 365.25 days, seasonic variation of observing conditions), 1 sidereal month ($\approx$ 27.32 days, variation of the angular distance between the Moon and the star being observed) and 1 synodic month ($\approx$ 29.53 days, variation of the Moon phase).

Section 4 will show that the annual errors of a few or even ten m/s is a common phenomenon in current planet search surveys, rather than exception. During the symposium, I have got from several participants a number of explanations of sources for such systematic errors, which acted in datasets that I analysed. They are as follows:

(*a*) Inaccurate barycentric correction of radial velocities.
(*b*) Incomplete exclusion of telluric lines during spectrum analysis.
(*c*) Variations in spectral line profiles having not well-established nature.

We can see that the annual errors may originate from very different sources having either physical or mathematical nature.

2.3. Non-linear effects

It is well known that estimations of coefficients of a linear model, provided by minimization of a χ^2 function, are unbiased (i.e., their expectations equal the true values of parameters). This property is important, because it allows us to hope that our estimations are related to the true values at all. Keplerian models are non-linear. In this case, the traditionally used χ^2 minimization or likelihood maximization principles provide, strictly speaking, biased estimations. However, this bias tends to zero when the number of observations grows (see references in the work by Baluev (2008a)).

Nevertheless, for real radial velocity time series, the biasing of Keplerian parameters and jitter may become practically significant and may exceed respective uncertainties in several times. For instance, we can never get negative estimation of orbital eccentricity; thus we always overestimate eccentricity for a circular orbit and probably we systematically overestimate any non-zero eccentricity. This means that estimations of orbital eccentricities of exoplanets are probably biased to larger values. For multiple planet systems we can quite be left with less than 10 RV measurements per free parameter. With such lack of observations, we may get biases exceeding formal uncertainties essentially.

3. Mathematical methods

3.1. Accounting for radial velocity jitter

Evidently, RV jitter is not an usual fitting parameter and cannot be estimated during usual χ^2 minimization. Information about the jitter may be extracted from the scattering of radial velocity measurements around best-fitting orbital solution. Mathematically, the jitter can be estimated using maximum likelihood principle. For N Gaussian observations v_i, the doubled log-likelihood function may be written down as

$$2 \ln \mathcal{L}(\theta, p) = - \sum_{i=1}^{N} \frac{(v_i - \mu(t_i, \theta))^2}{\sigma_i^2(p)} - \sum_{i=1}^{N} \ln \sigma_i^2(p) + \text{const}, \quad (3.1)$$

where $\mu(t, \theta)$ is the fit model having d 'usual' fitting parameters θ (planetary orbital elements, RV semi-amplitudes, etc.) and the RV uncertainty model $\sigma_i^2(p) = \sigma_{\text{mes},i}^2 + p$

introduces a 'variance' parameter – the jitter p. The first sum in the right hand side of eq. (3.1) represents the χ^2 function† which depends on the usual parameters θ and on the jitter p. The second term depends on the jitter only. The joint estimation of the parameters θ and p can be obtained in result of maximizing the full function (3.1). That is, the values of θ and p corresponding to the maximum of $\mathcal{L}(\theta, p)$ should be adopted as the 'maximum-likelihood estimations'. The maximum value of the likelihood function itself measures the quality of the fit found. It is important that the resulting estimation of p is automatically accounted for in the resulting estimations of θ (and vice versa).

To write down (3.1), we have assumed Gaussian distributions for RV errors. If this assumption is wrong, the function (3.1) is not a likelihood function and parameter estimations obtained during its maximization may be somewhat shifted with respect to the true maximum-likelihood estimations. For large datasets, however, the central limit theorem implies that the influence of non-Gaussian errors should tend to zero as $N \to \infty$. The non-gaussianity may affect the uncertainties of our estimations (by about $\mathcal{O}(1/\sqrt{N})$) and their bias (by $\mathcal{O}(1/N)$), but does not introduce a constant shift. These effects are calculated (till the first order) by Baluev (2008b) and are shown to be negligible or tolerable in usual practical situations.

3.2. Bias reduction

For the RV jitter, I applied an analytic way of bias reduction. Its first step is a built-in reduction achieved by slight modification of the function (3.1) as

$$2 \ln \tilde{\mathcal{L}}(\theta, p) = -\frac{N}{N-d} \sum_{i=1}^{N} \frac{(v_i - \mu(t_i, \theta))^2}{\sigma_i^2(p)} - \sum_{i=1}^{N} \ln \sigma_i^2(p). \quad (3.2)$$

This $\mathcal{O}(1/N)$ modification normally allows to correct a large fraction of the bias in jitter. This decreases the cross-influence on estimations of parameters θ (remind that these estimations depend on the jitter). The rest of the first-order (i.e., $\mathcal{O}(1/N)$) bias in $\sigma_\star^2$ can be calculated analytically as it is described by Baluev (2008b).

To correct the first-order bias in estimations, I used also the numerical algorithm proposed by Quenouille (1956). It is as follows:

(a) Calculate the basic (biased by $\mathcal{O}(1/N)$) estimation x of a given parameter ξ from the full time series of N observations.

(b) Construct N reduced time series with $i^{\text{th}} (i = 1, 2, \ldots N)$ measurement omitted. Therefore, each reduced time series should consist of $N-1$ data points.

(c) Calculate N new estimations $x'_i (i = 1, 2, \ldots N)$ of ξ by re-fitting with every of the reduced time series. The bias of x'_i will be about $\sim 1/(N-1)$, hence these new estimations will be shifted with respect to x by about $\sim (1/(N-1) - 1/N) = \mathcal{O}(1/N^2)$.

(d) Calculate the sum $b_1 = \sum_{i=1}^{N}(x'_i - x)$. The result $b_1 = \mathcal{O}(1/N)$ represents the first-order bias of x. That is, the corrected estimation $x - b_1$ should be biased by $\mathcal{O}(1/N^2)$ only.

The main advantage of this algorithm is that its implementation is model-independent and easy. This algorithm is valid for a wide variety of estimations (it is not restricted to, say, maximum-likelihood estimations only). Also, this algorithm does not require for the distribution of RV errors to be Gaussian (it is not restricted to any narrow family

† The χ^2 random quantity is defined as the sum of N standard (mean zero, variance one) independent Gaussian random quantities. For arbitrary values of θ and p, the first sum in (3.1) is not necessary a χ^2 quantity. But the words 'χ^2 function' denote (within this paper) the first sum in (3.1) as a function of parameters to be estimated. This should not confuse the reader.

of distributions). Unfortunately, it is rather time-consuming because it requires making many fits of non-linear Keplerian models.

3.3. Likelihood ratio periodograms

Routinely used normalizations of the Lomb-Scargle periodogram assume fixed weights for observations and, hence, the multiplicative model for RV uncertainties. For planet searches, it is preferred to use some extension of such periodogram with a built-in estimation of the square-additive RV jitter. I used a direct extension of the least-squares periodogram $z_3(f)$ from (Baluev 2008a). This extension is based on the likelihood ratio statistic and is described in more details in (Baluev 2008b).

3.4. Periodic systematic errors

To account for possible annual errors in radial velocities, I added to RV models a simple harmonic term $A\cos(2\pi(t-\tau)/1\mathrm{yr})$ of one year period. Usually, observations do not cover the full year but span a restricted observing season (always the same). Given the phase coverage $\sim 50\%$, we may not care much about possible deviations of systematic errors from a sinuous curve.

4. Preliminary results

4.1. Planetary systems with well-determined orbital solutions

To test the above described algorithms of jitter estimation, it is worth applying them to a number of planetary systems with well-determined parameters and large numbers of observations. For this purpose, I have selected planetary systems around stars 51 Peg, 70 Vir, 14 Her, HD83443, HD69830, μ Ara and 55 Cnc. The respective maximum-likelihood solutions are given in the first seven records in Table 1. We can see that the presence of annual periodicity in data is a rather frequent (although, cases free from it are not seldom). In ELODIE, CORALIE, and HET data, the semi-amplitude of this periodicity may reach $10-15$ m/s. Sometimes these periodicities produce clear peaks on periodograms and could be detected directly. In less bright examples, these peaks possess a high significance (given their frequency is known a priory), but may look unremarkable in comparison with the noise in a wide frequency range (see Fig. 1). On contrary, HARPS seems to be free from systematic errors. Data from Lick and Keck observatories show statistically significant annual periodicity in rare cases and with less semi-amplitude.

Addition of an annual term solves some inconsistencies between different surveys. For instance, a shallow linear RV trend detected by Butler *et al.* (2006) for 51 Peg, is not seen in the ELODIE data without accounting for the annual term with $A \sim 10$ m/s.

We can clearly see that the best-fitting values of RV jitter for one and the same star can differ largely for different instruments. RV jitters for ELODIE and CORALIE usually exceed those for other spectrographs, even with annual errors taken into account. This may indicate some extra systematic errors of unclear nature. Vice versa, HARPS jitters are perfectly less than for other instruments. Data from Keck, Lick and AAT demonstrate intermediate cases. It is surprising that the jitter estimation of HJST data for 14 Her is definitely *negative* ($\sigma_\star^2 < 0$), though no annual term was added to this dataset and the orbital solution is well constrained by data from other instruments. This indicates that

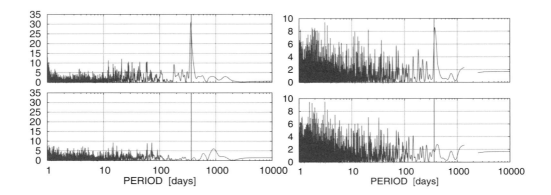

Figure 1. Left-Top: the modified likelihood ratio periodogram of RV residuals to the ELODIE radial velocities for the star 51 Pegasi. The RV variation induced by the planet 51 Peg b was included in the base RV model. Left-Bottom: a similar periodogram, but the annual term and the linear trend were included in the base RV model also. Right: similar periodograms for the ELODIE radial velocities of 14 Herculis (base RV models: 'planet b + quadratic trend' for the top panel and 'planet b + quadratic trend + annual harmonic' for the bottom panel). Positions of the annual period are marked by extra vertical lines.

the RV uncertainties stated by Wittenmyer *et al.* (2007) for 14 Her *overestimate* real errors by about 20%.

4.2. *Planetary system orbiting HD74156*

The period of the third planet in this system (recently announced based on HET observations, Bean *et al.* (2007)) is consistent with one year. The corresponding RV semi-amplitude is only 12 m/s. This oscillation is opposite in phase to that of ELODIE data for the same star and is similar to the annual periodicity in HET data for 55 Cnc. Therefore, it is quite possible that the planet HD74156 d is a false-detection inspired by instrumental systematic errors. At least, parameter estimations of this putative planet may be highly distorted by the systematic errors.

4.3. *Planetary system orbiting 47 UMa*

Wittenmyer *et al.* (2007) could not clearly confirm existence of the second planet with a period of about 2500 days. Instead, they found an extra signal having semi-amplitude 10 m/s and a very long period ~ 8000 days. In fact, this signal represents a shallow quadratic RV trend which is probably produced by a distant companion of poorly constrained period and mass. Addition of this parabolic trend to the model allowed to extract and confirm extra 2500 days periodicity of the planet 'c'. However, its semi-amplitude drops to $6-8$ m/s, what implies the minimum mass $m_c \sin i_c \approx 0.5\ M_{Jup}$ only.

4.4. Planetary system orbiting HD11964

The Extrasolar Planets Encyclopaedia by J. Schneider states that this system has two planets with periods of about 38 days (planet 'b') and 2000 days (planet 'c'), both discovered in 2005. However, there is no paper concerning the planet 'b' in this Encyclopaedia. Butler *et al.* (2006) don't mention the 38 days planet, but report on the detection of the outer (2000 days) planet and refer to it as planet 'b'. Likelihood periodogram of these data (with RV signal of the outer planet excluded) shows *two* almost equal and independent peaks near periods of 38 days and 365 days. In this case, the annual periodicity peak can be clearly seen on the periodogram. This annual periodicity should be accounted for in RV models in order to obtain more reliable estimations of the putative 38 days planet.

4.5. Planetary system orbiting HD37124

This system was a subject of many studies. The main problem is that the best-fitting orbital solution is far from dynamical stability due to a large eccentricity $e_c \approx 0.5$ of the outermost planet. It is possible to account for the stability requirement during data analysis (Goździewski *et al.* 2008). However, this implies too large loss in the fit quality. Moreover, these stable configurations occupy mainly boundaries of stability regions. This makes us to suppose that something is wrong with every of these orbital configurations.

Keck data alone yield an orbital fit not discussed in the discovery paper by Vogt *et al.* (2005). This fit has somewhat less period P_c and larger period P_d, so that the outer planets become trapped into 2/1 mean-motion resonance. This fit can be easily missed by a fitting algorithm due to larger values of r.m.s and χ^2 (or lower likelihood $\mathcal{L}$). However, if we enlarge fitting model by including some extra free terms (linear trend or annual periodicity), the picture becomes opposite. The main advantage of the resonant fit is that both orbital eccentricities become ill-determined and strongly biased to larger values. Therefore, it becomes possible to find a stable solution (having moderate eccentricities) with tolerable decreasing in fit quality.

Table 1 contains biased maximum-likelihood estimations of parameters for this system. Quenouille's bias reduction decreases eccentricities to ill-determined values $e_c = 0.3 \pm 0.2$ and $e_d = 0.3 \pm 0.1$. Hence, the most likely structure of this system is a pair of resonant (2/1) outer planets significantly perturbed by the inner (non-resonant) one.

4.6. Planetary system orbiting GJ876

Rivera *et al.* (2005) mentioned that periodogram of RV residuals to the best-fitting dynamical model for this three-planet system is not clean but contains at least three uncomfortable peaks at periods about 9, 13 and 125 days. I have re-fitted their radial velocities with free annual term added to Keck RVs and plotted a periodogram of resulting residuals. The 125 days peak rised up whereas other peaks retained at low levels. This 125 days periodicity may be interpreted as a fourth, Neptune-mass, planet in the system (although significantly negative value of Keck jitter estimation inspires some doubts and requires further studies). Dynamical behaviour of such four-planet system is remarkable due to the triple resonance 4/2/1 of outer planets.

Table 1. Revised data for several planetary systems.

Planet	P [days]	$\tilde{K}$ [m/s]	l_0 [°]	e	ω [°]	**Instr.**[5]	A [m/s]	τ [days]	$\sigma_\star$ [m/s][1]
51 Peg[3] b	4.230791(31)	55.80(48)	59.74(60)	0.0089(88)	31(53)	ELD	11.2(1.2)	364.7(9.3)	6.56(83)
						LCK			0.5(2.8)
70 Vir b	116.6901(44)	288.8(1.0)	79.48(18)	0.4009(28)	359.16(44)	ELD			−2.8(1.8)[2]
						LCK	5.1(1.6)	315(16)	3.58(79)
14 Her[3] b	1771.3(2.4)	83.61(45)	306.23(42)	0.3683(45)	21.88(85)	ELD	6.4(1.6)	185(12)	6.45(96)
						LCK			1.23(45)
						HJS			−4.44(86)[2]
HD83443[3] b	2.985746(36)	57.12(60)	144.0(1.7)	0.013(10)	157(48)	COR	5.8(1.4)	99.7(9.5)	6.40(48)
						AAT			6.0(1.7)
						KCK			2.70(58)
HD69830 b	8.6674(23)	3.49(15)	224.6(6.5)	0.101(42)	340(24)	HRP			0.22(26)
c	31.564(41)	2.64(18)	264.6(8.5)	0.126(62)	220(35)				
d	196.6(2.3)	2.20(20)	289(13)	0.069(84)	225(65)				
μ Ara b	644.7(1.4)	36.84(57)	18.22(63)	0.101(30)	33(14)	COR			5.67(95)
c	3900(700)	21.5(2.3)	70(11)	0.070(34)	100(90)	AAT			2.37(25)
d	9.6388(23)	2.88(22)	185.5(5.3)	0.172(77)	212(26)	HRP			1.52(15)
e	310.0(1.2)	15.1(1.1)	166.8(2.6)	0.083(19)	195(14)				
55 Cnc[4] b	14.6491(15)	71.78(64)	149.3(1.0)	0.0058(99)	116(87)	ELD	10.4(3.4)	86(19)	8.9(1.5)
c	44.472(50)	9.84(70)	24.8(5.9)	0.186(23)	39(11)	LCK			5.18(50)
d	4870(120)	46.09(97)	86.2(2.6)	0.235(30)	208.3(7.5)	HET	13.6(2.5)	229.7(7.8)	5.87(77)
90(fixed) e	2.79579(13)	4.93(68)	61.9(8.4)	0.26(14)	186(30)				
HD74156 b	51.6413(26)	87.7(1.1)	79.39(60)	0.6333(73)	176.58(89)	ELD	21.0(5.8)	31.1(7.3)	−4.7(2.2)[2]
c	2452.3(9.3)	105.4(2.2)	201.87(82)	0.3901(10)	267.8(2.4)	COR			11.6(1.9)
						HET	12.1(1.7)	255.5(8.0)	5.74(54)
47 UMa[3] b	1073.7(1.5)	49.91(97)	28.0(1.0)	0.018(17)	108(50)	ELD	10.9(3.7)	98(19)	7.7(1.6)
c	2520(110)	7.38(89)	77.0(7.8)	0(fixed)	0(fixed)	LCK			2.96(78)
						HJS			4.8(1.6)
						HET	5.2(1.4)	207(14)	4.66(44)
HD11964 b	38.002(59)	3.73(56)	86.5(9.6)	0(fixed)	0(fixed)	KCK	3.37(51)	16(17)	2.64(36)
c	1893(40)	10.21(84)	300.2(3.8)	0.051(63)	184(78)				
HD37124 b	154.53(12)	27.43(73)	308.6(3.0)	0.108(29)	134(14)	ELD	17.5(3.5)	68(20)	7.4(2.3)
c	1777(24)	15.2(1.0)	250.5(6.0)	0.563(56)	251.9(6.6)	COR			12.3(4.2)
d	902.3(8.7)	15.1(1.1)	119.9(6.5)	0.441(65)	276(13)	KCK	4.5(1.4)	122(19)	1.24(71)
GJ876[4] b	60.815(19)	213.43(56)	203.65(22)	0.0339(30)	5.1(4.4)	KCK	4.44(81)	284.7(8.7)	−1.13(76)[2]
c	30.482(19)	85.18(65)	42.1(1.0)	0.2521(55)	3.85(95)	LCK			−12.6(5.0)[2]
53.9(3.0) d	1.937836(47)	6.44(50)	347.3(6.3)	0(fixed)	0(fixed)				
GJ876[4] b	60.801(17)	213.51(50)	203.58(18)	0.0385(26)	2.9(4.0)	KCK	4.81(71)	282.4(6.9)	−2.32(25)[2]
c	30.491(15)	84.81(56)	42.12(94)	0.2520(43)	4.20(92)	LCK			−11.9(5.9)[2]
d	1.937826(42)	6.39(44)	348.6(5.3)	0(fixed)	0(fixed)				
53.8(2.3) e	125.6(1.1)	3.46(45)	305.5(5.9)	0.095(24)	209(15)				

Notes:
Orbital parameters are the usual Keplerian elements except for the modified semi-amplitude $\tilde{K} = K\sqrt{1-e^2}$. Mean longitudes l_0 and values of τ (and all other parameters for dynamical fits) are given for the epoch $JD = 2453000$ for the stars HD69830, μ Ara, HD74156, HD11964, and $JD = 2451000$ for other stars. The quantities $A, \tau, \sigma_\star$ are defined in the text. No bias correction applied here, except for analytic bias corrections of jitter estimations.

[1] Uncertainties of jitter estimations are calculated in the asymptotic Gaussian approximation as $\delta = \varepsilon/(2\sigma_\star)$, where ε is the uncertainty of the estimation of $p = \sigma_\star^2$ (ε was estimated simultaneously with p). When $\sigma_\star$ is comparable with δ, its distribution is far from Gaussian. Then it is necessary to return back to the quantity $p = \sigma_\star^2$ (not affected by the degeneracy) and to its uncertainty $\varepsilon = 2\delta\sigma_\star$.
[2] Negative values of $\sigma_\star$ reflect symbolically that the corresponding estimations of $\sigma_\star^2$ are negative.
[3] A linear/quadratic trend was included in the model.
[4] Dynamical co-planar fits (i.e., mutual gravitational interactions of planets are accounted for, all orbital planes are assumed co-planar). Fitted orbital inclinations of the systems (in degrees) are shown in the first column.
[5] ELD = ELODIE, COR = CORALIE, HRP = HARPS, LCK = Lick Observatory, KCK = Keck Telescope, AAT = Anglo-Australian Telescope, HET = Hobby-Eberly Telescope, HJS = Harlan J. Smith Telescope.

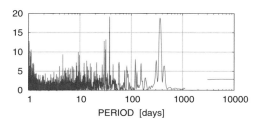
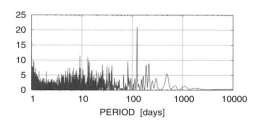

Figure 2. Left: periodogram of RV residuals for the star HD11964 with 2000-day planet included in the base model. Right: the periodogram of RV residuals for GJ876 to the best-fitting base model (dynamical three-planet RV curve + annual periodicity).

5. Conclusions

In this contribution, existing algorithms of analysis of radial velocity time series are improved for the purposes of detecting and characterizing extrasolar planets. A maximum-likelihood estimation of RV jitters is built in the usual RV curve fitting algorithms. Resulting algorithms often allow much higher accuracies than the jitter estimation based on empirical dependencies on stellar activity. It is shown that often the data from planet search radial velocity surveys suffer from periodic annual systematic errors. These errors should be taken into account during data analysis. Mathematical tools for statistical bias correction are proposed. Their usage is important mainly for multiple planet systems.

Application of improved algorithms allows to revise orbital solutions for a number of planetary systems. Recently announced planet HD74156 d may be a false-detection due to the annual systematic errors. The second planet orbiting 47 UMa does exist but has a smaller mass than it was estimated initially. Two outer planets around HD37124 are probably trapped in the 2/1 resonance. Planetary system around GJ876 may contain a fourth planet 'e' orbiting in resonances 2/1 and 4/1 with known giant planets GJ876 b,c.

Acknowledgements

I would like to thank Drs. V.V. Orlov and K.V. Kholshevnikov for critical reading of this paper, useful suggestions, and linguistic corrections. Dr S. Ferraz-Mello is thanked for fruitful suggestions concerning terminology used in the paper. I am grateful to Drs. M. Mayor, J.T. Wright, and A. Quirrenbach for detailed discussions of possible sources of annual RV errors in present planet search surveys. This work was supported by the Russian Foundation for Basic Research (Grants 05-02-17408, 06-02-16795) and by the President Grant NS-4929.2006.2 for the state support of leading scientific schools.

References

Baluev, R. V. 2008a, *Mon. Not. R. Astron. Soc.*, (Online Early), doi:101111/j.1365-2966.2008.12689.x

Baluev, R. V. 2008b, *Mon. Not. R. Astron. Soc.*, submitted (arXiv/astro-ph: 0712.3862)

Bean, J. L., McArthur, B. E., Benedict, G. F., & Armstrong, A. 2007, *Astroph. Journ.*, 672, 1202

Butler, R. P., Wright, J. T., Marcy, G. W., Fischer, D. A., Vogt, S. S., Tinney, C. G., Jones, H. R. A., Carter, B. D., Johnson, J. A., McCarthy, C., & Penny, A. J. 2006, *Astroph. Journ.*, 646, 505

Goździewski, K., Migaszewski, C. & Musieliński, A. 2008, these proceedings

Quenouille, M. H. 1956, *Biometrika*, 43, 353

Rivera, E. J., Lissauer, J. J., Butler, R. P., Marcy, G. W., Vogt, S. S., Fischer, D. A., Brown, T. M., Laughlin, G., & Henry, G. W. 2005, *Astroph. Journ.*, 634, 625

Schneider, J. 2007, *http://exoplanet.eu*, The Extrasolar Planets Encyclopaedia

Vogt, S. S., Butler, R. P., Marcy, G. W., Fischer, D. A., Henry, G. W., Laughlin, G., Wright, J. T., & Johnson, J. A. 2005, *Astroph. Journ.*, 632, 638

Wittenmyer, R. A., Endl M., & Cochran, W. D. 2007, *Astroph. Journ.*, 654, 625

Wright, J. T. 2005, *Pub. Astr. Soc. Pacific*, 117, 657

Selection effects in Doppler velocity planet searches

Simon O'Toole,[1] Chris Tinney[2] and Hugh Jones[3]

[1] Anglo-Australian Observatory, PO Box 296, Epping NSW 1710, Australia
email: otoole@aao.gov.au

[2] School of Physics, University of NSW, 2006 Australia
email: cgt@phys.unsw.edu.au

[3] Centre for Astrophysical Research, University of Hertfordshire, Hatfield, AL10 9AB, UK
email:hraj@star.herts.ac.uk

Abstract. The majority of extra-solar planets have been discovered by measuring the Doppler velocities of the host star. Like all exoplanet detection methods, the Doppler method is rife with observational biases. Before any robust comparison of mass, orbital period and eccentricity distributions can be made with theory, a detailed understanding of these selection effects is required, something which up to now is lacking. We present here a progress report on our analysis of the selection effects present in Anglo-Australian Planet Search data, including the methodology used and some preliminary results.

Keywords. Extra-solar planets, numerical simulations

1. Motivation

Planet searches - as with all surveys - necessarily suffer from selection effects and observational biases. Without characterising and properly accounting for these effects we cannot compare the catalogue of detected planets with models of planet formation. It has long been known that Doppler velocity planet searches preferentially detect short period planets with large masses - as exemplified by 51 Pegasi b - rather than solar system analogues or short period low mass planets. As surveys progress, longer temporal baselines are allowing us to find planets with longer periods and higher precision is lowering mass detection thresholds. The effects of data sampling and variable data quality have not been quantified, however, meaning a potentially significant selection effect has so far been ignored. Any studies designed to optimise the cadence of observations will inevitably introduce their own biases. Up to now, no detailed Monte-Carlo-like simulations designed to quantify these effects have been carried out.

Despite the lack of selection function analyses, statistical analyses of the sample are now routinely undertaken. Several of these have simply assumed that observed distribution of exoplanets is in fact the true distribution, while making overly simplified assumptions about the observational biases inherent in the sample (e.g. Lineweaver & Grether 2003). Others have either ignored the effects of eccentricity or used an inadequate treatment of it (e.g. Cumming 2004). No one has simulated observational data looking for the effects of data quality and sampling on a star-by-star basis. We have begun a project to investigate the latter effects, while examining the entire parameter space including period, eccentricity, and planet mass.

2. Methodology

To derive selection functions for our observations, we first require an automated toolbox. The Lomb-Scargle (LS) periodogram has been increasingly used in Doppler velocity planet searches where, however, circular orbits (giving rise to sinusoidal velocity curves) are apparently rare (Butler et al. 2006). It therefore makes more sense to fit Keplerians to data instead of sinusoids as discussed by Cumming (2004). As orbital eccentricity is an important parameter in these functions, we have expanded the traditional LS periodogram to two dimensions: period and eccentricity; we call this the 2D Keplerian LS (2DKLS) periodogram. The method we use to calculate the 2DKLS periodogram was introduced in O'Toole et al. (2007). Briefly, we use a grid of fixed periods and eccentricities to calculate the 2DKLS, with $e = 0 - 0.95$ in steps of 0.05, while the list of periods is on a logarithmic scale, with log P=–0.3 to 3.7. A Keplerian is then fitted to the data using a non-linear least squares fitting routine with Levenberg-Marquadt minimisation from Press et al. (1986).

3. Detection Criteria

As discussed above, automation is one of the important practical criteria of our simulations. This extends to the development of an adequate set of criteria to decide whether a planet has been detected. Given the number of simulations to be analysed, one cannot simply examine each power spectrum or velocity curve by eye as is often done (initially at least) with real observations.

There are several methods currently used to determine the reality of a planet detection. Marcy et al. (2005) present an excellent discussion of two different approaches based on the False Alarm Probability or FAP.

Because of the large number of simulations we plan to carry out during this project, it is necessary we have a simple set of criteria that can quickly test the reliability of a detection. This automatically rules out several approaches that are in themselves computationally intensive; in particular the determination of a FAP would add considerably to the time budget of our simulation analysis. The criteria must be able to detect as many planets as possible robustly, without introducing too many false positives. An individual criterion can be statistical or physical. There is a certain level of arbitrariness involved in selecting detection criteria, as many different combinations will lead to similar results. Below we list the criteria we have used in this work.

- Fit period must be less than twice the time-span of the data - periods beyond this are not constrained;
- RMS of data must be greater than twice median uncertainty - based on an often-used flag for real data;
- Fit period must be greater than twice fit error - the period must be reasonably constrained; and
- χ^2 must be within 1σ of median of all χ^2 values.

They are by no means the only criteria we could use; however, they allow a swift determination of planet detection and give false positives and false negatives of less than 5%. We will refine them when necessary during the project.

4. Preliminary Results

Selection functions (SF) for eccentricity and period have been derived for each data set using the detection criteria described above. The functions we show here are integrated

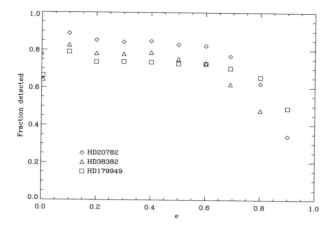

Figure 1. Eccentricity selection functions as a function of eccentricity for HD 20782, HD 38382 and HD 179949.

over semi-amplitude and period (eccentricity SF) and semi-amplitude and eccentricity (period SF). Currently the ordinate is simply the fraction of trials with a detection; this will be converted into a detection probability.

Figure 1 shows the eccentricity SFs for three of the objects we have simulated. There is a clear difference between each star, especially at high eccentricities. This difference does not appear to be dependent on the number of data points, as the HD 38382 simulations with 17 observations have only around 25-30% the number of observations as the HD 179949 simulations (56 observations). The sensitivity of our 35 observations of HD 20782 is greatest up to $e \sim 0.7$, but then drops off to below HD 179949. We suggest that data quality plays a significant role in determining the shape of the eccentricity SF. Below around $e = 0.5$ the sensitivity is approximately constant down to $e \sim 0.1$, where there appears to be a jump and then a drop at $e = 0.0$. The cause of this may be simply an artifact, and we are currently testing this hypothesis. The effects described above suggest that there is no appropriate parameterisation that can model the eccentricity SF, and that simulations such as these *must* be carried out on a star-by-star basis.

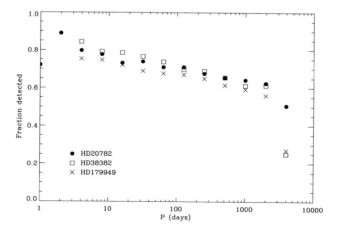

Figure 2. Period selection functions as a function of period for the three objects simulated.

The period SFs shown in the bottom left panel of Figure 2 are what we might have expected: a steady decrease in sensitivity as the input period increases. Two interesting points arise though. Firstly, the sensitivity at short periods (only shown for HD38382 in the figure) is high, even for periods around 2 days; it only drops off at very short periods, where data sampling becomes an issue. Secondly, the drop-off in sensitivity at very long periods is simply caused by the input period being longer than the length of the data string.

5. Conclusions and Future Work

One of the key conclusions of the preliminary selection function analysis presented in this poster is that carrying out the simulations to quantify selection effects on a star-by-star basis is of tremendous importance. We also find that Doppler velocity planet search observations are biased against finding planets in highly eccentric orbits. This does not mean that there are necessarily more of these planets to be found, just that we are not very sensitive to them. Finally, we find that our observations should be able to detect planets with orbital periods around 2 days, but we become less sensitive at shorter periods.

The ongoing and future work on this project includes:

- Continue simulations using simple noise model using Keter at UCL and Swinburne University supercomputer;
- Determine mass selection functions for stars already investigated; and
- Develop a more sophisticated noise model, incorporating stellar "jitter": magnetic activity $>\sim 2$ m/s (Wright 2005); solar-like oscillations ~ 0.5 m/s (O'Toole $et\ al.$ 2008); stellar convection and granulation noise, expected to be ~ 1-2 m/s, and possibly activity cycles.

References

Butler, R. P., et al. 2006, ApJ, 646, 505
Cumming, A. 2004, $MNRAS$, 354, 1165
Jones, H. R. A., et al. 2006, $MNRAS$, 369, 249
Lineweaver, C. & Grether, D. 2003, ApJ, 598, 1350
Marcy, G., et al. 2005, ApJ, 619, 570
O'Toole, S. J., et al. 2007, ApJ, 660, 1636
O'Toole, S. J., et al. 2008, $MNRAS$, submitted
Press, W., et al. 1986, $Numerical\ Recipes$, Cambridge University Press
Santos, N. C.. et al. 2001, $A\&A$, 379, 999
Tinney, C. G., et al. 2001, ApJ, 551, 507
Valenti, J. & Fischer, D. 2005, $ApJS$, 159, 141
Wright, J. 2005, $PASP$, 117, 657

Cadence optimisation and exoplanetary parameter sensitivity

Stephen R. Kane, Eric B. Ford and Jian Ge

Department of Astronomy, University of Florida, 211 Bryant Space Science Center,
Gainesville, FL 32611-2055, U.S.A.

Abstract. To achieve maximum planet yield for a given radial velocity survey, the observing strategy must be carefully considered. In particular, the adopted cadence can greatly affect the sensitivity to exoplanetary parameters such as period and eccentricity. Here we describe simulations which aim to maximise detections based upon the target parameter space of the survey.

Keywords. methods: data analysis – planetary systems – techniques: radial velocities

1. Introduction

Large-scale radial velocity surveys for extra-solar planets require a great deal of planning, particularly in terms of instrument considerations and the selection of targets. The duration of the survey will affect the sensitivity of the survey to different regions of period space. Additionally, the cadence of the observations affect the detection of short and long period planets and the overall planet yield. We present simulations of different observing strategies and demonstrate the change in sensitivity to a planetary period, mass, and eccentricity. These results are used to calculate the relative frequency of planet detections for various ranges of orbital parameters. By simulating a selection of cadence configurations, the optimal cadence for a given survey duration and observing constraints is estimated. The techniques presented here may be applied to a wide range of planet surveys with limited resources in order to maximise planet yield.

2. Simulation Framework

To investigate the detection efficiency properties of various radial velocity observing programs, we constructed a suite of simulated datasets using a FORTRAN code which also performs the analysis, as described by Kane, Schneider, & Ge (2007). The parameters of the initial simulation are based upon those of the Multi-object APO Radial-Velocity Exoplanet Large-Area Survey (MARVELS) (Ge *et al.* 2006). The stellar properties were estimated from Tycho-2 stars selected for observation by MARVELS for 60 separate fields. The noise model for the simulation was produced from the current and planned performance of the instrument and used to generate the radial velocity data. The cadences were defined by the number of observations per month during bright time. An example cadence can be conveniently expressed as 888, meaning 8 observations per month for 3 consecutive months. Since we are concerned with exoplanet parameter sensitivity, each dataset was injected with a planetary signature, the parameters of which were randomly chosen from a uniform distribution including mass, period, and eccentricity. In this way, > 4 million datasets were produced for each cadence simulation. Figure 1 represents an example simulated dataset, showing the radial velocity amplitudes and noise model.

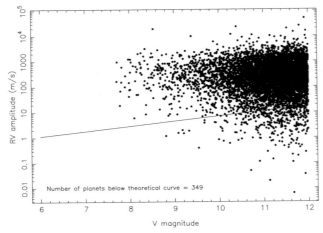

Figure 1. Radial velocity amplitudes of the planetary signatures along with the noise model for a small sample of datasets.

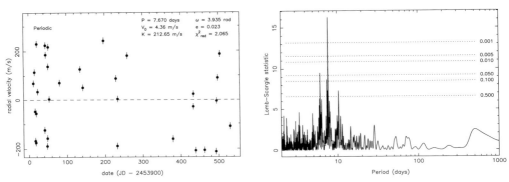

Figure 2. A simulated dataset for an 18 month cadence configuration (left) along with the accompanying periodogram (right).

3. Detection Criteria

The code written for the task of sifting planet candidates from the data uses a weighted Lomb-Scargle (L-S) periodogram to detect a periodic signal. The number of false detections resulting from this technique depends upon the periodic false-alarm probability threshold one adopts as the detection criteria. We selected this threshold for each cadence by producing a large number of datasets with no planets injected, then executing a Monte-Carlo simulation to determine that threshold which yields the required maximum number of false detections. In addition, we distinguish between those detections with unique and ambiguous periods based on the number of significant peaks in the periodogram.

Shown in Figure 2 (left) is a typical dataset from an 18 month cadence simulation. Figure 2 (right) also shows the corresponding periodogram where the dotted lines indicate various false-alarm probabilities.

4. Cadence Results and Planet Yield

For each cadence configuration, the simulated datasets were passed through the described detection algorithm and the results sorted by period, eccentricity, and sensitivity

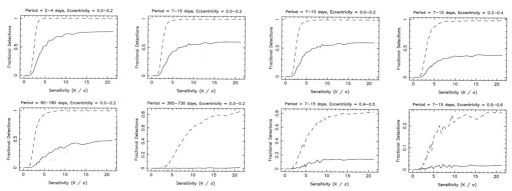

Figure 3. Detection efficiency results for the 888 cadence configuration, with the 4 plots on the left showing sensitivity variation with period, and the 4 plots on the right showing sensitivity variation with eccentricity.

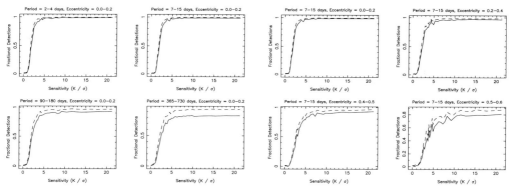

Figure 4. Detection efficiency results for the 7771111111000222222 cadence configuration, with the 4 plots on the left showing sensitivity variation with period, and the 4 plots on the right showing sensitivity variation with eccentricity.

(K/σ). The results for the 888 simulation are shown in Figure 3, where the dashed line indicates all detections and the solid line indicates only unique period detections. The four plots on the left show the period dependence for a circular orbit and the four plots on the right show the eccentricity dependence for a period range of 7–15 days. The detection efficiency of the 888 cadence performs moderately well for short-period planets with relatively circular orbits, but suffers greatly in the long-period and mid-high eccentricity regimes.

A far superior plan is to use slightly more measurements spread over a much longer time-scale. An example of this is an 18 month cadence with 33 measurements distributed in a cadence configuration described as 7771111111000222222. The results of this simulation are shown in Figure 4 in which it can be seen that the detection efficiency for both period and eccentricity fare significantly better than in the 888 case.

A large number of cadence configurations have been investigated in this manner. Given the parameter sensitivities derived from the cadence simulations, we can now use the known distribution of exoplanetary parameters to calculate the planet yield for each cadence. Table 1 shows the planet yield predictions, unique and total detections, from a subset of the cadences which, given uncertainties in stellar properties, provides a useful comparison. Included in this table is a simulation in which the measurements were

Table 1. Sample of cadence simulations performed with planet yields.

measurements	cadence	unique	total
15	111111111000111111	31	41
24	741111111000111111	88	94
33	771111111000222222	173	200
33	36 months uniform	184	187

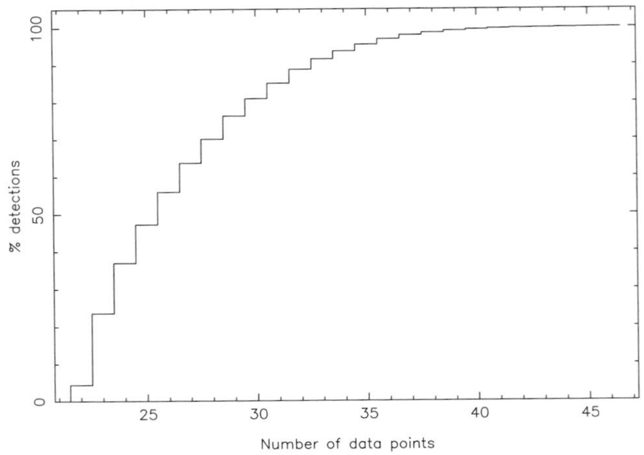

Figure 5. Cumulative histogram of the number of unique detections per measurement bin for a given cadence.

distributed uniformly over a 36 month period, avoiding monsoon seasons such as those experienced in Arizona.

There is a clear trade-off between the number of measurements and the number of unique detections. The choice of cadence therefore largely depends upon the amount of follow-up resources available. Figure 5 is a cumulative histogram of the number of unique detections for a given cadence. Beyond 30–35 data points, the fractional increase in unique detections becomes negligible, therefore suggesting that it would be most useful for increasing planet yield to change targets beyond this point.

5. Conclusions and Future Work

These simulations show that the choice of observing cadence can have a major impact on the exoplanetary parameter sensitivity. For example, reducing the number of measurements from 33 to 15 has a devastating impact on the planet yield. Furthermore, restricting the 33 measurements to 18 months rather than 36 months increases the sensitivity to short-period planets. The detection of mid-high eccentricity planets are biased against by the current algorithm, but this is being addressed by investigating the inclusion of higher order fourier terms. Since there is a continuum of possible cadence configurations, techniques to perform a more systematic search of cadence "parameter space" are being developed to determine optimal cadence solutions.

References

Ge, J., *et al.* 2006, *Proc. SPIE*, 6269, 75
Kane, S. R., Schneider, D. P., & Ge, J. 2007, *MNRAS*, 377, 1610

Exoplanets: Detection, Formation and Dynamics
Proceedings IAU Symposium No. 249, 2007
Y.-S. Sun, S. Ferraz-Mello and J.-L. Zhou, eds.

The astrometric data reduction software (ADRS) and error budget for PRIMA

N. M. Elias II[1,2], R. N. Tubbs[2], R. Köhler[1,2], S. Reffert[1], I. Stilz[1,2], R. Launhardt[2], J. de Jong[3], A. Quirrenbach[1], F. Delplancke[3], Th. Henning[2], D. Queloz[4] and ESPRI Consortium[1,2,3,4]

[1]Zentrum für Astronomie (ZAH) der Universität Heidelberg
Landesternwarte (LSW); Köenigstuhl 12; 69117 Heidelberg, Germany
email: n.elias@lsw.uni-heidelberg.de

[2]Max-Planck-Institut für Astronomie (MPIA)
Königstuhl 17; 69117 Heidelberg, Germany

[3]European Southern Observatory (ESO)
Karl-Schwarzschild-Strasse 2; 85748 Garching bei München, Germany

[4]Observatoire Astronomique de l'Université de Genève
1290 Sauverny, Switzerland

Abstract. The Astrometric Data-Reduction Software (ADRS) processes fringe, delay, environmental, and calibration data for PRIMA narrow-angle astrometry. It is automated software designed to provide fully-calibrated differential delays and separation angles. The ADRS is divided into on-line and off-line processing. The former deals with calibration and data compression, while the latter applies corrections and calculates science quantities. PRIMA is the first VLTI instrument that may require removal of long-term environmental trends. The trend identification and fitting routines are not part of the distributed on-line and off-line processing software. Instead, files containing fit parameters will be updated regularly. Coding is presently underway. The PRIMA error budget summarizes the principal sources of error in PRIMA astrometric observations.

Keywords. astrometry, stars: planetary systems, instrumentation: interferometers, techniques: interferometric, methods: data analysis, techniques: high angular resolution, infrared: stars

1. Introduction

PRIMA will add a dual-beam capability to the VLTI, allowing both narrow-angle astrometry with the 1.8 m Auxiliary Telescopes (ATs) and the ability to do phase-referenced imaging (further details of PRIMA can be found in Launhardt *et al.* (2008)). The Astrometric Data-Reduction Software (ADRS) processes PRIMA fringe, delay, environmental, and calibration data for astrometric observations. It is not an interactive package and it does not fit planetary orbits. The primary purpose of the ADRS is to automatically produce calibrated differential delays and separation angles. It is based upon the PRIMA error budget (Tubbs *et al.* (2008)) and follows the standard interferometric paradigm for data reduction: averaging, calculation of intermediate quantities, and calibration.

The top-level ADRS requirements are: 1) the code must be as simple as possible to perform the job; 2) intermediate quantities should be calculated and saved in tables and header keywords, to assess data quality and verify the mathematical assumptions; and 3) the code must be expandable, to respond to new information, ideas, and algorithms. The same software must run at ESO/Paranal (responding to the PRIMA instrumentation)

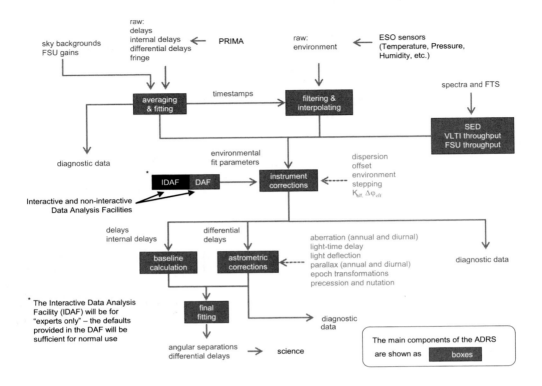

Figure 1. Flow diagram of the Astrometric Data-Reduction Software (ADRS)

and ESO/Garching (periodic reprocessing as fitted trends improve). Users can also run the ADRS, but we expect that the results produced by ESO will already be of the high quality required for scientific publications.

2. Data flow

The ADRS is divided into two main systems, on-line (pipeline) processing and off-line processing. The output files are divided into data levels, with the highest data level corresponding to science-grade data.

The on-line processing is responsible for reducing calibration data, averaging fringe data, fitting delay data, and interpolating environmental data (top of Figure 1). To eliminate the need for additional interpolation, all fringe, delay, and environmental data are placed onto a fixed ≈ one-second time grid. A number of intermediate diagnostic quantities are also produced, such as total photon numbers, squared visibilities, and phase errors. On-line recipes include Fourier-transform spectroscopy, gain corrections, sky backgrounds, laboratory darks, and fringe/delay/environment data compression.

The off-line processing corrects delays and differential delays for environmental and astrometric effects (middle and bottom of Figure 1). The delays are corrected for dispersion, and the differential delays are corrected for tracking errors and the arbitrary zero point. Environmental effects will likely consist of temperature, pressure, humidity, seeing, etc. Astrometric effects include light-time delay, light deflection, aberration, annual and diurnal parallax, and Earth orientation. The fringe data are used for dispersion and tracking corrections. The delay data are used to calculate the baseline vectors. The final data product consists of differential delays and star separations.

We expect long-term trends in delays and differential delays due to environmental effects. The PRIMA Consortium will use an IDL-based software package called the Interactive Data Analysis Facility (IDAF). It consists mostly of plotting and FITS-file routines, and it may eventually be expanded to include automated trend-identification routines.

After finding trends with the IDAF, the Data Analysis Facility (DAF) is used to select data and calculate fits that are used by the off-line processing recipes. The data selection part of the DAF is based upon the Organization, Classification, and Association (OCA) package. It is a rule-based interpreter that selects FITS files based on keywords in the primary headers. The fits, as well as instrument bandpasses and astrometric catalogs are saved in a group of files called the Correction Collection (CoCo). They are updated on a regular basis.

3. Error budget

The measurement errors in a single 30-minute PRIMA astrometric observation can be separated into errors in the knowledge of the narrow-angle baseline, and errors in the measurement of the differential delay using this baseline. The AT telescopes were designed before PRIMA, and two aspects of their design may limit the astrometric performance: the narrow-angle baseline stability and the differential delay introduced by the roughness of AT mirrors which lie close to an image plane. We are currently developing modeling and improvements for the ATs in order to overcome these problems. The PRIMA Error Budget is summarized in more detail in Tubbs et al. (2008).

4. Conclusion

The overall top-level design of the ADRS was finished in May 2007 at the Final Design Review (FDR). File formats have been created, but additional data may be saved as needed. Most recipes have been designed using work-breakdown structures, and preliminary versions of half of the recipes have already been written.

Additional functionality, which has already been planned, will be implemented in the near future. The beta versions will be delivered to ESO early in 2008. The commissioning versions will be delivered summer/fall 2008, where they will be debugged along with the hardware at the VLTI on Paranal.

Work on optimizing the operation of PRIMA and assessing the expected astrometric error is ongoing.

References

Launhardt, R., Henning, Th., Queloz, D., Quirrenbach, A., Delplancke, F., Derie, F., Elias, N., Pepe, F., Reffert, S., Segransan, D., Setiawan, J., Tubbs, R., & ESPRI Consortium 2008, The ESPRI project: Narrow-angle astrometry with VLTI-PRIMA (these proceedings).

Tubbs, R., Elias, N., Launhardt, R., Reffert, S., Delplancke, F., Quirrenbach, A. Henning, Th., Queloz, D., & ESPRI Consortium 2008. ESPRI data-reduction strategy and error budget for PRIMA. *In: Proc. IAUS 248 (in press)*.

Unveiling exoplanet families

Simone Marchi and Sergio Ortolani

Dipartimento di Astronomia, University of Padova,
I-35122 Padova, Italy
email: `simone.marchi@unipd.it, sergio.ortolani@unipd.it`

Abstract. The discovery of an increasing number of extrasolar planets (EPs) prompts the development of a planetary taxonomy. Such analysis, as in many other fields of research, is useful to identify groups of objects sharing similar traits. When applied to extrasolar planets, the taxonomy may provide a valid support for disentangling the role of the several physical parameters (semimajor axis, metallicity etc.) involved in the planetary formation processes and subsequent evolution. We present the state-of-the-art for exoplanets taxonomy obtained with hierarchical algorithms and the definition of robust clusters of planets (this is an update of the taxonomy published in Marchi 2007). The physical relevance of the exoplanet clusters along with their implications for the formation theories are also discussed. Finally, we comment on the future improvements of such analysis taking into account new algorithms and new input variables.

Keywords. planetary systems, planetary systems: formation

1. Multivariate analysis and dimensional scaling of EPs

The results presented here are an update of the taxonomy presented in Marchi (2007). Before going into the details of the present results, we recall few basic concepts about the method.

The inputs to our model are the following: planetary projected mass (M_p), semimajor axis (a), eccentricity (e), stellar mass (M_s), stellar metallicity ([Fe/H])†. Only objects having simultaneously estimates for $\{M_p, a, e, M_s, [Fe/H]\}$ have been used.

We consider 209 EPs (updated at 13 June 2007). To them, the Solar System planet Jupiter has been added. The first step is to perform a statistical analysis in order to find out if there are useless -or less significant- input variables. This is done using principal component analysis (Everit & Dunn 2001). The basic idea of principal component analysis (PCA) is to combine the input variables in such a way as to show the most important ones. This is done by describing the data with a number of new variables pc_i, ordered in terms of decreasing variance.

On the basis of the variance attained by each pc_i we may reject some of them. This procedure has the advantage of using only the variables which are important, allowing a simpler description of the data set with only a minor loss of information. According to general criteria, it seems reasonable to keep only the first three principal components which account for 81% of the total variance (notice we use the logarithms of M_p and a since this gave an higher variance of the first principal components).

† They are obtained from the interactive extrasolar planets catalog mantained by J. Schneider, see http://exoplanet.eu/

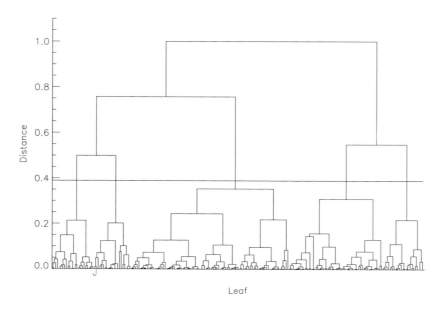

Figure 1. Dendrogram of the best solution. The vertical line correspond to the best cutoff which identify the five cluster (see text for futher details).

2. Cluster analysis

The choice of the clustering technique is quite arbitrary and it relies mostly on the kind of description of the data we are interested in. When the number of clusters are not known a priory, like in our case, hierarchical clustering is more suitable (Everit et al. 2001). One of the advantages of this technique is that it provides a classification which consists of a series of nested partitions, which is well illustrated by a two-dimensional diagram known as *dendrogram*. However, there are a number of possible ways to perform the analysis, and an accurate step-by-step evaluation of the process has to be performed.

First we decided not to standardize the clustering variables as this may reduce the difference among members, making the identification of clusters more difficult. Moreover, as a general rule, the same metrics should be used for the proximity matrix and the inter-group proximity measures. We explored different metrics and the effects of different algorithms of inter-group merging. The full set of possibilities has been investigated by using the traditional means (e.g. cophenetic coefficient) and analyzing the corresponding dendrograms.

We therefore identify a robust solution which is closely nested for small heights and stable against errors (e.g. observational errors) on the position of the EPs in the clustering space. The best solution was obtained with the Pearson correlation distance† and weighted centroid merging. It has five robust clusters.

It is interesting to note that traditional metrics (like the Euclidean) and traditional cluster merging (like single linkage) produce in general bad results. It turns out that they are not able to find structures for EPs.

† The distance between two items i, j is defined as $d_{ij} = [(1-\phi)/2]^{1/2}$, where ϕ is the Pearson correlation coefficient.

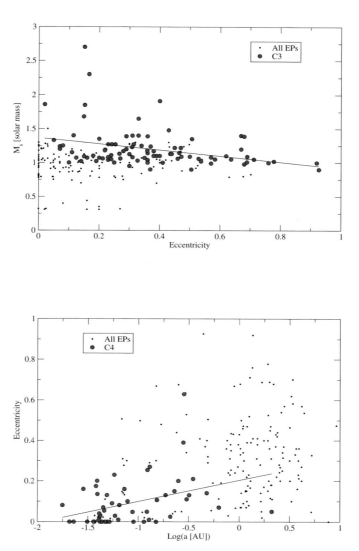

Figure 2. Examples of correlation found within clusters. Here we show the $e - M_s$ correlation for $\mathcal{C}3$ and $a - e$ correlation for $\mathcal{C}4$. We also overplot all the EPs.

Moreover, we tested the solution against the absence of clusters with Monte Carlo simulations. We also tested the solution with respect to the presence of observational errors. We find that the solution is stable with respect to both tests.

3. Analysing the clusters

In this section we present a briefly overview of the properties of the five clusters identified.

A first important point is that, although the clusters have been identified using pc_i, we may nevertheless find differences among the input variables. We may summarize these differences in this way: $\mathcal{C}1$ has sub-solar [Fe/H] and super-jovian M_p (notice that Jupiter belong to this cluster); $\mathcal{C}2$ has sub-solar [Fe/H] and sub-solar M_s; $\mathcal{C}3$ has super-solar [Fe/H] and M_s, $\mathcal{C}4$ has sub-jovian M_p and low e; $\mathcal{C}5$ has super-solar [Fe/H] and M_s. Moreover, the hot jupiters (HJs; i.e. planets with period less than 10 days) belongs to $\mathcal{C}4$ and $\mathcal{C}5$: we thus identify two main types of HJs. Most notably all the so-called hot Neptunes belong to cluster $\mathcal{C}4$. In other words, our cluster analysis is able to identify two types of hot "planets", possibly indicating different origin and evolution. We shall deal with this aspect in a forthcoming paper. As for the multiplicity, we find that all clusters have a similar number of multiple planet systems (MPS) and multiple stellar systems (MSS) except for $\mathcal{C}1$ which has no MSS and only a few MPS.

Finally, we also checked for inter-correlation among the input variables within each cluster. The reason for that, is that commonly accepted that planets may form in different ways (core accretion vs disk instability), and that their evolution is affected by several parameters (disk density, stellar types, opacity etc). The EP database may reflect such complexity, however the signature of these processes may be blurred in statistical analyses which deal with the whole EP dataset. On the contrary, if cluster separation has something to do with the formation and evolutionary processes and not being just a mere classification, it becomes important to look for trends within each cluster. In the following we report only highly significant (i.e. having a 2-tailed probability less then 5%) intra-cluster correlations. These correlations are: $a - M_s$, M_p-[Fe/H] for $\mathcal{C}1$; $M_p - M_s$ for $\mathcal{C}2$; $e - M_s$ for $\mathcal{C}3$ (see fig. 2); $a - e$ for $\mathcal{C}4$ (see fig. 2); $a - M_s$ for $\mathcal{C}5$. Therefore the stellar mass seems to play an important role in several clusters, while the [Fe/H] seems to be by far less important. This is somehow in contadiction with Marchi (2007), where both variable where found to be important.

It is important to realize that such correlations rise considering the present population of discovered exoplanets. In this respect two important points have to be considered. First, some of the correlations may be affected by observational biases. On the other hand, some of them are probably affected by planetary evolution. This is clearly the case for the cluster $\mathcal{C}4$, where the $a - e$ correlation is likely the results of tidal circularization. For the purpose of the present proceeding we do not tackle with the physical interpretation of the clusters and the correlations, which will be the topic of a detailed forthcoming paper.

Another relevant point, is that we find a significant correlation between [Fe/H] and M_s for several clusters and also for the whole EPs sample. This clearly implies that the two parameters are linked in the field stars selected for planetary search purpose. Indeed for a given age this is expected from the theoretical models of stellar evolutions (Girardi et al. 2002): we estimated an increase of 0.5 $M_\odot$/dex at 5 Gyr for a star of about 1 $M_\odot$ (see fig. 3).

4. Discussion and conclusion

In this paper we develop the basis for an extrasolar planet taxonomy. We identify the best procedure to follow: a multivariate statistical analysis (PCA) to find the most important variables, and then hierarchical clustering analysis. The best result is achieved with non traditional metric and merging algorithms, namely the Pearson correlation metric and weighted centroid cluster merging. We reject the absence of clustering structure with Monte Carlo simulations, and also tested the stability of the solution against of ~ 100 K, comparable to Jupiter. However, if the heat deposited on the day side can be efficiently redistributed, the planet's entire outer envelope would reach an isotropic photospheric temperature of $\sim T_{equil}/2^{1/4} \sim 10^3$ K. The actual day-night temperature distribution is likely to be between these extreme limits.

Transiting planets are ideal objects for studying planetary atmospheres. Observations include the investigation of primary eclipses with both the Spitzer (Richardson et al. 2006) and the Hubble (Brown et al. 2001) Space Telescopes that reveal crucial information on planetary radii, and IR spectroscopy (Richardson et al. 2003, 2007; Swain et al. 2007; Knutson et al. 2007a) conducted by Spitzer that allows us to probe the composition and thermal structure. In addition, the Spitzer observations of secondary eclipses (Deming et al. 2005; Charbonneau et al. 2005; Deming et al. 2006), and phase monitoring (Harrington et al. 2006; Cowan et al. 2007; Knutson et al. 2007b) allow for the determination of planetary surface temperatures. These observations indicate that at least a fraction of the incident stellar flux is advected from the day to the night side.

Recent observational data also raises several important new puzzles. Among the > 26 observed radii of transiting extrasolar planets there exist those that are both too large and too small to be explained by conventional formation scenarios. Several solution have been proposed to account for this diversity: 1) the stellar irradiation reduces the temperature gradient and the radiative flux in the planet's outer layers significantly slowing the Kelvin-Helmholtz contraction of the planet (Burrows et al. 2000); 2) a dispersion in the planets' metallicity, opacity, and internal structure may lead to diversity in their mass-radius relation (Burrows et al. 2007a); 3) the kinetic heating resulting from the dissipation of the gas flow within the atmosphere induced by the day-night pressure gradient (Showman & Guillot 2002); 4) the additional energy provided by tidal dissipation within the planet's envelope (Bodenheimer et al. 2001) caused by the circularization of its orbit may inflate its radius, and 5) the dissipation of the obliquity tides in a Cassini-state resonance between the planet's spin precession and its orbital precession (Winn & Holman 2005). In each scenario, the efficiency of radiation transfer and atmospheric dynamics play crucial roles.

In this section we present the results of full 3-D, hydrodynamical, radiative transfer models of hot-Jupiter's for a variety of opacities. We cast our numerical model in spherical coordinates and advance the equations of continuity, motion, and energy, including the Coriolis and centrifugal forces. We concentrate on the upper atmosphere $(1.0 - 1.2 R_J)$ and simulate the azimuth in its entirety. We utilize the flux-limited radiation transfer approximation of Levermore & Pomraning (1981) to calculate the heat transfer carried by the radiative flux, allowing for a self-consistent treatment of radiation transfer in both the optically thick and optically thin limits. The radiative portion of the energy equation is solved *implicitly*, while the advection scheme, described in Hawley et al. (1984) and Kley & Hensler (1987), is an extension of the first-order van Leer (van Leer 1977) method known as the 'mono-scheme'. We use an ideal gas equation of state for the gas pressure and a fixed molecular weight.

For this study we adopted a $0.63 M_J$ planet (HD 209458b) that is synchronously spinning with its 3 day orbit around a 4.5Gyr old solar-type star. In Fig. 6, we illustrate the effective temperature distribution (at the planet's photosphere) of a representative model. For this model, a clear day-night delineation is apparent, with the night-side characterised by temperatures from 270 to 400K. Slightly hotter regions near the terminators, associated with jets from the day-side, are apparent with temperatures reaching ~ 480K. Although the spin period of this planet is much longer than that of Jupiter, the associated Coriolis force alters the resulting flow dynamics and has implications concerning the ability of the planet to fully synchronise its spin. The left-hand plot of Fig. 7

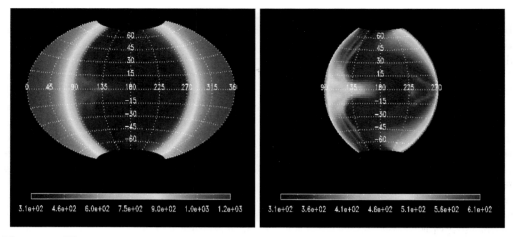

Figure 6. The temperature at the photosphere of a planet rotating with a period of 3days. The left panel shows the entire photosphere, while the right panel highlights the structure on the night-side. Although some higher temperature regions exist associated with jets, a clear day-night delineation persists, despite complicated dynamical structure.

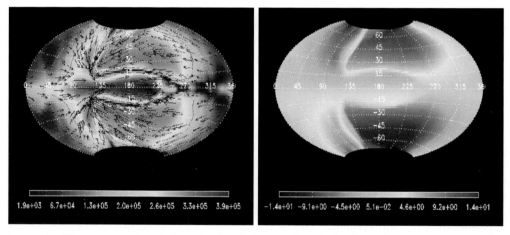

Figure 7. The magnitude of the velocity (left panel) and the latitudinal component of the Coriolis force (right panel) at the photosphere of a planet spinning at 3days. The rotation of the planet causes eastward and westward moving fluid to behave quite differently.

shows the velocity magnitude at the photosphere. Material near the terminator is moving quite rapidly, reaching speeds of over 3km/s. Eastward moving material appears to be funnelled into an equatorial jet near $\phi = \frac{\pi}{2}$, while westward moving material is pushed toward the poles near $\phi = \frac{3\pi}{2}$. The right-hand plot in Fig. 7 shows the latitudinal component of the Coriolis force. The asymmetry imposed by the rotation causes the fluid moving in eastward and westward directions to behave significantly different then in the non-rotating case. These flow structures, and the advected heat they carry account for the structure of the hotter regions seen on the night-side.

Also evident in Fig. 7 is the marked decrease in velocity where the eastward and westward flows converge. It is clear that neither flow is able to instigate circumplanetary flow at the surface. The convergent points, both at the equator and higher latitudes, differ significantly from π, so upon meeting one of the flows has undergone substantially more cooling then its counterpart. Thus, when the eastward flow encounters the westward flow at the equator, it is cooler and sinks below. The opposite is true at higher latitudes, with the westward flow experiencing more cooling and sinking below the eastward flow. The flow pattern is likely to generate potential vorticity and excite baroclinic instabilities. Nevertheless, the dynamics will be dominated by the large scale anisotropic heating imposed by the stellar irradiation.

Our results differ significantly from previous dynamical models (Showman & Guillot 2002; Cho *et al.* 2003; Menou *et al.* 2003; Cooper & Showman 2005; Cho *et al.* 2006). These investigations solve either the primitive equations or the equivalent-barotropic equations. The primitive equations are widely used in meteorology and replace the vertical equation of motion with the hydrostatic equation. The equivalent-barotropic equations are derived by vertically integrating the primitive equations. To represent radiative energy transfer, Showman & Guillot (2002) and Cooper & Showman (2005) also utilised a Newtonian radiative scheme to approximate stellar irradiation. In this prescription, the temperature is relaxed toward some predefined equilibrium distribution on some radiative timescale. As noted by the authors, this approximation is crude, but allows rapid computation of a large number of models. Our models add two crucial components, including the flux-limited diffusion treatment for radiation and the full 3D treatment. Our results indicate that throughout the entire planet and for a range of opacities the radiative timescale can be quite long in comparison to the dynamical timescale, highlighting the need for a more accurate treatment of radiation. In addition our full 3D simulation flow accurately models the flow in the radial direction. Disallowing this flow would significantly alter the resulting dynamics and leads to much more uniform temperature distributions. In fact, the interpretation (Burrows *et al.* 2007b) of recent Spitzer observations by (Knutson *et al.* 2007a) have confirmed a temperature inversion similar to that predicted in ().

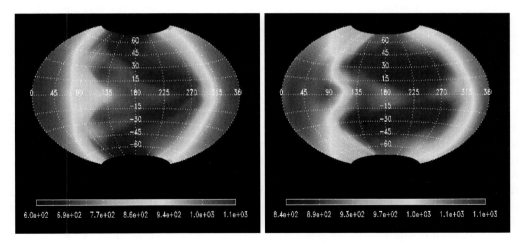

Figure 8. The temperature distribution at the atmosphere of planets whose opacities are reduced by a factor of 100 (left panel) and a factor of 1000 (right panel). Differences in opacity between different planets may play a key role in the observed diversity, and the ability to self–consistently model the radiative transfer across a wide range in opacity is a crucial component to any dynamical model of these short period planets.

The opacity of the planet regulates the efficiency of both the absorption and re-radiation of the incident stellar energy, plays a large role in determining the night-side temperature distribution, and may play a role in the overall evolution of the planetary structure. Our current models are based on the interstellar opacities that are most certainly an upper limit to the actual opacities of these atmospheres. Atmospheres containing grains with an abundance and size distribution comparable to that of the interstellar medium will undergo only shallow heating on the day-side; the resulting circulation will not effectively transmit heat to the night-side, which thus cools well below the day-side. However, as the abundance of grains in the atmosphere is reduced, the stellar radiative flux penetrates more deeply into the atmosphere on the day-side, and the higher density atmospheric circulation carries a larger flux of heat over to the night-side. Our numerical simulations show a similar behaviour. In Fig. 8 we show the effects of reducing the atmospheric opacity by factors of 100 (left-hand panel) and 1000 (right-hand panel). In comparison to Fig. 6, the night-side in both these simulations is considerable warmer, reaching $\sim 900K$ for the lowest opacity.

5. Conclusion

In this proceeding we have explored several phenomena that may help to explain some of the unusual observed features of short period gas giants. In particular we concentrate on the observed pile-up of planets around $2-3$ days, and the large diversity in both structural and atmospheric properties inferred from recent transit observations. The proximity of the host stars suggests that interactions between the star and planet likely play a large role in the evolution of the orbit and the atmosphere.

To explain the pile-up of planets at $2-3$ days we present the results from an analysis of the ohmic dissipation within a planet due to the stellar magnetic field. The magnetic field of young T-Tauri stars may be as large as several kilogauss, which will easily penetrate into the planetary interior to a depth that depends on the planets interior conductivity. A slightly misaligned or non-syncronized magnetic field will lead to field modulation and energy dissipation within the planet. The associated expansion of the envelope with instigate Roche lobe overflow through the inner Lagrange point, providing an outward torque on the orbit. Such interactions ultimately lead to smaller mass planets and stalled migration.

Recent transit observations allow for the computation of mean densities and reveal a large diversity in the structure of gas giants. When compared to the conventional sequential accretion model, there are a number of planets too dense, but also a number that are too large. Such diversity, taken together with the formation constraints imposed by observations of disk lifetimes, suggests that post formation processes may be responsible for the large range in observed properties. We have presented coupled analysis of the rate and nature of planetesimal/embryo bombardment and the resulting structural changes due to the larger of such impacts. SPH simulations show that the angle of giant impacts plays a large role in determining the retention efficiency; head on collisions result in near total assimilation and significant envelope expansion, while in glancing collisions the core of the impactor is preferentially retained perhaps explaining those planets observed to have very large cores.

Finally, continuing observations of close in planets has been revealing a diversity in the atmospheric temperature structures both across the surface of the planet and with depth. We present the results of a three dimensional study of planetary atmospheres irradiated by their host stars. We highlight the importance of using such fully three dimensional radiative hydrodynamical simulation to capture the actual dynamics and energy transfer. In addition we demonstrate the importance of opacity in determining the resulting efficiency of heat transport. Changes in opacity due to pollution via methods similar to those described above or the settling/upwelling of high opacity material may help to explain the observed diversity.

References

Bodenheimer, P., Lin, D. N. C., & Mardling, R. A. 2001, ApJ, 548, 466
Brown, T. M., Charbonneau, D., Gilliland, R. L., Noyes, R. W., & Burrows, A. 2001, ApJ, 552, 699
Burrows, A., Guillot, T., Hubbard, W. B., Marley, M. S., Saumon, D., Lunine, J. I., & Sudarsky, D. 2000, ApJL, 534, L97
Burrows, A., Hubeny, I., Budaj, J., & Hubbard, W. B. 2007a, ApJ, 661, 502
Burrows, A., Hubeny, I., Budaj, J., Knutson, H. A., & Charbonneau, D. 2007b, ApJL, 668, L171
Campbell, C. G. 1983, MNRAS, 205, 1031
Charbonneau, D., Allen, L. E., Megeath, S. T., Torres, G., Alonso, R., Brown, T. M., Gilliland, R. L., Latham, D. W., Mandushev, G., O'Donovan, F. T., & Sozzetti, A. 2005, ApJ, 626, 523
Cho, J. Y.-K., Menou, K., Hansen, B., & Seager, S. 2006, ArXiv Astrophysics e-prints
Cho, J. Y.-K., Menou, K., Hansen, B. M. S., & Seager, S. 2003, ApJL, 587, L117
Cooper, C. S., & Showman, A. P. 2005, ApJL, 629, L45
Cowan, N. B., Agol, E., & Charbonneau, D. 2007, MNRAS, 552
Deming, D., Harrington, J., Seager, S., & Richardson, L. J. 2006, ApJ, 644, 560, secondary Ecclipse Discovery Paper for HD189733b
Deming, D., Seager, S., Richardson, L. J., & Harrington, J. 2005, Nature, 434, 740
Dermott, S. F. 1970, MNRAS, 149, 35
Dobbs-Dixon, I., & Lin, D. N. C. 2008, ApJ, 673, 513
Dobbs-Dixon, I., Lin, D. N. C., & Mardling, R. A. 2004, ApJ, 610, 464
Goldreich, P., & Lynden-Bell, D. 1969, ApJ, 156, 59
Gu, P.-G., Bodenheimer, P. H., & Lin, D. N. C. 2004, ApJ, 608, 1076
Guillot, T., Stevenson, D. J., Hubbard, W. B., & Saumon, D. 2004, The interior of Jupiter (Jupiter. The Planet, Satellites and Magnetosphere), 35–57
Harrington, J., Hansen, B. M., Luszcz, S. H., Seager, S., Deming, D., Menou, K., Cho, J. Y.-K., & Richardson, L. J. 2006, Science, 314, 623
Hawley, J. F., Wilson, J. R., & Smarr, L. L. 1984, ApJS, 55, 211
Ida, S., & Lin, D. N. C. 2004, ApJ, 604, 388
Johns-Krull, C. M. 2007, ApJ, 664, 975
Kley, W., & Hensler, G. 1987, A&A, 172, 124
Knutson, H. A., Charbonneau, D., Allen, L. E., Burrows, A., & Megeath, S. T. 2007a, ArXiv e-prints, 709
Knutson, H. A., Charbonneau, D., Allen, L. E., Fortney, J. J., Agol, E., Cowan, N. B., Showman, A. P., Cooper, C. S., & Megeath, S. T. 2007b, Nature, 447, 183
Kretke, K. A. Lin, D. N. C., Turner, N. J. 2008, in: Y.-S. Sun, S. Ferraz-Mello & J.-L. Zhou, (eds.), *Exoplanets: Detection, Formation and Dynamics*, Proc. IAU Symposium No. 249 (Suzhou,China), p. 293
Koenigl, A. 1991, ApJL, 370, L39
Laine, R., & Lin, D. 2008, in preparation
Levermore, C. D., & Pomraning, G. C. 1981, ApJ, 248, 321
Li, S., & Lin, D. 2008, in preparation
Lin, D. N. C., Bodenheimer, P., & Richardson, D. C. 1996, Nature, 380, 606
Marcy, G., Butler, R. P., Fischer, D., Vogt, S., Wright, J. T., Tinney, C. G., & Jones, H. R. A. 2005, Progress of Theoretical Physics Supplement, 158, 24
Mardling, R. A., & Lin, D. N. C. 2004, ApJ, 614, 955
Mayor, M., & Queloz, D. 1995, Nature, 378, 355
Menou, K., Cho, J. Y.-K., Seager, S., & Hansen, B. M. S. 2003, ApJL, 587, L113

Nagasawa, M., & Lin, D. N. C. 2005, ApJ, 632, 1140
Novak, G. S., Lai, D., & Lin, D. N. C. 2003, 294, 177
Ogilvie, G. I., & Lin, D. N. C. 2007, ArXiv Astrophysics e-prints
Pollack, J. B., Hubickyj, O., Bodenheimer, P., Lissauer, J. J., Podolak, M., & Greenzweig, Y. 1996, Icarus, 124, 62
Richardson, L. J., Deming, D., Horning, K., Seager, S., & Harrington, J. 2007, ArXiv Astrophysics e-prints
Richardson, L. J., Deming, D., & Seager, S. 2003, ApJ, 597, 581
Richardson, L. J., Harrington, J., Seager, S., & Deming, D. 2006, ApJ, 649, 1043
Setiawan, J., Henning, T., Launhardt, R., Muller, A., Weise, P., & Kurster, M. 2008, Nature, 451, 38
Showman, A. P., & Guillot, T. 2002, A&A, 385, 166
Swain, M. R., Bouwman, J., Akeson, R., Lawler, S., & Beichman, C. 2007, ArXiv Astrophysics e-prints
van Leer, B. 1977, Journal of Computational Physics, 23, 276
Winn, J. N., & Holman, M. J. 2005, ApJL, 628, L159
Zhang, X. J., Kretke, K. A. Lin, D. N. C. 2008, in: Y.-S. Sun, S. Ferraz-Mello & J. -L. Zhou, (eds.), *Exoplanets: Detection, Formation and Dynamics*, Proc. IAU Symposium No. 249 (Suzhou,China), p. 309
Zhou, J. -L., Aarseth, S. J., Lin, D. N. C., & Nagasawa, M. 2005, ApJL, 631, L85
Zhou, J.-L., & Lin, D. N. C. 2007, ApJ, 666, 447
Zhou, J.-L., Lin, D. N. C., & Sun, Y.-S. 2007, ApJ, 666, 423

Internal waves driven by stellar irradiation in a non-synchronized hot Jupiter

Pin-Gao Gu[1] and Gordon I. Ogilvie[2]

[1]Institute of Astronomy & Astrophysics, Academia Sinica, Taipei 106, Taiwan
email: gu@asiaa.sinica.edu.tw

[2]Dept. of Applied Mathematics & Theoretical Physics, University of Cambridge,
CB3 0WA, UK
email: gio10@cam.ac.uk

Abstract. We investigate the dynamical response of a non-synchronized hot Jupiter to stellar irradiation. In our current model, the stellar radiation acts like a diurnal thermal forcing from the top of a radiative layer of a hot Jupiter. If the thermal forcing period is longer than the sound speed crossing time of the planet's surface, the forcing can excite internal waves propagating into the planet's interior. When the planet spins faster than its orbital motion, these waves carry negative angular momentum and are damped by radiative loss as they propagate downwards from the upper layer of the radiative zone. As a result, the upper layer gains the angular momentum from the lower layer of the radiative zone. Simple estimates of angular momentum flux are made for all transiting planets.

Keywords. diffusion, hydrodynamics, waves, planets and satellites: general

1. Introduction

Infrared observations of transiting planetary systems with the Spitzer Space Telescope have revealed that the peak infrared brightness occurs before the secondary eclipse (Knutson *et al.* 2007). One interpretation of this phenomenon is that the atmosphere of a hot Jupiter is not tidally locked but exhibits a super-rotation state which advects the stellar heating downstream from the substellar point. Since a super-rotating atmosphere experiences day-night changes, internal waves can be excited by the periodic thermal forcing in the super-rotating flow. Furthermore, internal waves can carry angular momentum and therefore serve as a reasonable candidate for maintaining such a super-rotating flow. Here we employ linear analyses and investigate the possibility of wave excitation in a non-synchronized surface layer of a hot Jupiter driven by stellar irradiation.

2. Thermal tides in the non-rotating plane-parallel atmosphere

We initially consider a non-rotating plane-parallel atmosphere with uniform gravity $\mathbf{g} = -g\,\mathbf{e}_z$. The basic equations are

$$\frac{\partial \mathbf{u}}{\partial t} + \mathbf{u} \cdot \nabla \mathbf{u} = -\frac{1}{\rho}\nabla p + \mathbf{g}, \qquad (2.1)$$

$$\frac{\partial \rho}{\partial t} + \nabla \cdot (\rho \mathbf{u}) = 0, \qquad (2.2)$$

$$\frac{\partial p}{\partial t} + \mathbf{u} \cdot \nabla p + \gamma p \nabla \cdot \mathbf{u} = -(\gamma - 1)\nabla \cdot \mathbf{F}, \qquad (2.3)$$

$$\mathbf{F} = -\frac{16\sigma T^3}{3\kappa\rho}\nabla T, \tag{2.4}$$

$$p = \frac{R\rho T}{\mu}. \tag{2.5}$$

For simplicity we also assume that κ, γ and μ are constant. We use the radiative diffusion approximation (2.4) throughout the atmosphere and apply the 'Marshak' boundary condition (cf. Pomraning 1973)

$$\sigma T^4 = \tfrac{1}{2}F_z + F_i \tag{2.6}$$

at $z = +\infty$, where F_i is the irradiating flux. For the equilibrium state we have

$$\frac{dp}{d\tau} = \frac{g}{\kappa}, \tag{2.7}$$

$$\frac{d}{d\tau}(\sigma T^4) = \tfrac{3}{4}F_z. \tag{2.8}$$

where F_z is the intrinsic radiative flux density of the planet and τ is the optical depth.

The thermal tide is driven by a variation of the irradiating flux. Therefore we linearize equations (2.1)-(2.5) with Eulerian perturbations of the form $\text{Re}\,[\mathbf{u}'(z)\exp(\mathrm{i}k_x x - \mathrm{i}\omega t)]$, etc., where k_x is a real horizontal wavenumber and ω a real frequency determined by the tidal forcing (thermal tides). The perturbed boundary condition at $\tau = 0$ is $4\sigma T^3 T' = \tfrac{1}{2}F_z' + F_i'$, where F_i' is the thermal forcing term. The linearized equations admit a solution in the form of a internal wave in the limit of large τ. In the case of diurnal thermal forcing, the magnitude of the horizontal wavelength $2\pi/|k_x|$ is just the circumference of the planet's surface $2\pi R_p$, where R_p is the planet's radius.

We employ the following parameters for HD 209458b to be the input parameters for solving the linearized equations: $R_p = 1.32 R_J$, $M_p = 0.69 M_J$, $a = 0.0474$ AU, and $T_* = 6099$K. We adopt $F_z = 7 \times 10^6$ erg/cm^2 s (Bodenheimer et al. 2003), $\kappa = 0.01$ gm/cm^2 (molecular opacity with no grains), $\mu = 2$ gm/mol., and $\gamma = 1.4$ for the gas in the radiative surface layer of the planet. The Brunt–Väisälä frequency N ranges from

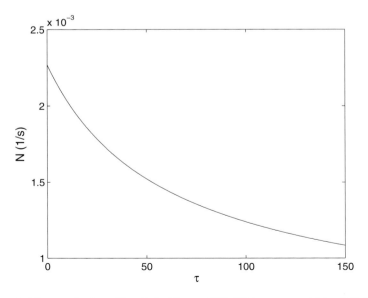

Figure 1. The vertical profile of the Brunt–Väisälä frequency N for HD 209458b.

0.0022/s at $\tau \approx 0$ to 0.001/s at $\tau \approx 150$ (see Fig 1) and is far larger than the thermal forcing frequency ω which is some fraction of the orbital frequency $n = 2\pi/(\text{a few days})$. The solutions are insensitive to the value of F_z because F_z is much smaller than the stellar irradiation F_i. We focus on the case that the planet is rotating faster than its orbital motion. Therefore the thermal forcing and the thermal tides propagate in a retrograde sense in the frame of the rotating planet.

Figure 2 shows the vertical structure of T' in units of K for different thermal forcing periods $2\pi/\omega$. The real and imaginary parts of T' are denoted by a solid and a dotted curve respectively. When the forcing periods are short (e.g. 0.1 and 1 day as shown in the top two panels), T' decays with the depth and the solutions behave like those to the thermal diffusion problem with the heat diffusing from the top of the atmosphere to a depth $\propto 1/\sqrt{\omega}$. Comparing the 0.1-day to the 1-day case, Figure 2 shows that T' can penetrate deeper in the 1-day case as a result of a longer forcing period and therefore a longer diffusion length.

On the other hand, when the forcing periods are long (e.g. 10 and 50 days as shown in the bottom panels of Figure 2), the vertical profiles of T' exhibit wavy-like solutions, meaning that waves are excited from the top of the atmosphere and propagate in. These waves are known as internal waves (i.e. g-mode).

3. Thermal tides in a rotating planetary atmosphere

We now consider the linearized dynamics of the form $\text{Re}[u'(z)H_{\nu,m=1,n}(\theta)e^{im\phi-i\omega t}]$ in a thin uniformly rotating atmosphere (Ω =constant & no winds), where $H_{\nu,m=1,n}$ is the Hough function with $m = 1$ (diurnal heating). Then the linearized equations remain the same except that k is replaced by λ/R_p^2, where λ is the Hough eigenvalue (Ogilvie & Lin 2004). λ is a function of $2\Omega/\omega$. We find that damped wave solutions are possible for even smaller ω ($2\pi/\omega > 50$ days) in the rotating case than in the non-rotating case ($2\pi/\omega > 10$ days).

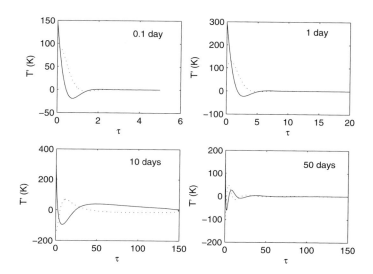

Figure 2. Temperature perturbation T' as a function of τ for the cases of 4 different forcing periods: $2\pi/\omega = 0.1$, 1, 10, and 100 days. The real and imaginary parts of T' are denoted by a solid and a dotted curve respectively.

Assuming that $2\pi/\omega = 100$ days, we find that $\lambda \approx 0.11$ for the most dominant Hough heating. The linearized equations can then be solved. Fig 3 shows the angular momentum flux carried by inward propagating Hough waves through the equator ($\theta = 90°$) as a function of τ for all transiting planets (Burrows et al. 2007). The positive value of angular momentum flux means that the angular momentum is transported outwards by Hough waves. The plot illustrates that for a given forcing period, the hot Jupiters which receive more stellar irradiation have larger angular momentum flux.

Although differential rotation and gravitational tides are not included in our analysis, we consider the following highly hypothetical scenario for the planetary spin based on our simple model. We assume that while the interior of the planet evolves to a synchronous state by both gravitational and thermal tides, the outer layer of the planet can maintain a super-synchronous state by the dissipation of Hough waves (thermal tides). Then as illustrated in Fig 3, the planet gas located below the photosphere ($\tau \gtrsim 1$) keeps losing angular momentum to the planetary atmosphere and is therefore spinning down. We further assume that the spun-down gas below the photosphere due to the Hough-wave torque τ_{Hough} can spin down the whole planetary interior against gravitational tides; i.e.,

$$\tau_{Hough} = \frac{9GM_*^2 R_p^5}{2a^6 Q'_p}\left(1 - \frac{\Omega_{p,interior}}{n}\right), \qquad (3.1)$$

where M_* is the stellar mass, Q'_p is the tidal quality factor of the planet, and $\Omega_{p,interior}$ is the spin rate of the planetary interior. Then the gravitational tidal heating rate $\dot{E}_{grav}$ due to the sub-synchronous rotation of the planetary interior is generated (e.g. Eggleton et al. 1998):

$$\dot{E}_{grav} = \frac{9n^3 a^2 M_*}{2Q'_p}\left(\frac{R_p}{a}\right)^5 \left(1 - \frac{\Omega_{p,interior}}{n}\right)^2. \qquad (3.2)$$

The heating rates $\dot{E}_{grav}$ in the case of $2\pi/\omega = 100$ days & $R_p = 1.67 R_J$ for all transiting planets are illustrated in Fig 4: the resulting tidal heating rates $\sim 10^{27}$ erg/s

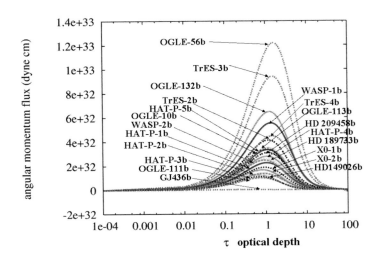

Figure 3. The vertical profile of angular momentum flux at the equator of the transiting planets when $2\pi/\omega = 100$ days.

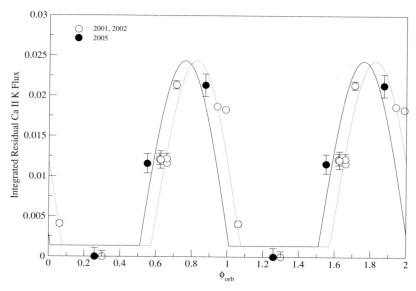

Figure 1. Integrated flux of the K-line residuals from a normalized mean spectrum of HD 179949 as a function of orbital phase for the 2001 and 2002 data (open circles) published in Shkolnik et al. (2003) and 2005 data (filled circles). The grey line is a best-fit spot model to the earlier data whose thickness reflects the error in the phase shift. The black line is the same fit slightly shifted in phase by -0.07 to better fit the 2005 data (See Shkolnik et al. 2008). This small shift relative to the earlier data is not signficant. Error bars in the integrated residual K flux are ± the intranight residual RMS. Note that the data points are repeated for two cycles.

periods of the planets are well known and uniquely established by the PRV and transit discovery methods, but the rotation periods of the stars are much harder to determine in part due to stellar differential rotation. For studies of SPI, differentiating between rotational and orbital modulation of the chromospheric emission is key.

In Shkolnik et al. (2005) we presented evidence of planet-induced heating on HD 179949. The effect lasted for over a year and peaked only once per orbit, suggesting a magnetic interaction. In the simplest configuration, a magnetic interaction would occur near the sub-planetary point, when the planet is in front of the star relative to the line-of-sight, which defines orbital phase $\phi_{orb} = 0$. Reproduced in Figure 1, we fitted a truncated, best-fit spot model to our 2001 and 2002 data with $P = P_{orb} = 3.092$ d corresponding to the change in projected area of a bright spot on the stellar surface before being occulted by the stellar limb. The fit to the 2001 and 2002 data peaks at $\phi_{orb} = 0.83 \pm 0.04$ with an amplitude of 0.027. We over-plot new data from 2005 which is fit remarkably well by the same model with only an insignificantly small relative phase shift of -0.07.

This phase lead may help identify the nature of the interaction. For example, the offset from the sub-planetary point of a starspot or group of starspots can be a characteristic effect of tidal friction, magnetic drag or reconnection with off-center stellar magnetic field lines. For further discussion on such mechanisms see papers by Gu et al. (2005), Preusse et al. (2006) and McIvor et al. (2006). In any case, the phasing, amplitude and period of the activity have persisted for over 4 years.

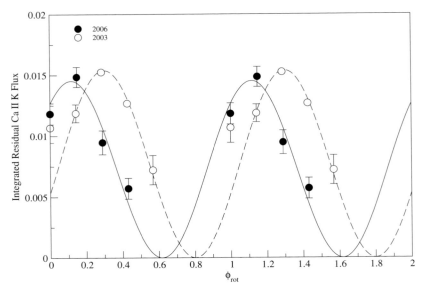

Figure 2. Integrated flux of the K-line residuals from a normalized mean spectrum of HD 179949 for 2003 and 2006 data plotted on a 7-day rotation period with phases relative to the first night of each run. The points are vertically shifted such that the minimum of each curves is zero. Error bars are ± the intranight residual RMS. The curves are best-fit spot models to the two data sets.

Ca II data acquired in 2003 and 2006 of HD 179949 do not phase with the planet's orbit, but both phase well with a 7-day period, likely the rotation period of the star. In Figure 2 we fit data from each year separately with a rotation curve because the effects of differential rotation and the appearance and disappearance of new spots over the three years would produce variations in phase, amplitude and period in the observed modulation. Note that the amplitude of the rotational activity is only 0.6 of that of induced by SPI. Indirect indications of the rotation rate of HD 179949 imply $P_{rot} \approx 9$ days and are presented in Shkolnik et al. (2003) and Saar et al. (2004). Wolf & Harmanec (2004) weakly detect (1.5 σ) a photometric rotation period for HD 179949 of 7.07 d with an amplitude of only 0.008 mag. While more photometry is needed to determine a rotation period conclusively, the modulated Ca II emission of this star in both 2003 and 2006 strongly suggests a rotation period of 7 days.

Similarly, previous Ca II data of v And indicated possible SPI (Figure 8 of Shkolnik et al. 2005), yet our September 2005 data appears to vary with the rotation. Again, the rotation period is not well known. Henry et al. (2000) quotes both 11 and 19 days, with a probable 11.6-day period from the $\langle S_{HK} \rangle$ index. Unlike data from 2002 and 2003, the 2005 data phase much better with P_{rot}=11.6 d than with a planetary orbit.

This on/off characteristic of SPI observed in the HD 179949 and v And systems is explained by the models of Cranmer & Saar (2007). They model the Ca II H & K light curve of a sun-like star with a hot Jupiter interacting with the field geometry at various stages of the empirically derived solar magnetic field at annual steps of the 11-year solar cycle. They conclude that due to the complex nature of the multipole fields, the Ca II K light curves due to SPI do not repeat exactly from orbit to orbit, and at

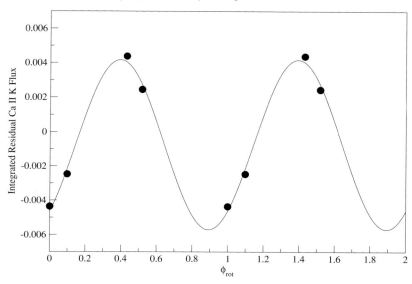

Figure 3. Integrated flux of the K-line residuals from a normalized mean spectrum of υ And for September 2005 data plotted on a 12-day rotation period. Error bars are within the size of the of the points. The curve is a best-fit spot model.

times the planet-induced enhancement may disappear altogether leaving only rotationally modulated emission. This may explain the 2003 and 2006 disappearance of the strong orbital modulation seen in 2001, 2002 and 2005 for HD 179949. Their models also show that for sparsely sampled data, the apparent phase shift between the peak Ca II emission and the sub-planetary point may fall between 0.2 and +0.2 (or ± 72°), consistent with the -0.17 phase shift we detected repeatedly for HD 179949.

2.2. Night-to-night activity correlates with planet's magnetic moment

Sánchez-Lavega (2004) looked at the internal structure and the convective motions of giant extrasolar planets in order to calculate their dynamo-generated surface magnetism. Given the same angular frequency (which is a reasonable approximation for the short-period planets in question), the magnetic dipole moment, and hence the magnetospheric strength, increases with planetary mass. This is observed for the magnetized planets in our own solar system, where the magnetic moment grows proportionally with the mass of the planet (Stevens 2005), and more specifically, with the planet's angular momentum ($L \propto M_p R_p^2 P_{p,rot}^{-1}$, Arge et al. 1995). Since only lower limits exist for the masses of most hot Jupiters and at such small semi-major axes they should be tidally locked ($P_{p,rot} = P_{orb}$), we plot $M_p \sin i / P_{orb}$ against $\langle MADK \rangle$, the average of the integrated MAD of the K line residuals per observing run, in Figure 4 of our complete sample of 13 systems. (The targets with system parameters are listed in Tables 1 of Shkolnik et al. 2005 and 2008.) Of our sample, τ Boo has the most massive planet and yet falls well below the correlation. This is consistent with the proposed Alfvén wave model where the near zero relative motion due to the tidal locking of both the star and the planet ($P_{*,rot} = P_{p,rot} = P_{orb}$) produces minimal SPI because of the weak Alfvén waves generated as the planet passes through the stellar magnetosphere, thereby transporting little excess energy to the stellar surface along the magnetic field lines (Gu *et al.* 2005). If this correlation between short-term activity and planetary magnetic moment holds for more hot Jupiter systems engaging in SPI, this could provide an empirical tool with which to estimate the strength of extrasolar planetary magnetic fields.

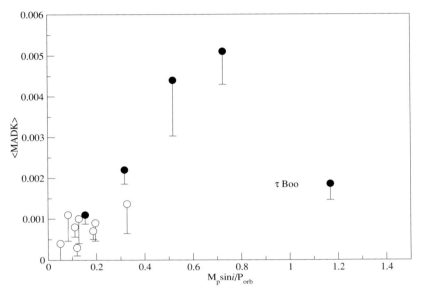

Figure 4. The ratio of the minimum planetary mass (in Jupiter masses) to the orbital period (in days) plotted against the average MAD of the K-line per observing run for all 13 stars observed. The x-axis quantifies the planet's magnetic moment assuming tidal locking, such that $P_{rot} = P_{orb}$. The filled-in circles are of stars which exhibit significant night-to-night variability in the Ca II K line: HD 73526, v And, HD 179949 and HD 189733 (this work). The tidally-locked system of τ Boo does not follow the correlation traced out by the others. The error bars are one-sided due to the positive contribution of integrated MAD immediately outside the Ca II emission core and reflect the S/N obtained for each target.

3. Summary

For our prime target, HD 179949, we now have a total of six observing runs spanning 5 years. During four runs (Aug 2001, July 2002, Aug 2002 and Sept 2005) the Ca II emission varied with the orbital period of 3.092 days, with consistent amplitude and peak phase indicative of a magnetic interaction between the star and planet. The peak activity on HD 179949 in these epochs occurs at $\phi_{orb} \approx 0.8$, leading the sub-planetary longitude by some 70°. The phase lead can provide information on the field geometries (i.e. Parker spiral) and the nature of the effect such as tidal friction, magnetic drag or reconnection with off-center magnetic fields.

HD 179949 data from the other two runs (Sept 2003 and June 2006) clearly vary with the rotation period of 7 days. A similar effect is seen on v And where one of four epochs appears to be modulated by rotation rather than the planet's motion. This *on/off* behavior has been modeled by Cranmer & Saar (2007) to be an effect of magnetic reconnection with the stellar field as it varies with the star's long-term activity cycle.

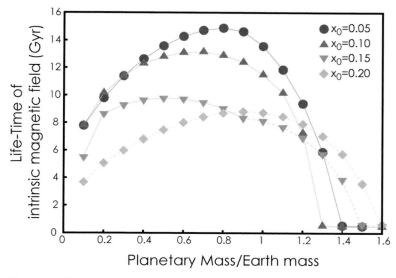

Figure 2. Life-time of intrinsic magnetic field for various sized planets.

For the case of $x_0 = 10\mathrm{wt}\%$, an inner core forms at 2.2Gyr. Since then reduction rate of core heat flux becomes moderate because of gravitational energy and latent heat released at the surface of inner core. These results indicate that the inner core nucleation enhance the subsequent heat flux from the core. Core heat flux exceeds the minimum flux to drive the convection within the outer core until about 12.0 Gyr, i.e. planet can drive dynamo action and can sustain its intrinsic magnetic field for 12.0 Gyr.

We carry out the numerical simulation for various sized planets to estimate the life-time of magnetic field as a function of planetary mass. Figure 2 shows the relationship between life-time of the intrinsic magnetic field and planetary mass. Each curve represent the life-time of intrinsic magnetic field for different initial impurity concentration in the core. Generally, these curves show convex upward configuration. Note that solid curve shows the cases with inner core formation and dashed curve shows the cases without inner core formation.

When planetary mass is less than 0.5 Earth mass, life-time of the field becomes shorter for smaller planet. This is because of the smaller planet cools down more quickly and thermal energy of the core decrease within short period. On the other hand, when planetary mass is more than Earth-mass, life time of the field becomes shorter with larger planet. This is because of depression of heat transport efficiency. Since the viscosity of the mantle depends on the pressure, the viscosity at the bottom of the mantle becomes higher for heavier planet. In this case such a viscous mantle can not sustain large heat flux from the core to invoke vigorous convection within outer core while the core remains hot.

According to our numerical results, the planetary mass must be between 0.1 and 1.4 Earth mass to sustain the intrinsic magnetic field for more than 4.5Gyr. If existence of intrinsic magnetic field were a key factor to make the planet habitable, the mass range above indicates that super-Earths would not be habitable.

References

Alfé, D., Price, G. D., & Gillan, M. J. 2002, *Phys. Rev.* B 65 (16),165118
Ida, S. & Lin, D. N. C. 2004, *ApJ*, 604, 388
Ida, S. & Lin, D. N. C. 2004, *ApJ*, 616, 567
Mayer, M. & Queloz, D. 1995, *Nature*, 378, 355
Sasaki, S., & Nakazawa, K. 1986, *J, Geophys. Res.* 91, 9231
Stevenson, D. J., Spohn, T., & Schubert G. 1983, *Icarus* 54 (3), 466
Stixrude, L., & Lithgow-Bertelloni, C. 2005, *Geophys. J. Int.* 162, 610
Valencia, D., OConnell, R. J., & Sasselov, D. 2006, *Icarus* 181, 545
Yukutake, T. 2000, *Phys. Earth Planet. Inter.* 121, 103

On uncertainty of Jupiter's core mass due to observational errors

Yasunori Hori[1], Takayoshi Sano[2], Masahiro Ikoma[1] and Shigeru Ida[1]

[1]Department of Earth and Planetary Sciences, Tokyo Institute of Technology,
Ookayama, Meguro-ku, Tokyo 152-8551, Japan
email: hori@geo.titech.ac.jp

[2]Institute of Laser Engineering, Osaka University,
Yamadaoka, Suita, Osaka 565-0871, Japan

Abstract. The origins of extrasolar gas giant planets have been discussed, based on our understanding of the gas giant planets in the solar system, Jupiter and Saturn. However, how Jupiter and Saturn formed is still uncertain because of the uncertainty in their interiors, especially the core mass (M_c). The uncertainty in M_c is partly due to those in observational data such as gravitational moments (J_{2n}), equatorial radius (R_{eq}) and 1-bar temperatures (T_{1bar}). New frontiers mission to Jupiter by NASA (JUNO) launched in 2011 is expected to reduce the observational errors. However, it is not necessarily clear yet which observational uncertainty dominates and how accurate observation is needed to constrain M_c enough to know the origin of Jupiter. Thus, modeling the interior of Jupiter, we evaluate each effect on M_c and required precision. We have found that the observational error of 5% in T_{1bar} yields an error of several $M_\oplus$ in M_c. We have also found that the values of J_6 of our successful models are confined in a narrow range compared to its observational error. This implies that comparison between the values of J_6 of our successful models and the J_6 value obtained from JUNO mission helps us to know whether the present theoretical model is valid.

Keywords. gas giant planet, interior structure, core mass, gravitational moment

1. Introduction

To date, more than 250 exoplanets have been detected (e.g., Butler *et al.* 2006). The estimated planetary masses suggest that at least 90% of them should be gas giant planets. The interior structure, formation and evolution of those planets have been discussed, based on our knowledge of the gas giant giants, Jupiter and Saturn, in our solar system.

However, the interior structures of Jupiter and Saturn, especially the masses of their cores (M_c), are still uncertain; thus it is not yet possible to determine how they formed. The interior structure of the gas giant planets is determined by finding theoretically proper models that satisfy observational constraints such as equatorial radii (R_{eq}), 1bar-temperatures (T_{1bar}) and even gravitational moments (J_{2n}). Errors of the observational data and uncertainty in equation of state (EOS) for hydrogen and helium result in the uncertainty in M_c: Estimated core masses of Jupiter and Saturn are 0 to $14M_\oplus$ and 15 to $25M_\oplus$, respectively (Saumon & Guillot 2004). On the other hand, the core accretion model, which is a widespread scenario for gas giant formation, requires cores larger than about $10M_\oplus$ as seeds to form gas giant planets (e.g., Pollack *et al.* 1996). Thus, the prediction by theories of the interior structure, especially of Jupiter, is not consistent with that by formation theories. We need to limit the possible range of M_c more tightly by reducing the uncertainties in EOS and observational data.

Fortunately, more precise observational constraints will be obtained in the near future by JUNO launched in 2011, while it might not be easy to reduce the uncertainties in EOS soon. However, it is not necessarily clear in the literature which observational uncertainty dominates and how accurate observation is required to constrain M_c more tightly. For example, the results by Saumon & Guillot (2004) include all of the uncertain factors. Therefore, in this study, we evaluate each effect on M_c of Jupiter and required observation precision by focusing on errors of the observational data.

2. Modeling

We model the static structure of gas giant planets in a similar way to previous studies (e.g., CEPAM; Guillot & Morel 1995). We assume that the interior is three-layered, namely, a icy/rocky core, metallic-hydrogen and molecular-hydrogen layers outward; each layer is chemically homogeneous. The interior is assumed to be fully convective (i.e. isoentropic). We also assume that the planetary rotation is rigid and axisymmetrical. The equations of state used in this study are Saumon & Chabrier EOS (Saumon et al. 1995) for hydrogen and helium and Hubbard & Marley's (1989) EOS for ice and rock.

The input parameters and their ranges of values are listed in Table 1. Those include 1-bar temperature (T_{1bar}), the mass fraction of elements heavier than hydrogen in the molecular-hydrogen layer (Y_z^{mole}), and the ratio of ice to rock in the core (f_{ice}). The range of T_{1bar} corresponds to the current observational errors (see Table 2). We assume large Y_z^{mole} because in situ measurements by the Galileo probe suggest the heavy elements are enriched in the atmosphere of Jupiter compared to the solar abundance. In our calculations, f_{ice} is fixed at 0.5, because the effects of observational errors are of special interest.

For a given set of values of the parameters, we integrate the interior structure. Our successful models are such that the values of R_{eq}, J_2, J_4 and J_6 are within their observational errors given on Table 2. Changing values of M_c and the fraction of heavy elements (including helium) in the metallic layer (Y_z^{metal}), we look for successful models in an iterative way.

Table 1. Input parameters (Guillot 1999)

T_{1bar} [K]	Y_z^{mole}	f_{ice}
160 - 170	0.345 - 0.380	0.5

Table 2. Observational data (Guillot 2005)

	T_{1bar} [K]	$J_2 \times 10^2$	$J_4 \times 10^4$	$J_6 \times 10^4$	$R_{eq} \times 10^9$ [cm]
Jupiter	165 ± 5	1.4697 ± 0.0001	-5.84 ± 0.05	0.31 ± 0.20	7.1492 ± 0.0006
Saturn	135 ± 5	1.6332 ± 0.0010	-9.19 ± 0.40	1.04 ± 0.50	5.8210 ± 0.0006

3. Results

Figure 1 plots M_c and the total mass of heavy elements (i.e., elements other than hydrogen and helium) contained in the envelope (M_z) of successful interior models. We calculated M_z by using Y_z^{mole} and Y_z^{metal} under the constraint that the total mass fraction of helium of Jupiter must equal to that of the protosolar disk. The domain enclosed with dotted lines in Fig.1 contains all the successful models. Each small box

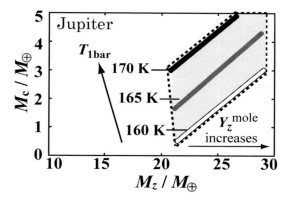

Figure 1. Possible ranges of the core mass (M_c) and the total mass of heavy elements contained in the envelope (M_z) of Jupiter that match the observed values, R_{eq}, J_2, J_4 and J_6 within 1σ.

corresponds to the case of a fixed value of T_{1bar}; the black, gray, and white boxes represent results for 170 K, 165 K, and 160 K respectively. The left end of each box corresponds to the result for $Y_z^{mole} = 0.345$ and Y_z^{mole} increases from left to right. Note that all the core masses of our successful models are less than 10 $M_\oplus$, which do not always mean that M_c of Jupiter could be less than $10 M_\oplus$. Those results arise because our calculations adopt only specific EOS models and a limited range of Y_z^{mole}. That is enough to see the sensitivity of M_c to the observational constraints.

In Fig. 1, one finds that the error of 5% in yields an error of several $M_\oplus$ in M_c. Higher T_{1bar} results in larger M_c because the entropy of the interior is also high, which causes a decrease in the interior density. This reduces the degree of concentration near the deep part. As a result, higher T_{1bar} eventually yields larger M_c to satisfy the constraint of J_2, namely, the inertia moment. The difference in M_z caused by that in T_{1bar} is so small that it would be of little significance from the viewpoint of planet formation.

The uncertainty in Y_z^{mole} also yields an error of a few $M_\oplus$ in M_c. An increase in Y_z^{mole} from 0.345 to 0.380 is found to result in a increase in M_z by about $10 M_\oplus$. Then, larger M_z reduces the degree of concentration near the deep part like the case of higher T_{1bar}. Larger M_c is thus required to satisfy the constraint of J_2.

Finally, we have found two implications concerning J_4 and J_6 for the interior structure of Jupiter. The uncertainty in M_c for fixed values of T_{1bar} and Y_z^{mole} is due to those in observational constraints, especially J_4 and J_6. That uncertainty is small compared to the one due to the errors of T_{1bar} and Y_z^{mole}. Our calculations suggest that the interior models that match the observed J_4 require a little difference between Y_z^{mole} and Y_z^{metal}. Also, we have found that the values of J_6 in our successful models are in a narrow area from 0.32×10^{-4} to 0.37×10^{-4} compared to the observational error of J_6.

4. Conclusions

We have realized that the error in T_{1bar} should be responsible for the uncertainty in M_c mostly. To avoid the uncertainty of several or a few Earth-mass in M_c, the error in T_{1bar} is reduced by less than 1% at least.

Furthermore, the importance of the uncertainty in T_{1bar} suggests that the thermal structure must be investigated in more detail. Although our calculations assumed that the interior is fully convective, there exists one possibility that the interior is super-adiabatic. In fact, Guillot et al. (2004) estimated temperature gradient based on the mixing theory and found that this possibility may be high. The resultant hot interior would yield larger M_c like in the case of higher $T_{1\text{bar}}$, which has the potential to solve the formation problem of Jupiter, i.e. the problem of the possibility that Jupiter may have a small core or be a coreless planet.

We have also reconfirmed the need for accurate data for J_4 and J_6. Our successful models prefer a little difference between Y_z^{mole} and Y_z^{metal} to satisfy the J_4 constraint. This suggests that J_4 is a key to know the composition of the metallic layer (i.e. Y_z^{metal}) that probes are unable to reach. In other words, if we obtain the accurate J_4 and Y_z^{mole} of Jupiter by in situ measurements such as JUNO, we will know the content of heavy elements of Jupiter. We also found that the resultant J_6 values are confined in a narrow range compared to its observational error. This implies that whether the J_6 value obtained by JUNO is in the narrow range predicted by interior models or not helps us to confirm the validity of the present interior model of Jupiter. Finally, the error of the observed J_6 is required to be within the narrow range at least.

References

Butler R. P., Wright, J. T., Marcy, G. W., Fischer, D. A., Vogt, S. S., Tinney, C. G., Jones, H. R. A., Carter, B. D., Johnson, J. A., McCarthy, C. & Penny, A. J. 2006, *ApJ*, 646, 505
Guillot, T. 1999, *Plan. Space. Sci.* 47, 1183
Guillot, T. 2005, *Ann. Rev. Earth Planet. Sci.*, 33, 493
Guillot, T. & Morel, P. 1995, *ApJS*, 109, 109
Guillot, T., Stevenson, D. J., Hubbard, W. B. & Saumon, D. 2004, F. Bagenal *et al.* ed., Cambridge University Press
Hubbard, W. B. & Marley, M. S. 1989, *Icarus*, 78, 102
Pollack, J. B., Hubickyj, O., Bodenheimer, P., Lissauer, J. J., Podolak, M. & Greenzweig, Y. 1996, *Icarus*, 124, 62
Saumon, D., Chabrier, G., & Van Horn, H. M. 1995, *ApJS*, 99, 713
Saumon, D. & Guillot, T. 2004, *ApJ*, 609, 1170

Silicate, ruby, opal – Why gas giants keep their jewels in the atmosphere

Christiane Helling

[1]SUPA, School of Physics & Astronomy, University of St Andrews, North Haugh, St Andrews, KY16 9SS, Scotland, UK
email: Christiane.Helling@st-andrews.ac.uk

Abstract. Giant gas-planets - and brown dwarfs - form dust clouds in their atmospheres which are made of a variety of gemstone-like and possible liquid materials. Our theoretical approach, where we calculate homogeneous nucleation, heterogeneous growth/evaporation, gravitational settling, and element consumption for composite dust grains, allows to access the evolution of the dust complex in the cloud, and hence also the elements remaining in the gas phase. The cloud formation process is imprinted into these remaining elements. Following a (T, p) trajectory into the atmosphere we observe that 1. metals disappear, 2. dust forms, 3. metals re-appear, 4. dust disappears. For the first time, our kinetic cloud formation approach is coupled with an 1D atmosphere simulation and, hence, synthetic spectra can be produced based on detailed cloud micro-physics. Results are demonstrated for metal-poor gas giants and the strong influence of the dust modelling on alkali-line profile is shown.

Keywords. astrochemistry, radiative transfer, methods: numerical, stars: atmospheres, stars: low-mass, brown dwarfs, planets: atmospheres

1. Introduction

Cloud formation has a distinct influence on the atmospheres of substellar objects like giant gas planets and brown dwarfs, which in fact, share the basic mechanisms of cloud formation and micro-physics. Cloud *formation* is chemically (and physically) a kinetic process where a first surface has to form out of a gas phase (*seed formation* rate J_* Fig. 1)†. This seed formation requires a considerable super-cooling, and a variety of other materials is already thermally stable at these temperatures. The seeds provide the surface where chemical surface reactions can grow a grain mantle made of these already thermally stable compounds, and the size of the grain increases rapidly due to gas-grain surface reactions. ($\langle a \rangle$ in Fig. 1). The bigger the grains, the faster they fall ($\langle v_{\rm dr} \rangle$ in Fig. 1) until the grain materials become thermally unstable at higher temperatures inside the atmosphere. The grain size decreases now due to evaporation and elements previously locked into the grains at higher altitudes are set free in deeper atmospheric layers. Convection is acting as transport mechanism on a certain time scale ($\tau_{\rm mix}$ in Fig. 1) to bring up fresh material which keeps the cycle of dust formation running, producing the cloud in a substellar atmosphere.

† The seed formation process can be neglected in terrestrial planets since here wind sweeps up dust from the ground into the atmosphere. However, the actual rate, which determines the size and the number of the droplets, is still debated.

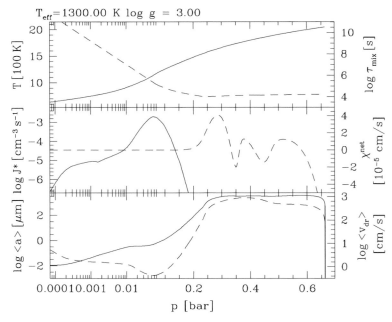

Figure 1. Dust cloud structure in a giant gas planet of $T_{\rm eff} = 1300$K, logg=3.0, [M/H]=0.0. **Top panel:** Prescribed $(T, p, \tau_{\rm mix})$-profile, **Middle panel:** nucleation rate J_* (left), net growth velocity $\chi^{\rm net}$ (right), **Bottom panel:** mean grain size $\langle a \rangle$ (left), drift velocity $\langle v_{\rm dr} \rangle$ (right).

2. Dust model

We model homogeneous seed formation (nucleation), heterogeneous growth, evaporation, and drift (gravitational settling) of dirty dust particles in a quasi-static atmosphere by using the moment method (Gail & Sedlmayr 1988; Dominik et al 1993; Woitke & Helling 2003, 2004; Helling & Woitke 2006; Helling, Woitke & Thi 2008). We consider the formation of compact spherical grains out of an oxygen-rich gas by the initial nucleation of TiO_2 seed particles, followed by the growth of a dirty mantle. The moment and elemental conservation equations are evaluated for given $(T, \rho, v_{\rm conv})$ either for a prescribed static model atmosphere structure (Helling, Woitke & Thi 2008) or inside a iterative solution of the radiative transfer problem (Helling et al. 2008, Dehn et al. 2007). Our dust model calculates the amount of condensates, the mean grain size $\langle a \rangle$, the parameterised grain size distribution function, and the volume fractions V_s of each materials as a function of height z in the atmosphere. 12 solids ($MgSiO_3$[s], Mg_2SiO_4[s], MgO[s], SiO_2[s], SiO[s], Fe[s], FeO[s], FeS[s], Fe_2O_3[s], Al_2O_3[s], $CaTiO_3$[s], TiO_2[s]) made of 8 elements (Ti, Si, Fe, Al, Mg, O, S, Ca) are considered to form the grain mantle by 60 chemical surface reactions. We solve 19 stiff differential equations already in the dust and element conservation complex. All details of the dust model can be found in the above references. Note that the equation used here represent the stationary dust formation case omitting every time-dependence and, hence, can be applied in the framework of classical stellar atmosphere simulations. The time-dependent version of our dust models has by now been tested in turbulence calculations of brown dwarf atmospheres (Helling & Woitke 2004, Helling 2005) and in winds of asymptotic giant stars (Woitke 2006).

spectra by modelling cloud characteristics like the number of dust particles, the material composition of the cloud grains/droplets, material composition, and the size of the individual cloud particles. In general, formation of clouds by condensation would be controlled by kinetic factors, for example induced by turbulence, and amorphous solids would form. Given sufficient time and energy, these amorphous solids will crystallise and form equilibrium solids known as minerals. Phase-equilibrium is adopted in the approaches of Ackerman & Marley (2001), Tsuji (2005), Allard et al. (2003, 2007), and the kinetic approach is taken by Woitke & Helling (2004) and Helling, Woitke & Thi (2008). The dust cloud models involved in the test calculations are summarised by the following simplified description:

Fixed grain size (Tsuji 2005;
– grain size a = const also Burrows et al.)
– cloud layer between T_{cond} and T_{crit}
– parameter: T_{crit}

Time scale comparison (Allard et al. 2003, 2007)
– $\tau_{growth} > \tau_{mix}$ ⇒ grain size obtained from $\tau_{growth} = \tau_{mix}$
– $\tau_{growth} < \tau_{mix}$ ⇒ number density of grains obtained from $\tau_{growth} = \tau_{mix}$
– convective advection transported into radiative layers according to Ludwig et al. (2002)

Diffusive transport (Ackerman & Marley 2001,
– static solution of diffusion equation Marley et al. 2007)
– constant mixing efficiency, homogeneous cloud particles
– parameter: settling efficiency f_{sed}

Top - Down approach (Woitke & Helling 2003, Helling & Woitke 2006,
– quasi-static solution of stationary Helling, Woitke, Thi 2008)
moment methods for nucleation, dirty growth/evaporation and element conservation including gravtiational settling
– convective over-shooting parameterized according to Ludwig et al. (2002)

We designed test cases to study difference arising from our different theoretical approach to model clouds in substellar atmosphere simulations. We disentangle our individual cloud models from the complete atmosphere problem (**Test case 1**) in order to exclude possible feedback amplifications in the entire model atmosphere codes. Only as a second step, we investigate possible differences in the spectral appearance of a given substellar object (**Test case 2**). The test cases are designed as follows and more details can be found under http://phoenix.hs.uni-hamburg.de/BrownDwarfsToPlanets1/:

Test case 1:
– compare results from cloud models only
– *local* quantities given : (T, p ,v_{conv}) structure
– different dust cloud treatment

Test case 2:
– compare results from complete 1D atmosphere simulation
– *global* quantities given: T_{eff} = 1800K, log g = 5.
– different model atmosphere codes including different dust cloud treatment

3. Results

We perform our tests for brown dwarfs of spectral class L and T. Figure 1 shows the dust content (dust-to-gas ratio ρ_d/ρ_g) of the clouds calculated by different model approaches (Sect. 2). We observe similarities amongst the different cloud model results like the location of the inner cloud edge and the approximate location of the maximum dust content inside the cloud. However, differences are apparent like e.g. the pressure range covered by the clouds. We find the same behaviour for other opacity relevant cloud quantities like grain size and cloud material composition. It is therefore not surprising that the resulting spectra (Fig. 2) show a very similar general behaviour, like the absolute flux level, but vary in details. Figure 2 also demonstrates that the model atmosphere simulations for $T_{\rm eff} = 1800$K, $\log g = 5.0$, [M/H]=0.0 fall into two groups: those producing a thin cloud layer (Tsuji with $T_{\rm crit}$=1900K, Homeier & Allard) and those producing a thick cloud layer (Tsuji with $T_{\rm crit}$=1700K, Marley & Ackerman, Dehn/Helling&Woitke (Dehn 2007)). It is interesting to realise that two entirely different mechanisms are responsible to move from one group to the other: Tsuji varies the geometrical cloud thickness by using a critical temperature $T_{\rm crit}$ for a constant grain size which results in an optically thinner/thicker cloud. The cloud models employed by Homeier & Allard, Marley & Ackerman, Dehn/Helling&Woitke produce different amounts of dust (Fig. 1) and different grain size distributions across the cloud height causing an optically thiner/thicker cloud (not shown).

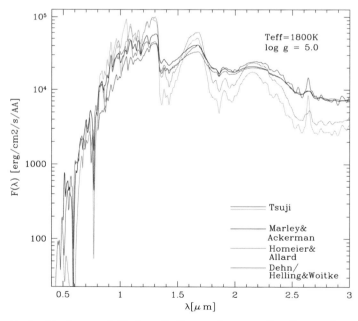

Figure 2. Results for **Test case 2** (L dwarf): Spectral energy distribution between 0.5–3.0μm from different atmosphere codes utilising different cloud-model approaches for ($T_{\rm eff} = 1800$K, $\log g = 5.0$, [M/H]=0.0). Note, that the two Tsuji-cases demonstrate the influence of the cloud thickness: $T_{\rm crit}$=1700K (brown) – thick cloud, $T_{\rm crit}$=1900K (orange) – thin cloud.

4. Conclusion

Our test case studies show that the results of our individual cloud models are comparable regarding general feature like the location of cloud base and the maximum dust-to-gas ratio. However, the cloud-model results differ if studied in more detail. It is therefore no surprise that the spectral energy distributions produced from our different 1D atmosphere simulations for a given parameter combination ($T_{\rm eff}$, logg, [M/H]) differ in almost the entire wavelength range. It remains to quantify these differences in order to provide a general range of applicability for e.g. stellar parameter determinations inside and beyond the substellar regime.

Acknowledgement:

We thank all the participants of the workshop **From Brown Dwarfs to Planets: Chemistry and Cloud formation** which was supported by the Lorentz Center of the University Leiden, Nederlandse Organisatie voor Wetenschappelijk Onderzoek, The Netherlands research School for Astronomy, the Scottish University Physics Alliance, and European Space Agency. ChH acknowledges an IAU travel grant.

References

Ackerman, A. & Marly, M. 2001, ApJ 556, 872
Allard, F., Allard, N. F., Homeier, D. et al. 2007, A&A 474, L21
Allard, F., Guillot, T., Ludwig, H.-G., Hauschildt, P. H. et al. 2003, IAU 211, 325
Dehn, M. 2007, PhD Thesis, University Hamburg
Helling, Ch., Woitke, P., Thi, W.-F. 2008, submitted
Helling, Ch., Woitke, P. 2006, A&A 455, 325
Ludwig, H.-G., Allard, F., Hauschildt, P.H. 2002, A&A 395, 99
Sudarsky D., Burrows A., Hubeny I. 2003, ApJ 588, 1121
Tsuji, T. 2005, ApJ 621, 1033
Woitke P., Helling Ch. 2004, A&A 414, 335
Woitke P., Helling Ch. 2003, A&A 388, 297

Tidal friction in close-in planets

Adrián Rodríguez[1], Sylvio Ferraz-Mello[1] and Hauke Hussmann[2]

[1] Instituto de Astronomia, Geofísica e Ciências Atmosféricas, University of São Paulo
Rua do Matão, 1226, CEP 05508-900, São Paulo, Brasil
email: sylvio@astro.iag.usp.br, adrian@astro.iag.usp.br

[2] Institut für Planetenforschung, DLR, Berlim-Adlershof, Germany
email: hauke.hussmann@dlr.de

Abstract. We use Darwin's theory (Darwin, 1880) to derive the main results on the orbital and rotational evolution of a close-in companion (exoplanet or planetary satellite) due to tidal friction. The given results do not depend on any assumption linking the tidal lags to the frequencies of the corresponding tide harmonics (except that equal frequency harmonics are assumed to span equal lags). Emphasis is given to the study of the synchronization of the planetary rotation in the two possible final states for a non-zero eccentricity : (1) the super-synchronous stationary rotation resulting from the vanishing of the average tidal torque; (2) the capture into a 1:1 spin-orbit resonance (true synchronization), which is only possible if an additional torque exists acting in opposition to the tidal torque. Results are given under the assumption that this additional torque is produced by a non-tidal permanent equatorial asymmetry of the planet. The indirect tidal effects and some non-tidal effects due to that asymmetry are considered. For sake of comparison with other works, the results obtained when tidal lags are assumed proportional to the corresponding tidal wave frequencies are also given.

Keywords. Tidal friction, Close-in exoplanets, Synchronization, Stationary rotation, Energy dissipation, Hot Jupiters

---

1. Introduction

It is well-known that tidal friction tends to synchronize the rotation period of a close-in companion (exoplanet or planetary satellite) with the period of its orbital motion around the primary. However, it is not always taken into account the fact that there are two possible final states: (1) the super-synchronous stationary rotation resulting from the vanishing of the average tidal torque (Goldreich, 1966); (2) the capture into a spin-orbit resonance (generally the synchronous one) as a result of the vanishing of the average sum of two torques, the tidal torque and an additional torque acting in opposition to it. The selection of one or another of the two final states is determined by the orbit eccentricity and by the existence or not of a permanent non-tidal equatorial asymmetry of the close-in planet. In a 'cold' solid quasi-elastic body, the existence of one important equatorial asymmetry is not superseded by the tidal deformations and the final state is the capture into a spin-orbit resonance which may be 1:1, as in the case of the Moon, or a different one, as in the case of Mercury (captured into a 3:2 spin-orbit resonance; see Lemaitre *et al.*, 2006). Otherwise, the tidal torque dominates and the final state is non-synchronous, at least while the orbit is not circularized. This is expected to be the case for hot Jupiters, since the giant planets in our Solar System do not show any measurable equatorial asymmetry. In this case, one classical result is that the system tends to a final state whose rotation speed is given by $\Omega_{\rm stat} = n(1 + \kappa\, e^2)$ where n and e are the orbital mean motion and eccentricity and κ is a positive constant whose actual value depends on the hypotheses done on the tidal phase lag. When it is assumed that each harmonic of the tide has a phase lag proportional to its frequency, we obtain $\kappa = 6$ (see Eq. 2.4); when the lag is assumed to be constant and frequency independent (MacDonald, 1964), the value corresponding to the vanishing of the average torque is $\kappa = 9.5$ (Goldreich, 1966). If, however, we just assume that equal frequencies give rise to equal lags, without assuming any particular law relating lags to frequencies, the result is expressed in a different way: $\varepsilon_0 = 12\,e^2 \varepsilon_2$ where ε_0 and ε_2 are the phase lags of the tide harmonics whose frequencies are $2\Omega - 2n$ and $2\Omega - n$ respectively (Ω is the planet's angular velocity of rotation).

2. Planetary Rotation

To study the planet rotation, we have to consider the torque $\mathcal{M}$ resulting from the misalignment of the tidal bulge created by the delayed response of the planet to the tide raising forces and use the angular momentum conservation. Initially, it is enough to consider only the interaction between the orbit and the tidally deformed planet: the change $\dot{\mathcal{L}}$ of angular momentum in the orbit may be compensated by a change $\dot{\mathcal{L}}_{\rm rot}$ in the angular momentum acting on the planet in the opposite direction. Hence $\dot{\mathcal{L}}_{\rm rot} = -\mathcal{M}$. The study of this equation is simplified by the fact that the torque $\mathcal{M}$ is, in this case, normal to the line of nodes ($M_x = 0$) thus allowing this equation to be decomposed into two parts which may be studied separately. We remind that the rotational angular momentum of the planet is given by $\mathcal{L}_{\rm rot} \simeq C\Omega \mathbf{k}$ where C is the moment of inertia of the planet with respect to the rotation axis and $\mathbf{k}$ is a unit vector along that axis. We obtain $C\dot{\Omega} = -M_z$ and $C\Omega\dot{J} = -M_y$, where M_y, M_z are the components of $\mathcal{M}$ and $\dot{J}$ is a variation in the obliquity of the planet. Computing the tidal forces and torques and averaging the torque over the orbital period, we obtain

$$<\dot{\Omega}> = -\frac{3k_d GM^2 R^5}{2Ca^6}\left[\varepsilon_0 - \xi(e^2, S^2)\right] \tag{2.1}$$

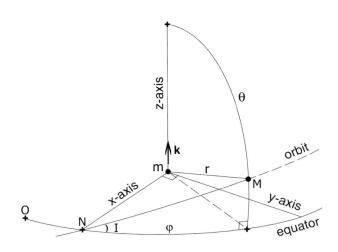

Figure 1. Spherical coordinates and reference system

where k_d is the dynamical Love number, G the gravitation constant, M the mass of the tide raising body (the star), R the mean radius of the planet, a the orbit semi-major axis and $\xi(e^2, S^2)$ a function whose terms are proportional to e^2 and S^2 (S is the sine of the obliquity) whose coefficients depend on the lags of the several harmonics in which the tide is decomposed. (For details, see Ferraz-Mello et al., 2007). Near synchronization,

$$\xi(e^2, S^2) = e^2(5\varepsilon_0 + 12\varepsilon_2) + S^2\varepsilon_0. \tag{2.2}$$

If we assume that the planet has no other significant deformation besides the tidal one, Eqn. (2.1) shows that, in the first approximation, the averaged acceleration $<\dot\Omega>$ is proportional to $|\varepsilon_0|$ with a sign contrary to $\Omega - n$, indicating that the system will evolve towards synchronization.

We say that the system reaches a state of stationary rotation when the average angular acceleration (or the average tidal torque component $\langle M_z\rangle$) vanishes. Solving the equation $<\dot\Omega> = 0$, we obtain, to the second order in e, S,

$$\varepsilon_0 = 12e^2\varepsilon_2. \tag{2.3}$$

In Darwin's theory, as well as in many theories inspired by it, the tidal lags ε_0 and ε_2 are assumed to be proportional to the frequencies of the corresponding tide harmonics ($2\Omega - 2n$ and $2\Omega - n$, respectively). With this assumption, the above result becomes

$$\Omega_{\text{stat}} = n(1 + 6e^2) \tag{2.4}$$

(see eqn. 6.1), which means that, when $e \neq 0$, the rotation stabilizes at a slightly super-synchronous value.

The second equation resulting from the angular momentum conservation gives the variation of the planet's obliquity:

$$<\dot J> = -\frac{3k_d S G M^2 R^5}{2C\Omega a^6}\varepsilon_2. \tag{2.5}$$

The part of the equation corresponding to the vanishing of the torque along the nodal line indicates that no precession of the nodes occurs due to tidal friction.

3. Spin-orbit synchronization

The spin-orbit synchronization condition is given by $\Omega = n$ or, if we assume the lags proportional to the frequencies of the corresponding harmonics, $\varepsilon_0 = 0$, in which case eqns. (2.1) and (2.2) give

$$<\dot\Omega> = \frac{18k_d G M^2 R^5}{C a^6} e^2 \varepsilon_2, \tag{3.1}$$

which cannot vanish if $e \neq 0$. It is worth noting that S does not affect the above condition (at least when terms of the fourth order in e, S are neglected).

3.1. Synchronous Asymmetric Planets

The previous conclusions were derived from the assumption that the planet has no other significant deformation besides the tidal one. If, for instance, it has a permanent non-tidal (solid-like) equatorial ellipticity, different results follow. Indeed, in such case, we have to add to the potential due to the tide, the non-tidal quadrupolar term associated with the equatorial ellipticity:

$$U_{22} = \frac{GmR^2}{r^3} J_{22} P_{22}(\cos\theta) \cos 2(\varphi - \varphi_{22}) \qquad (3.2)$$

where m is the mass of the planet, r, θ, φ are the radius vector, co-latitude and longitude of the star in the planetocentric frame (see Fig. 1) and P_{22} is an associated Legendre function. J_{22} and φ_{22} are the two parameters characterizing the ellipticity of the equator (see Beutler, 2005). If J_{22} is large enough, the torque due to U_{22} will be dominant and the synchronization will end with capture into an exact 1:1 spin-orbit resonance (Goldreich, 1966). In this case, we have, instead of eqn. (2.3),

$$\varepsilon_0 = 0, \qquad (3.3)$$

and the equations obtained in the previous section cannot be used since the torque forcing the motion to be synchronous must also be considered. Since the tidal friction in a synchronous companion moving in an eccentric orbit tends to accelerate its motion, the body will rotate faster, but, being asymmetric, the advance with respect to the symmetric configuration will create a torque in the contrary direction, which will compensate the tidal torque. The equilibrium is obtained with the body displaced of an angle δ forward. The average torque due to this displacement is

$$<(M_{22})_z> = \frac{6GMmR^2 J_{22} \sin 2\delta}{a^3} \left(1 - \frac{5}{2}e^2 - \frac{1}{2}S^2\right). \qquad (3.4)$$

The condition $<\dot{\Omega}> = 0$ allows the offset angle δ to be determined. At the order of approximation adopted in this paper,

$$\sin 2\delta \simeq \frac{3MR^3}{mJ_{22}a^3} k_d \varepsilon_2 e^2. \qquad (3.5)$$

This equation shows that when J_{22} is too small, the right-hand side is large and the synchronization cannot be reached (δ becomes larger than the critical limit allowing synchronization)†. Nevertheless, the concomitant process of circularization of the orbit (see Sec. 5) will continue and, at some moment in the evolution, e will become enough small to allow capture into a 1:1 spin-orbit resonance.

The torque due to U_{22} also contributes with a precession of the equatorial plane of the synchronous companion. However, at the order of approximation adopted here, the only effects do not depend on any misalignment $\delta \neq 0$ and are thus not of tidal origin. They are due to the figure of the companion captured into a 1:1 spin-orbit resonance.

4. Work done by the tidal force and energy release

The work done by the tidal forces in a displacement ds is given by $dW = \mathbf{F} \cdot d\mathbf{s}$, or $\dot{W} = \mathbf{F} \cdot \mathbf{v}$ where $\mathbf{v}$ is the velocity vector. In the case of close-in planets in stationary rotation, the introduction of the condition given by eqn. (2.3) gives

$$<\dot{W}> = -\frac{3nk_d GM^2 R^5}{2a^6}(7e^2 + S^2)\varepsilon_2. \qquad (4.1)$$

It is worth emphasizing that this result does not depend on any hypothesis linking lags to frequencies (the only assumptions are that equal frequencies lead to equal lags).

† When the tidal phase lag is assumed to be frequency independent (MacDonald theory), the resulting critical value of J_{22} is proportional to e^4 (Goldreich, 1966) instead of e^2.

If the same calculations were done using $\varepsilon_0 = 0$, instead of the condition given by eqn (2.3), we would obtain for the work the value

$$-\frac{3nk_d GM^2 R^5}{2a^6}(19e^2 + S^2)\varepsilon_2. \tag{4.2}$$

However, the latest value is not the net work done because a synchronous rotation cannot exist without additional forces acting in the system when $e \neq 0$. As the additional forces act in opposition to the tidal forces, the work done by them has a different sign. From (3.2), we indeed obtain

$$<\dot{W}_{22}> = \frac{6GMmR^2 nJ_{22}\sin 2\delta}{a^3}\left(1 - \frac{5}{2}e^2 - \frac{1}{2}S^2\right) \simeq \frac{18nk_d GM^2 R^5}{a^6}e^2\varepsilon_2, \tag{4.3}$$

where the rightmost expression was obtained introducing the value of $\sin 2\delta$ given by eqn. (3.5).

When this term is added to the direct tidal work given by eqn. (4.2), the result becomes equal to that given by eqn. (4.1). This is the same result found by several authors (Segatz et al. 1988; Levrard, 2008; Wisdom, 2008) and may also be obtained from the results of Peale and Cassen (1978) on energy dissipation due to tidal friction. In the comparisons, the phase lag ε_2 of the tide harmonic whose frequency is $2\Omega - n$, is related to the planet's quality factor through $Q = 1/\varepsilon_2$.

4.1. Energy Dissipation

The total energy variation of the system must be equal to zero. Besides the orbital energy, whose variation is given above, we have the rotational energy of the deformed body and the thermal energy dissipated in the body. This balance equation allows us to calculate the energy dissipated in the planet due to the tides raised on it by the star.

The variation of the rotational energy is given by $C\Omega\dot{\Omega}$ and vanishes when $\dot{\Omega} = 0$. Therefore, in the studied cases, $|\dot{W}|$ may be roughly equal to the thermal energy released inside the companion (they are equal when we neglect the change in the equilibrium rotation speed due to the variation of the mean motion, which is of the order of $(R/a)^2$.)

5. Variation in the orbital elements

The variation of the mean motion, semi-major axis, eccentricity and obliquity can be known from the variations of the energy and angular momentum (see Ferraz-Mello et al., 2007) using classical two-body dynamics equations. For the mean-motion, we obtain

$$<\dot{n}> = -\frac{3n}{2a}<\dot{a}> = -\frac{3na}{GMm}<\dot{W}> \tag{5.1}$$

or

$$<\dot{n}> = \frac{9n^2 k_d M R^5}{2ma^5}(7e^2 + S^2)\varepsilon_2. \tag{5.2}$$

We note that in some papers (e.g. Mardling and Lin, 2004), the term $19\,e^2$ appears in the equation giving $<\dot{n}>$ instead of $7e^2$. That result is only partial because it does not take into account that a torque counteracting the tidal torque is necessary to allow the synchronization to be achieved.

For the eccentricity, we obtain

$$<\dot{e}> = -\frac{21 n e k_d M R^5}{2 m a^5} \varepsilon_2. \tag{5.3}$$

At the order considered here, the torque due to U_{22} does not contribute to the variation of the eccentricity. Its only contribution is a motion of the node whose secular part is

$$<\dot{ON}> = \frac{3 J_{22} n R^2}{a^2}\left(1 - 2e^2 - \frac{1}{4} S^2\right)\cos 2\delta. \tag{5.4}$$

However, this precession exists even when $\delta = 0$ and, thus, it is not an effect due to the tidal misalignment of U_{22}, but an effect due to the trapping of the planet into the 1:1 spin-orbit resonance.

The obliquity is the angle between the equator and the orbital plane and we have to consider the possible variation of both planes. The torque of the forces acting on the main body gives the variation of the orbital plane with respect to a fixed frame; its reaction, acting on the companion gives the variation of the equatorial plane. If we call I the angle fixing the orbital plane and J the angle fixing the equator (w.r.t. an inertial plane), we obtain, for close-in synchronous or near-synchronous planets: $<\dot{I}>/<\dot{J}> \propto (R/a)^2 \ll 1$. This means that we may neglect the variation of the obliquity due to the motion of the orbital plane when compared to the variation due to the motion of the equator.

6. Linear theories with a constant time lag

For the sake of comparison with theories where the phase lags ε_i are assumed to be proportional to the frequencies of the corresponding tide harmonics, with τ (time lag) as the coefficient of proportionality, we give below the resulting equations.

For the parameters related to the rotation of the close-in companion we have

$$<\dot{\Omega}> = \frac{3 n k_d G M^2 R^5}{C a^6}\left[\left(1 + \frac{27}{2} e^2 - \frac{1}{2} S^2\right) - \left(1 + \frac{15}{2} e^2 - \frac{1}{2} S^2\right)\frac{\Omega}{n}\right]\tau \tag{6.1}$$

$$<\dot{J}> = -\frac{3 n k_d G M^2 R^5}{C \Omega a^6} S\left(1 - \frac{1}{2}\frac{\Omega}{n}\right)\tau. \tag{6.2}$$

For the equations giving the variation of the orbital elements, we have

$$<\dot{n}> = -\frac{3n}{2a}<\dot{a}> = \frac{9 n^3 k_d M R^5}{m a^5}\left[(1 + 23 e^2) - \left(1 + \frac{27}{2} e^2 - \frac{1}{2} S^2\right)\frac{\Omega}{n}\right]\tau, \tag{6.3}$$

$$<\dot{e}> = -\frac{27 n^2 e k_d M R^5}{m a^5}\left(1 - \frac{11}{18}\frac{\Omega}{n}\right)\tau, \tag{6.4}$$

These equations are the same given by Hut (1981) and several papers on tidal friction on close-in exoplanets using Hut's approach (e.g. Mardling and Lin, 2004; Dobbs-Dixon et al., 2004). It is worth noting that the equations in the cited papers are expanded to a much higher order in eccentricity and that the quality factor adopted there is related to the time lag used here through $Q = 1/n\tau$.

To obtain the equations corresponding to the free stationary rotation, it is enough to replace Ω/n by $(1 + 6e^2)$.

Sometimes, the condition $\Omega = n$ is introduced to obtain the equations corresponding to the case of a planet locked in a 1:1 spin-orbit resonance. However, the results thus obtained are inconsistent. For instance, eqn. (6.1) gives, in such case,

$$<\dot{\Omega}> = \frac{18 n k_d G M^2 R^5}{C a^6} e^2 \tau, \tag{6.5}$$

that is, a non-zero acceleration (when $e \neq 0$) showing that the condition $\Omega = n$ cannot subsist. The results, obtained with such condition are only partial and to become correct need that the contributions coming from M_{22} (or from another forcing torque) be added to the right-hand sides.

7. Conclusion

This communication presents some consequences of Darwin's theory for bodily tides. As expected, the main results formally coincide with the general results given by Kaula (1964) and Hut (1981) to the order of approximation adopted here, when we do the same hypotheses on the physics of the lagging done by them. Emphasis is given in the paper to companions having reached one of the two possible end states of the planet synchronization. If tides are the only source of perturbations in the system, the perfect synchronization cannot be achieved while the orbit is not completely circularized. If a remnant eccentricity exists, the final state is rather non-synchronous and the rotation reaches a stationary state with a rotation velocity slightly larger than the orbital mean-motion. In order to have a true synchronous rotation, it is necessary to provide the system with a torque counteracting the tidal torque. This additional torque can, for instance, be due to a permanent non-tidal solid-like equatorial asymmetry of the companion. Tides drive the figure of the body to a small misalignment which creates a torque acting in opposition to the tidal torque. At the order considered here, the results for the tidal variations of the orbital elements are the same in the stationary (free) and synchronous (forced) cases but, in the synchronous case, we have some additional non-tidal effects coming from the figure of the body captured into a 1:1 spin-orbit resonance, which also affects the evolution of the system and may be considered.

Acknowledgements

The authors acknowledge the support of CNPq, DFG, FAPESPand the CAPES/SECYT cooperation programme. We thank the referee for his clever comments.

References

Darwin, G. H., 1880, *Philos. Trans.* 171, 713 (repr. Scientific Papers, Cambridge, Vol. II, 1908).
Dobbs-Dixon, I., Lin, D. N. C. and Mardling, R. A. 2004, *Astrophys. J.* 610, 464
Ferraz-Mello, S., Rodríguez, A. & Hussmann, H., 2007, *Cel. Mech. Dynam. Astron.* (submitted). ArXiv: astro-ph 0712.1156
Goldreich, P. 1966, *Astron. J.* 71, 1
Hut, P., 1981, *Astron. Astrophys.* 99, 126
Kaula, W. M. 1964, *Rev. Geophys.* 3, 661
Lemaitre, A., D'Hoedt, S., & Rambaux, N., 2006, *Cel. Mech. Dynam. Astron.* 95, 213
Levrard, B., 2008, *Icarus* (in press)
MacDonald, G. F., 1964, *Rev. Geophys.* **2**, 467.
Mardling, R. A. & Lin, D. N. C., 2004 *Astrophys. J.* 614, 955

Peale, S. J. & Cassen, P., 1978, *Icarus*, 36, 245-269.
Segatz, M., Spohn, T., Ross, M. N., & Schubert, G., 1988, *Icarus*, **75**, 187-206.
Wisdom, J., 2008, *Icarus* (in press).

Tidal evolution of close-in extra-solar planets

Brian Jackson[1], Richard Greenberg[1] and Rory Barnes[1]

[1]Lunar and Planetary Laboratory, University of Arizona,
1629 E University Blvd
Tucson, Arizona 85721-0092 USA
email: bjackson@lpl.arizona.edu

Abstract. The distribution of eccentricities e of extra-solar planets with semi-major axes $a > 0.2$ AU is very uniform, and values for e are generally large. For $a < 0.2$ AU, eccentricities are much smaller (most $e < 0.2$), a characteristic widely attributed to damping by tides after the planets formed and the protoplanetary gas disk dissipated. We have integrated the classical coupled tidal evolution equations for e and a backward in time over the estimated age of each planet, and confirmed that the distribution of initial e values of close-in planets matches that of the general population for reasonable tidal dissipation values Q, with the best fits for stellar and planetary Q being $\sim 10^{5.5}$ and $\sim 10^{6.5}$, respectively. The current small values of a were only reached gradually due to tides over the lifetimes of the planets, *i.e.*, the earlier gas disk migration did not bring all planets to their current orbits. As the orbits tidally evolved, there was substantial tidal heating within the planets. The past tidal heating of each planet may have contributed significantly to the thermal budget that governed the planet's physical properties, including its radius, which in many cases may be measured by observing transit events. Here we also compute the plausible heating histories for a few planets with anomalously large measured radii, including HD 209458 b. We show that they may have undergone substantial tidal heating during the past billion years, perhaps enough to explain their large radii. Theoretical models of exoplanet interiors and the corresponding radii should include the role of large and time-variable tidal heating. Our results may have important implications for planet formation models, physical models of "hot Jupiters", and the success of transit surveys.

Keywords. celestial mechanics, planetary systems: formation, protoplanetary disks

1. Introduction

As the number of known extra-solar planetary orbits has increased, interesting patterns are emerging. Figure 1 shows the semi-major axes a and eccentricities e. Eccentricities of extra-solar planets with $a > 0.2$ AU average 0.3 and are broadly distributed up to near 1. The distribution of eccentricities is fairly uniform over a. For example, a Kolmogorov-Smirnov (K-S) test shows that the e distribution for a between 0.2 and 1.0 AU matches that for a between 1.0 and 5.0 AU at the 96% confidence level. However, for close-in extra-solar planets (by which we mean $a < 0.2$ AU), e values are smaller, with an average of 0.09, but still large compared to our solar system. For $a < 0.2$ AU, the K-S test shows agreement at only the 0.1% level compared with planets further out.

Here we consider the conventional idea that close-in planets began with a distribution of e similar to that of planets farther out. Because the magnitude of tidal effects falls off very rapidly with increasing a, Rasio *et al.* 1996 suggested that tides could have reduced e for close-in planets and not for those farther out.

The implications of this hypothesis for planetary masses M_p, radii R_p, and heating rates has been considered in previous studies (Rasio *et al.* 1996; Trilling 2000; Bodenheimer *et al.* 2003). Several assumptions have been common. First, the tidal dissipation parameter Q_p for the planet is usually based on model-dependent estimates based on the tidal evolution of the Jovian satellite system (Yoder & Peale 1981) or on physical theories of tidal dissipation (Goldreich & Nicholson 1977; Ogilvie & Lin 2004). Second, effects of tides raised on the star are only partially taken into account. Often the orbit-circularizing effects of tides raised on the host star are ignored, although they can be important. Third, some studies (*e.g.* Ford & Rasio 2006) assumed that orbital angular momentum is conserved during tidal evolution, which is not true when tides on the star are taken into account. Fourth, previous work ignored the strong coupling of the tidal evolution of e with that of a, generally describing the circularization of an orbit in terms of an implicitly exponential "timescale". In fact, the coupling with a means that evolution is more complex than a simple exponential damping. Moreover, tidal evolution has resulted in significant inward migration of many close-in planets.

In our study, we avoid these assumptions. We use the full tidal evolution equations to determine possible past orbital change. The equations account for tides raised on the stars by the planet and on the planet by the star, each of which affects changes in both e and a. We employ conservative assumptions about M_p and R_p, to test the tidal

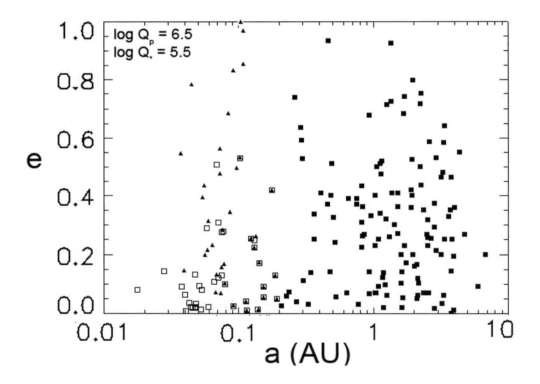

Figure 1. Distributions of orbital elements. Squares (filled and open) represent the currently observed orbital elements, with the open squares (with $a < 0.2$ AU) being candidates for significant tidal evolution. Triangles show the initial orbital elements ($e_{initial}$ and $a_{initial}$) determined by integrating the equations of tidal evolution backward in time to the formation of the planet, using Q values that gave the best fit of the e distribution to that of the other planets.

circularization hypothesis. And we consider a wide range of possible Q values, allowing us to determine which pair of Q_p and Q_* values yields the most plausible evolution history.

These calculations also yield the corresponding past tidal heating history for each planet. During the course of the tidal evolution, tidal distortion of the figure of the planet can result in substantial amounts of internal heating at the expense of orbital energy, so the heating rate as a function of time is coupled to the evolution of the orbit. In a typical case, tidal heating might have begun modest, but then increased as tides reduced a. As the tides became stronger, they would circularize the orbit, which in turn would shut down the tidal heating mechanism. The relative strength and timing of these two effects would determine a planet's history, typically with a gradual increase in the heating rate followed by a decrease.

The thermal history of a planet is critical to determining its physical properties. For example, studies of extra-solar planets have considered the effects of heating on their radii, which can be measured directly by transit observations. Heat sources that have been considered in these models include the energy of planetary accretion and radiation from the star, as well as tidal heating (Bodenheimer *et al.* 2003; Mardling 2007). In many cases the theoretical predictions match the observations reasonably well (Burrows *et al.* 2007). However, there are notable exceptions. HD 209458 b has been observed by Knutson *et al.* 2007 to have a radius of 1.32 Jupiter radii (R_J), which is 10-20% larger than predicted by theoretical modeling Guillot 2005. Similarly, HAT-P-1 b is 10-20% larger than predicted by theory (Bakos *et al.* 2007a). On the other hand, HAT-P-2 b seems to be smaller than theory would predict (Bakos *et al.* 2007b), while for GJ 436 b, theory predicts a radius consistent with observation (Gillon *et al.* 2007).

Theoretical models to date have not taken into account the history of tidal heating for close-in planets, and of course those are the planets most likely to have radii measurable by transits. The tidal heating histories reported here and in Jackson *et al.* 2008b provides motivation and a basis for construction of improved physical models.

2. Method

To test the hypothesis that tides have been responsible for reducing e, we numerically integrated the canonical tidal evolution equations of Goldreich & Soter 1966 and Kaula 1968 backwards in time for all close-in planets for which we have adequate information (see Jackson *et al.* 2008a for details). For each planet, we began the integration with the current best estimates of e and a ($e_{current}$ and $a_{current}$) and integrated backwards over the estimated age of the host star to find the orbital elements, $e_{initial}$ and $a_{initial}$. We assumed tidal evolution dominated the orbital evolution after the protoplanetary disk dissipated and collisional effects became negligible, which probably happened only a few Myr after the star's formation. Hence $e_{initial}$ and $a_{initial}$ may describe the orbits at that time. We repeated the integration for 289 combinations of Q_p and Q_*, each Q ranging from 10^4 to 10^8.

Our study involves a number of assumptions and approximations which are detailed and discussed by Jackson *et al.* 2008a. In particular, our model assumes host stars rotate much more slowly than their close-in companion planet revolves, an assumption largely corroborated by observation (Trilling 2000; Barnes 2001). As a result, tides raised on the star by the planet tend to decrease both e and a, as do tides raised on the planet. Our results will inevitably need to be revisited and refined as improved data and physical models become available. Of necessity, we considered only planets for which the reported e is non-zero, and for which there is some estimate available for the age of the system. Even with these restrictions, we can still study about 40% of all known close-in planets.

Stellar masses and radii come from a variety of sources: Bakos *et al.* 2007a; Bakos *et al.* 2007b; Da Silva *et al.* 2006; Fischer & Valenti 2005; Moutou *et al.* 2006; Saffe *et al.* 2006; Gorda & Svechnikov 1996; and Takeda *et al.* 2007. For planetary masses, we use the radial-velocity minimum mass. For planetary radii, if $M_p > 0.3$ Jupiter's mass, we fix $R_p = 1.2\ R_J$ since, for Jovian planets in this range of mass, the radius is insensitive to mass (Hubbard 1984). This value is near the average radius of almost all observed transiting extra-solar planets. For planets with $M_p <0.3$ Jupiter's mass, we assume the planet has Jupiter's density and scale R_p accordingly. This assumption agrees fairly well with the observed radius for GJ 436 b (Gillon *et al.* 2007; Deming *et al.* 2007). Where R_p and M_p have been determined directly from transit observations, we instead use those values. Planetary data were taken from a variety of sources: Bakos *et al.* 2007a; Bakos *et al.* 2007b; Butler *et al.* 2006; Da Silva *et al.* 2006; Deming *et al.* 2007; Gillon *et al.* 2007; Johnson *et al.* 2006; Knutson *et al.* 2007; Laughlin *et al.* 2005; Lovis *et al.* 2006; McArthur *et al.* 2004; Maness 2007; Mayor 2004; Moutou *et al.* 2006; Rivera *et al.* 2005; Udry *et al.* 2002; Valenti & Fischer 2005; Vogt *et al.* 2005; Wright *et al.* 2006; and Zucker *et al.* 2004.

Based upon the above model, we can also calculate the tidal heating rate for planets undergoing tidal circularization. The tidal heating of the planet results from the reduction of orbital energy (and hence a) due to energy dissipation in the planet. By tracking changes in a due to tides raised on the planet throughout the process of tidal circularization, we can estimate the tidal heating rate over the whole lifetime of the planet (Peale *et al.* 1979). The magnitude and, indeed, the very shape of the past heating curve over time depends sensitively upon the assumed current orbital elements. Accordingly, we have calculated multiple plausible heating curves for each planet, corresponding to the range of observationally allowed values for $e_{current}$ and $a_{current}$.

3. Orbital Evolution

The 289 combinations of Q_p and Q_* that we tested gave a wide variety of distributions of initial eccentricities for the close-in planets. (Remember "initial" refers to the time, shortly after formation, that a planet's orbital evolution begins to be dominated by tides.) In addition to the current orbits (squares), Figure 1 shows the computed distributions of $e_{initial}$ and $a_{initial}$ (filled triangles) for the case of $Q_p = 10^{6.5}$ and $Q_* = 10^{5.5}$. We compared the $e_{initial}$ distributions for the close-in planets with the standard e distribution observed for $a > 0.2$ AU. In this case, the agreement is excellent, with a K-S score of 90%. These Q values are well within plausible ranges. Reasonable fits can also be obtained with other values of Q_* as long as $Q_p \sim 10^{6.5}$. Other good fits (K-S $\sim$ 70%) have $Q_* = 10^{4.25}$ or $> 10^{6.75}$.

Figure 2 shows the evolutionary track of a and e over time for each of the planets, in the case of the best-fit Q values. The current orbital elements are at the lower left end of each trajectory in (a, e) space. These points correspond to empty squares in Figure 1. The tick marks show the orbital elements at intervals of 500 Myr, going back in time, for 15 Gyr, from the present toward the upper right. Black dots have been placed at a point representing the best age estimate for the planetary system. These same points appear as the triangles in Fig. 1.

The evolutionary histories derived here include substantial changes in semi-major axis coupled with the changes in eccentricity. For many close-in planets, Figures 1 and 2 show that initial a values were significantly higher than the currently observed values. These initial values of a likely represent their locations at the termination of gas disk migration in each early planetary system. Given the extent of tidal migration for observed close-in

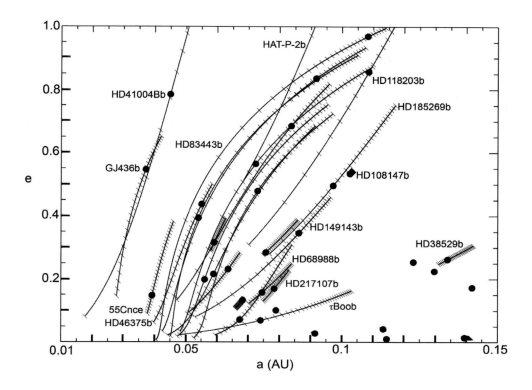

Figure 2. Tidal evolution of e and a for the sample of known close-in extra-solar planets using our best-fit values of $Q_* = 10^{5.5}$ and $Q_p = 10^{6.5}$. Solid curves represent the trajectories of orbital evolution from current orbits (lower left end of each curve) backward in time (toward the upper right). On the trajectories, tick marks are spaced every 500 Myr to indicate the rate of tidal evolution. Tidal integrations were performed for 15 Gyr for all planets, but the filled circles indicate the initial values of orbital elements at the beginning of each planet's life. Due to space restrictions, most planets are not labeled, however they can be identified by the (e, a) values at the lower left end of each trajectory. Jackson et al. 2008a includes a table of $e_{current}$ and $a_{current}$ (Table 1) which can be used to identify planets in this figure.

planets, planets that formed inward of 0.04 AU could have subsequently fallen into their host stars. Consideration of such hypothesized scenarios will be the subject of future work.

4. Tidal Heating

Using the results of our tidal evolution calculations, we modeled the past tidal heating rates for numerous planets. In Figure 3, we illustrate plausible heating histories for three interesting examples: HD 209458 b, HAT-P-1 b and GJ 436 b, using $Q_p = 10^{6.5}$ and $Q_* = 10^{5.5}$. Results for other planets are presented in Jackson et al. 2008b. Note that in these examples the planetary radii have been held constant, even though the changing heating rates would most likely produce a changing radius.

For HD 209458 b, Burrows *et al.* 2007 suggest that a heating rate of about 4×10^{19} W would be required to yield the observed planetary radius, which is much larger than any allowable current tidal heating rate. However, the history plotted in Figure 3 shows that the required heating rate was available as recently as 1 billion years ago. If the lag in the response of the planet to the heating rate were on the order of a billion years, this heating rate may explain the observed large radius. Such a lag seems reasonable based on the long duration of the influence of heat of formation on the planet's radius in the modeling by Burrows *et al.* 2007.

Like HD 209458 b, HAT-P-1 b's observed radius of 1.36 R_J (Bakos *et al.* 2007a) is larger than expected from theoretical modeling that did not include tidal heating (Guillot 2005). Similar to HD 209458 b, Figure 3 shows that its heating rate $\sim$ 1 Gyr ago was substantially higher than the present tidal heating. For both HD 209458 b and HAT-P-1 b, the substantial heating rate $\sim$ 3-4 x 10^{19} W about 1 Gyr ago may help account for the

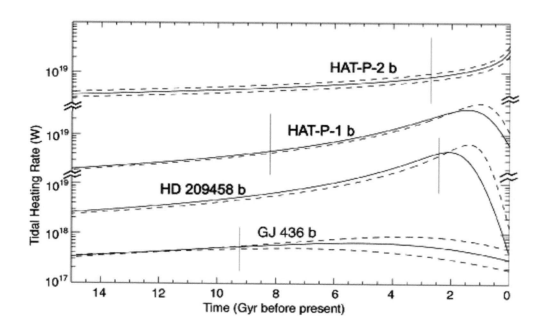

Figure 3. The tidal heating rates for planets HAT-P-2 b, HAT-P-1 b, HD 209458 b, and GJ 436 b as a function of time. The present time ($t = 0$) is at the right. The solid vertical line through each of the heating curves represents the estimated time of formation for that planet. Note that the vertical scales has been shifted for HAT-P-2 b and HAT-P-1 b to make its curves more visible. (The vertical scale that corresponds to each curve is the scale intersected by that curve.) The solid curve for each planet is based on the current nominal eccentricity value. The dashed lines represent upper and lower limits on the heating, *i.e.* they assume the extreme values of current a and e consistent with observations that give the maximum and minimum current tidal heating rates, respectively. For HAT-P-1 b and HD 209458 b, observations could not exclude a current eccentricity of zero, so the lower bound on heating rates is formally zero. Hence in those cases only one dashed line is shown, representing the upper limit.

discrepancy between the large observed planetary radii and the predictions of physical modeling.

In the case of HAT-P-2 b there also has been a substantial amount of tidal heating. The current heating rate is similar to the maximum rate attained by HD 209458 b and HAT-P-1 b, so again we might expect a larger radius than predicted by theory that ignored tidal heating. In this case, however, the measured radius is actually about 12% smaller than predicted (Bakos et al. 2007b). Thus there is a discrepancy between theory and observation even if tidal heating is neglected. The fact that there is likely a high rate of tidal dissipation makes the problem even worse. On the other hand, a key factor in the reconciliation may be that, while the current tidal heating rate is high and increasing, in the recent past the heating rate was much lower. HAT-P-2 b is still on the increasing part of the heating curve, which is unusual among planets considered here, most of which have passed their peaks. The fact that the heating rate was several times smaller a billion years ago than it is now may help explain the small radius.

Figure 3 also illustrates tidal heating histories for GJ 436 b. This planet has a measured radius consistent with theoretical models, independent of tidal heating (Gillon et al. 2007). The tidal heating history shown in Figure 3 is consistent with that result. Compared with the previous two cases, the maximum heating rate was two orders of magnitude less, small enough perhaps not to affect the radius.

5. Discussion

This investigation supports the hypothesis that tidal interactions between a star and a planet are responsible for the relatively small e values of close-in planets (Rasio et al. 1996), although our calculations introduce important corrections to previous studies. Because even close-in planets evidently formed with much larger e, the processes that governed their early dynamics were probably similar to other planets. A plausible mechanism for producing the initial e distribution of extra-solar planets is planet-planet scattering (Rasio & Ford 1996; Weidenschilling & Marzari 1996), although some modification to the original model are needed (Barnes & Greenberg 2007).

If the tidal circularization hypothesis is borne out, then our study also provides constraints on Q values. We find that agreement between eccentricity distributions requires $Q_p \sim 10^{6.5}$, with $Q_* \sim 10^{4.25}$, $10^{5.5}$ (best fit), or $10^{6.75}$. This Q_p is in good agreement with other constraints (Yoder & Peale 1981; Ogilvie & Lin 2004), and this Q_* agrees well with studies of binary star circularization (Mathieu 1994). Of course, it is likely that individual planets and stars have unique Q values, owing to variation in their internal structures. These Q values are only meant to be representative values, good for the population as a whole. Corrections to our tidal model might also result in different suitable Q values.

Significant reductions in semi-major axes have accompanied the changes in eccentricity, with important implications. First, models of protoplanetary migration in the primordial gas disk need not carry "hot Jupiters" in as far as their current positions. Lin et al. 1996 proposed that migration in the gas disk halted near the inner edge of the disk, a boundary determined by clearing due to the host star's magnetosphere. Our results show that the inner edge was probably farther out than indicated by the current semi-major axes of the planets, which were only reached during tidal migration long after the nebula had dissipated. In order to evaluate where migration due to the gas disk halted, (and thus where the inner edge of the nebula was) models should account for the subsequent tidal evolution.

The tidal changes in orbital semi-major axes also have implications for observations of planetary transits, such as surveys of young open galactic clusters (Bramich *et al.* 2005; von Braun *et al.* 2005; Burke *et al.* 2006). The probability to observe a planetary transit increases for smaller semi-major axes, but decreases as orbits become more circular (Borucki & Summers 1984; Barnes 2007). Tidal evolution means that the probability of an observable transit depends on a star's age, but the exact relation depends on the particular evolutionary path through (a, e) space. As our understanding of the statistics of tidal evolution paths improves, the observed frequency of transits in the field and in open clusters may eventually help to constrain planetary formation scenarios, distinguishing, for example, between the relative roles of embedded migration and of gravitational scattering, which set up the initial conditions for tidal evolution. Transit statistics may not yet be refined enough to be sensitive to detect this effect (Pepper & Gaudi 2006), but such systematic effects may show up in future surveys.

The tidal heating calculations here suggest that past tidal heating may well have played an important role in the evolution of the physical properties of many extra-solar planets, specifically the planetary radius. We caution that the specific calculations displayed here depend on numerous assumptions and several uncertain parameters. The heating rates correspond to the orbital evolution trajectories computed by Jackson *et al.* 2008a, and various caveats are discussed in detail there. It is quite likely that the actual thermal history of any particular planet was different to some degree from what we show here. In particular, when analyzing radial-velocity observations to solve for a planet's orbit, it is difficult to rule out a completely circular orbit, in which case the tidal heating rate would have been zero. However, the unavoidable point is that past tidal heating may be significant for many planets and should be considered as a factor in theoretical modeling of physical properties of exoplanets.

For every planet whose tidal evolution we modeled, we have calculated corresponding tidal heating histories (Jackson *et al.* 2008b). For the cases presented here, see Fig. 3, we see that past tidal heating may provide a previously unconsidered source of heat for planets with larger-than-predicted radii. However, it may make things worse in cases where measured radii seemed to fit the current models. Theoretical models of tidal evolution and planetary interiors will generally need to be adjusted and improved so as to yield a match between predicted and observed planetary radii.

To conclude, we find that the distribution of orbital eccentricities for exoplanets was once strikingly uniform across all semi-major axes. By varying tidal dissipation parameters, we can match the original distribution of close-in planetary eccentricities to that of planets far from their host star for stellar and planetary Q's $\sim 10^{5.5}$ and $\sim 10^{6.5}$, which are consistent with previous estimates. After the formation of the close-in exoplanets, tides raised on the host star and on the planet acted over Gyrs to reduce orbital eccentricities and semi-major axes of the close-in exoplanets. This reduction in e as well as a has important implications for the thermal histories of close-in exoplanets and for transit studies.

We also find that tidal heating in the past was significantly larger than current heating. For example, about 1 Gyr ago, HD 209458 b may have undergone tidal heating 100 times the present value. This substantial heating may help resolve the mystery of the anomalously large radii observed for many transiting planets today. If the lag in response of the planetary radii to tidal heating is of order a Gyr, then past tidal heating must be included in models of exoplanetary radii. Previous studies suggest such a lag is reasonable. However, further studies are required to elucidate this effect.

References

Bakos, P., *et al.* 2007a, *ApJ*, 656, 552.
Bakos, P., *et al.* 2007b, *ApJ*, 670, 826.
Barnes 2001, *ApJ*, 561, 1095.
Barnes, R. & Greenberg, R. 2007, *ApJL*, 659, L53.
Barnes, J. 2007, *PASP*, 119, 986.
Bodenheimer, P., Laughlin, G., & Lin, D.N.C 2003, *ApJ*, 592, 555.
Borucki, W. J. & Summers, A. L. 1984, *Icarus*, 58, 121.
Bramich, D. M. *et al.* 2005, *Mon. Not. R. Astron. Soc.*, 359, 1096.
Butler, R. P. *et al.* 2006, *ApJ*, 646, 505.
Burke, C. J. *et al.* 2006, *ApJ*, 132, 210.
Burrows, A., Hubeny, I., Budaj, J., & Hubbard, W. B. 2007, *ApJ*, 661, 514.
Da Silva, R. *et al.* 2006, *A&A*, 446, 717.
Deming, D. *et al.* 2007, *ApJL*, 667, L199.
Fischer, D. & Valenti, J. 2005, *ApJ*, 622, 1102.
Ford, E. & Rasio, F. 2006, *ApJ*, 638, L45.
Gillon, M. *et al.* 2007, *A&A*, 472, L13.
Goldreich, P. & Nicholson, P. 1981, *Icarus*, 30, 301.
Goldreich, P. & Soter, S. 1966, *Icarus*, 5, 375.
Gorda, S. & Svechnikov, M. A. 1996, *Astron. Reports*, 42, 793.
Guillot, T. 2005, *Ann. Rev. Earth and Planet. Sci.*, 33, 493.
Hubbard, W. B. 1984, Planetary Interiors, (New York: Van Nostrand Reinhold Co).
Jackson, B., Greenberg, R., & Barnes, R. 2008a *ApJ*, accepted
Jackson, B., Greenberg, R., & Barnes, R. 2008b, *ApJ*, submitted.
Johnson, J. A. *et al.* 2006 *ApJ*, 652, 1724.
Kaula, W. 1968, An Introduction to Planetary Physics, Wiley, NY.
Knutson, H., Charbonneau, D., Noyes, R., Brown, T., & Gilliland, R. 2007, *ApJ*, 655, 564.
Laughlin, G. *et al.* 2005 *ApJL*, 629, L121.
Lin, D. N. C. *et al.* 1996, *Nature*, 380, 606.
Lovis, C. *et al.* 2006 *Nature*, 441, 305.
Maness, H. L. *et al.* 2007, *PASP*, 119, 90.
Mardling, R. 2007, *Mon. Not. R. Astron. Soc.*, in press.
Weidenschilling, S. J. & Marzari, F. 1996, *Nature*, 384, 619.
Mathieu, R. 1994, *Annu. Rev. Astron. Astrophys.*, 32, 465.
Mayor, M. *et al.* 2004, *A&A*, 415, 391.
McArthur, B. *et al.* 2004, *ApJL*, 614, L81.
Moutou, C. *et al.* 2006, *A&A*, 458, 327.
Ogilvie, G. & Lin, D. N. C. 2004, *ApJ*, 610, 477.
Peale, S. J., Cassen, P. & Reynolds, R. T. 1979, *Science*, 203, 892.
Pepper, J. & Gaudi, B. S. 2006, *Acta Astronomica*, 56, 183.
Rasio, F. A. & Ford, E. B. 1996, *Science*, 279, 954.
Rasio, F. A., Tout, C. A., Lubow, S. H., & Livio, M. 1996, *ApJ*, 470, 1187.
Rivera, E. *et al.* 2005, *ApJ*, 634, 625.
Saffè, C. *et al.* 2006, *A&A*, 443, 609.
Takeda, G. *et al.* 2007, *ApJS*, 168, 297.
Trilling, D. 2000, *ApJL*, 537, L61.
Udry, S. *et al.* 2002, *ApJ*, 634, 625.

Valenti, J. & Fischer, D. 2005, *ApJ*, 159, 141.
Vogt, S. *et al.* 2005, *ApJ*, 632, 638.
von Braun, K. *et al.* 2005, *PASP*, 117, 141.
Wright, J. T. *et al.* 2007, *ApJ*, 657, 533.
Yoder, C. & Peale, S. 1981, *Icarus*, 47, 1.
Zucker, S. *et al.* 2004, *A&A*, 426, 695.

Modeling of evolution of the rotational axis of "hot Jupiter" planets under tidal perturbations

Irina Kitiashvili

Kazan state university, Dept. of Astronomy
420008, Kremlevskaya str., 18, Kazan, Russia
email: gabbiano@mail.ru

Abstract. In this report, we present results of analytical and numerical calculations of evolution the axis of rotation of planets moving at very close orbits. We consider the evolution of the axis of rotation caused by tidal perturbations of a parent star and obtain estimates of the principal moment of inertia and the dynamical flattening for nine exoplanets. From analysis of evolutionary equations, we obtain the critical values of the kinetic momentum vector, $\vec{L}$, for different values of orbital eccentricity. We find a general tendency of vector $\vec{L}$ to evolve to the direction perpendicular to the orbital plane.

Keywords. exoplanets, spin evolution, tidal perturbations

1. Introduction

Investigations of the spin motion of celestial bodies were started by d'Alembert, Euler, Lagrange and Laplace. Their results led to the theories of precession and nutation of the Earth's axis and libration of the Moon, and to the general perturbation theory.

Discoveries of extra-solar planets have inspired intensive developments of the planetary dynamics. Unfortunately, the current technical capabilities do not allow us to measure dynamical and physical characteristics of exoplanets, such as the spin period, dynamical flattening, elastic properties, etc. However, the problems of planetary and stellar spins and physical properties of exoplanets can be studied for observed planetary systems through modeling. In particular, the planets in close-in orbits are of great interest for planetary dynamics because of the extreme orbital characteristics, which are quite different from properties of planetary orbits in the Solar System.

In this report, we discuss the spin evolution of hot Jupiters under the action of tidal perturbations. In the next section, we will obtain estimates of the principal moment of inertia of a planet, and calculate the main moments of inertia for nine exoplanets.

2. Estimates of the principal moment of inertia and dynamical flattening of exoplanets

The basic properties of evolution of the axis of rotation of a planet, caused by tidal effects are illustrated in Fig. 1. The tidal deformations lead to changes of the planet rotational axis. If the planet has a permanent volume and is liquid with small viscosity, then the tidal bulge is always directed along a line connecting the centers of masses of the star and planet. Otherwise, the tidal bulge lags behind this line by angle γ_π (Fig. 1). For quantitative estimates of tidal perturbations, the dissipative factor, $Q^{-1} = \sin \gamma_\pi$, is often used. The deformation of planets may be non-elastic, and, hence, it can be

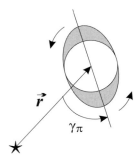

Figure 1. Illustration of the star-planet tidal interaction.

accompanied by significant dissipation of energy. In the case when a central star with mass $M_\star$ distorts a spherical exoplanet, dynamic flattening of the planet due to to the tides can be described by the following expression (Alfven & Arrhenius 1979):

$$\chi = \frac{15 M_\star R^3}{4 M_p a^3}, \qquad (2.1)$$

where χ is dynamical flattening, $M_\star$ and M_p are the masses of the star and the planet correspondingly, R is the planetary radius, a is the semi-major axes.

It is obvious that for planets with very close orbits the spin evolution strongly depends on the tidal forces of the parent star. For numerical estimates we consider the spin evolution of nine exoplanets, listed in Table 1. Assuming that the planets are uniform bodies, we calculate the principal momenta of inertia, I, and the values of their dynamical flattening, χ. Then using these data we estimate the main moments of inertia A and B, shown in Table 2.

In the next section, we consider the basic equations describing the spin evolution of exoplanets under the action of tidal perturbation of a parent star and calculate the basic evolutionary effects.

Table 1. Some properties of close-in planets.

Planet	M, M_J	R, R_J	a, AU	P, days	e	M, M_{Sun}
OGLE-TR-10b	0.63	1.26	0.042	3.10	0	1.18
OGLE-TR-56b	1.29	1.3	0.023	1.212	0	1.17
OGLE-TR-111	0.53	1.067	0.047	4.01	0	0.82
OGLE-TR-113b	1.32	1.09	0.023	1.43	0	0.78
OGLE-TR-132b	1.14	1.18	0.031	1.69	0	1.26
TrES-1b	0.61	1.081	0.039	3.03	0.135	0.87
TrES-2b	1.98	1.22	0.037	2.47	0	0.98
HAT-P-1b	0.53	1.36	0.055	4.47	0.09	1.12
HD 209458b	0.69	1.32	0.045	3.52	0.07	1.01

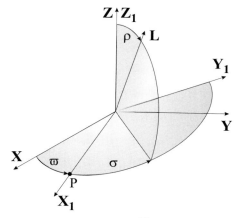

Figure 2. Orientation of kinetic moment vector, $\vec{L}$, relative to the pericentre of the orbit, P.

3. Spin evolution of planets under action tidal perturbations

For investigation of the spin evolution of the hot Jupiter planets we use the reference frame, $X_1 Y_1 Z_1$, which is connected to the orbital pericentre, P. In this case, axes X_1 and Y_1 are rotated relative to the orbital reference frame, XYZ, by an angle of the longitude of the pericentre, $\tilde{\omega}$ (Fig. 2). The orientation of the kinetic momentum vector $\vec{L}$ relative to the pericentre of the orbit of a planet is described by angles ρ and σ.

The perturbed rotation of a planet can be described by the following system of evolutionary equations in the case when the angular spin velocity of a planet is significantly higher than its angular orbital velocity (Beletskii & Khentov 1995):

$$\frac{d\rho}{dt} = \frac{1}{L \sin \rho} \left(\frac{\partial U}{\partial \psi} \cos \rho - \frac{\partial U}{\partial \sigma} \right) + \frac{M_1}{L} - K_\Omega \sin i \cos(\sigma + \omega_\pi), \qquad (3.1)$$

Table 2. Estimates of the surface gravity, dynamical flattening and the moment of inertia for "hot Jupiters".

Planet	Surface gravity m s^{-2}	χ	$I \times 10^{42}$, kg m^2	$A \times 10^{42}$ kg m^2	$C \times 10^{42}$ kg m^2	$F_{tidal} \times 10^{-10}$, N/kg
OGLE-TR-10b	10.07	0.020	3.79	3.82	3.74	6.49
OGLE-TR-56b	19.38	0.072	8.26	8.47	7.85	40.72
OGLE-TR-111	11.82	0.007	2.29	2.29	2.28	3.13
OGLE-TR-113b	28.20	0.026	5.94	6	5.84	25.74
OGLE-TR-132b	20.78	0.026	6.02	6.07	5.91	17.43
TrES-1b	13.25	0.012	2.70	2.71	2.68	5.68
TrES-2b	33.77	0.0075	11.17	11.2	11.11	7.86
HAT-P-1b	7.27	0.013	3.72	3.73	3.68	2.65
HD 209458b	10.05	0.015	4.56	4.58	4.511	4.39

Note:
[1] For comparison: the tidal force of the Jupiter on Io is equal to 9×10^{-37} N/kg; for the Moon, the tidal action of the Earth is 3.6×10^{-39} N/kg.

$$\frac{d\sigma}{dt} = \frac{1}{L\sin\rho}\frac{\partial U}{\partial \rho} + \frac{M_2}{L\sin\rho} + K_\Omega\left(\sin i \cot\rho \sin(\sigma+\omega_\pi) - \cos i\right), \tag{3.2}$$

$$\frac{dL}{dt} = \frac{\partial U}{\partial \psi} + M_3, \tag{3.3}$$

$$\frac{d\vartheta}{dt} = L\sin\vartheta \sin\phi\cos\phi\left(\frac{1}{A} - \frac{1}{B}\right) + \frac{1}{L\sin\vartheta}\left(\cos\vartheta\frac{\partial U}{\partial \psi} - \frac{\partial U}{\partial \phi}\right) +$$
$$+ \frac{M_2 \cos\psi - M_1 \sin\psi}{L}, \tag{3.4}$$

$$\frac{d\psi}{dt} = L\left(\frac{\sin^2\phi}{A} + \frac{\cos^2\phi}{B}\right) - \frac{1}{L}\left(\frac{\partial U}{\partial \vartheta}\cot\vartheta + \frac{\partial U}{\partial \rho}\cot\rho\right) - \tag{3.5}$$
$$- \frac{M_1 \cos\psi + M_2 \sin\psi}{L}\cot\vartheta - \frac{M_2}{L}\cot\rho - K_\Omega \frac{\sin i}{\sin\rho}\sin(\sigma+\omega_\pi),$$

$$\frac{d\phi}{dt} = L\cos\vartheta\left(\frac{1}{C} - \frac{\sin^2\phi}{A} - \frac{\cos^2\phi}{C}\right) + \frac{1}{L\sin\vartheta}\frac{\partial U}{\partial \vartheta} + \tag{3.6}$$
$$+ \frac{M_1 \cos\psi + M_2 \sin\psi}{L\sin\vartheta},$$

where angles ρ, σ describe the orientation of vector $\vec{L}$ in the orbital reference frame (Fig. 1); ψ, ϕ and ϑ are the Euler's angles, U is the potential force function of perturbations, M_1, M_2 and M_3 are projections of the nonpotential perturbation forces on components of vector $\vec{L}$, and A, B and C are the main moments of inertia of the planet.

In a first approximation we assume that the planet is a dynamically symmetrical body ($A = B \neq C$) and divide all variables in the system (3.1 - 3.6) into fast and slow. We define fast variables as the variables, which change in the unperturbed motion. The variables, which are constants for the unperturbed motion but vary in the case of perturbations, are called slow variables. Hence, in the case of a dynamically symmetrical planet the slow variables are angles ρ and σ, which describe position of vector $\vec{L}$ in space (Fig. 2), the value of $\vec{L}$, and the angle of spin rotation, ϕ. Thus, for the tidal perturbations the spin evolution of "hot Jupiters" can be described by the following dynamical system:

$$\frac{d\rho}{dt} = \frac{M_1}{L}, \tag{3.7}$$

$$\frac{d\sigma}{dt} = \frac{M_2}{L\sin\rho}, \tag{3.8}$$

$$\frac{dL}{dt} = M_3, \tag{3.9}$$

$$\frac{d\phi}{dt} = L\cos\vartheta\left(\frac{1}{C} - \frac{1}{A}\right) + \frac{M_1 \cos\psi + M_2 \sin\psi}{L\sin\vartheta}. \tag{3.10}$$

It is known, that if a celestial body has fast non-resonant rotation then the moment of the tidal interaction in the first approximation is described as

$$\vec{M} = \frac{l_1 l_3}{r^6}\left\{[(\vec{\omega}_{spin} - \vec{\omega}_{rot}) \times \vec{e}_r] \times \vec{e}_r\right\}, \tag{3.11}$$

where $\vec{e}_r$ is the unit vector of the radius-vector $\vec{r}$, $l_1 = 1.51 l_2 \gamma_g M_\star^2 R_{pl}^5$ is a constant coefficient, l_2 is a coefficient that depends on elastic properties of the planet (for Jupiter $l_2 = 5/2$).

If the tidal perturbations can be described by equation (3.11), then evolutionary equations (3.7 - 3.10) become (Beletskii 1981):

$$\frac{d\rho}{dt} = \frac{\delta}{2LP^6} \sin\rho \{-2\omega\nu_1 + \alpha\beta L \cos\rho (\nu_2 + 0.5\nu_3 \cos 2\sigma)\}, \quad (3.12)$$

$$\frac{d\sigma}{dt} = -\frac{\delta}{4P^6} \alpha\beta\nu_3 \sin 2\sigma, \quad (3.13)$$

$$\frac{dL}{dt} = \frac{\delta}{P^6} \{\omega\nu_1 \cos\rho - \alpha\beta L (\nu_2(1 - 0.5\sin^2\rho) - 0.5\nu_3 \sin^2\rho \cos 2\sigma)\}, \quad (3.14)$$

$$\frac{d\phi}{dt} = L \cos\vartheta \left(\frac{1}{C} - \frac{1}{A}\right). \quad (3.15)$$

where $\alpha = \sin^2\vartheta/A + \cos^2\vartheta/C$, $\beta = (1-e^2)^{3/2}$, P is the semilatus rectum, $\delta = l_1 l_3$, $\omega = \sqrt{GM_\star/a^3}$ is the angular velocity of the orbital motion, G is the gravity constant, e is eccentricity, $\nu_1 = 1 + 7.5e^2 + 5.625e^4 + ...$, $\nu_2 = 1 + 3e^2 + 0.375e^4$, $\nu_3 = 1.5e^2 + 0.25e^4$.

Equation (3.15) that describes evolution of the angle of the planet's spin ϕ, depends only on the vector of kinetic moment $\vec{L}$ of a planet and the values of the main moments of inertia (we assume that the angle of nutation, ϑ, is constant). Consequently we have only three independent equations (3.12 - 3.14). In addition, equation (3.13) is independent from the other two. Therefore, we can reduce the evolutionary equations to a system of two equations (3.12, 3.14). After the change of variables:

$$x = \rho, \qquad y = L, \qquad \dot{x} = d\rho/dt, \quad \dot{y} = dL/dt,$$
$$a = \delta/(2P^6), \quad b = \omega\nu_1, \qquad c = \alpha\beta, \qquad d = \nu_2 + 0.5\nu_3 \cos 2\sigma$$

we obtain the following system:

$$\dot{x} = \frac{a}{y} \sin\rho(-2b + cdy \cos x), \quad (3.16)$$
$$\dot{y} = 2a \left[b \cos x - 0.5cy \left(\nu_2(1 + \cos^2 x) - \nu_3 \cos 2\sigma \sin^2 x\right)\right].$$

We investigate stability of the dynamic system (3.16) from the point of view the Lyapunov's stability by methods of the qualitative analysis (Kitiashvili & Gusev 2008). The dynamical system (3.16) has six equilibrium states (ESs) with coordinates:

$$x_{1,2} = 0, \pi, \qquad y_{1,2} = \pm\frac{b}{c\nu_2} \quad (3.17)$$

$$x_{3,4} = \pm\arccos\xi, \qquad y_{3,4} = \frac{2b}{cd}\xi \quad (3.18)$$

$$x_{5,6} = \pm\arccos(-\xi), \qquad y_{3,4} = -\frac{2b}{cd}\xi, \quad (3.19)$$

where

$$\xi = \sqrt{\frac{0.5\nu_3 \cos 2\sigma}{\nu_3 \cos 2\sigma - \nu_2}}$$

Table 3. Critical values of parameters for planets at close-in orbits ($e = 0$).

Planet	ES 1-2: L for $\rho = 0^0, 180^0$	Planet	ES 1-2: L for $\rho = 0^0, 180^0$
OGLE-TR-10b	1.39×10^{37}	OGLE-TR-113b	4.72×10^{37}
OGLE-TR-56b	7.51×10^{37}	OGLE-TR-132b	4.05×10^{37}
OGLE-TR-111	6.6×10^{36}	TrES-2b	5.21×10^{37}

Table 4. Critical values of parameters for planets with $e \neq 0$.

Planet	ES 1-2: $L \times 10^{37}$	ES 3-4: ρ	ES 3-4: $L \times 10^{35}$	ES 5-6: ρ	ES 5-6: $L \times 10^{35}$
HD 209458b	1.52	$\pm (1.56^0 - 1.53^0)$	$3.14 - 13.08$	$\pm (1.58^0 - 1.61^0)$	$3.14 - 13.08$
HAT-P-1b	1.00	$\pm (1.56^0 - 1.52^0)$	$2.64 - 10.99$	$\pm (1.58^0 - 1.63^0)$	$2.64 - 10.99$
TrES-1b	1.14	$\pm (1.56^0 - 1.49^0)$	$4.43 - 18.45$	$\pm (1.59^0 - 1.65^0)$	$4.43 - 18.45$

The six equilibrium states (3.17 - 3.19) are possible only at $e \neq 0$. In the case of a circular orbit, there are only two ESs. Numerical estimates of angle ρ and the value of the kinetic moment $\vec{L}$ for the ESs are presented in Table 3 for the planets with $e = 0$, and in Table 4 for the planets with $e \neq 0$. The phase trajectories in the vicinity of ES 1 and 2 are of the steady node type, while the trajectories for other ES have the type of steady focus. Thus, the value of $\vec{L}$ evolves to the value of $b/(c\nu_2)$ ($\sim 1.14 \times 10^{37}$) or $2b\xi/(cd)$ ($\sim 4.43 \times 10^{35}$ for TrES-1b in Table 4. The orientation of the kinetic moment vector, $\vec{L}$, tend to evolve to the direction perpendicular to the orbital plane of a planet, $\rho = 0^0$ or $\rho \sim 1.6^0$ for TrES-1b. In Table 4, we show the numerical estimates of the critical values of angle ρ and $\vec{L}$. Possible variations of the ρ and $\vec{L}$ values, which depend on the assumed angle σ, are shown in the brackets.

According to our results, under the action of the tidal forces between a hot Jupiter planet and a parent star the vector of the kinetic moment $\vec{L}$ of the planet evolves to the position perpendicular to the orbital plane in the case of a circular orbit or close to this position for elliptical orbits.

4. Conclusions

In this report, we have considered possible scenarios of the spin evolution of exoplanets at close orbits. We have found the effect of the evolution of orientation of the exoplanet spin to the position perpendicular to the orbital plane for circular orbits or close to this orientation in the case of elliptical orbits. We have obtained the dynamical estimates of the principal moment of inertia of some hot Jupiter planets, and their dynamical flattening under the action of tidal forces from the parent stars.

References

Alfven, H., & Arrhenius, G. 1976, *Evolution of the Solar system*. Scientific and technical information office NASA, Washindton, D.C., USA.
Beletskii, V.V. 1981, *Celestial Mechanics*, 23, 371.
Beletsky, V.V. & Khentov, A.A. 1995 *Resonance rotation of celestial bodies*. Nizhnii Novgorod, Nizhegorodskii gumanitarnii centr, Russia (in Russian).
Kitiashvili, I. & Gusev, A. 2008, *Celestial Mecanics and Dynamical Astronomy*, 100 (in press).

Astrobiological effects of F, G, K and M main-sequence stars

M. Cuntz[1], L. Gurdemir[1], E. F. Guinan[2] and R. L. Kurucz[3]

[1] Department of Physics, University of Texas at Arlington,
Arlington, TX 76019-0059, USA
email: cuntz@uta.edu, gurdemir@uta.edu

[2] Department of Astronomy and Astrophysics, Villanova University,
Villanova, PA 19085, USA
email: edward.guinan@villanova.edu

[3] Harvard-Smithsonian Center for Astrophysics,
Cambridge, MA 02138, USA
email: rkurucz@cfa.harvard.edu

Abstract. We focus on the astrobiological effects of photospheric radiation produced by main-sequence stars of spectral types F, G, K, and M. The photospheric radiation is represented by using realistic spectra, taking into account millions or hundred of millions of lines for atoms and molecules. DNA is taken as a proxy for carbon-based macromolecules, assumed to be the chemical centerpiece of extraterrestrial life forms. Emphasis is placed on the investigation of the radiative environment in conservative as well as generalized habitable zones.

Keywords. Astrobiology, stars: atmospheres, stars: late-type

1. Introduction and Methods

The centerpiece of all life on Earth is carbon-based biochemistry. It has repeatedly been surmised that biochemistry based on carbon may also play a pivotal role in extraterrestrial life forms, if existent. This is due to the pronounced advantages of carbon, especially compared to its closest competitor (i.e., silicon), which include: its relatively high abundance, its bonding properties, and its ability to form very large molecules as it can combine with hydrogen and other molecules as, e.g., nitrogen and oxygen in a very large number of ways (Goldsmith & Owen 2002).

In the following, we explore the relative damage to carbon-based macromolecules in the environments of a variety of main-sequence stars using DNA as a proxy by focussing on the effects of photospheric radiation. The radiative effects on DNA are considered by applying a DNA action spectrum (Horneck 1995) that shows that the damage is strongly wavelength-dependent, increasing by more than seven orders of magnitude between 400 and 200 nm. The different regimes are commonly referred to as UV-A, UV-B, and UV-C. The test planets are assumed to be located in the stellar habitable zone (HZ). Following the concepts by Kasting *et al.* (1993), we distinguish between the conservative and generalized HZ. Stellar photospheric radiation is represented by using realistic spectra taking into account millions or hundred of millions of lines for atoms and molecules (Castelli & Kurucz 2004, and related publications). We also consider the effects of attenuation by an Earth-type planetary atmosphere, which allows us to estimate attenuation coefficients appropriate to the cases of Earth as today, Earth 3.5 Gyr ago, and no atmosphere at all (Cockell 2002).

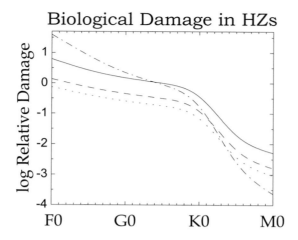

Figure 1. Biological damage to DNA for a planet at an Earth-equivalent position without an atmosphere (solid line), an atmosphere akin to Earth 3.5 Gyr ago (dashed line) and an atmosphere akin to Earth today (dotted line). The dash-dotted line refers to a planet without an atmosphere at a distance of 1 AU.

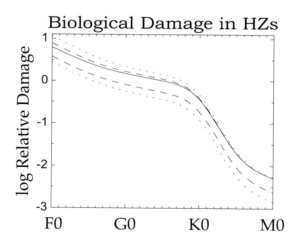

Figure 2. Biological damage to DNA for a planet (no atmosphere) at an Earth-equivalent position (solid line), at the limits of the conservative HZ (dashed lines) and at the limits of the generalized HZ (dotted lines).

2. Results and Conclusions

Our results are presented in Figs. 1, 2, and 3. The first two figures show the relative damage to DNA due to stars between spectral type F0 and M0, normalized to today's Earth. We also considered planets at the inner and outer edge of either the conservative or generalized HZ as well as planets of different atmospheric attenuation.

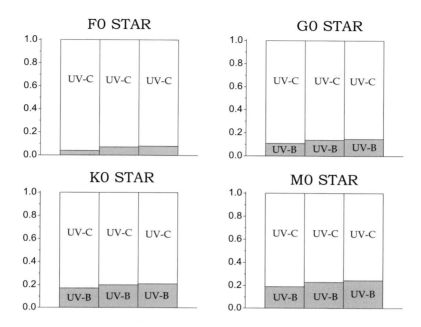

Figure 3. Relative significance of UV-A, UV-B, and UV-C for the damage to DNA for a planet without an atmosphere (left), an atmosphere akin to Earth 3.5 Gyr ago (center) and an atmosphere akin to Earth today (right) for different types of main-sequence stars. Note that the fraction due to UV-A is unidentifiable.

Based on our studies we arrive at the following conclusions: (1) All main-sequence stars of spectral type F to M have the potential of damaging DNA due to UV radiation. The amount of damage strongly depends on the stellar spectral type, the type of the planetary atmosphere and the position of the planet in the habitable zone (HZ); see Cockell (1999) for previous results. (2) The damage to DNA for a planet in the HZ around an F-star (Earth-equivalent distance) due to photospheric radiation is significantly higher (factor 5) compared to planet Earth around the Sun, which in turn is significantly higher than for an Earth-equivalent planet around an M-star (factor 180). (3) We also found that the damage is most severe in the case of no atmosphere at all, somewhat less severe for an atmosphere corresponding to Earth 3.5 Gyr ago, and least severe for an atmosphere like Earth today. (4) Any damage due to photospheric stellar radiation is mostly due to UV-C. The relative importance of UV-B is between 5% (F-stars) and 20% (M-stars). Note that damage due to UV-A is virtually nonexistent (see Fig. 3).

Our results are of general interest for the future search of planets in stellar HZ (e.g., Turnbull & Tarter 2003). They also reinforce the notion that habitability may at least in principle be possible around M-type stars, as previously discussed by Tarter *et al.* (2007). Note however that a more detailed analysis also requires the consideration of chromospheric UV radiation, especially flares (e.g., Robinson *et al.* 2005), as well as the detailed treatment of planetary atmospheric photochemistry, including the build-up and destruction of ozone, as pointed out by Segura *et al.* (2003, 2005) and others.

References

Castelli, F., & Kurucz, R. L. 2004, in: N.E. Piskunov, W.W. Weiss & D.F. Gray (eds.), *Modelling of Stellar Atmospheres*, IAU Symp. 210 (San Francisco: ASP), CD-ROM, Poster 20

Cockell, C. S. 1999, *Icarus*, 141, 399

Cockell, C. S. 2002, in: G. Horneck & C. Baumstark-Khan (eds.), *Astrobiology: The Quest for the Conditions of Life* (Berlin: Springer), p. 219

Goldsmith, D. & Owen, T. 2002, *The Search for Life in the Universe*, 3rd Ed. (Sausalito: University Science Books)

Horneck, G. 1995, *J. Photochem. Photobiol. B: Biology*, 31, 43

Kasting, J. F., Whitmire, D. P., & Reynolds, R. T. 1993, *Icarus*, 101, 108

Robinson, R. D., et al. 2005, *ApJ*, 633, 447

Segura, A., et al. 2003, *Astrobiology*, 3, 689

Segura, A., et al. 2005, *Astrobiology*, 5, 706

Tarter, J. C., et al. 2007, *Astrobiology*, 7, 30

Turnbull, M. C., & Tarter, J. C. 2003, *ApJ (Suppl.)*, 145, 181

Part 3

Planet Formation

Testing planet formation theories with giant stars

Luca Pasquini[1], M.P. Döllinger[1], A. Hatzes[2], J. Setiawan[3], L. Girardi[4], L. da Silva[5], J.R. de Medeiros[6] and A. Weiss[7]

[1] ESO
Garching, Germany
email: lpasquin@eso.org, mdoellin@eso.org

[2] Thüringer Landessternwarte
Tautemburg, Germany
email: artie@tls-tautenburg.de

[3] MPiA
Heidelberg, Germany
email: setiawan@mpia-hd.mpg.de

[4] INAF-OaPD
Padova, Italy
email: leo.girardi@oapd.inaf.it

[5] ON
Rio de Janeiro, Brazil
email: licio@on.br

[6] UFRN
Natal, Brazil
email: renan@dfte.ufrn.br

[7] MPA
Garching , Germany
email: weiss@mpa-garching.mpg.de

Abstract. Planet searches around evolved giant stars are bringing new insights to planet formation theories by virtue of the broader stellar mass range of the host stars compared to the solar-type stars that have been the subject of most current planet searches programs. These searches among giant stars are producing extremely interesting results. Contrary to main sequence stars planet-hosting giants do not show a tendency of being more metal rich. Even if limited, the statistics also suggest a higher frequency of giant planets (at least 10 %) that are more massive compared to solar-type main sequence stars.

The interpretation of these results is not straightforward. We propose that the lack of a metallicity-planet connection among giant stars is due to pollution of the star while on the main sequence, followed by dillution during the giant phase. We also suggest that the higher mass and frequency of the planets are due to the higher stellar mass. Even if these results do not favor a specific formation scenario, they suggest that planetary formation might be more complex than what has been proposed so far, perhaps with two mechanisms at work and one or the other dominating according to the stellar mass. We finally stress as the detailed study of the host stars and of the parent sample is essential to derive firm conclusions.

Keywords. planetary systems: formation, stars: abundance, stars: fundamental parameters.

1. Introduction

Out of the more than 200 exoplanets known, only an handful orbit around evolved giants stars. These stars, however, are interesting targets for planet searches. Unlike the host stars of most known exoplanets which have a mass distribution that peaks at 0.8 ±0.3 solar masses, giant stars can have masses several times this value. The radial velocity (RV) method, the most succesful technique at finding exoplanets, is insentive to more massive, early-type stars. These hot stars have few spectral lines that are often broadened by high stellar rotation rates. The typical RV error for a hot early-type star might be 100 – 1000 ms^{-1}, much higher than the $\approx$ 10 ms^{-1} needed to detect sub-stellar companions. On the other hand, early-type main sequence stars that have evolved off the main sequence and have become giants have a plethora of narrow asorption lines that are amenable to RV planet searches. Figure 1 clearly shows this difference, where the spectral region of a main sequence A star is compared to the same spectral region of a giant star with approximately the same mass. While in the main sequence star only one broad feature is visible, in the giants many, narrow lines are present.

• The first interest in searching planets around giants is therefore to enlarge the stellar mass range surveyed, including higher masses.

Giants differ from main sequence stars in their structure: they have much larger radii (typical of 10 R$_\odot$, see (da Silva *et al.* 2006) and they have a much deeper convective zone compared to solar-type stars (cfr. discussion later). Differences in stellar radius and depth of the convection zone are expected also along the main sequence, but their variations are small compared to the large differences occurring when the stars evolved into giants. So it shall be much more evident and easy to detect those effects, such chemical pollution, whose signatures vary with internal structure changes.

• The second interest in searching planets around giants is therefore that, thanks to their different structures, giants provide ideal testbed to understand which stellar or planetary system characteristics depend from some stellar parameters, such as depth (mass) of the convective zone or stellar radius, which change radically when passing from the main sequence to the giant phase.

Interestingly, giants have been early recognized as radial velocity variable, and they could be broadly characterized by two variability timescales:

(*a*) Short timescales , of the order of a few days or fraction of day. These are most likely due to pulsations or solar-type oscillations (Hatzes & Cochran 1994, Frandsen *et al.* 2002)

(*b*) Long timescales, of the order of one hundred or a few hundred days, either due to planetary companions or to stellar surface structure. (Hatzes & Cochran 1993).

Because giant stars have higher intrinsic "noise" due to stellar oscillations, these stars have been so far neglected by planet searches. Although the situation is rapidly evolving, as testified by the contributions to this conference (Quirrenbach; Liu; Masasshi, reports in IAUS249 2007; Niedzielski 2008). In addition, one program is dedicated at Lick Observatory to follow up evolved A stars (Johnson *et al.* 2007a).

2. The FEROS and the Tautenburg Surveys

Our group started to obtain accurate radial velocity of giant stars in 1999, with the thesis work of J. Setiawan (Setiawan *et al.* 2003b, 2004) who used the FEROS spectrograph at the ESO 1.5m telescope. Seventy-six stars were observed for four years, and the most promising were followed up in the last years after the spectrograph was moved at the 2.2 m telescope, using time allocated to the Max Planck Institute for Astronomy. We will refer in the following to this sample as the 'southern' sample.

In 2004 M. Döllinger started a survey with the 2m Alfred Jensch telescope in Tautenburg on 67 stars in the northern sky.

In both cases the selected stars are giants (class III) with accurate Hipparcos parallaxes, known to better than 10%. For the Tautenburg sample we added the constraint that the stars should be circumpolar so that observations could be obtained throughout the year. Intrinsically bright and thus cool giants were not selected for our samples to avoid the possible occurance of pulsating AGB stars which may have large RV amplitude variations and the fact that it is difficult to characterize their stellar parameters (cfr. next sections).

Figure 2 shows the colour-magnitude diagram of all our sample stars. Clearly we sample well the RGB, the clump and the beginning of the AGB. Very few subgiants were included.

3. Characterization of the Stars

Shortly after the first exoplanet has been discovered (Mayor and Queloz 1995) it was clear that it is essential to characterize the host stars. Age, mass and metallicity are fundamental parameters to understand the planetary formation mechanism and the relationship between stars and protoplanetary disks. The early discovery of the 'planet-metallicity' connection (Gonzalez 1997, Santos et al. 2000, 2001) was exciting and triggered more observational studies and theoretical efforts. When dealing with giants, the characterization of the stars is particularly important and also more difficult because of the presence of two degeneracies. First, while stars of different masses are quite well separated on the main sequence, while ascending the RGB (and in the clump) they occupy more or less the same region of the color-magnitude diagram. Second, evolutionary tracks of metal poor giants are bluer than for the tracks of metal rich giants. Thus an old metal poor star will occupy the same region of the CMD as a younger metal rich giant. This is the well known age-metallicity degeneracy. Thus to fully characterize a giant star (mass,

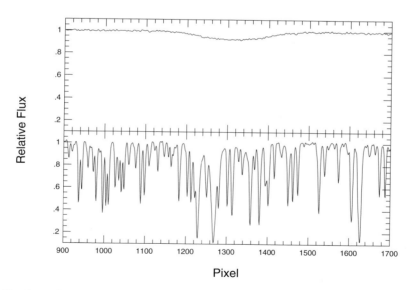

Figure 1. Portion of spectrum in the optical region of two stars: one massive main sequence star and one cool giant. This comparison shows that as soon as the star evolves from the main sequence migrating in the giant domain, it develops a very crowded line system, suitable for precise Radial Velocity (RV) measurements.

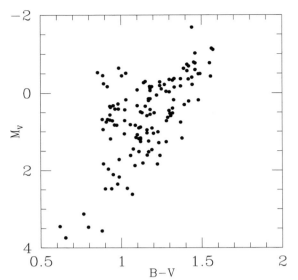

Figure 2. Color-magnitude diagram of the sample stars, observed with FEROS and at Tautenburg. The stars well sample the Red Giant Branch (RGB), the clump and the beginning of the AGB. Only a few subgiants are present, and a few cool, luminous stars.

radius, etc.) using evolutionary tracks requires the analysis of high resolution spectra from which effective temperature and metallicity are derived.

Effective temperature and metallicity were derived using a classical abundance analysis in a plane parallel atmospheres and LTE conditions. From these values the age and mass have been derived using a modified version of the Jorgensen and Lindegren (2005) method which computes the probability distribution function (PDF) of each quantity using theoretical tracks (cfr. da Silva et al. 2006) The exact shape of the PDF differs from star to star because that depends on the parameters errors and mostly on the position of the star in the CMD. In most instances the PDFs are sharply peaked and thus result in a well-determined parameter (Figure 3). Some stars, however, can have more uncertain parameters due to a broader, or even double-peaked PDF. da Silva et al. (2006) have performed several sanity checks on the recovered stellar parameters by comparing the derived colors and radii with observed ones. They found excellent agreement with the values derived with our PDF method.

We have further measured the effective temperatures of all the stars of the southern sample using the line-depth ratio method (Gray 1991, Biazzo et al. 2007a). They agreed very well with the temperatures derived with the abundance analysis. There was a small zero point difference (5 K) and a dispersion of less 70 K, as shown in Figure 4. Finally, applying the same PDF method to six giants of the open cluster IC4651 we obtain for the stars of this cluster a mass of 2.0 ±0.2 solar mass and an age of 1.2 ±0.2 Gyrs, very similar to what is quoted in the WEBDA database (Biazzo et al. 2007b).

We are confident that we are able to produce reliable parameters for our sample stars, and we have applied the same analysis also to the Northern sample (Döllinger et al. 2008a).

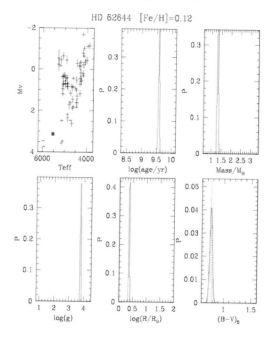

Figure 3. Probability Distribution Functions for one star of our sample. For this object all the PDFs are very well peaked and give unambiguous results (from da Silva et al. 2006).

4. Observational Results

The number of giants hosting planets is not large enough to allow refined analysis, but it can start to provide strong indications on several aspects such as planetary frequency, dependence on metallicity, planets characteristics, which can be compared with similar properties in main sequence stars.

4.1. Radial Velocity Statistics of the Surveys

After several years of observations, we can summarize the general statistics we have obtained separately for the southern and northern samples. Table 1 summarizes our main findings in terms of RV behavior.

A few aspects of this table require some comments. In the introduction we have mentioned that, when analyzing giant stars, special care should be taken in excluding other sources of RV variability. Indeed out of the 32 RV variable in the Southern hemisphere, 5 stars show associated chromospheric variability and/or bisector variability in phase with the RV period. An example is given in Figure 5, which shows 'planet-like' RV variability that is correlated with Ca II core variations. Therefore these RV variations are likely caused by chromospheric activity rather than due a planetary companion. For the 'binaries' bin we included all stars which showed RV variability of several $km\,s^{-1}$, but for which we have insufficient data to calculate for them a full orbital solutions. For the southern sample they were simply removed from future RV follow-up measurements. For the northern sample we will continue to obtain RV measurements, but with less frequent sampling, in order to derive the orbit.

Clearly, most numbers agree very well between the two surveys. The only real discrepancy is in the number of 'RV constant stars', whose percentage is much larger in the southern sample. However, at issue here is what is defined as a "constant star". The

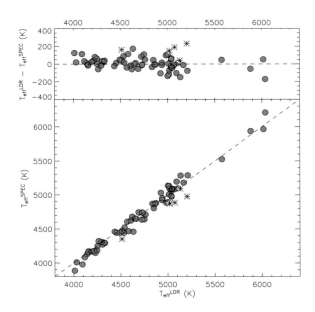

Figure 4. Comparison between effective temperatures obtained from the abundance analysis and effective temperature as obtained by the Line Depth Ratio method for the stars of the southern sample. The agreement is excellent (From Biazzo et al. 2007b)

FEROS program had a measurement precision four times worse than for the northern sample. Not surprisingly, this program revealed more "constant stars". Several of the northern sample giant stars have RV variations of ≈ 20 m s^{-1} which were clearly variable, but would have qualified as a constant star in the southern program. With increased measurement precision many of the constant stars measured by FEROS may turn out to be variable. We therefore confirm the first conclusions obtained by e.g. Walker et al. (1989) that G and K giants are a new class of Radial Velocity variable.

The second important result is that giant planets around giant stars seem to be common. Their frequency is of at least 10%, since our findings probably represent a lower limit. We also note that the long orbital periods require observations which extend over a baseline of several years. Continued monitoring may reveal more stars in our sample as hosting giant planets. The planets span a range in periods between 150 and 1000 days, and in mass between 3 and 15 Jupiter masses. We are indeed sensitive only to quite massive planets, given the limited precision of our observations and the fairly high intrinsic RV variability of evolved stars. Thus a 10% frequency of very massive planets around more massive stars than the sun may represent a minimum value. These results seem to agree well with other surveys around evolved stars, which find massive planets and even a good rate of brown dwarfs orbiting around giant stars (see i.e. Quirrenbach 2008).

Out of the 13 exoplanets so far discovered in our surveys, only 4 have been published (Setiawan et al. 2003a, Setiawan et al. 2004, Setiawan et al. 2005, Döllinger et al. 2007), while the others are in preparation. We show in Figure 6 the RV curves of the 6 planet candidates from the Northern sample.

Table 1. Overview of the Radial Velocity variability obtained in our surveys.

	Northern Sample (76)	Southern Sample (62)
Binaries	15 (20%)	13 (21%)
Variables	32 (43%) 5 Activity Modulation	22 Long Period (34%) 19 Short Period (31%)
Planets	7 (10%)	6 (10%)
Constant	21 (27%)	2 (3%)
Precision	22 m/sec	∼5 m/sec

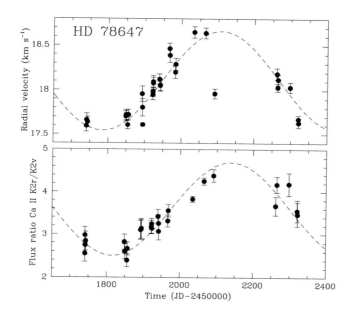

Figure 5. Variations of chromospheric flux (Ca II K line core) in phase with the radial velocity period of one giant. These variations indicate that the associate Radial Velocity orbit is not due to the presence of planetary companion, rather by variable chromospheric activity.

4.2. Stellar Metallicity and Planets

Gonzales et al. (1997, 2001) first showed that stars hosting exoplanets tend to have higher metallicities than stars without giant planets. Subsequent investigations showed this to be a real effect and not an observational bias. (e.g. Santos et al. 2004, 2006, Fisher and Valenti 2005).

The first determinations of abundances of planet-hosting giant stars suggested that such a metallicity-planet connection may not hold for evolved stars (Sadakane et al. 2005 Schuler et al. 2005). This was based, however, on a small sample so no reliable conclusions could be drawn.

Using the first results from our Southern and Northern surveys plus literature data Pasquini et al. (2007) have shown that this suspected trend was correct: giant planet-hosting stars have the same metallicity distribution as giant stars without planets, and have a different abundance distribution from planet-hosting main sequence stars. Figure 7

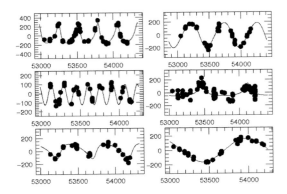

Figure 6. Radial Velocity variations for the 6 stars of the northern sample hosting planets. (Döllinger et al. 2008a,b in preparation)

shows the age-metallicity distribution for the giants of our samples, and for the giants hosting planets (including data from literature), and it is clear that the hosting planets giants do follow the same age-metallicity relationship of the sample stars.

Admittedly, the mixing of stars from literature and our own sample may introduce some bias, because we do not know the parent sample from which the literature data are taken, or if the quoted abundances are on the same scale as the one of our sample. We have therefore computed the metallicity distribution of the planet hosting (13) and non planet-hosting (126) stars of our samples, and this is shown in Figure 8. Clearly the two distributions overlap and there is no indication of metal excess in the planet-hosting stars.

Our result are indirectly supported by the first 5 planets included in the survey of evolved A stars by Johnson et al. (2007a), which have a rather uniform metal distribution. We cannot derive more firm conclusions from this sample because we do not know the distribution of the parent sample and being a young population we do expect that its age distribution should be quite metal rich.

For sake of completeness we report that Hekker & Melendez (2007) claim a possible metal excess in their anaysis of giants hosting planets. We however notice two potential problems in their analysis: the metal excess is entirely due to the presence of two subgiants with exceptionally high metallicity. We have basically no subgiants in our sample. The second and most important problem is that the planet host sample used is not derived by their parent sample: the planet sample is taken from other surveys, whose metal distribution is not known. Clearly this is a flaw, and it is clear from their figures , which show that no star is present in the parent sample in the highest metallicity bins. Without a knowledge of the parent sample from which the planet hosts stars have been derived, a comparison is subject to possible bias and is thus inconclusive.

Concerning metallicity the evidence so far collected seems clear: giant stars hosting planets do not show any tendency towards being metal rich.

4.3. Planets Characteristics

The planets found around giant stars have long periods (longer than 150 days), and relatively large masses. In this section we briefly investigate some of the planets' characteristics.

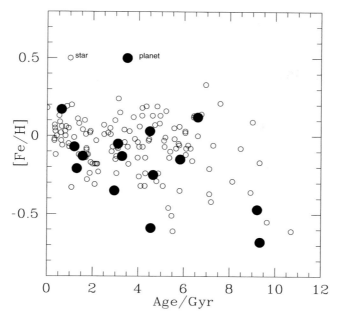

Figure 7. Age-metallicity relationship for the giants of our Southern and Northern samples and for giants with planets, either from our survey or from literature. Clearly there is no difference between the two distributions. (From Pasquini *et al.* 2007)

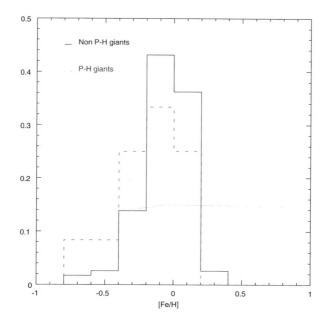

Figure 8. Metallicity distribution of the (126) Southern and Northern giants from our sample not hosting planets (Blue continuous line). With the red dashed line are indicated the 13 giants hosting planets, found in the same samples. No data from literature are included. This comparison clearly shows that hosting planet giants do not prefer metal rich systems.

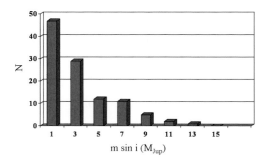

Figure 9. Mass distribution of the planets around stars with masses below 1.1 $M_\odot$ showing the strong increase for small planet mass.

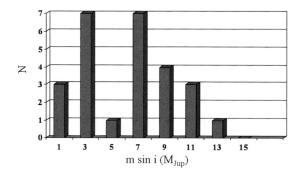

Figure 10. Mass distribution of the planets around stars with masses above 1.1 $M_\odot$ (dominated by the giants) and for stars with smaller masses. The distribution is clearly different, with more massive stars showing a very high frequency of massive planets.

Planet masses for giant stars are typically larger than what has been observed around main sequence stars. Figure 9 shows the planet mass distribution for main sequence stars with masses below 1.1 $M_\odot$ while Figure 10 is the same but for stars (mostly giants) with masses larger than 1.1$M_\odot$. Clearly the distribution of planet masses for low mass main sequence stars increases towards lower mass planets, and it is definitely different from the one of more massive stars. Of course we know that there are biases in the giants' sample, in that small mass planets are presently out of reach of surveys due to their limited precision and to the intrinsic variability of the stars. Nevertheless, if we assume that the difference between the two distribution is due to observational biases and the distribution was the same than the low mass stars, the plot of Figure 10 would imply an extremely high planet occurrence for the massive stars.

We shall therefore consider as an observational evidence that planets around giants (more massive stars) are more massive than those around lower mass stars. If they were sharing a similar overall mass distribution, the frequency so far observed among giants would imply a very high planetary frequency around massive stars.

As far as orbital parameters are concerned, only long period planets are found around giants. This is quite expected since the stars have radii typically as large as 10 $R_\odot$, therefore short period planets would be swallowed by the stars. It would be interesting to study the case of migration and planetary evolution in giants, in the presence of radius growth and mass losses, as occurring along the RGB and AGB phases. We note that the short period RV variations in giant stars may mask the detection of possible short period planets that may still reside outside the photosphere of the star. These would have orbital periods of many days, or similar to the periods for stellar oscillations. Such variability if found by RV surveys may be dismissed as due to stellar oscillations rather than a short period companion.

The distribution of eccentricity is very similar to what is observed among main sequence stars and ranges from planets in nearly circular orbits to those in highly eccentric orbits. It seems therefore that eccentricity is not affected by the stellar characteristics such as the mass. We note that there could be also here a hidden bias in the eccentricity distribution: orbits with high eccentricity are much more difficult to be mimiced by other phenomena, like oscillations or stellar surface structure so such RV variations are more easily attributed to Keplerian motion. A nearly sinsoidal RV variations can also arise from stellar surface structure and is thus more suspicious. Additional tests (e.g. lack of photometric, Ca II H&K, or line bisector variations) are needed to confirm these as planetary companions.

5. Interpretation

It is natural to interpret the differences between giants' and main sequence stars starting from their three main differences: giants are on average more massive, have larger convective zones and larger radii.

Are giants hosting planets indeed more massive than their main sequence stars counterparts? Given a volume-limited samples this would be a natural result of stellar evolution, and this difference is confirmed in Figure 11, which shows the mass-metallicity distribution for main sequence stars and giants hosting planets. Even if the giants we studied are not really originating from very massive stars, their mass is, on average, higher than that of main sequence stars.

As far as the tendency for planet-hosting giant stars not to be metal rich, we favor the hypothesis that the difference between main sequence and giants is due to pollution: the most external parts of main sequence atmospheres are polluted by engulfed debris and small planets which have a higher metal content. Because of the relatively shallow convection zone of stars on the main sequence, these metals are well mixed only in a small fraction of the star. During the giant phase the convective zone of the stars greatly deepens and the more efficient mixing of the atmosphere dilutes the atmospheric abundance of metals. This scenario has been debated at length in literature, and most analyses conclude against this hypothesis (Santos *et al.* 2004, 2006, Fischer and Valenti 2005, Ecuvillon *et al.* 2006). We find that the counter arguments are not particularly strong, mostly because it is not very well known the real depth of the mixing zone in main sequence stars and its variation with stellar mass (cfr. Vauclair 2006). By comparison this effect is greatly enhanced (a factor $\sim$30 or more) as soon as a star evolves and becomes a red giant. A similar effect should happen at the very low mass end of the stellar mass distribution, where the stars become fully convective. If highly convective M stars show a non-dependence on metallicity similar to the giants, as seems from the first analysis

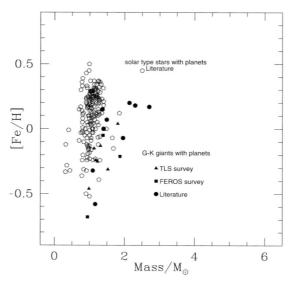

Figure 11. Mass-Metalicity distribution of hosting - planets dwarfs and giants. As expected, evolved stars are on average more massive than main sequence stars.

(Bonfils *et al.*, these proceedings), then the evidence would be even stronger. The main observational argument against this interpretation stems in the subgiants analyzed by Fischer and Valenti (2005), which do not observe any difference between subgiants in the blue and red part of the CMD. Stars on the two sides of the Hertzpung gap (i.e. before and after the deep convective zones have developed) show no differences in metallicity. We can offer no explanation for this point. We already noticed as the metal distribution of subgiants shows a tail at very high metallicity which is not present in the giant samples analyzed so far, and we also note that Murray *et al.* (2001) reached opposite conclusions than Fischer and Valenti. The issue is still open and we believe that subgiants are worth a dedicated study.

As far the trend that giant stars might have a higher frequency of more massive planets, stellar mass seems the most likely governing parameter, possibly because massive stars may develop more massive proto-planetary disks. Similar conclusions have been reached recently by two papers (Lovis and Mayor 2007, Johnson *et al.* 2007b), which discussed stars in clusters and in field low mass stars. It is also highly suggestive that while Paulson *et al.* (2004) did not find any planet around almost 100 low mass stars surveyed in the Hyades, one planet has been found around one of the 2 single giants of this cluster (Sato *et al.* 2007). As far to this topic, we can observe that all analyses so far agree with an enhanced frequency (and higher mass) of planets around more massive stars. We note also that a similar behavior is present in the samples of Fischer and Valenti (2005) and Santos *et al.* 2006, but both authors conclude that this dependence was uniquely the result of a selection bias, and that metallicity is the relevant parameter, because metallicity and stellar mass were strongly correlated in their samples. Since giants do not show a dependence of planet formation on metallicity and widen the range of stellar masses observed, we propose that there may be a strong dependence of planet formation on stellar mass. Clearly, more studies are needed to disentangle the possible effects of various stellar parameters (e.g. mass, abundance) on planet formation.

6. Giants and Planet Formation Theories

Considering the 'classical' planet formation schemes: core accretion (Ida & Lin 2004) and disk instability (Boss 1997), we cannot conclude that the characteristics of the planets around giant stars favor either of the two schemes. It is on the other hand accepted that the metal dependence of massive planet formation in main sequence stars is considered a great success of the core accretion scenario (Ida & Lin 2004).

It is very possible that reality is more complex than firstly envisaged; could it be that the two mechanisms are both at work, and one is favored with respect to the other depending, for instance, on the stellar mass? Something in this line has been proposed by Matsuo et al. (2007).

On the other hand, a more complex core accretion scenario, which involves the interaction with the stellar magnetosphere and the position of the dust evaporation line for intermediate mass stars, in addition of the snow line considered in less massive stars, could also explain the observations (Lin, private communication).

We do not know yet the answer, but clearly the extension of planet search to giants is bringing new insides, which should be strongly pursued.

7. Next steps

In this last section we would like to express our vision for the continuation of this exciting research.

Obviously, the results from other surveys are needed to confirm our results, and to better quantify them. Reaching smaller mass planets would also be essential, to determine the planet mass distribution among intermediate mass stars. The results presented at this conference by other groups are extremely encouraging, and promise to bring a wealth of new results in a few years.

We stress that a proper analysis of the hosting stars is essential, if we aim at obtaining results which can express more about physics. The parent sample require to be analyzed as carefully as the planet hosting stars.

Open clusters, where stars are supposed to share the same age and chemical compositions, could bring direct new results. Similarly subgiants are worth to have a dedicated study. In principle it would be great to compare equivalent samples of field giants and dwarfs, but this seems quite difficult, when all the variables (including age) are considered.

Since ultimately we will aim at comparing stars in different evolutionary status, one important point is to be sure that the analysis methods do not suffer of biases and peculiarities when applied at stars of different evolutionary status. Open clusters could be again privileged testbeds: analyzing stars in a well populated open cluster from main sequence to tip of RGB would be a crucial test to ensure that the results obtained do not suffer of any bias (see e.g. the analysis of Pasquini et al. 2004 of the open cluster IC4651).

8. Acknowledgements

The observations have been collected at ESO and the 2m Alfred Jensch Telescope of the Thüringer Landessternwarte Tautenburg.

References

Biazzo , K. et al. 2007a, AN 328, 938
Biazzo, K., Pasquini, L. et al. 2007b, A&A 475, 981

Boss, A. P. 1997, Science, 276, 1836
da Silva, L., Girardi, L., et al. 2006, A&A 458, 603
Döllinger. M. P., et al. 2007a: A&A, 472, 649
Döllinger. M. P., et al. 2007b: A&A, in preparation
Döllinger. M. P., et al. 2007c: A&A, in preparation
Ecuvillon, A., Israelian, G., et al. 2006, A&A, 449, 809
Fischer, D., & Valenti, J. 2005, ApJ, 622, 1102
Frandsen, S., Carrier, F., Aerts, C. et al. 2002, A&A, 394, L5.
Gonzalez, G. 1997, MNRAS, 285, 403
Gonzalez, G. 1998, A&A, 334, 221
Gonzalez, G., Laws, C. et al. 2001, AJ, 121, 432
Gray, D. Johnes, H. 1991 PASP 103, 409
Hatzes, A. P., Cochran, W. D. 1993, ApJ, 413, 339
Hatzes, A. P., Cochran, W. D. 1994, ApJ, 422, 366
Hekker, Melendez 2007 astro-ph 07091145
Ida, S., & Lin, D. N. C. 2004, ApJ, 616, 567
Johnson, J. et al. 2007a ApJ 665, 785
Johnson, J. et al. 2007b ArXiv 0707.0518
Jørgensen, B.R., Lindegren, L. 2005, A&A, 436, 127
Laughlin, G., & Adams, F. C. 1997, ApJ, 491, L51
Lovis, C., Mayor, M. 2007 A&A 472, 657
Matsuo, T., et al. 2007, ApJ, in press (astro-ph 0703237)
Mayor, M., & Queloz, D. 1995, Nature, 378, 355
Murray N., Chaboyer B. 2001, ApJ 555, 801
Pasquini, L., Randich, S. et al. 2004, A&A 424, 951
Pasquini, L., Döllinger, M. Weiss, A. et al. 2007, A&A 473, 979
Paulson, D.B., Cochran, B., Hatzes, A.P. 2004, AJ 127, 3579
Pollack, J. B., Hubickyj, O., et al. 1996, Icarus, 124, 62
Sadakane, K., Ohnishi, T., Ohkubo, M., Takeda, Y. 2005, PASJ 57, 127
Santos, N. C., Israelian, G., Mayor, M. 2000, A&A, 363, 228
Santos, N. C., Israelian, G., Mayor, M. 2001, A&A, 373, 1019
Santos, N. C., Israelian, et al. 2003, A&A, 398, 363
Santos, N. C., Israelian, G., Mayor, M. 2004, A&A, 415, 1153
Santos, N. C., Israelian, G., et al. 2005, A&A, 437, 1127
Sato, B., Ando, H., Kambe, E. 2003, ApJ, 597, L157
Sato, B., Izumiura, H., et al. 2007, ApJ, 661, 527
Schuler, S., Kim, J.H. et al. 2005, ApJ, 632, L131
Setiawan, J., Hatzes, A.P., et al. 2003a, A&A, 398, L19
Setiawan, J., Pasquini, L., et al. 2003b, A&A, 397, 1151
Setiawan, J., Pasquini, L., et al. 2004, A&A 421, 241
Setiawan, J., Rodman, J., et al. 2005, A&A, 437, L31
Vauclair S. 2004, ApJ 605, 874
Walker, G.A.H., Yang, S., Campbell, B., Irwin, A.W. 1989, ApJ, 343, L21.

Orbital migration and mass-semimajor axis distributions of extrasolar planets

Shigeru Ida[1] and D. N. C. Lin[2,3]

[1] Tokyo Institute of Technology, Ookayama, Meguro-ku, Tokyo 152-8551, Japan
email: ida@geo.titech.ac.jp

[2] UCO/Lick Observatory, University of California, Santa Cruz, CA 95064, USA; Kavli Institute of Astronomy & Astrophysics, Peking University, Beijing, China
email: lin@ucolick.org

Abstract. Here we discuss the effects of type-I migration of protoplanetary embryos on mass and semimajor axis distributions of extrasolar planets. We summarize the results of Ida & Lin (2008a, 2008b), in which Monte Carlo simulations with a deterministic planet-formation model were carried out. The strength of type-I migration regulates the distribution of extrasolar gas giant planets as well as terrestrial planets. To be consistent with the existing observational data of extrasolar gas giants, the type-I migration speed has to be an order of magnitude slower than that given by the linear theory. The introduction of type-I migration inhibits in situ formation of gas giants in habitable zones (HZs) and reduces the probability of passage of gas giants through HZs, both of which facilitate retention of terrestrial planets in HZs. We also point out that the effect of magneto-rotational instability (MRI) could lead to trapping of migrating protoplanetary embryos in the regions near an ice line in the disk and it significantly enhances formation/retention probability of gas giants against type-I migration.

Keywords. planetary systems: formation, solar system: formation, stars: statics

1. Introduction

1.1. Difficulty of too rapid type-I migration

In the core accretion scenario for planet formation (e.g., Hayashi et al. 1985), migration of planetary embryos due to tidal interactions of embedded embryos with surrounding disk gas (e.g., Goldreich & Tremaine 1980, Ward 1986) is one of the most serious problems. Analytic studies suggest that isolated embryos lose angular momentum to the disk exterior to their orbits faster than that they gain from the disk interior to their orbits. This torque imbalance leads to "type-I" migration.

The migration timescales for a planet with mass M_p and orbital radius r (orbital frequency Ω_K) in a disk with gas surface density Σ_g ($\propto r^{-p}$) and sound velocity c_s around a host star with mass M_* is given by (Tanaka et al. 2002)

$$\begin{aligned}\tau_{\mathrm{mig1}} &= \frac{r}{\dot{r}} \simeq \frac{1}{2.728+1.082p}\left(\frac{c_s}{a\Omega_K}\right)^2 \frac{M_*}{M_p}\frac{M_*}{a^2\Sigma_g}\Omega_K^{-1} \\ &\simeq 7\times 10^4 \times \left(\frac{\Sigma_g}{\Sigma_{g,\mathrm{MMSN}}}\right)^{-1}\left(\frac{M_p}{M_\oplus}\right)^{-1}\left(\frac{r}{1\mathrm{AU}}\right)^p\left(\frac{M_*}{M_\odot}\right)^{3/2} \text{yrs},\end{aligned} \quad (1.1)$$

where $\Sigma_{g,\mathrm{MMSN}} = 1700(r/10\mathrm{AU})^{-3/2}\mathrm{g\,cm^{-2}}$ is Σ_g of the minimum-mass solar nebula (MMSN) model (Hayashi 1981). For an Earth-mass planet at $\sim 1\mathrm{AU}$ and a 10 Earth-mass core at $\sim 5\mathrm{AU}$, $\tau_{\mathrm{mig1}} \sim 10^5$ years in a disk similar to MMSN, which is much shorter than the observationally inferred lifetime of protoplanetary disks ($\sim 10^6$–10^7 years). Since more than several $M_\oplus$ may be required for a core to start runaway gas accretion onto the core (see eq. [1.2]), the short migration timescale implies the difficulty in formation of gas giants. However, it is apparently inconsistent with observationally postulated probability ($\gtrsim 10\%$) of solar type stars harboring gas giants (e.g., Marcy *et al.* 2005, Mayor *et al.* 2005). Furthermore, the mostly refractory compositions of the terrestrial planets in the present-day Solar system suggest that they probably did not migrate to their present locations from regions well beyond the ice line.

1.2. *Preservation of solid materials in terrestrial planet regions*

The difficulty for formation of terrestrial planets is less serious if oligarchic growth model (Kokubo & Ida 1998, Kokubo & Ida 2002) is considered. The oligarchic growth model suggests that growth of planetary embryos is stalled at their isolation mass that is about a Mars mass in the terrestrial planet regions in the disk similar to MMSN, until Σ_g is depleted to 10^{-3} times of that of MMSN and coagulation between the embryos start (Kominami & Ida 2002). It prevents rapid depletion of solid materials in full amount of disk gas, since the migration timescale of embryos is $\tau_{\text{mig1}} \sim 10^6$ years (eq. [1.1]), which is not significantly shorter than lifetime of protoplanetary disks. With slight reduction of type-I migration speed or slightly enhanced dust to gas ratio of the disk, enough amount of materials for the formation of Earth-mass planets can be retained (McNeil *et al.* 2005, Daisaka, Tanaka & Ida 2006).

Through Monte Carlo simulations on the basis of a comprehensive treatment of the sequential planet formation scenario, Ida & Lin (2008a) found that although uninhibited type-I migration leads to efficient self-clearing of embryos, materials of a few times $M_\oplus$ remain at $a \gtrsim 1\text{AU}$, almost independent of initial surface density of planetesimals (Σ_d) and migration strength. At $a \gtrsim 1\text{AU}$, embryos start type-I migration well before they attain isolation mass, so after they have migrated, new-generation embryos continue to form from residual planetesimals. This repeated generation and migration slow down when Σ_d declines to values comparable to or smaller than that of the MMSN model and type-I migration is no longer an effective disruption mechanism for Mars-mass embryos. With this reservoir, there is an adequate inventory of residual embryos to subsequently assemble into Earth-mass rocky planets.

1.3. *Formation and retention of cores for giant planets*

For a core to start runaway gas accretion and form a gas giant within the disk lifetime ($\sim 10^6$–10^7 years), a core of at least several $M_\oplus$ must accrete, because the Kelvin-Helmholtz contraction timescale (τ_{KH}) of the gas envelope is given by

$$\tau_{\text{KH}} \simeq 10^{k1} \left(\frac{M_\text{p}}{M_\oplus}\right)^{-k2} \text{ yrs,} \qquad (1.2)$$

where $k1 = 8$–10 and $k2 = 3$–4 (see Ida & Lin 2004a, Ida & Lin 2004b and references therein). Note, however, that more efficient cooling of the envelope by low dust contents in inflow gas and/or non-spherically symmetric structure of the envelope can reduce τ_{KH}. Equation (1.1) shows that for the several $M_\oplus$ cores to start runaway gas accretion and form gas giants, significant reduction of type-I migration speed is required.

Cores formed after the disk gas has been severely depleted may withstand the disruption by the declining type-I migration speed (Thommes & Murray 2006), since $\tau_{\text{mig1}} \propto \Sigma_g^{-1}$ (eq. [1.1]). Incorporating gas accretion onto cores and the effect of type-II migration, Ida & Lin (2008a) found that cores repeatedly form and migrate inward and those formed in "late" stage can survive for slightly reduced type-I migration. But, they showed that the formation probability of gas giant planets and hence the predicted mass and semimajor axis distributions of extrasolar gas giants are sensitively determined by the strength of type-I migration. They suggested that the observed fraction of solar-type stars with gas giant planets can be reproduced only if the actual type-I migration timescale is an order of magnitude longer than that deduced from linear theories (§2.3). Note that this late formation scenario is conceptually consistent with that inferred from the noble gas enrichment in Jupiter (Guillot & Hueso 2006).

1.4. *Retardation of type-I migration*

Retardation processes for this type-I migration are explored by many authors. One possibility for the retardation is non-linear fluid dynamical effects: self-induced unstable flow (Koller *et al.* 2004, Li *et al.* 2005), non-linear radiative and hydrodynamic feedbacks (Masset *et al.* 2006a), and entropy gradient (Baruteau & Masset 2008). Magnetic field can affect wave propagation (Fromang, Terquem & Nelson 2005, Muto & Inutsuka 2008). Random torque due to density fluctuations caused by MRI can also overcome monotonic torque for type-I migration (Laughlin, Steinacker & Adams 2004, Nelson & Papaloizou 2004).

Another possibility is surface density gradient of disk gas. As shown in eq. (1.1), if p is negative locally, type-I migration is slowed down or even reversed there. Masset *et al.* (2006b) considered an inner cavity in a gas disk and showed that migrating embryos are captured near the outer edge of the cavity. Kretke & Lin (2007) pointed out protoplanetary disks are composed of an inactive neutral "dead zone" near the mid plane, sandwiched together by partially ionized surface layers where MRI is active. Because the main agents for removing electrons from the gas are grains (e.g., Sano et al. 2000), due to a transition in the surface density of the icy dust grains across the ice line, the thickness of the active layer decreases abruptly outside the ice line, resulting in local positive surface density. Zhang, Lin & Kretke (2008) showed that type-I migration is stalled near the ice line with the transition. Ida & Lin (2008b) incorporated these effects into the Monte Carlo simulation and found that the mass and semimajor axis distribution of extrasolar giant planets consistent with the observed data are reproduced with much less reduction of type-I migration strength than that required in the case without this effect.

In the below, we summarize the results of Ida & Lin (2008a) and Ida & Lin (2008b). In section 2, we describe the results in the disks with Σ_g that has a power-law radial dependence (Ida & Lin 2008a). In section 3, we show the cases in which Σ_g has the non-uniform radial dependence due to a coupling effect of MRI and the ice line (Ida & Lin 2008b).

2. Disks with surface density having a power-law radial dependence

In a series of papers (Ida & Lin 2004a, Ida & Lin 2004b, Ida & Lin 2005, Ida & Lin 2008a), a numerical scheme to simulate the anticipated mass-semimajor distribution of planets was constructed based on a comprehensive treatment of the sequential planet formation scenario. In Ida & Lin (2004a), we presented calculations for a solar-type stars by neglecting the effect of type-I migration. With the same assumptions, we simulated the distribution for stars with a range of metallicity, $[\text{Fe/H}] = \log_{10}((\text{Fe/H})_*/(\text{Fe/H})_\odot)$, and mass ($M_*$) in Ida & Lin (2004b) and Ida & Lin (2005), respectively.

The prescriptions for planetesimal accretion, gas accretion onto cores and type-II migration for gas giant planets are essentially the same in these papers. In Ida & Lin (2008a) and Ida & Lin (2008b), the effect of type-I migration was included. For the details of the prescription, see Ida & Lin (2008a).

2.1. Disk model

Ida & Lin (2008a) introduced multiplicative factors (f_d and f_g) to scale surface density of gas (Σ_g) and solid components (Σ_d) such that

$$\begin{cases} \Sigma_d = \Sigma_{d,10} 10^{[\text{Fe/H}]} \eta_{\text{ice}} f_d (r/10\text{AU})^{-3/2}, \\ \Sigma_g = \Sigma_{g,10} f_g (r/10\text{AU})^{-p}, \end{cases} \quad (2.1)$$

where normalization factors $\Sigma_{d,10} = 0.32 \text{g/cm}^2$ and $\Sigma_{g,10} = 75 \text{g/cm}^2$ correspond to 1.4 times of Σ_g and Σ_d at 10AU of the MMSN model, and the step function $\eta_{\text{ice}} = 1$ inside the ice line and 4.2 for $r > a_{\text{ice}}$ (Hayashi 1981). We assume that dust to gas ratio is proportional to [Fe/H] and [Fe/H] does not affect the distribution of Σ_g.

Assuming an equilibrium temperature in optically thin disk regions (Hayashi 1981),

$$T = 280 \left(\frac{r}{1\text{AU}}\right)^{-1/2} \left(\frac{L_*}{L_\odot}\right)^{1/4} \text{K}, \quad (2.2)$$

where L_* and $L_\odot$ are the stellar and solar luminosity. The ice line is determined by this temperature distribution as

$$a_{\text{ice}} = 2.7 \left(\frac{L_*}{L_\odot}\right)^{1/2} \text{AU}. \quad (2.3)$$

Note that a_{ice} is modulated by opacity of the disk and viscous heating (Davis 2005, Garaud & Lin 2007).

Here, we assume that planetesimals have been formed from dust grains and Σ_d is identified as surface density of planetesimals. f_d at a given location r continuously decreases with time from its initial value $f_{d,0}$ as planetesimals are accreted by embryos that in term undergo orbital decay. Note that here semimajor axis a is identified as orbital radius r, since we neglect evolution of orbital eccentricities. For gas component, we adopt exponential decay with decay constant τ_{dep},

$$f_g = f_{g,0} \exp(-t/\tau_{\text{dep}}). \quad (2.4)$$

2.2. Surface density evolution due to type-I migration

The results on the Σ_d reduction due to type-I migration are shown in Figures 1. They show Σ_d at $t = 10^5, 10^6$, and 10^7 years (dashed, dotted, and solid lines) with $f_{g,0} = f_{d,0}$ and $p = 1.0$: (a) $C_1 = 1$, $f_{g,0} = 3$, (b) $C_1 = 1$, $f_{g,0} = 30$, and (c) $C_1 = 0.1$, $f_{g,0} = 3$, where C_1 is a reduction factor for type-I migration speed ($\dot{r}$) from the linear theory (eq. [1.1]). Smaller C_1 means slower migration. These results correspond to surviving protoplanets at $t \sim \tau_{\text{dep}}$ for $\tau_{\text{dep}} = 10^5, 10^6$, and 10^7 years, although depletion of f_g on time scales τ_{dep} is not taken into account in this result (in the Monte Carlo simulations, the exponential decay is assumed).

The results show that Σ_d is depleted in an inside-out manner. Ida & Lin (2008a) found through analytical argument that significant depletion occurs at

$$r \lesssim a_{\text{dep,mig}} \simeq C_1^{-1/8} \left(\frac{f_{g,0}}{3}\right)^{1/40} \left(\frac{t}{10^6 \text{yrs}}\right)^{1/4} \left(\frac{M_*}{M_\odot}\right)^{3/8} \text{AU}, \quad (2.5)$$

for $C_1 \gtrsim 0.1$. This boundary is in agreement with the critical location within which Σ_d has reduced from its initial values by an order of magnitude. Note that the dependences of $a_{\text{dep,mig}}$ on C_1 and $f_{g,0}$ are very weak. This is because for larger C_1 and/or $f_{g,0}$, embryos start migration at smaller mass and more number of embryos must be generated to clear the surface density.

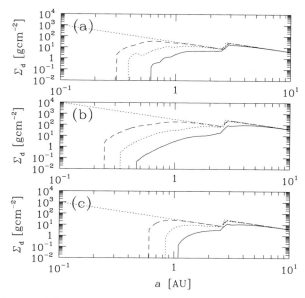

Figure 1. The evolution of Σ_d due to accretional sculpting by type-I migration for (a) $C_1 = 1$ and $f_{g,0} = 3$, (b) $C_1 = 1$ and $f_{g,0} = 30$, and (c) $C_1 = 0.1$ and $f_{g,0} = 3$. In this plot, we assume constant f_g. The distributions at $t = 10^5, 10^6$, and 10^7 years are expressed by dashed, dotted, and solid lines. The initial distribution is also shown by dotted lines.

During the early epoch of disk evolution, embryos form and migrate repeatedly to clear out the residual planetesimals. For larger $f_{g,0}$, the clearing is faster initially. Eventually, the disk gas is so severely depleted that relatively massive embryos no longer undergo significant amount of type-I migration. As shown in Figures 1, this self-regulation process retains the total mass of surviving embryos in terrestrial planet regions that is comparable to or smaller than that of the MMSN model. The typical individual mass of the surviving embryos is $\sim 0.1 M_\oplus$. These embryos can coalesce through giant impacts during and after the severe depletion of the disk gas (Kominami & Ida 2002).

Type-I migration leads to clearing of planetesimals close to their host stars and sets the inner edge of the embryos' population at $\sim$ 1AU. N-body simulations show that the most massive terrestrial planets tend to form in inner regions of the computational domain where the isolated embryos are initially placed. The lack of planets inside the Mercury's orbit in our Solar system might also be attributed to this result. The self-regulated clearing process ensures the formation of Mars to Earth-sized terrestrial planets in habitable zones. Even in the limit of large $f_{d,0}$, the reduction of Σ_d at $\lesssim$ a few AU inhibits *in situ* formation of gas giants interior to the ice line and facilitates the formation and retention probability of habitable terrestrial planets.

Although the formation of Mars to Earth-sized habitable planets depends only weakly on type-I migration speed, it is critical for the formation of cores of gas giant planets. For giant planets to actually form, sufficiently massive cores that accrete gas on timescales at least shorter than a few folding time of $\tau_{\rm dep}$ must be formed. For $\tau_{\rm dep} \sim 10^6$–10^7 years, we deduce, from eq. (1.2) with $k1 = 9$ and $k2 = 3$, that gas giant formation is possible only for $M_c \gtrsim$ several times $M_\oplus$. But, type-I migration suppresses the emergence of such massive cores in disk regions with relatively large Σ_g. For $C_1 \gtrsim 0.1$, the asymptotic core masses are generally much smaller than that needed to launch efficient gas accretion even though there is little decline in the magnitude of Σ_d.

2.3. Mass - semimajor axis distributions

In the Monte Carlo simulations, we first generate a 1,000 set of disks with various surface density of gas (Σ_g) and planetesimals (Σ_d). For each disk, 15 semimajor axes of the protoplanetary seeds are selected from a log uniform distribution in the ranges of 0.05–50AU, assuming that averaged orbital separation between planets is 0.2 in log scale (the averaged ratio of semimajor axes of adjacent planets is $\simeq 1.6$). We also assume that $\tau_{\rm dep}$ and M_* have log uniform distributions in the range of 10^6–10^7 yrs and 0.8–$1.25M_\odot$, respectively. (We focus on the effects of type-I migration but not dependences on stellar mass, so we consider solar-type stars). Corresponding to mm observations of T Tauri disks (e.g., Beckwith & Sargent 1996), it is assumed that $f_{g,0}$ has a log normal distribution which is centered on the value of $f_{g,0} = 1$ with a dispersion of 1 ($\delta \log_{10} f_{g,0} = 1.0$) for $M_* = M_\odot$. We assume $L_* \propto M_*^4$ and the distribution of $f_{g,0}$ is shifted in proportion to M_*^2 (Ida & Lin 2005). The choice of the M_*-dependences do not affect the results, because the range of M_* that we consider here is relatively narrow.

Figures 2a show the predicted M_p–a distributions for $C_1 = 0.03, 0.1, 0.3$ and 1. We also plot data of extrasolar planets around stars with $M_* = 0.8$–$1.25M_\odot$ discovered by radial velocity surveys (http://exoplanet.eu/). The planet masses M_p is a factor of $4/\pi$ times the values of $M_p \sin i$ determined from radial velocity measurements, assuming a sample of planetary systems with randomly oriented orbital planes.

Formation probability of gas giants dramatically changes with C_1. In order to quantitatively compare with observations, we determine the fraction (η_J) of stars with planets within the detectability limit by the magnitude of radial velocity ($v_r > 10$m/s) and orbital periods ($T_K < 4$ years). Because we artificially terminate type-I and II migrations near disk inner edge at a 2 day period ($\simeq 0.03$AU for $M_* = 1 M_\odot$) and we have not specified a survival criterion for the close-in planets, we exclude close-in planets with $a < 0.05$AU in the evaluation of η_J. The values of η_J are 12.8, 3.7, 0.4, and 0% for the models with $C_1 = 0.03, 0.1, 0.3$, and 1, respectively. Only for $C_1 \lesssim 0.03$, the predicted η_J can be comparable to the observed data (Fischer & Valenti 2005). In models with higher C_1, only the low-mass cores can survive type-I migration. The envelope contraction time scales for these low-mass cores are generally much longer (eq. [1.2]) than the gas depletion time scales. Consequently, η_J is very small. Ida & Lin (2008a) showed that a mass function of close-in planets is also sensitively dependent on the magnitude of C_1 and it will be able to be calibrated from an observed mass function of close-in planets.

In these models, we approximate the gas accretion rate with $(k1, k2) = (9, 3)$ in eq. (1.2). In view of uncertainties in the gas accretion rate, we also simulated models with $k1 = 8$ and 10, with $k2 = 3$ for all cases. For models with $C_1 = 0.03$–0.1 in which type-I migration marginally suppresses the formation of gas giants, the magnitude of η_J depends sensitively on the minimum mass for the onset of dynamical gas accretion (which is represented by $k1$). A smaller value of $k1(= 8)$ can lead to a significant increase in η_J because smaller mass cores can initiate the runaway gas accretion within $\tau_{\rm dep} \sim 10^6$–10^7 years (eq. [1.2]).

The survival of terrestrial planets depends on their post-formation encounter probability with migrating giant planets. In the absence of any type-I migration, this probability is modest. But the inclusion of a small amount of type-I migration significantly reduces the fraction of stars with massive close-in gas giants because the retention of the progenitor cores becomes possible only at the late stages of disk evolution when the magnitude of Σ_g is reduced and type-II migration is no more efficient. Repeated migration of gas giants is less common in models with $C_1 \gtrsim 0.01$ than those with $C_1 = 0$. The low type-II migration probability reduced the need for efficient disruption of largely accumulated

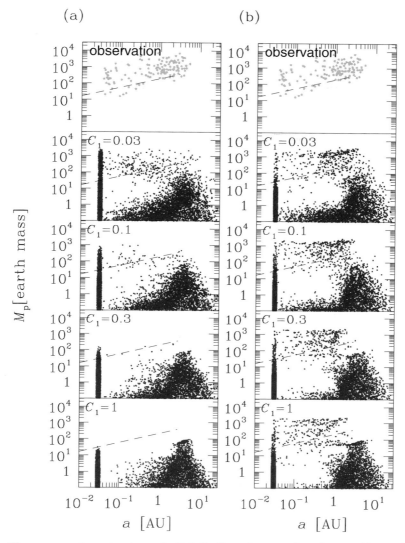

Figure 2. The mass and semimajor axis distribution of extrasolar planets. Units of the mass (M_p) and semimajor axis (a) are Earth mass ($M_\oplus$) and AU. (a) Disks with Σ_g of a power-law radial dependence and (b) Disks with a bump in Σ_g due to the coupling effect of MRI activity and the ice line. The top panels are observational data of extrasolar planets (based on data in http://exoplanet.eu/) around stars with $M_* = 0.8$–$1.25 M_\odot$ that were detected by the radial velocity surveys. The determined $M_p \sin i$ is multiplied by $1/\langle \sin i \rangle = 4/\pi \simeq 1.27$, assuming random orientation of planetary orbital planes. The other panels are theoretical predictions with $M_* = 0.8$–$1.25 M_\odot$ for various values of C_1. The dashed lines express observational limit with radial-velocity measure precision of $v_r = 10$m/s. In these models, the magnitude of the metallicity [Fe/H] = 0.1 and the contraction time scale parameters in eq. (1.2) are assumed to be $(k1, k2) = (9, 3)$.

close-in planets. It also ensures that most of the terrestrial planets formed in the HZs are not removed by the migrating gas giants. Note that type-I migration also inhibits *in situ* formation of gas giants near 1AU (§2.2). Thus, a small amount of type-I migration facilitates formation and retention of terrestrial planets in HZs in extrasolar planetary systems, rather than inhibits them.

Ida & Lin (2004b) showed through a model without the effects of type-I migration that η_J increases with [Fe/H]. The correlation reflects the fact that high [Fe/H] enhances formation of cores large enough for runaway gas accretion. The effect of type-I migration enhances the η_J–[Fe/H] correlation, because type-I migration is more efficient in metal-poor disks (Ida & Lin 2008a). The steep dependence is in a better agreement with the observed data (Fischer & Valenti 2005).

The dependence on M_* was studied by Ida & Lin (2006), using a simple model without the effects of type-I migration. They predicted that gas giants are much more rare around M dwarfs than around FGK dwarfs while super-Earths are abundant around M dwarfs. These conclusions are not changed by the effects of type-I migration (Ida & Lin 2008a). Around M dwarfs, super-Earths at 1–3AU, which are inferred to be abundant by microlensing survey, survive type-I migration.

3. Disks with non-uniform surface density

3.1. *Disk model*

The surface density of planetesimals is the same as eq. (2.1). However, in order to take into account the effect of spatial non-uniformity of the alpha parameter (α) of disk viscosity (Shakura & Sunyaev 1973), that of gas is modeled as

$$\Sigma_g = \frac{\dot{M}}{3\pi\nu} \qquad (3.1)$$
$$\dot{M} = 3 \times 10^{-9} f_{g,0} \exp(-t/\tau_{\rm dep}) \, [M_\odot/{\rm yr}],$$

where $\nu = \alpha c_s h$. Since the variation of Σ_g due to change in $\eta_{\rm ice}$ across $a_{\rm ice}$ is important, in this section the change in $\eta_{\rm ice}$ is smoothed out by a tanh function with width of scale height h.

Assuming the disk temperature given by eq. (2.2), $\alpha_{\rm active} = 10^{-3}$ and $\alpha_{\rm dead} = 10^{-4}$, and surface density of the surface MRI active layer (Ida & Lin 2008a),

$$\Sigma_A = \min\left(6\eta_{\rm ice}^{-1}\left(\frac{r}{1{\rm AU}}\right)^3 \, [{\rm g/cm^2}], \Sigma_g\right), \qquad (3.2)$$

the equilibrium Σ_g distribution is given as a function of $\dot{M}$ in Figure 3. When Σ_g declines to the values comparable to Σ_A near the ice line, the Σ_g distribution has a positive gradient near the ice line. As mentioned in §1.4, type-I migration would be halted in the positive gradient regions.

With these prescriptions, we carried out the Monte Carlo simulations to predict mass and semimajor axis of extrasolar planets. The assumed distributions of $f_{g,0}$, $\tau_{\rm dep}$, and M_* are the same as those in §2.3. The mass accretion rate ($\dot{M}$) in this range of $f_{g,0}$ corresponds to late T Tauri stage (Calvet, Hartmann & Strom 2000). Figures 2b show the predicted M_p–a distributions for $C_1 = 0.03, 0.1, 0.3$ and 1. The values of η_J are 16.2, 14.3, 11.9, and 9.5% for the models with $C_1 = 0.03, 0.1, 0.3$, and 1, respectively. Compared with the case without the ice line effect (Figures 2a), η_J is enhanced, in particular for relatively large C_1, and the dependence on C_1 is significantly weakened. Even for $C_1 \simeq 0.3$–1, the predicted η_J can be comparable to the observed data. Cores are trapped near the ice line, almost independent of the strength of type-I migration, and they accrete planetesimals until they can start runaway gas accretion.

The active layer thickness given by eq. (3.2) is ten times higher than that used in Kretke & Lin (2007), because it has large uncertainty and we intended to highlight the importance of the effect of the ice line. Figures 2 suggest that the coupling effect of MRI

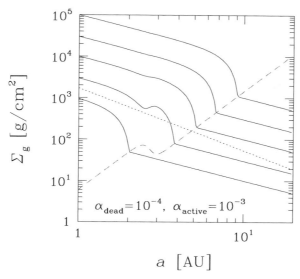

Figure 3. The evolution of Σ_g with the coupling effect of MRI and the ice line. The solid lines express the Σ_g distributions for $\dot{M} = 10^{-7}, 3 \times 10^{-8}, 10^{-8}, 3 \times 10^{-9}, 10^{-9} M_\odot/\mathrm{yr}$ from top to bottom. The dashed and dotted lines are Σ_A and Σ_g of the MMSN model.

activity and the ice line can enhance formation and retention rates of gas giants without significant reduction of the strength of type-I migration.

Since the formation and retention of gas giants are facilitated only in the regions beyond the ice line, *in situ* formation of gas giants in HZs are still inhibited by type-I migration. Figures 2 suggest that the formation probability of hot jupiters is apparently low in the cases with the ice line effect compared with the cases without the effect for similar η_J. Hence, the property that type-I migration facilitates formation and retention of terrestrial planets in HZs is preserved.

The effect of the ice line also preserves the dependence on M_* and the correlation between η_J and [Fe/H]. But, the correlation is now similar to the case without type-I migration, which is weaker than that in the cases with type-I migration but without the ice line effect. Because the results shown here are limiting cases in which the ice line effect is maximally efficient, realistic cases may be in between the results here and the results without the ice line effect.

4. Summary

We have investigated the effects of type-I migration of protoplanetary embryos on mass and semimajor axis distributions of extrasolar planets, through Monte Carlo simulations. In disks with gas surface density having a power-law radial dependence, the type-I migration speed has to be an order of magnitude slower than that given by the linear theory, to be consistent with the existing observational data of extrasolar gas giants. However, a bump in gas surface density can be produced by change in the thickness of MRI active layer near the ice line. The bump can trap migrating protoplanetary embryos. Even with the type-I migration strength similar to that predicted by the linear theory, the formation/retention probability of gas giants can be comparable to the observational data.

The introduction of type-I migration preserves the dependence on M_* that gas giants are much more rare around M dwarfs than around FGK dwarfs while super-Earths are abundant around M dwarfs. The correlation that η_J increases with [Fe/H] is enhanced by the effect of type-I migration. It is in a better agreement with the observed data, although the enhancement is weak in the disk with the bump.

The introduction of type-I migration inhibits *in situ* formation of gas giants in HZs and reduces the probability of passage of gas giants through HZs, both of which facilitate retention of terrestrial planets in HZs. The strength of type-I migration can be constrained by a mass function of close-in planets, which will be tested by transit observation from space.

References

Baruteau, C. & Masset, F. 2008, this proceedings
Beckwith, S. V. W. & Sargent, A. I., 1996, Nature, 383, 139
Calvet, N., Hartmann, L. & Strom, S. E. 2000, in Protostars and Planets IV, ed. V. Mannings, A. P. Boss and S. S. Russell (Tucson:Univ. of Arizona Press), 377
Daisaka, K. J., Tanaka, H. & Ida, S. 2006, Icarus, 185, 492
Davis, S. S. 2005, ApJ, 620, 994
Fischer, D. A. & Valenti, J. A. 2005, ApJ, 622, 1102
Fromang, S., Terquem, C. & Nelson , R. P. 2005, MNRAS, 363, 943
Garaud, P. & Lin, D. N. C. 2007, ApJ, 654, 606
Goldreich, P., & Tremaine, S. 1980, ApJ, 241, 425
Guillot, T. & Hueso, R. 2006, MNRAS, 367, L47
Hayashi, C. 1981, Prog. Theor. Phys. Suppl., 70, 35
Ida, S. & Lin, D. N. C. 2004a, ApJ, 604, 388
Ida, S. & Lin, D. N. C. 2004b, ApJ, 616, 567
Ida, S. & Lin, D. N. C. 2005, ApJ, 626, 1045
Ida, S. & Lin, D. N. C. 2008, ApJ, in press
Ida, S. & Lin, D. N. C. 2008, in preparation
Kokubo, E. & Ida, S. 1998, Icarus, 131, 171
Kokubo, E. & Ida, S. 2002, ApJ, 581, 666
Kominami, J. & Ida, S. 2002, Icarus, 157, 43
Koller, J., Li, H. 2004, in The Search For Other Worlds. AIP Conf. Proc., 713, pp. 63-66
Kretke, K. A. & Lin, D. N. C. 2007, ApJ, 664, L55
Laughlin, G., Steinacker, A. & Adams, F. C. 2004, ApJ, 309, 846
Li, H. *et al.* 2005, ApJ, 624, 1003
McNeil, D., Duncan, M. & Levison, H. F. 2005, AJ, 130, 2884
Marcy, G. *et al.* 2005, Prog. Theor. Phys. Suppl., 158, 24
Masset, F. S., D'Angelo, G. & Kley, W., 2006, ApJ, 652, 730
Masset, F. S., Morbidelli, A., Crida, A., Ferreira, J., 2006, ApJ, 642, 478
Mayor, M., Pont, F. & Vidal-Madjar, A. 2005, Prog. Theor. Phys. Suppl., 158, 43
Muto, T. & Inutsuka, S. 2008, this proceedings
Nelson, R. P. & Papaloizou, J. C. B. 2004, MNRAS, 350, 849
Sano, T., Miyama, S. M., Umebayashi, T., & Nakano, T., 2000, ApJ, 543, 486
Shakura, N. I. & Sunyaev, R. A. 1973, A&A, 24, 337
Tanaka, H., Takeuchi, T. & Ward, W. 2002, ApJ, 565, 1257
Thommes, E. & Murray, N. 2006, ApJ, 644, 1214
Ward, W. 1986, Icarus, 67, 164
Zhang, X., Lin, D. N. C. & Kretke, K. A. 2008,in: Y.-S. Sun, S. Ferraz-Mello & J.-L. Zhou, (eds.), *Exoplanets: Detection, Formation and Dynamics*, Proc. IAU Symposium No. 249 (Suzhou,China), p. 309

Terrestrial planet formation in extra-solar planetary systems

Sean N. Raymond[1]

[1]NASA Postdoctoral Fellow, Center for Astrophysics and Space Astronomy and Center for Astrobiology, University of Colorado, Campus Box 389, Boulder, CO, 80309-0389, USA
email: raymond@lasp.colorado.edu

Abstract. Terrestrial planets form in a series of dynamical steps from the solid component of circumstellar disks. First, km-sized planetesimals form likely via a combination of sticky collisions, turbulent concentration of solids, and gravitational collapse from micron-sized dust grains in the thin disk midplane. Second, planetesimals coalesce to form Moon- to Mars-sized protoplanets, also called "planetary embryos". Finally, full-sized terrestrial planets accrete from protoplanets and planetesimals. This final stage of accretion lasts about 10-100 Myr and is strongly affected by gravitational perturbations from any gas giant planets, which are constrained to form more quickly, during the 1-10 Myr lifetime of the gaseous component of the disk. It is during this final stage that the bulk compositions and volatile (e.g., water) contents of terrestrial planets are set, depending on their feeding zones and the amount of radial mixing that occurs. The main factors that influence terrestrial planet formation are the mass and surface density profile of the disk, and the perturbations from giant planets and binary companions if they exist. Simple accretion models predicts that low-mass stars should form small, dry planets in their habitable zones. The migration of a giant planet through a disk of rocky bodies does not completely impede terrestrial planet growth. Rather, "hot Jupiter" systems are likely to also contain exterior, very water-rich Earth-like planets, and also "hot Earths", very close-in rocky planets. Roughly one third of the known systems of extra-solar (giant) planets could allow a terrestrial planet to form in the habitable zone.

Keywords. astrobiology, methods: n-body simulations, solar system: formation, planetary systems: formation, planetary systems: protoplanetary disks, stars: late-type

1. Introduction

Recent research has developed a model for the growth of the terrestrial planets in the Solar System via collisional accumulation of smaller bodies (e.g., Wetherill 1990). Several distinct dynamical stages have been identified in this process, from the accumulation of micron-sized dust grains to the late stage of giant impacts. In this article, I first review the current state of knowledge of the stages of terrestrial planet formation (§ 2). Second, I explore the effects of external parameters on the accretion process, including new work showing that simple accretion models predict that low-mass stars should preferentially harbor low-mass terrestrial planets in their habitable zones (§ 3). These include the orbit of a giant planet and the surface density profile of the protoplanetary disk. Third, I examine the effects of the migration of a Jupiter-mass giant planet on the accretion of terrestrial planets (§ 4), and also apply accretion models to the known set of extra-solar planets to show that roughly one third of the known systems could have formed an Earth-like planet in the habitable zone (§ 5). Finally, conclusions and avenues for future study are presented (§ 6).

2. Stages of Terrestrial Planet Growth

The process of planet formation starts when a parcel of gas with a giant molecular cloud becomes gravitationally unstable, and completes on a $\sim 10^8$ year timescale with the formation of a complete planetary system. I cannot hope to encompass all relevant processes in this short article. Therefore, I will start from a protoplanetary disk in which dust has settled to the thin "mid-plane" of the disk, a process which requires $\sim 10^4$ years in a non-turbulent disk (Weidenschilling 1980). I will review three distinct dynamical stages of terrestrial planet growth: (i) the formation of km-sized planetesimals starting from dust grains; (ii) the formation of $\sim$1000-km sized protoplanets (also called "planetary embryos"); and (iii) the formation of full-sized terrestrial planets. For more detailed reviews, the reader is referred to Lissauer (1993), Chambers (2004), and Papaloizou & Terquem (2006).

Models suggest that the surface density of solid material Σ in disks roughly follow a power law with radial distance r, i.e.,

$$\Sigma = \Sigma_1 r^{-\alpha}, \qquad (2.1)$$

where Σ_1 is the surface density at 1 AU and α controls the radial distribution of solids (not to be confused with the viscosity parameter of the same name). In the "minimum-mass solar nebula" (MMSN) model of Weidenschilling (1977a) and Hayashi (1981), $\alpha = 3/2$. However, new analyses of the MMSN model derive values for α between $1/2$ and 2 (Davis 2005; Desch 2007). In addition, current models and observations generally favor $\alpha \approx 1$ (Dullemond *et al.* 2007; Andrews & Williams 2007; Garaud & Lin 2007).

From dust to planetesimals. The growth of km-sized planetesimals from micron-sized dust grains presents a significant modeling challenge, which can be constrained to some degree by observations of dust populations in disks around young stars (Dullemond & Dominik 2005). Collisional growth of micron-sized grains, especially if they are arranged into fluffy aggregates, appears efficient for relatively small particle sizes and impact speeds of $\sim 1\,m\,s^{-1}$ or slower (Dominik & Tielens 1997; Wurm & Blum 2000; Poppe *et al.* 2000; Benz 2000; see Dominik *et al.* 2007 and references therein). However, there is a constant battle between disk turbulence, which increases random velocities, and drag-induced settling, which reduces them (Cuzzi *et al.* 1993). Growth of particles in such collisions appears effective until they reach roughly 1 m in size. At that point, continued growth may be suppressed by collision velocities of $\gtrsim 10\,m\,s^{-1}$ (Dominik *et al.* 2007), as well as rapid inspiralling of m-sized bodies due to aerodynamic drag (Weidenschilling 1977b). Indeed, the timescale for infall can be as short as 100 years, leading to what has been called the "meter-size catastrophe". Collisional growth models must quickly cross the barrier at meter-sizes (Weidenschilling & Cuzzi 1993; Benz 2000; Weidenschilling 2000).

If solids can be sufficiently concentrated in the disk relative to the gas, then local gravitational instability may occur, leading to the top-down formation of planetesimals (Goldreich & Ward 1973; Youdin & Shu 2002). Concentration of particles has been proposed to occur at local maxima in the disk pressure (Haghighipour & Boss 2003), which can exist as a result of turbulence in the disk (Johansen *et al.* 2007). It is interesting to note that m-sized particles are the fastest to drift toward pressure maxima (Johansen *et al.* 2006), for the same reason as the "meter-size catastrophe." An alternate location for planetesimal formation via gravitational instability are regions with an increased local density of solids (Goodman & Pindor 2000), via concentration due to drag-induced inspiralling (Youdin & Chiang 2003), in vortices (Tanga *et al.* 1996), or even via photo-evaporative depletion of the gas layer (Throop & Bally 2005).

Thus, the details of planetesimal formation are not fully resolved. The current best understanding invokes a combination of collisional sticking of grains at small sizes until they reach ~ 1 m in size, concentration of these boulders in pressure maxima, and subsequent growth, via either gravitational instability or collisional accretion, into km-sized planetesimals. These objects are the building blocks of terrestrial planets.

From planetesimals to protoplanets. While gas is still present in the disk, eccentricities of planetesimals are damped quickly due to aerodynamic gas drag (Adachi et al. 1976). While velocities remain low, bodies that are slightly larger than the typical size can accelerate their growth due to gravitational focusing (Safronov 1969; Greenberg et al. 1978):

$$\frac{dM}{dt} = \pi R^2 \left(1 + \frac{v_{esc}^2}{v_{rand}^2}\right), \quad (2.2)$$

where R represents the body's physical radius, v_{esc} is the escape speed from the body's surface ($= \sqrt{2GM/R}$), and v_{rand} represents the velocity dispersion of planetesimals. While random velocities are small, gravitational focusing can increase the growth rates of bodies by a factor of hundreds, such that $dM/dt \sim M^{4/3}$, leading to a phase of "runaway growth" (Greenberg et al. 1978; Wetherill & Stewart 1989, 1993; Ida & Makino 1992; Kokubo & Ida 1996; Goldreich et al. 2004). The length of this phase depends on the timescale for random velocities of planetesimals to approach the escape speed of the larger bodies. For small (~ 100 m-sized) planetesimals, gas drag is stronger such that runaway growth can be prolonged and embryos may be larger and grow faster (Chambers 2006).

Eventually, the random velocities of planetesimals are increased by gravitational interactions with the larger bodies that have formed, in a process called "viscous stirring" (Ida & Makino 1992a). During this time, the random velocities of large bodies are kept small via dynamical friction with the swarm of small bodies (Ida & Makino 1992b). Gravitational focusing is therefore reduced, and the growth of large bodies is slowed to the geometrical accretion limit, such that $dM/dt \sim M^{2/3}$ (Ida & Makino 1993; Rafikov 2003). Nonetheless, large bodies continue to grow, and jostle each other such that a characteristic spacing of several mutual Hill radii $R_{H,m}$ is maintained ($R_{H,m} = 0.5[a_1 + a_2][M_1 + M_2/3M_\star]^{1/3}$, where a_1 and M_1 denote the orbital distance and mass of object 1, etc; Kokubo & Ida 1995). This phase of growth is often referred to as "oligarchic growth", as just a few large bodies dominate the dynamics of the system, with reduced growth rates and increased interactions between neighboring protoplanets (Kokubo & Ida 1998; Leinhardt & Richardson 2005).

Oligarchic growth tends to form systems of protoplanets with roughly comparable masses and separations of 5-10 mutual Hill radii (Kokubo & Ida 1998, 2000; Weidenschilling et al. 1997). Masses and separations of protoplanets depend on the total mass and surface density distribution of the disk (Kokubo & Ida 2002). Typical protoplanet masses in a solar nebula model are a few percent of an Earth mass, i.e., roughly lunar to Mars-sized (Kokubo & Ida 2000; Collis & Sari 2007). Fig 1 shows three distributions of protoplanets: each contains 10 $M_\oplus$ between 0.5 and 5 AU (roughly twice the minimum-mass disk), with 7 $R_{H,m}$ between adjacent protoplanets (from Raymond et al. 2005b). Note that for surface density profiles steeper than r^{-2}, the protoplanet mass decreases with orbital distance.

It is important to realize that terrestrial protoplanets are the same objects as giant planet cores, assuming giant planets to form via the bottom-up, "core-accretion" scenario (Mizuno 1980; Pollack et al. 1996; Ida & Lin 2004; Alibert et al. 2005). Models of core-accretion estimate the growth rate of an isolated core, then calculate the accretion of gas

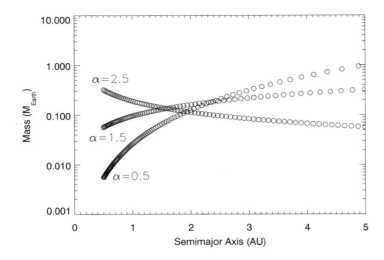

Figure 1. Three sets of protoplanets at the end of oligarchic growth, assuming they formed with separations of 7 mutual Hill radii. Each set of protoplanets contains 10 $M_\oplus$ between 0.5 and 5 AU, differing only in the radial surface density exponent α. From Raymond et al. (2005b).

onto the core, generally neglecting the potentially important effects of nearby cores. Note also that protoplanets may excite spiral density waves in the gaseous disk and undergo inward type 1 migration on a 10^{4-5} year timescale (Goldreich & Tremaine 1980; Ward 1986; Masset et al. 2006). This can reduce the efficiency of terrestrial planet growth by removing a significant fraction of material (McNeil et al. 2005; Ida & Lin 2007).

From protoplanets to terrestrial planets. Once the local mass in planetesimals and protoplanets (i.e., "mall" and "large" bodies, respectively) is comparable, feeding zones of neighboring protoplanets overlap, and giant collisions between protoplanets begin to occur (Wetherill 1985; Kenyon & Bromley 2006). This is the start of the final phase of terrestrial planet growth, sometimes called "late-stage accretion" (Wetherill 1996; Chambers & Wetherill 1998). During this stage, planets grow by accreting other protoplanets as well as planetesimals (Chambers 2001; Raymond et al. 2006a; O'Brien et al. 2006). Eccentricities of protoplanets are generally smaller than for planetesimals because of dynamical friction, but not small enough to prevent protoplanet collisions. Growing planets clear their nearby zones of material, and their feeding zones widen and move outward in time (Raymond et al. 2006a). During the very late stages of accretion, there can be substantial mixing across the inner planetary system, with important implications for terrestrial planet compositions (see below).

Giant collisions between protoplanets may not always be accretionary: high-speed or off-center collisions can actually erode the target mass (Agnor & Asphaug 2004; Asphaug et al. 2006; Genda et al. 2007, in preparation). High-speed collisions may alter the planetary composition, by preferentially removing more volatile materials; this is a proposed explanation for Mercury's large iron core (Wetherill 1988; Benz et al. 1988). In addition, low-speed, off-center collisions have the potential to create a circum-planetary disk of debris from which a large moon may accrete (Benz et al. 1987; Ida et al. 1997; Canup & Asphaug 2001; Canup 2004). Note, however, that such low-speed collisions are rare, especially at later times when a planet's nearby zone has been cleared out (Agnor et al. 1999).

Note that giant planets are constrained to form in the 1-10 Myr lifetimes of the gaseous component of protoplanetary disks (Haisch et al. 2001; Briceño et al. 2001; Pascucci et al. 2006). Thus, if giant planets form in a planetary system, as Jupiter and Saturn did in the Solar System, then they can have an effect on the late-stage accretion of terrestrial planets. Indeed, it has been shown that giant planets have a large impact on the final assembly of terrestrial planets (Levison & Agnor 2003; Raymond et al. 2004; Raymond 2006; O'Brien et al. 2006). For example, in systems with less massive giant planets, induced eccentricities are smaller, reducing the width of feeding zones and causing the formation of more, smaller terrestrial planets as compared with a more massive giant planet. In addition, in systems with more massive or eccentric giant planets, the innermost terrestrial planet tends to be the most massive (Levison & Agnor 2003).

Figure 2 shows snapshots in time of the accretion of terrestrial planets from a simulation by Raymond et al. (2006a), including 9.9 $M_\oplus$ of material from 0.5-5 AU (roughly twice the mass of the MMSN model). The simulation included a Jupiter-mass giant planet on a circular orbit at 5.5 AU, which is somewhat different than its current orbit, although it is consistent with the 'Nice' model of the giant planets' orbital evolution (Tsiganis et al. 2005). Several mean motion resonances are clearly visible as vertical spikes in eccentricity in the '0.1 Myr' panel. Eccentricities are driven in the inner disk via interactions between protoplanets, and in the outer disk via secular and resonant perturbations from the giant planet. Large bodies grow more quickly closer to the Sun, because of the faster orbital timescales; by 10 Myr the planet at ~ 1 AU has reached 1 $M_\oplus$. However, large-scale mixing between zones does not happen until about 20 Myr, when the feeding zones of all three final planets overlap and encompass the entire terrestrial zone, out to $\sim$4 AU (Raymond et al. 2007a). The final configuration of three planets contains a 1.5 $M_\oplus$ planet at 0.55 AU, a 2 $M_\oplus$ planet at 0.98 AU, and a 0.95 $M_\oplus$ planet at 1.93 AU (see Raymond et al. 2006a for details). The orbits of the planets have slightly higher eccentricities than the current-day terrestrial planets; in similar simulations, O'Brien et al. (2006) formed terrestrial planet systems with eccentricities even lower than those of Earth, Venus and Mars.

Formation times of Earth-like planets from dynamical simulations are 10-100 Myr, comparable to timescales derived from Hf/W and other isotopic systems for the last core-forming event on Earth (e.g., Kleine et al. 2002; Jacobsen 2005). Although this formation timescale is far longer than the gaseous disk lifetime, Interactions with the dissipating gaseous disk may affect this final stage of accretion. Important processes include tidal gaseous drag (also called "type 1" damping; Ward 1993; Agnor & Ward 2002; Kominami & Ida 2002, 2004) and secular resonance sweeping, which can be induced by the changing gravitational potential as the disk dissipates (Ward 1981; Nagasawa et al. 2005).

The final compositions of terrestrial planets are determined mainly during this final stage of terrestrial accretion. A planet's composition is mainly determined by its feeding zone, i.e., the sum of all the material incorporated during formation (as well as physical processes during and after accretion). If planets form locally, then their compositions are a simple reflection of the composition of the local building blocks. Thus, eccentricity-driven mixing between different radial zones during accretion is perhaps the key process that determines the final planetary composition.

The Earth's water content is anomalously high: nebular models suggest that the local temperature at 1 AU was too hot to allow for hydration of planetesimals (Boss 1998). Thus, it is thought that Earth's water was "delivered" from more distant regions, in the form of hydrated asteroids (Morbidelli et al. 2000; Raymond et al. 2004, 2007b) or comets (Owen & Bar-Nun 1995). The D/H ratio of Earth's water is virtually identical to that of carbonaceous chondrites, which are linked to C type asteroids in the outer

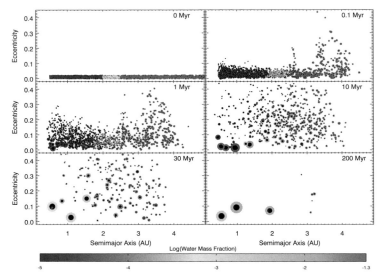

Figure 2. Snapshots in time from a simulation of the final growth of terrestrial planets, starting from 1885 sub-isolation mass objects (from Raymond *et al.* 2006a). The size of each body is proportional to its mass$^{1/3}$, the dark circle represents the relative size of each body's iron core (in the black and white version, iron cores are shown only for bodies larger than 0.05 $M_\oplus$), and the color corresponds to its water content (red = dry, blue = 5% water; in the black and white version, white = dry and black = 5% water; see color bar). For a movie of this simulation, go to http://lasp.colorado.edu/~raymond/ and click on "movies and graphics".

main belt (Robert & Epstein 1982; Kerridge 1985; see Table 1 of Morbidelli *et al.* 2000). Comets, though poorly sampled, appear to have D/H ratios two times higher than Earth (Balsiger *et al.* 1995; Meier *et al.* 1998; Bockelee-Morvan *et al.* 1998), suggesting that primitive outer asteroid material may be the best candidate for the source of Earth's water. In Fig. 2, material is given a starting water content that matches the values for primitive meteorites (Abe *et al.* 2000; see Fig. 4 of Raymond *et al.* 2004). Radial mixing during formation delivers water from the primitive outer asteroid belt (beyond 2.5 AU) to the growing terrestrial planets. The amount of water delivered in planetesimals vs. protoplanets is comparable (Raymond *et al.* 2007a). Thus, we expect that water delivery from planetesimals is statistically robust, while the amount of additional water from a small number of water-rich protoplanets can vary significantly from system to system.

3. Effects of External Parameters on Terrestrial Accretion

Dynamical simulations of terrestrial planet growth display a wide range in outcomes, in terms of planet masses, orbits and compositions (Wetherill 1996; Agnor *et al.* 1999; Chambers 2001; Raymond *et al.* 2004). The stochastic nature of accretion is due to the importance of individual scattering events during the very late phases of accretion. Indeed, Quintana & Lissauer (2006) showed that varying the initial conditions by moving a single embryo by just 1 meter in its same orbit could cause a significant change in the outcome. Nonetheless, systematic trends are seen between certain system parameters and the properties of the planets that form. Many simulations with a given starting condition are needed to suppress stochastic effects and to understand the systematic trends.

To date, systematic trends have been established with the following system parameters: 1) the total disk mass; 2) the disk's surface density profile; 3) the stellar mass; 4) the mass and 5) orbit of external giant planets; and 6) the mass and 7) separation of a binary stellar companion. Here I summarize the effect of these parameters on terrestrial planet formation, in terms of the terrestrial planets' final masses, orbits, and compositions (specifically water contents, in the context of water delivery from an asteroidal source, as described in § 2).

Disk Mass. There exists a range of roughly two orders of magnitude in observed circumstellar disk masses (Andre & Montmerle 1994; Eisner & Carpenter 2003; Andrews & Williams 2005; Scholz et al. 2006). An increase in the disk mass allows more massive embryos to form than a fiducial case (Kokubo & Ida 2002). These embryos have stronger gravitational interactions, leading to larger eccentricities and therefore wider feeding zones. Thus, a disk with a larger total mass in solids will tend to form a smaller number of more massive terrestrial planets than a less massive disk (Raymond et al. 2005a, 2007; Kokubo et al. 2006). The planet mass, M_p, scales with the disk mass, M_d, as $M_d^{1.1}$ (Kokubo et al. 2006; Raymond et al. 2007b). The super-linear scaling is due to the increase in feeding zone width. In addition, a wider feeding zone implies increased radial mixing, such that more massive disks tend to form more water-rich planets in the habitable zone (Raymond et al. 2007b).

Disk Surface Density Profile. The surface density profile of protoplanetary disks is uncertain; estimates range from $\Sigma \sim r^{-1/2}$ to r^{-2} (i.e., from eqn. 2.1, $\alpha = 0.5\text{-}2$; e.g., Weidenschilling 1977a; Davis 2005; Kuchner 2004; Desch 2005; Andrews & Williams 2007). Systematic simulations of disks with varying surface density profiles were run by Raymond et al. (2005b) and Kokubo et al. (2006). They found that for steeper profiles (higher values of α), the terrestrial planets (1) are more numerous, (2) form more quickly, (3) form closer to the star, (4) are more massive, (5) have higher iron contents, and (6) have lower water contents. However, the possibility of forming potentially habitable planets (water-rich planets in the habitable zone) does not vary strongly with α.

Stellar Mass. A very rough correlation appears to exist between protoplanetary disk mass and the mass of the host star, with a scatter of about 2 orders of magnitude for a given stellar mass (e.g., Fig. 3 of Scholz et al. 2006). Indeed, eqn. 2.1 can be rewritten as:

$$\Sigma = \Sigma_1 \, f \, Z \, r^{-\alpha} \left(\frac{M_\star}{M_\odot}\right)^h, \tag{3.1}$$

where f represents the relative disk mass, Z represents the disk metallicity on a linear scale (i.e., how much of the disk mass is in solids), and $M_\star$ is the stellar mass (Ida & Lin 2005; Raymond et al. 2007b). Current best-fit values are $\alpha \sim 1$ (as discussed above) and $h \sim 1$, while f and Z are unique to each disk and have inherent distributions (e.g., Andrews & Williams 2005; Nordstrom et al. 2004). Given that terrestrial planets are thought to form interior to the snow line, the location of terrestrial planet formation depends on the stellar luminosity, which is a strong function of the stellar mass (e.g., Hillenbrand & White 2004). Indeed, the habitable zones of low-mass stars are very close-in (Kasting et al. 1993), where the amount of material available to form terrestrial planets is limited.

Figure 3 shows the expected mean mass of terrestrial planets in the habitable zone as a function of stellar mass in the fiducial model of Raymond et al. (2007b), with $fZ = 1$, $h = 1$, and $\alpha = 1$. The data come from 20 simulations of the late-stage accretion of terrestrial planets from protoplanets, with the error bars signifying the range of outcomes. The curve represents a very simple model, in which the mass of a terrestrial planet scales linearly with the local mass (as expected from results of Kokubo et al. 2006). Both the curve and data are calibrated to match 1 Earth mass for a Solar-mass star. The "habitable

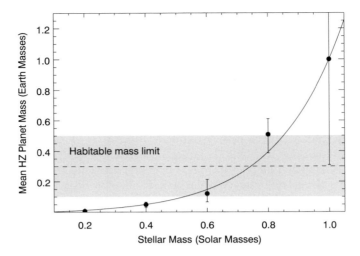

Figure 3. An estimate of the mass of terrestrial planets formed via accretion in the habitable zone as a function of stellar mass, assuming $f = \alpha = h = 1$ (see eqn.3.1). From Raymond et al. (2007b).

planet limit" of 0.1-0.5 $M_\oplus$ represents the range of expected planet masses capable of sustaining life for long timescales (Williams et al. 1997; Raymond et al. 2007b).

Raymond et al. (2007b) therefore show that a simple *in situ* accretion model predicts that low-mass stars should typically harbor low-mass terrestrial planets in their habitable zones. This result is contingent on the parameters of eqn. 3.1, especially the value of h. In addition, this study neglects interactions with the gaseous disk as well as tidal effects, both of which could be important for terrestrial planets around low-mass stars (Ida & Lin 2005; Barnes et al. 2007). If many large terrestrial planets are discovered in the habitable zones of low-mass stars, the study of Raymond et al. (2007b) provides evidence that these planets did not form via *in situ* accretion, but rather migrated to their current locations (see, e.g., Terquem & Papaloizou 2007). In fact, it may be possible to differentiate planet formation models in systems with close-in terrestrial planets if the planet transits its host star, given the relation between planet size and composition (Valencia et al. 2007; Seager et al. 2007; Fortney et al. 2007; Selsis et al. 2007a; Raymond et al. 2007c).

Giant Planet Mass. The effects of giant planet mass mimic those of the disk mass in some ways. A more massive giant planet excites larger eccentricities among protoplanets, leading to wider feeding zones and therefore fewer, more massive terrestrial planets (Levison & Agnor 2003; Raymond et al. 2004). In addition, large eccentricities among embryos often causes the innermost terrestrial planet to be the most massive (Levison & Agnor 2003). However, giant planets in virtually all configurations are destructive to asteroidal water delivery (Raymond 2007, in preparation). The larger eccentricities induced by a more massive giant planet cause a larger fraction of water-rich material (2-4 AU) to become unstable and be removed from the system, usually via ejection (Raymond 2007, in preparation).

Giant Planet Orbit. The current explanation for the deficit of mass in the asteroid belt is that the material was removed because of the dynamical effects of Jupiter (Wetherill 1992; Petit et al. 2001; Bottke et al. 2005). There exists a minimum separation between a giant planet and the closest terrestrial planet that can form, of a factor of 3-5 in orbital period (Raymond et al. 2005b; Raymond 2006). Thus, terrestrial planet formation is inhibited in the vicinity of a giant planet, such that a giant planet's orbital distance provides information as to where terrestrial planets could have formed in the system (see § 5 and Raymond 2006). In addition, for giant planets on more distant orbits, more water-rich material is available to the terrestrial planet region, allowing planets to be somewhat more water-rich, although this effect is small compared with the disk mass (Lunine 2001; Raymond et al. 2004, 2007b).

The orbital eccentricity of a Jupiter-like planet has important consequences for water delivery to terrestrial planets. An eccentric giant planet destabilizes primitive, water-rich asteroidal material and prevents it from colliding with growing planets in the inner system (Chambers & Cassen 2002; Raymond et al. 2004; Raymond 2006; O'Brien et al. 2006). Thus, dry terrestrial planets are probably correlated with eccentric giant planets.

Binary Companion Mass and Separation. Terrestrial planet formation has been studied in binary star systems, both orbiting both components of a close binary (P-type orbits) or one star with a more distant companion (S-type orbits). Perturbations from the binary can increase the relative velocities of planetesimals, making it difficult to form protoplanets (Thebault et al. 2002). However, gas drag can act to align the orbital apses of planetesimals, thereby reducing collision velocities and allowing protoplanets to grow (Thebault et al. 2006). This effect is also seen for protoplanet formation in the presence of a giant planet (Kortenkamp et al. 2001).

Terrestrial planets on P-type orbits are statistically the same if the binary's maximum separation is less than 0.2 AU (Quintana & Lissauer 2006). For larger separations, fewer terrestrial planets form, especially in the inner disk. Water delivery in these systems has not been studied. However, for an equal mass binary of two 0.5 $M_\odot$ stars, the habitable zone is actually much closer in than for the Sun. If the habitable zone scales with the stellar flux, i.e., as the square root of the stellar luminosity, then the total luminosity of two 0.5 $M_\odot$ stars is only 6% of the Solar flux, using the mass-luminosity relation of Scalo et al. (2007), which is a fit to data of Hillenbrand & White (2004). Thus, if the Sun's habitable zone is at 0.8-1.5 AU, then the habitable zone of this binary system is at 0.2-0.37 AU, such that it is far more vulnerable to the effects of the binary than a naïve contemplation.

As for the case of a giant planet's orbital distance, there exists a correlation between the orbit of a binary companion for S-type orbits and the most distant terrestrial planet that can form. For an equal-mass binary, there exists a ratio of roughly 5 between the orbital periastron of the binary and the more distant terrestrial planet that can form around the primary (Quintana et al. 2007; Haghighipour & Raymond 2007). Indeed, Quintana et al. (2002) showed that terrestrial planets could accrete in the α Centauri binary system. Thus, binaries with separations larger than 5-10 AU do not directly impede the accretion of terrestrial planets, although they may act to preferentially remove water-rich material as it lies at larger orbital distances and is therefore more susceptible to binary perturbations. In cases with a giant planet on a Jupiter-like orbit and a binary companion, the giant planet can act to transfer angular momentum between the binary star and inner terrestrial material (Haghighipour & Raymond 2007). The binary's orbit can induce a large eccentricity in the giant planet and cause it to deplete water-rich material as discussed in the case of an isolated, eccentric giant planet.

4. Giant Planet Migration and Terrestrial Planet Formation

Close-in giant planets are thought to have formed at larger orbital distances and migrated in to their current locations (Lin et al. 1996; Bodenmeimer et al. 2000). Planets more massive than roughly a Saturn-mass have feeding zones that are wider than the disk height, and so carve annular gaps in the disk (Lin & Papaloizou 1986; Takeuchi et al. 1996; Rafikov 2002; Crida et al. 2006). Such planets are then tied to the viscous evolution of the disk, and "type 2" migrate, usually inward, on the viscous timescale of $\sim 10^{5-6}$ years (Lin & Papaloizou 1986; Ward 1997; D'Angelo et al. 2003; Ida & Lin 2004). On its way to becoming a hot Jupiter or falling into the star, the giant planet must pass through the terrestrial planet zone.

Until recently, the effects of this migration on the accretion of Earth-like was debated without any direct calculations. Several authors made unfounded, pessimistic assumptions about the existence of terrestrial planets (Ward & Brownlee 2000; Gonzalez et al. 2001; Lineweaver et al. 2004). Others made limited calculations with mixed results, both pessimistic (Armitage 2003; Mandell & Sigurdsson 2003) and optimistic (Edgar & Artymowicz 2004; Raymond et al. 2005a; Lufkin et al. 2006). Since 2005, two groups have conducted realistic simulations of the migration of a giant planet through a region of accreting terrestrial bodies, with similar results. These papers include the migration of a single giant planet, a disk containing more than a thousand terrestrial particles, and effects of the gaseous disk on the terrestrial bodies (Fogg & Nelson 2005, 2007a, 2007b; Raymond et al. 2006b; Mandell et al. 2007).

Figure 4 shows snapshots of a simulation that was integrated for 200 Myr, including the migration of a Jupiter-mass giant planet from 5.2 to 0.25 AU in the first 100,000 years, from Raymond et al. (2006b). During the giant planet's migration, material interior to the giant planet is shepherded inward by mean motion resonances (MMRs), especially the 2:1 MMR (Tanaka & Ida 1999). This shepherding occurs because eccentricities are increased by the MMR and subsequently damped by gas drag and dynamical friction, causing a net loss of orbital energy. In this simulation, roughly half of the total material was shepherded inward, while the other half underwent a close encounter with the giant planet and was thrown outward, exterior to the giant planet. This material had very high eccentricities and inclinations, but these were damped on Myr timescales by gas drag (Adachi et al. 1976) and dynamical friction. On 100 Myr timescales, accretion continued in the outer disk, including the formation of a 3 $M_\oplus$ planet at 0.9 AU. This planet orbits its star in the habitable zone, and has roughly 30 times more water than other simulations designed to reproduce the Earth (Raymond et al. 2004). Thus, this planet is likely a "water world" or "ocean planet", with deep oceans covering its surface (Léger et al. 2004). Its formation time was significantly longer than Earth's, which may have consequences for its geological properties.

Systems with migrated giant planets tend to contain terrestrial planets both interior and exterior to their final orbits, although close-in terrestrial planets may not always survive (Fogg & Nelson 2005). "Hot Earth" planets tend to be formed interior to strong MMRs, especially the 2:1 MMR (Mandell et al. 2007). These may be favorable for detection via transits (Croll et al. 2007) or transit timing (Agol et al. 2005; Holman & Murray 2005). Exterior terrestrial planets form on long timescales from scattered material and tend to be very water-rich for two reasons: 1) material scattered by the giant planet has very large eccentricities, causing large-scale radial mixing, and 2) gas drag causes more distant, icy planetesimals to spiral inward and delivery vast amounts of water to these planets (Raymond et al. 2006b; Mandell et al. 2007). If the lifetime of the gaseous disk does not extend long past the end of migration (as in the scenario of Trilling et al. 1998), then exterior terrestrial planets could still form, but eccentricity damping would be slower (causing even longer formation times) and the amount of planetesimal in-spiralling would be reduced (somewhat reducing their water contents; Mandell et al. 2007). Nonetheless, hot Jupiter systems may be good places to look for additional planets, including potentially habitable terrestrial planets.

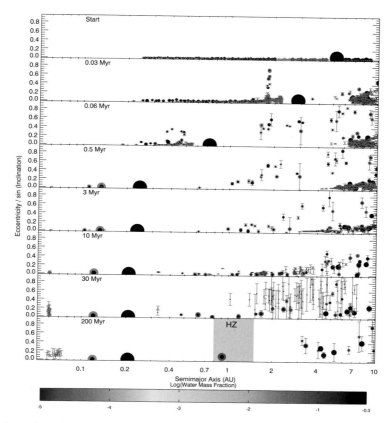

Figure 4. Snapshots in time from a simulation of the growth of terrestrial planets in the presence of giant planet migration. The giant planet is shown in black, and note the logarithmic scale of the x axis. As in Fig. 2, the size of each body is proportional to its mass$^{1/3}$, the dark circle represents the relative size of each body's iron core (color version only), and the color corresponds to its water content (red = dry, dark blue = 50% water; in the black and white version white = dry and black = 50% water; see color bar). For bodies with inclinations larger than 5°, inclinations are shown via vertical error bars with lengths of sin (inc.), readable on the y axis. Reproduced from Raymond *et al.* (2006b). For a movie of this simulation, go to http://lasp.colorado.edu/~raymond/ and click on "movies and graphics".

5. Terrestrial Planet Formation in the Known Exoplanet Systems

Given the large number of simulations of terrestrial planet formation that have been done while varying the characteristics of giant planets, it is possible to apply certain limits to the known extra-solar planet population. Specifically, we can constrain which of the systems of known (giant) planets may have formed currently-undetected terrestrial planets. To accomplish this, I will follow the procedure employed in Raymond *et al.* (2006b) and described in detail in section 4 of Mandell *et al.* (2007), which divides giant planets into two classes: those interior and those exterior to the habitable zone. I assume the habitable zone to be a simple function of the stellar luminosity, calibrated to a value of 0.8-1.5 AU for a Solar-mass star, in rough agreement with Kasting *et al.* (1993; for a more detailed treatment of the habitable zone, see Selsis *et al.* 2007b). The stellar luminosity is calculated using the mass-luminosity relation of Scalo *et al.* (2007), which is a fit to the data of Hillenbrand & White (2004). Table 1 lists the inner and outer giant

Table 1. Giant Planet Limits for Potentially Habitable Systems

M_* ($M_\odot$)	Sp. Type[1]	Hab Zone (AU)[2]	Inner Limit (AU)	Outer Limit (AU)
0.1	M6	0.024-0.045	0.015	0.075
0.4	M3	0.10-0.19	0.06	0.32
0.7	K6	0.28-0.52	0.17	0.87
1.0	G2	0.8-1.5	0.5	2.5
1.3	F8	2.3-4.3	1.45	7.2
1.6	F0	6.5-12.3	4.1	20.5
2.0	A5	25-47	15.7	78.3

Notes:
[1] Spectral types from Table 8.1 of Reid & Hawley (2000) and Appendix E from Carroll & Ostlie (1996). Note that spectral types of low-mass stars are age-dependent.
[2] Habitable Zones calibrated to 0.8-1.5 AU for a solar-mass star.

planet limits on the formation of a terrestrial planet in the habitable zone, reproduced from Mandell *et al.* (2007).

For giant planets exterior to the habitable zone, I assume that terrestrial accretion occurred in a similar fashion to the Solar System, with the late stages of accretion governed in part by fully-formed giant planets. In Raymond (2006), I ran 460 low-resolution simulations of terrestrial accretion under the influence of a single, Jupiter-mass giant planet. Systems with giant planets beyond 2.5 AU formed terrestrial planets larger than 0.3 $M_\oplus$ within the habitable zone (0.8-1.5 AU), while only systems with giant planets beyond 3.5 AU formed $> 0.3 M_\oplus$ water-rich terrestrial planets in the habitable zone. These limits depend strongly on the eccentricity of the giant planet (as shown in Raymond 2006), and certainly also on the giant planet's mass, although this is not currently accounted for. By assuming this separation (in terms of ratios of orbital periods) between the most distant terrestrial planet and the giant planet to be characteristic and independent of stellar mass, it was showed that only 5-6 % of the known systems (as of 2006) could form Earth-like planets in the habitable zone in Solar System-like configurations (Raymond 2006). Note that this prescription assumes that the giant planet acquired its large eccentricity before the bulk of terrestrial planet formation occurred. If eccentricity was acquired later then the situation is more complicated: the giant planet's orbit may have been more favorable during terrestrial accretion, but the eccentricity-causing instability has the potential to destabilize the terrestrial planets (Veras & Armitage 2006).

For giant planets interior to the habitable zone, limits are calibrated to simulations including giant planet migration (see § 4). From the 8 relevant simulations of Mandell *et al.* (2007), the spacing between the close-in giant planet and the innermost large terrestrial planet varies considerably, from 3.3 to 43 in terms of orbital period ratios, with a median value of 9. Thus, we adopt the median as a characteristic spacing, although there clearly exists a range of outcomes. For the most optimistic case, a giant planet at 0.7 AU would allow a terrestrial planet to form just inside the outer edge of the habitable zone at 1.5 AU. Note that we have limited calculations for inner giant planets on eccentric orbits. We therefore require inner giant planets to have eccentricities less than 0.1, which was the largest eccentricity of any giant planet in the simulations of Mandell *et al.* (2007).

With knowledge of extra-solar planet host stars' masses, we can apply the limits from Table 1 to the know extra-solar planets. I use the planet population as of August 1 2006, with values taken from Butler *et al.* (2006) and exoplanet.eu and references therein. This sample includes 207 planets in 178 systems, with 21 multiple planet systems. Figure 5 shows the results of applying our limits to this sample of giant planets: 65 out of 178 systems (37%) meet our criteria and appear able to have formed a terrestrial planet in the habitable zone. Of these, 17 met our outer limits and could have Solar System-like

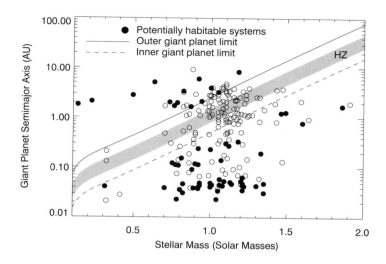

Figure 5. Extra-solar planetary systems that could harbor an Earth-like planet in the habitable zone. The habitable zone is shaded, and our inner and outer giant planet limits, listed in Table 1 for specific values of the stellar mass, are shown by the dashed (inner) and solid (outer) lines. Each circle represents a known planet; those that fulfill either the inner or outer limit are filled in black. Note that many planets that appear to meet the limits are not filled; this is because their orbital eccentricities were too large, although this is not shown explicitly. From Mandell et al. (2007).

architectures, while 49 could have giant planets interior to the terrestrial planets. For a list of all systems that met our limits, see Mandell et al. (2007).

One multiple planet system, 55 Cancri, met both our inner and outer giant planet limits. Indeed, detailed simulations had already shown that an Earth-sized terrestrial planet could accrete *in situ*, in the presence of the four giant planets (Raymond et al. 2006c). Interestingly, a fifth planet in 55 Cancri was recently discovered, although it is far larger than a terrestrial planet (Fischer et al. 2007). In addition, our list was compiled prior to the discovery of the 5 $M_\oplus$ planet Gliese 581 c (Udry et al. 2007), whose orbit lies just interior to the habitable zone (although the more distant, 8 $M_\oplus$ planets Gliese 581 d may be within the outer boundary; Selsis et al. 2007b). Indeed, based on the orbit of Gliese 581 b, the star was on our list of candidate systems for Earth-like planets.

This is an interesting exercise in terms of its predictive power, an indication of which known systems are good candidates to search for Earth-like planets. In addition, it is interesting to compare our results for where terrestrial planets can form with studies looking at where terrestrial planets' orbits would be stable. Several authors have examined the stability of test particles, used as proxies for terrestrial planets, in the known systems (e.g., Menou & Tabachnik 2003; Jones et al. 2005). These studies find that about half of the known systems could have a stable terrestrial planet in the habitable zone, somewhat higher than our estimate that about 1/3 could form an Earth-like planet.

It is important to realize that the existence of a stable region does not imply that a planet must occupy it. For example, a planet of up to 5 $M_\oplus$ would be stable at 3 AU in the Solar System's asteroid belt (Lissauer et al. 2001), but it could not have formed there. However, a contrasting example was recently discovered among the extra-solar planets, in the system HD 74156. Two giant planets were known to exist in HD 74156 as of 2003 (Naef et al. 2004). Detailed studies showed that both test particles and Saturn-mass planets were stable between the orbits of the two known planets, in a relatively narrow stable region (Barnes & Raymond 2004; Raymond & Barnes 2005). However, simulations showed that accretion of smaller bodies could not occur in this stable zone, making it somewhat analogous to the Solar System's asteroid belt (Raymond et al. 2006c). Recently, and with no prior knowledge of the theoretical results, Bean et al. (2007) discovered a $\sim$ Saturn-mass planet on the orbit predicted by Raymond & Barnes (2005). Thus, there exists an interesting contrast between the Solar System's asteroid belt (no planet accrete form there and there is no planet) with the previously-identified stable zone of HD 74156 (no planet could accrete there but there is a planet). Perhaps this is simply due to the order of planet discoveries, and the stable zone where HD 74156 d is located has little meaning. On the other hand, perhaps we have much more to learn about planet formation and dynamics.

6. Conclusions

In this proceedings, I have summarized the current state of knowledge about terrestrial planet formation in extra-solar planetary systems. In § 2, I reviewed the stages of terrestrial planet formation, from micron-sized grains to planetesimals, protoplanets, and full-sized planets. In § 3 I described the effect of external parameters such as the disk and giant planet properties on the terrestrial accretion process as well as radial mixing and water delivery. In § 4, I presented recent results showing that terrestrial planets can form in systems with close-in giant planets, assuming those to have migrated to their final locations. In fact, close-in giant planets should be accompanied by both very close-in "hot Earths" and exterior ocean planets (Raymond et al. 2006b). In § 5, I used results from previous work to derive limits designed to predict, based on the observed orbits of giant planets, where to search for other Earths among the known extra-solar systems. Approximately one third of the known systems of giant planets are good candidates for harboring an Earth-like planet (Mandell et al. 2007).

Many issues remain to be resolved in each topic I described. The details of planetesimal formation are still poorly known, although recent results suggest that several process including turbulence and migration of meter-sized bodies acting in tandem might be the solution (Johansen et al. 2007). The details and consequences of giant impacts are not well understood, in terms of the fate of collisional debris and compositional changes induced by the impacts (Genda & Abe 2005; Asphaug et al. 2006; Canup & Pierazzo 2006). In addition, some of the effects of external parameters are perhaps less well understood than we would like to think. For example the giant planet eccentricity - terrestrial planet water content correlation (Chambers & Cassen 2002; Raymond et al. 2004) inherently assumes that eccentric giant planets acquire their eccentricities early, before terrestrial accretion. If this assumption is wrong, then one can imagine a scenario in which terrestrial planets tend to form with relatively circular giant planets; this is a beneficial situation in terms of water delivery. However, a late, impulsive eccentricity increase can destabilize the orbits of terrestrial planets and remove them from the system entirely (Veras & Armitage 2006). Indeed, there are many known extra-solar planets that would be favorable for terrestrial planet formation if their orbits were more circular (see Fig. 5). Thus, if giant planet eccentricity is acquired late, many systems may undergo "planetary system suicide", forming an Earth-like planet and subsequently destroying it. A more complete, holistic view of planet formation and evolution is needed to distentangle these effects.

7. Acknowledgments

I want to thank the organizers of the symposium for an excellent meeting in a beautiful place. The research presented here was made possible by contributions from several of my collaborators, including Tom Quinn, Jonathan Lunine, Avi Mandell, Rory Barnes, Steinn Sigurdsson, John Scalo, Vikki Meadows, Nate Kaib, Eric Gaidos, and Nader Haghighipour. I was supported by an appointment to the NASA Postdoctoral Program at the University of Colorado Astrobiology Center, administered by Oak Ridge Associated Universities through a contract with NASA. I dedicate this paper to my son Owen Zahler Raymond, born January 1, 2008.

References

Abe, Y., Ohtani, E., Okuchi, T., Righter, K., & Drake, M. 2000, *Origin of the earth and moon, edited by R.M. Canup and K. Righter and 69 collaborating authors. Tucson: University of Arizona Press., p.413-433*, 413.
Adachi, I., Hayashi, C., & Nakazawa, K. 1976, *Progress of Theoretical Physics* **56**, 1756.
Agnor, C. & Asphaug, E. 2004, *Astrophys. J.* **613**, L157.
Agnor, C.B., Canup, R.M., & Levison, H.F. 1999, *Icarus* **142**, 219.
Agol, E., Steffen, J., Sari, R., & Clarkson, W. 2005, *Monthly Notices of the Royal Astronomical Society* **359**, 567.
Alibert, Y., Mordasini, C., Benz, W., & Winisdoerffer, C. 2005, *Astron. Astroph.* **434**, 343.
Andre, P. & Montmerle, T. 1994, *Astrophys. J.* **420**, 837.
Andrews, S. M. & Williams, J. P. 2007, *Astrophys. J.* **659**, 705.
Andrews, S. M. & Williams, J. P. 2005, *Astrophys. J.* **631**, 1134.
Armitage, P. J. 2003, *Astrophys. J.* **582**, L47.
Asphaug, E., Agnor, C. B., & Williams, Q. 2006, *Nature* **439**, 155.
Balsiger, H., Altwegg, K., & Geiss, J. 1995, *Journal of Geophysical Research* **100**, 5827.
Barnes, R. & Raymond, S. N. 2004, *Astrophys. J.* **617**, 569.
Bean, J. L., McArthur, B. E., Benedict, G. F., & Armstrong, A. 2007, *ArXiv e-prints* **709**, arXiv:0709.1656.
Benz, W., Slattery, W. L., & Cameron, A. G. W. 1988, *Icarus* **74**, 516.
Benz, W., Slattery, W. L., & Cameron, A. G. W. 1987, *Icarus* **71**, 30.
Benz, W. 2000, *Space Science Reviews* **92**, 279.
Bockelee-Morvan, D., Gautier, D., Lis, D. C., Young, K., Keene, J., Phillips, T., Owen, T., Crovisier, J., Goldsmith, P. F., Bergin, E. A., Despois, D., & Wootten, A. 1998, *Icarus* **133**, 147.
Bodenheimer, P., Hubickyj, O., & Lissauer, J. J. 2000, *Icarus* **143**, 2.
Boss, A. P. 1998, *Annual Review of Earth and Planetary Sciences* **26**, 53.
Bottke, W. F., Durda, D. D., Nesvorný, D., Jedicke, R., Morbidelli, A., Vokrouhlický, D., & Levison, H. 2005, *Icarus* **175**, 111.
Briceño, C., Vivas, A. K., Calvet, N., Hartmann, L., Pacheco, R., Herrera, D., Romero, L., Berlind, P., Sánchez, G., Snyder, J. A., & Andrews, P. 2001, *Science* **291**, 93.
Butler, R. P., Wright, J. T., Marcy, G. W., Fischer, D. A., Vogt, S. S., Tinney, C. G., Jones, H. R. A., Carter, B. D., Johnson, J. A., McCarthy, C., & Penny, A. J. 2006, *Astrophys. J.* **646**, 505.
Canup, R. M. & Pierazzo, E. 2006, *37th Annual Lunar and Planetary Science Conference* **37**, 2146.
Canup, R. M. 2004, *Icarus* **168**, 433.
Canup, R. M. & Asphaug, E. 2001, *Nature* **412**, 708.
Carroll, B. W. & Ostlie, D. A. 1996, *An Introduction to Modern Astrophysics*. New Yorl: Addison-Wesley.
Chambers, J. E. 2001, *Icarus* **152**, 205.
Chambers, J. E. & Cassen, P. 2002, *Meteoritics and Planetary Science* **37**, 1523.
Chambers, J. E. & Wetherill, G. W. 1998, *Icarus* **136**, 304.
Chambers, J. 2006, *Icarus* **180**, 496.

Chambers, J. E. 2004, *Earth and Planetary Science Letters* **223**, 241.
Collins, B. F. & Sari, R. 2007, *ArXiv e-prints* **706**, arXiv:0706.1079.
Crida, A., Morbidelli, A., & Masset, F. 2006, *Icarus* **181**, 587.
Croll, B., Matthews, J. M., Rowe, J. F., Kuschnig, R., Walker, A., Gladman, B., Sasselov, D., Cameron, C., Walker, G. A. H., Lin, D. N. C., Guenther, D. B., Moffat, A. F. J., Rucinski, S. M., & Weiss, W. W. 2007, *Astrophys. J.* **658**, 1328.
Cuzzi, J. N., Dobrovolskis, A. R., & Champney, J. M. 1993, *Icarus* **106**, 102.
D'Angelo, G., Kley, W., & Henning, T. 2003, *Astrophys. J.* **586**, 540.
Davis, S. S. 2005, *Astrophys. J.* **627**, L153.
Desch, S. 2007. *Astrophys. J.*, in press.
Dominik, C., Blum, J., Cuzzi, J. N., & Wurm, G. 2007, *Protostars and Planets V*, 783.
Dominik, C. & Tielens, A. G. G. M. 1997, *Astrophys. J.* **480**, 647.
Dullemond, C. P. & Dominik, C. 2005, *Astron. Astroph.* **434**, 971.
Dullemond, C. P., Hollenbach, D., Kamp, I., & D'Alessio, P. 2007, *Protostars and Planets V*, 555.
Edgar, R. & Artymowicz, P. 2004, *Monthly Notices of the Royal Astronomical Society* **354**, 769.
Eisner, J. A. & Carpenter, J. M. 2003, *Astrophys. J.* **598**, 1341.
Fischer, D., *et al.* 2007, *Astrophys. J.*, in press.
Fogg, M. J. & Nelson, R. P. 2007, *Astron. Astroph.* **472**, 1003.
Fogg, M. J. & Nelson, R. P. 2007, *Astron. Astroph.* **461**, 1195.
Fogg, M. J. & Nelson, R. P. 2005, *Astron. Astroph.* **441**, 791.
Fortney, J. J., Marley, M. S., & Barnes, J. W. 2007, *Astrophys. J.* **659**, 1661.
Gaidos, E., Haghighipour, N., Agol, E., Latham, D., Raymond, S., & Rayner, J. 2007, *Science* **318**, 210.
Garaud, P. & Lin, D. N. C. 2007, *Astrophys. J.* **654**, 606.
Genda, H. & Abe, Y. 2005, *Nature* **433**, 842.
Goldreich, P. & Tremaine, S. 1980, *Astrophys. J.* **241**, 425.
Goldreich, P., Lithwick, Y., & Sari, R. 2004, *Astrophys. J.* **614**, 497.
Goldreich, P. & Ward, W. R. 1973, *Astrophys. J.* **183**, 1051.
Gonzalez, G., Brownlee, D., & Ward, P. 2001, *Icarus* **152**, 185.
Goodman, J. & Pindor, B. 2000, *Icarus* **148**, 537.
Greenberg, R., Hartmann, W. K., Chapman, C. R., & Wacker, J. F. 1978, *Icarus* **35**, 1.
Haghighipour, N. & Boss, A. P. 2003, *Astrophys. J.* **598**, 1301.
Haghighipour, N. & Raymond, S. N. 2007, *Astrophys. J.* **666**, 436.
Haisch, K. E., Jr., Lada, E. A., & Lada, C. J. 2001, *Astrophys. J.* **553**, L153.
Hillenbrand, L. A. & White, R. J. 2004, *Astrophys. J.* **604**, 741.
Holman, M. J. & Murray, N. W. 2005, *Science* **307**, 1288.
Ida, S., Canup, R. M., & Stewart, G. R. 1997, *Nature* **389**, 353.
Ida, S. & Lin, D. N. C. 2005, *Astrophys. J.* **626**, 1045.
Ida, S. & Lin, D. N. C. 2004, *Astrophys. J.* **604**, 388.
Ida, S. & Makino, J. 1993, *Icarus* **106**, 210.
Ida, S. & Makino, J. 1992, *Icarus* **98**, 28.
Ida, S. & Makino, J. 1992, *Icarus* **96**, 107.
Jacobsen, S. B. 2005, *Annual Review of Earth and Planetary Sciences* **33**, 531.
Johansen, A., Oishi, J. S., Low, M.-M. M., Klahr, H., Henning, T., & Youdin, A. 2007, *Nature* **448**, 1022.
Jones, B. W., Underwood, D. R., and Sleep, P. N. 2005, *Astrophys. J.* **622**, 1091.
Kasting, J. F., Whitmire, D. P., & Reynolds, R. T. 1993, *Icarus* **101**, 108.
Kenyon, S. J. & Bromley, B. C. 2006, *Astronomical Journal* **131**, 1837.
Kerridge, J. F. 1985, *Geochimica et Cosmochimica Acta* **49**, 1707.
Kleine, T., Münker, C., Mezger, K., & Palme, H. 2002, *Nature* **418**, 952.
Kokubo, E. & Ida, S. 2002, *Astrophys. J.* **581**, 666.
Kokubo, E. & Ida, S. 2000, *Icarus* **143**, 15.
Kokubo, E. & Ida, S. 1998, *Icarus* **131**, 171.
Kokubo, E. & Ida, S. 1996, *Icarus* **123**, 180.
Kokubo, E., Kominami, J., & Ida, S. 2006, *Astrophys. J.* **642**, 1131.

Kortenkamp, S. J., Wetherill, G. W., & Inaba, S. 2001, *Science* **293**, 1127.
Kuchner, M. J. 2004, *Astrophys. J.* **612**, 1147.
Léger, A., Selsis, F., Sotin, C., Guillot, T., Despois, D., Mawet, D., Ollivier, M., Labèque, A., Valette, C., Brachet, F., Chazelas, B., & Lammer, H. 2004, *Icarus* **169**, 499.
Leinhardt, Z. M. & Richardson, D. C. 2005, *Astrophys. J.* **625**, 427.
Levison, H. F. & Agnor, C. 2003, *Astronomical Journal* **125**, 2692.
Lin, D. N. C., Bodenheimer, P., & Richardson, D. C. 1996, *Nature* **380**, 606.
Lin, D. N. C. & Papaloizou, J. 1986, *Astrophys. J.* **309**, 846.
Lineweaver, C. H., Fenner, Y., & Gibson, B. K. 2004, *Science* **303**, 59.
Lissauer, J. J. 1993, *Annual Review of Astronomy and Astrophysics* **31**, 129.
Lissauer, J. J. 2007, *Astrophys. J.* **660**, L149.
Lissauer, J. J., Quintana, E. V., Rivera, E. J., & Duncan, M. J. 2001, *Icarus* **154**, 449.
Lufkin, G., Richardson, D. C., & Mundy, L. G. 2006, *Astrophys. J.* **653**, 1464.
Lunine, J. I. 2001, *Proceedings of the National Academy of Science* **98**, 809.
Mandell, A. M., Raymond, S. N., & Sigurdsson, S. 2007, *Astrophys. J.* **660**, 823.
Mandell, A. M. & Sigurdsson, S. 2003, *Astrophys. J.* **599**, L111.
Masset, F. S., D'Angelo, G., & Kley, W. 2006, *Astrophys. J.* **652**, 730.
McNeil, D., Duncan, M., & Levison, H. F. 2005, *Astronomical Journal* **130**, 2884.
Meier, R., Owen, T. C., Matthews, H. E., Jewitt, D. C., Bockelee-Morvan, D., Biver, N., Crovisier, J., & Gautier, D. 1998, *Science* **279**, 842.
Menou, K. & Tabachnik, S. 2003, *Astrophys. J.* **583**, 473.
Mizuno, H. 1980, *Progress of Theoretical Physics* **64**, 544.
Morbidelli, A., Chambers, J., Lunine, J. I., Petit, J. M., Robert, F., Valsecchi, G. B., & Cyr, K. E. 2000, *Meteoritics and Planetary Science* **35**, 1309.
Naef, D., Mayor, M., Beuzit, J. L., Perrier, C., Queloz, D., Sivan, J. P., & Udry, S. 2004, *Astron. Astroph.* **414**, 351.
Nagasawa, M., Lin, D. N. C., & Thommes, E. 2005, *Astrophys. J.* **635**, 578.
Nordström, B., Mayor, M., Andersen, J., Holmberg, J., Pont, F., Jørgensen, B. R., Olsen, E. H., Udry, S., & Mowlavi, N. 2004, *Astron. Astroph.* **418**, 989.
O'Brien, D. P., Morbidelli, A., & Levison, H. F. 2006, *Icarus* **184**, 39.
Owen, T. & Bar-Nun, A. 1995, *Icarus* **116**, 215.
Papaloizou, J. C. B. & Terquem, C. 2006, *Reports of Progress in Physics* **69**, 119.
Pascucci, I., Gorti, U., Hollenbach, D., Najita, J., Meyer, M. R., Carpenter, J. M., Hillenbrand, L. A., Herczeg, G. J., Padgett, D. L., Mamajek, E. E., Silverstone, M. D., Schlingman, W. M., Kim, J. S., Stobie, E. B., Bouwman, J., Wolf, S., Rodmann, J., Hines, D. C., Lunine, J., & Malhotra, R. 2006, *Astrophys. J.* **651**, 1177.
Petit, J.-M., Morbidelli, A., & Chambers, J. 2001, *Icarus* **153**, 338.
Pollack, J. B., Hubickyj, O., Bodenheimer, P., Lissauer, J. J., Podolak, M., & Greenzweig, Y. 1996, *Icarus* **124**, 62.
Poppe, T., Blum, J., & Henning, T. 2000, *Astrophys. J.* **533**, 454.
Quintana, E. V., Adams, F. C., Lissauer, J. J., & Chambers, J. E. 2007, *Astrophys. J.* **660**, 807.
Quintana, E. V. & Lissauer, J. J. 2006, *Icarus* **185**, 1.
Quintana, E. V., Lissauer, J. J., Chambers, J. E., & Duncan, M. J. 2002, *Astrophys. J.* **576**, 982.
Rafikov, R. R. 2003, *Astronomical Journal* **125**, 942.
Rafikov, R. R. 2002, *Astrophys. J.* **572**, 566.
Raymond, S. N. 2006, *Astrophys. J.* **643**, L131.
Raymond, S. N. & Barnes, R. 2005, *Astrophys. J.* **619**, 549.
Raymond, S. N., Barnes, R., & Kaib, N. A. 2006c, *Astrophys. J.* **644**, 1223.
Raymond, S. N., Barnes, R., & Mandell, A. M. 2007c, MNRAS 384, 663–674
Raymond, S. N., Mandell, A. M., & Sigurdsson, S. 2006b, *Science* **313**, 1413.
Raymond, S. N., Quinn, T., & Lunine, J. I. 2007a, *Astrobiology* **7**, 66.
Raymond, S. N., Quinn, T., & Lunine, J. I. 2006a, *Icarus* **183**, 265.
Raymond, S. N., Quinn, T., & Lunine, J. I. 2005b, *Astrophys. J.* **632**, 670.
Raymond, S. N., Quinn, T., & Lunine, J. I. 2005a, *Icarus* **177**, 256.

Raymond, S. N., Quinn, T., & Lunine, J. I. 2004, *Icarus* **168**, 1.
Raymond, S. N., Scalo, J., & Meadows, V. S. 2007b, *Astrophys. J.* **669**, 606.
Reid, N. & Hawley, S. L. 2000, *New light on dark stars : red dwarfs, low mass stars, brown dwarfs / Neill Reid and Suzanne L. Hawley. New York : Springer, 2000. (Springer-Praxis series in astronomy and astrophysics)*.
Robert, F. & Epstein, S. 1982, *Geochimica et Cosmochimica Acta* **46**, 81.
Safronov, V. S. 1969, *1969.*.
Scalo, J., Kaltenegger, L., Segura, A. G., Fridlund, M., Ribas, I., Kulikov, Y. N., Grenfell, J. L., Rauer, H., Odert, P., Leitzinger, M., Selsis, F., Khodachenko, M. L., Eiroa, C., Kasting, J., & Lammer, H. 2007, *Astrobiology* **7**, 85.
Scholz, A., Jayawardhana, R., & Wood, K. 2006, *Astrophys. J.* **645**, 1498.
Seager, S., Kuchner, M., Hier-Majumder, C. A., & Militzer, B. 2007, *Astrophys. J.* **669**, 1279.
Selsis, F., Chazelas, B., Bordé, P., Ollivier, M., Brachet, F., Decaudin, M., Bouchy, F., Ehrenreich, D., Grießmeier, J.-M., Lammer, H., Sotin, C., Grasset, O., Moutou, C., Barge, P., Deleuil, M., Mawet, D., Despois, D., Kasting, J. F., & Léger, A. 2007a, *Icarus* **191**, 453.
Selsis, F., Kasting, J. F., Levrard, B., Paillet, J., Ribas, I., & Delfosse, X. 2007b, *ArXiv e-prints* **710**, arXiv:0710.5294.
Takeuchi, T., Miyama, S. M., & Lin, D. N. C. 1996, *Astrophys. J.* **460**, 832.
Tanaka, H. & Ida, S. 1999, *Icarus* **139**, 350.
Tanga, P., Babiano, A., Dubrulle, B., & Provenzale, A. 1996, *Icarus* **121**, 158.
Terquem, C. & Papaloizou, J. C. B. 2007, *Astrophys. J.* **654**, 1110.
Thébault, P., Marzari, F., & Scholl, H. 2006, *Icarus* **183**, 193.
Thébault, P., Marzari, F., & Scholl, H. 2002, *Astron. Astroph.* **384**, 594.
Throop, H. B. & Bally, J. 2005, *Astrophys. J.* **623**, L149.
Trilling, D. E., Benz, W., Guillot, T., Lunine, J. I., Hubbard, W. B., & Burrows, A. 1998, *Astrophys. J.* **500**, 428.
Tsiganis, K., Gomes, R., Morbidelli, A., & Levison, H. F. 2005, *Nature* **435**, 459.
Udry, S., Bonfils, X., Delfosse, X., Forveille, T., Mayor, M., Perrier, C., Bouchy, F., Lovis, C., Pepe, F., Queloz, D., & Bertaux, J.-L. 2007, *Astron. Astroph.* **469**, L43.
Valencia, D., Sasselov, D. D., & O'Connell, R. J. 2007, *Astrophys. J.* **665**, 1413.
Veras, D. & Armitage, P. J. 2006, *Astrophys. J.* **645**, 1509.
Ward, P. & Brownlee, D. 2000, *Rare earth : why complex life is uncommon in the universe / Peter Ward, Donald Brownlee. New York : Copernicus, c2000.*.
Ward, W. R. 1997, *Icarus* **126**, 261.
Ward, W. R. 1986, *Icarus* **67**, 164.
Ward, W. R. 1981, *Icarus* **47**, 234.
Weidenschilling, S. J. 1984, *Icarus* **60**, 553.
Weidenschilling, S. J. 1977, *Astrophysics and Space Science* **51**, 153.
Weidenschilling, S. J. 1977, *Monthly Notices of the Royal Astronomical Society* **180**, 57.
Weidenschilling, S. J. & Cuzzi, J. N. 1993, *Protostars and Planets III*, 1031.
Weidenschilling, S. J., Spaute, D., Davis, D. R., Marzari, F., & Ohtsuki, K. 1997, *Icarus* **128**, 429.
Weidenschilling, S. J. 2000, *Space Science Reviews* **92**, 295.
Wetherill, G. W. 1985, *Science* **228**, 877.
Wetherill, G. W. 1996, *Icarus* **119**, 219.
Wetherill, G. W. & Stewart, G. R. 1993, *Icarus* **106**, 190.
Wetherill, G. W. & Stewart, G. R. 1989, *Icarus* **77**, 330.
Wetherill, G. W. 1992, *Icarus* **100**, 307.
Wetherill, G. W. 1990, *Annual Review of Earth and Planetary Sciences* **18**, 205.
Williams, D. M., Kasting, J. F., & Wade, R. A. 1997, *Nature* **385**, 234.
Wurm, G. & Blum, J. 2000, *Astrophys. J.* **529**, L57.
Youdin, A. N. & Chiang, E. I. 2004, *Astrophys. J.* **601**, 1109.
Youdin, A. N. & Shu, F. H. 2002, *Astrophys. J.* **580**, 494.
Zhou, J.-L., Aarseth, S. J., Lin, D. N. C., & Nagasawa, M. 2005, *Astrophys. J.* **631**, L85.

Planet formation in binary stars

Wilhelm Kley

Institut für Astronomie & Astrophysik and Kepler Center for Astro and Particle Physics,
Universität Tübingen, Auf der Morgenstelle 10, D-72076 Tübingen, Germany
email: wilhelm.kley@uni-tuebingen.de

Abstract. As of today more than 30 planetary systems have been discovered in binary stars. In all cases the configuration is circumstellar, where the planets orbit around one of the stars. The formation process of planets in binary stars is more difficult than around single stars due to the gravitational action of the companion. An overview of the research done in this field will be given. The dynamical influence that a secondary companion has on a circumstellar disk, and how this affects the planet formation process in this challenging environment will be summarized. Finally, new fully hydrodynamical simulations of protoplanets embedded in disks residing in a binary star will be presented. Applications with respect to the planet orbiting the primary in the system γ Cephei will be presented.

Keywords. Planet formation, protoplanetary disk, binary stars, hydrodynamics

1. Observational Data

Planet formation is obviously a process that may occur around single as well as in multiple stars, which is indicated by the detection of about 35 planetary systems that reside in a binary or even multiple star configurations. All of the observed systems display a so called S-type configuration in which the planets orbit around one of the stars and the additional star, the companion or secondary star, acts as a perturber to this system. In this review we shall refer to the secondaries as single objects, even if they may be multiple. As indicated in Table 1 the distances of the secondaries to the host star of the planetary system range from very small values of about 20 AU for Gl 86 and γ Cep to several thousand AU. As the influence of the secondaries on the planet formation process will obviously be smaller for larger distances, the mere existence of the last three systems, below the horizontal separation line in the table, represent a special challenge to any kind of planet formation process. Interestingly, all the additional systems have separations larger than 100 AU, and some several 1000 AU, but note that the list is very incomplete for larger separations. Despite the actual detection of planets in binary systems there is additional circumstantial evidence of debris disks (which are thought of as a byproduct of the planet formation process) in binary systems as indicated by recent Spitzer data. Here, for S-type configurations it is found that disks around an individual star of the binary exist mainly for binary separations larger than 50 AU, while P-type circumbinary debris disks are detected only in very tight binaries with a_{bin} smaller than about 3 AU (Trilling *et al.* 2007).

2. Constraints on the planet formation process in binary stars

In a binary star system the formation of planets is altered and effectively handicapped due to the dynamical action of the companion and the subsequent change in the internal structure of the protoplanetary disks. The tidal torques of the companion generate strong spiral arms in the circumstellar disk of the primary and angular momentum will be

Star	a_{bin} [AU]	a_{pl} [AU]	$M_p \sin i$ [M_{Jup}]	e_p	Remarks
HD 40979	6400	0.811	3.32	.23	
Gl 777 A	3000	3.65	1.15	.48	
HD 80606	1200	0.439	3.41	.93	
55 Cnc B	1065	0.1-5.9	0.8-4.05	.02-.34	multiple
16 Cyg B	850	1.66	1.64	.63	
υ And	750	0.06-2.5	0.7-4.0	.01-.27	multiple
HD 178911 B	640	0.32	6.3	.12	
HD 219542 B	288	0.46	0.30	.32	
τ Boo	240	0.05	4.08	.02	
HD 195019	150	0.14	3.51	.03	
HD 114762	130	0.35	11.03	.34	
HD 19994	100	1.54	1.78	.33	
HD 41004A	23	1.33	2.5	.39	Quadruple
γ Cep	20.2	2.04	1.60	.11	new data
Gl 86	20	0.11	4.0	0.046	White Dwarf

Table 1. A representative selection of planets in binary systems (see Eggenberger et al. 2004).

transferred to the binary orbit which in turn leads to a truncation and restructuring of the disk. Using analytical and numerical methods Artymowicz & Lubow (1994) have shown how the truncation radius r_t of the disk depends on the binary separation a_{bin}, its eccentricity e_{bin}, the mass ratio $q = M_2/M_1$ (where M_1, M_2 denote are the primary and secondary mass, respectively), and the viscosity ν of the disk. For typical values of $q \approx 0.5$ and $e_{bin} = 0.3$ the disk will be truncated to a radius of $r_t \approx 1/3 a_{bin}$ for typical disk Reynolds numbers of 10^5 (Artymowicz & Lubow 1994, Armitage et al. 1999). For a given mass ratio q and semi-major axis a_{bin} an increase in e_{bin} will reduce the size of the disk while a large ν will increase the disks radii.

Whether these changes in the disk structure have an influence on the likelihood of planet formation in such disks has been a matter of debate. The dynamical action of the secondary induces several consequences which are adverse to planet formation: *i*) it changes the stability properties of orbits around the primary, *ii*) it reduces the lifetime of the disk, and *iii*) it increases the temperature in the disk.

In a numerical study Nelson 2000 investigated the evolution of an equal mass binary with a 50 AU separation and an eccentricity of 0.3. He argued that both main scenarios of planet formation (i.e. through core instability or gravitational instability) are strongly handicapped, because the eccentric companion will induce a periodic heating of the disk up to temperatures possibly above 1200 K. Since the condensation of particles as well as the occurrence of gravitational instability require lower temperatures, planet formation will be made more difficult. Clearly the strength of this effect will depend on the binary orbital elements, i.e. a_{bin} and e_{bin} and its mass ratio q.

However, recent numerical studies of the early planetesimal formation phase in rather close binaries with separations of only 20-30 AU show that it is indeed possible to form planetary embryos in such systems (Lissauer et al. 2004, Turrini et al. 2005, Quintana et al. 2007).

Already in ordinary planet formation around single stars the lifetime of the disk represents a limiting factor in the formation of planets from such a disk. It has been suspected that the dynamical action of a companion will limit the lifetime of disks substantially and place even tighter constraints on the possibility of planet formation. However, a recent analysis of the observational data of disks in binary stars finds no or very little change in the lifetimes of the disks at least for separations larger than about 20 AU (Monin et al. 2007).

3. Evolution of protoplanets in disks

In this section we present models of the evolution and growth of planetary cores in a binary system where the orbital elements of the binary have been chosen to match the system γ Cep quite closely. The data for γ Cep have been taken from Hatzes et al. (2003) which do not include the most recent improvements by Neuhäuser et al. (2007). This system is the tightest binary system known to contain a Jupiter-sized protoplanet. For this reason, it has attracted much attention in past years. Several studies looked at the stability and/or the possibility of (additional) habitable planets in the system (e.g. Dvorak et al. 2004, Turrini et al. 2004, Haghighipour 2006, Verrier & Evans 2006). Due to the enhanced influence of the secondary on the disk this system presents a strong challenge to any kind of planet formation process. Following other studies (Thébault et al. 2004), we assume that that the formation of planetesimals within the protoplanetary disk around the primary has been possible and start our investigation at the point where the planetary core has been grown to about 30 M_{Earth}. From this point on we follow the orbital evolution and growth of the planetary core in the disk.

3.1. Model definition

For the present study we have chosen a binary with $M_1 = 1.59 M_\odot$, $M_2 = 0.38 M_\odot$, $a_{bin} = 18.5$ AU and $e_{bin} = 0.36$, which translates into a binary period of $P = 56.7$ yrs. The primary star is surrounded by a flat circumstellar disk, where the binary orbit and the disk all lie in one plane, i.e. they are coplanar. The typical dynamical timescale in the disk, the orbital period at a few AU, is substantially shorter than the binary period, but in a numerical simulation the system's evolution can only be followed on these short dynamical time scales. To simplify the simulations we assume that the disk is vertically thin and perform only 2D hydrodynamical simulations of an embedded planet in a circumstellar disk which is perturbed by the secondary. We assume that the effects of the intrinsic turbulence of the disk can be described approximately through the viscous Navier-Stokes equations, which are solved by a finite volume method which is second order in space and time. To substantiate our results we have utilized two different codes RH2D (Kley 1999, Kley 1989) and NIRVANA (Nelson et al. 2000, Ziegler & Yorke 1997). The disk is assumed to be locally isothermal where the ratio of the vertical thickness H to the distance r from the primary is constant, here $H/r = 0.05$. For the viscosity an α type parameterization is used with $\alpha = 0.02$.

In the runs presented the computational domain covers a radial range from 0.5 to 8 AU, and 0 to 2π in azimuth. This is covered with an equidistant 300×300 grid. To allow for parameter studies we have found it highly useful to increase the performance of the code and have implemented the FARGO-algorithm to the code RH2D which is particularly designed to model differentially rotating flows (Masset 2000). For our chosen radial range and grid resolution we find a speed-up factor of about 7.5 over the standard case. Then, applying a Courant number of 0.75 still about 160,000 timesteps are required for only 10 binary orbits for our radial range and a standard resolution of 300×300.

3.2. Disk equilibration and effects on the planetary orbit

Before inserting the planet into the disk it is necessary to first relax the disk in the binary to its equilibrium configuration in the presence of the secondary. This makes sure that the calculated planetary evolution is not spoiled by long term transients due to

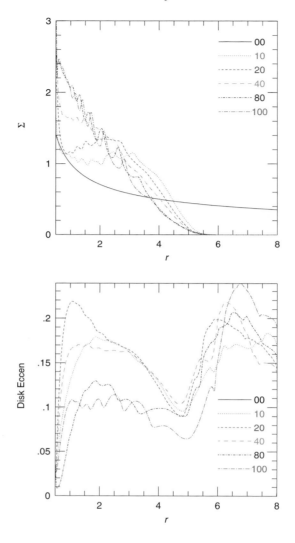

Figure 1. Azimuthally averaged disk structure at different evolutionary times given in binary orbits. On the left the surface density $\Sigma(r)$ is displayed and on the right the mean eccentricity $e(r)$.

the influence of the secondary star. In Fig. 1 we display such an initial settling of the disk with no planet. The initial surface density profile $\Sigma_0(r) \propto r^{-1/2}$ is given by the dark solid line, and the other lines represent the azimuthally averaged density during the subsequent evolution. As can be seen from the figure the disk is truncated very soon in the simulations (in fact after one or two orbits) and then re-adjusts as a whole on longer, viscous timescales to reach equilibrium at around 60-70 binary orbits. The disk is still perturbed periodically at each orbit due to the eccentric orbit of the secondary. At around each periastron strong spiral arms appear in the disk which are then damped until apoastron. However, the azimuthally averaged density structure at $t = 80$ and $t = 100$ does not differ anymore. During the process of equilibration the average eccentricity of the disk, e_{disk}, settles to a value of about 0.1 in the bulk part of the disk. Only in the

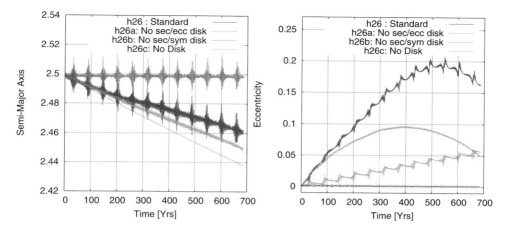

Figure 2. Evolution of the semi-major axis and eccentricity for fiducial models where either the disk or the binary have been switched off individually to test their influence separately.

outer low density parts of the disk the eccentricity remains high due to the disturbance of the secondary. This eccentric disk shows a coherent slow retrograde precession with a period of about 700 yrs or 14 binary orbits, see also Fig. 4 below.

The final two-dimensional density structure at the time $t = 100$ is then used as the initial condition for the embedded planet simulations. The total mass of the disk is rescaled to $3 M_{Jup}$ and the planet of $30 M_{Earth}$ is placed on a circular orbit at a given semi-major axis (distance) from the primary star, ranging from 2.0AU to 3.5AU.

Before performing a full run where the planet may evolve and accrete material over a long time scale evolution we study a few test cases where individual effects are switched off selectively in order to estimates its influences. After inserting the protoplanet on a circular orbit at 2.5 AU we generally expect that, in addition to the typical planet-disk interaction, its orbital elements will change due to the gravitational influence of the binary and the distorted disk. To differentiate the different contributions we have decided to check the origin of the dynamical behavior, through a variation of physical conditions. The standard model resembles the true physical situation where the planet feels the full influence of the binary and the disk which is perturbed by the binary. In the other setups we switch the various contributions on and off. Not to add additional complications to the problem the protoplanet is not allowed to accrete here such that its mass remains fixed at $30 M_{Earth}$. The results are displayed in Fig. 2, where the semi-major axis and eccentricity of the planet are shown for four different setups. The standard model (h26, dark solid line) refers to the full model (including binary and disk) as just described, in model (h26a, dashed bright line) the mass of the secondary has been switched off to test its influence, and the planet evolves in the initially eccentric disk which becomes more circular during the evolution because of the lacking secondary. In the next model (26b, dark dotted line), in addition to having no secondary the density in the disk has been azimuthally symmetrized keeping the radial distribution intact. Hence, this model suits as a reference of what happens in the single star case. In the last model (26c, fine dotted line) the secondary is present but the disk mass has been reduced, such that effectively only a 3-body problem is solved. The curves with the periodic bumps in Fig. 2 refer to the cases (26,26c) including the secondary while the smoother curves describe the situation where the secondary star has been excluded (26a,26b). The bumps occur with the binary period and indicate the 'kick' the planet experiences due to the interaction with the spiral arms in the disk. The results show that the main contribution to the *initial* growth of planetary eccentricity e_p is in fact the eccentric disk. The eccentricity e_p for models (h26 and h26a) rise initially with the same speed but due to the fact that without the secondary star the disk slowly circularizes, such that the model (h26a) falls behind the full model (which reaches $e_p = 0.2$ at $t \approx 500$) and then e_p begins to drop off after a time of about 400 yrs after insertion of the protoplanet. The maximum eccentricity reached for the reduced case (with no binary) is only $e_p = 0.1$. In the symmetrized disk case without secondary (26b) the planet behaves as expected for single star disks, it displays inward migration (the fastest for all cases) and the eccentricity remains zero. The case with only the binary (h26c) does not show any change in the semi-major axis a_p but rather a slow secular rise in e_p. This initially linear rise represents the first part of the long term periodic variation of e_p induced by the binary which has a period of approximately $T_{sec} = 4/3\, M_*/M_{bin}\, (a_p/a_{bin})^{-3} P_p \sim 2200 P_p$, where P_p is the period of the planetary orbit. Additional test cases with similar setup but different starting distances $a_p = 3.0$ and 3.5 AU display similar behavior. For the largest initial distance 3.5AU there are clear jumps in e_p evolution visible at each individual periapse.

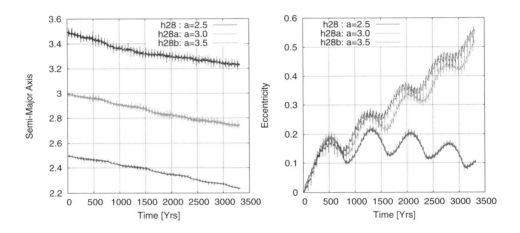

Figure 3. The evolution of the semi-major axis and eccentricity for planets released at different distances from the binary, i.e. 2.5, 3.0 and 3.5 AU.

3.3. *Evolving planets without mass accretion*

Planetary cores form in the outer cooler regions of protoplanetary disks beyond the so called ice-line. However, in a binary star system the outer disk is affected most by the secondary, and to find possible restrictions on the planet forming regions in the disk it is important to analyze the evolution of cores near the outer parts of the disk. To study the effect of initial position we start our embryos at 3 different locations in the disk 2.5, 3.0 and 3.5 AU always on a circular orbit, and choose again non-accreting cores. Because the initial characteristic growth time of the cores may be long, even in comparison to the orbital period of the binary these set of runs constitute a test suite to estimate the orbital evolution of protoplanets in the disk. The results for the semi-major axis and eccentricity evolution of the planet are displayed in Fig. 3, where the only difference in the three cases is the release distance (2.5, 3.0 and 3.5 AU) of the planet. From all three locations the

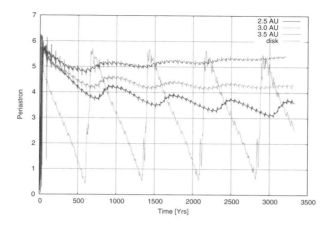

Figure 4. Time evolution of the argument of pericenter (periastron) for an embedded planet starting at 3 different locations and the protoplanetary disk.

planets migrate inwards at approximately the same rate with the tendency for a slow down for the two outer cases. However, the different initial starting radii lead to a very different eccentricity evolution. While only the innermost planet (starting at 2.5 AU) shows a finite eccentricity evolution the two outer cases display a very strong increase in their eccentricity beyond $e_p = 0.5$ after about 55 binary orbits. Clearly the strongly disturbed disk in the outer regions at around 4 AU significantly perturbs the orbits of the protoplanet and induces initially such high eccentricities that the resulting elongated orbits become more and more influenced by the action of the binary. This increases the eccentricities to such high values that the orbits will certainly be unstable eventually. The region of stability in this orbital domain has been analyzed through simple N-body simulations (e.g. Dvorak *et al.* 2004, Turrini *et al.* 2004) which match very well to our results here.

As the planets move on non-circular orbits in an eccentric disk and binary, a temporal change of the apsidal line may be expected. In Fig. 4 the evolution of the argument of pericenter of the planets and the disk are plotted versus time. The innermost planet has on average a periastron angle of about 200 deg while the outer planets have a larger angle. For the binary the angle of periastron lies fixed at 0 degrees, and the disk is slowly precessing retrograde as pointed out above and indicated in Fig. 4 by the tightly dashed line. The influence of the disk on the planets is visible by the slight oscillations of the periastron angle about the mean with the same period as the oscillations in the eccentricity. The innermost planet has approximately a phase shift of 180 deg with respect to the binary and is nearly in an anti-symmetric state while the other planets are lagging behind this configuration.

3.4. Evolution with mass accretion

To estimate the influence of accretion on the planetary evolution we have performed models where the mass of planets is allowed to grow due to accretion from the ambient disk. This accretion process is modelled by taking out mass within a given radius r_{acc} around the planet at a prescribed rate. Whenever the center of a gridcell is closer than about half of the planetary Hill radius, $r_{acc} = 1/2 R_{Hill}$, the density in that cell is reduced by a factor $1 - f_{acc} * \Delta t$, where Δt the the actual timestep of the computation and f_{acc} is a model dependent reduction factor. The rate is doubled whenever the distance is smaller

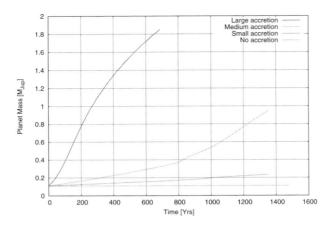

Figure 5. Mass growth of a protoplanet released at a distance of 2.5 AU with an initial mass of about 30 M_{Earth} for different imposed accretion rates.

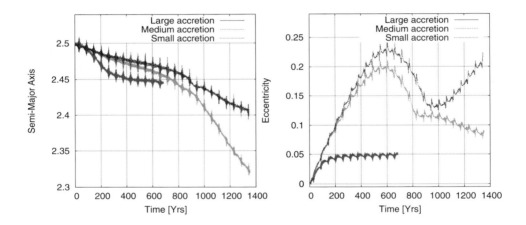

Figure 6. The evolution of the semi-major axis and eccentricity for planets released at an initial distance of 2.5 AU for different mass accretion rates as given in Fig. 5.

than 1/4 r_{acc} (for details see Kley 1999). In Fig. 5 we display the mass evolution of a planetary core released at a distance of 2.5 AU for different accretion rates. The largest accretion rate refers to a value of $f_{acc} = 1.0$, and the medium and small to 10^{-1} and 3×10^{-2}, respectively. For the large accretion rate the mass of the planet increases very rapidly and reaches about $2 M_{Jup}$ after 800 yrs while the other models take accordingly longer.

The migration rate is initially similar for all accretion rates but then accelerates as the mass of the planet increases (left panel of Fig. 6), and finally it slows down because the mass reservoir of the disk becomes exhausted. For the same reason (faster reduction of disk material) the final eccentricity of the planet is smaller for higher accretion rates. Hence, the detailed evolution of the orbital elements of the planet depends on the rate of mass accretion onto the planet. The efficacy of the accretion process cannot be determined straightforwardly but is given for example by thermal processes in the vicinity of the growing planet. In our simulations we did not find a single case of outward or highly reduced migration in the case of smaller masses (accretion rates). However, the migration rate may also be affected by thermal processes in the disk. These will be studied in a future investigation.

4. Summary

In this contribution we have concentrated on the planetary formation process in the system γ Cep which places, due the relatively tight binary star, severe constraints on the formation process. To study the effect of the binary we have followed the evolution of planetary embryos interacting with the ambient protoplanetary disk which is perturbed by the secondary star.

As suspected, the perturbations of the disk, in particular the periodic creation of strong tidally induced spiral density arms, lead to non-negligible effect on the planetary orbital elements. While embryos placed in the disk at different initial distances from the primary star continue to migrate inwards at approximately the same rate, the eccentricity evolution is markedly different for the individual cases. If the initial distance is beyond about $a \gtrsim 2.7$ AU the eccentricity of the embryo continues to rise to very high values and apparently only due to the damping action of the disk the orbit remains bound. The main excitation mechanism of the initial rise of the eccentricity is the perturbed disk and the spiral arms near the outer edge of the disk, and for larger e_p the eccentric binary will pump it up to even larger values. This finding is consistent with the stability analysis of planetary orbits in γ Cep in the case of pure 3-body simulations.

For a disk mass of $3M_{Jup}$ a $1.6M_{Jup}$ planet can easily be grown, and the final semi-major axis and eccentricity are also in the observed range of the γ Cep planet for suitable accretion rates onto the planet. One of the major problems in forming a planet in this binary system via the core instability model is the problem of the formation of the planetary core in the first place. Due to the large relative velocities induced in a planetesimal disk especially for objects of different sizes the growth process is also problematic in itself, see however new results by Xie & Zhou (2008) in this volume.

Hence, the formation of the Jupiter sized planet observed in γ Cep via the standard scenarios remains difficult but may not be impossible. Future research will have to concentrate on additional physical effects such as radiative transport, three-dimensional effects and self-gravity of the disk.

5. Acknowledgements

This review is based on joint work with Richard Nelson from London (GB).

References

Armitage, P. J., Clarke, C. J., & Tout, C. A. 1999, *MNRAS*, 304, 425
Artymowicz, P. & Lubow, S. H. 1994, *ApJ*, 421, 651
Dvorak, R., Pilat-Lohinger, E., Bois, E., *et al.* 2004, in *Revista Mexicana de Astronomia y Astrofisica Conference Series*, Vol. 21, ed. C. Allen & C. Scarfe, 222–226
Eggenberger, A., Udry, S., & Mayor, M. 2004, *A&A*, 417, 353
Haghighipour, N. 2006, *ApJ*, 644, 543
Hatzes, A. P., Cochran, W. D., Endl, M., *et al.* 2003, *ApJ*, 599, 1383
Kley, W. 1989, *A&A*, 208, 98
Kley, W. 1999, *MNRAS*, 303, 696

Lissauer, J. J., Quintana, E. V., Chambers, J. E., Duncan, M. J., & Adams, F. C. 2004, in *Revista Mexicana de Astronomia y Astrofisica Conference Series*, 99–103

Masset, F. 2000, *A&AS*, 141, 165

Monin, J.-L., Clarke, C. J., Prato, L., & McCabe, C. 2007, in *Protostars and Planets V*, ed. B. Reipurth, D. Jewitt, & K. Keil, 395–409

Nelson, A. F. 2000, *ApJ*, 537, L65

Nelson, R. P., Papaloizou, J. C. B., Masset, F. S., & Kley, W. 2000, *MNRAS*, 318, 18

Neuhäuser, R., Mugrauer, M., Fukagawa, M., Torres, G., & Schmidt, T. 2007, *A&A*, 462, 777

Quintana, E. V., Adams, F. C., Lissauer, J. J., & Chambers, J. E. 2007, *ApJ*, 660, 807

Thébault, P., Marzari, F., Scholl, H., Turrini, D., & Barbieri, M. 2004, *A&A*, 427, 1097

Trilling, D. E., Bryden, G., Beichman, C. A., et al. 2007, ArXiv Astrophysics e-prints

Turrini, D., Barbieri, M., Marzari, F., Thebault, P., & Tricarico, P. 2005, *Memorie della Societa Astronomica Italiana Supplement*, 6, 172

Turrini, D., Barbieri, M., Marzari, F., & Tricarico, P. 2004, *Memorie della Societa Astronomica Italiana Supplement*, 5, 127

Verrier, P. E. & Evans, N. W. 2006, *MNRAS*, 368, 1599

Xie, J. -W. & Zhou, J. -L. 2008, in: Y.-S. Sun, S. Ferraz-Mello & J.-L. Zhou, (eds.),*IAU symposium 249: Exoplanets: Detection, Formation and Dynamics*, Proc. IAU Symposium No. 249 (Suzhou,China), p. 419

Ziegler, U. & Yorke, H. 1997, *Computer Physics Communications*, 101, 54 diffused into the gap can actually reach its Roche lobe. Consequently, the accretion rate derived from Equation (2.1) is substantially smaller than that obtained when neglecting the effect of the azimuthal tidal barrier.

We can apply the appropriate boundary condition in protostellar disks to determine the asymptotic mass of protoplanets. Using the 2D mass accretion rate given by Equation (2.1), together with the approximate global depletion formula $\Sigma = \Sigma_0 \exp(-t/\tau_{\rm dep})$ (Ida & Lin 2004), we deduce an asymptotic mass for a protoplanet of

$$\frac{M_p}{M_*} = \frac{3}{|\beta|^{3/2}} \left(\frac{H}{a}\right)^3 \left[\ln \frac{8\pi|\beta|}{9} \left(\frac{aM_{\rm gap}\tau_{\rm dep}}{HM_*P}\right)\right]^{3/2} \tag{2.2}$$

where $M_{\rm gap} = \pi\epsilon\Sigma_o a^2$ is the characteristic mass associate with the gap region. For $\beta \sim 1$ and $\epsilon \sim 10^{-3}$, the above asymptotic mass would yield $R_R \sim H$. To calculate the actual values of β and f_{sub}, we must utilise numerical simulations.

3. Numerical Results

The 2D numerical scheme we use is a fully parallel hydrodynamical code with which the continuity and momentum equations are solved on a fixed Eulerian grid in 2D cylindrical coordinates (r, ϕ). We solve the standard continuity and momentum equations for a global 2D inviscid disk. We have adopted an isothermal equation of state and solve the equations in a frame rotating at the same angular frequency Ω as the planet. Following Tanigawa & Watanabe (2002), a $1M_{jup}$ planet is held fixed at $1AU$ and mass is accreted from the region $r_{sink} \leqslant 0.5R_R$ at a rate given by $\Sigma(t + \Delta t) = \Sigma(t)[1 - \Delta t/\tau_{sink}]$.

We present the results of two models in which $H/a = 0.04$ and 0.07 respectively. Figure (1) shows both the total mass and accretion timescale as a function of orbital period for these models. The accretion rates approach a steady state on the time scale of 100 orbital periods. Scaled to $5AU$, we see that the growth timescale becomes comparable to the gas depletion timescale. The results in Figure (1) clearly indicate that the mass of a planet embedded in a disk with a lower aspect ratio levels off at significantly lower mass than one embedded in a disk with larger (H/a). Based on these results, we compute, from Equation (2.1), the value of

$$\beta \simeq \left(\frac{R_R^2}{H_2^2} - \frac{R_R^2}{H_1^2}\right)^{-1} \ln\left(\frac{\dot{M}_1 H_2}{\dot{M}_2 H_1}\right) \tag{3.1}$$

where the subscripts 1, 2 refer to the values for the two models. We find the value of β increases from a very small value to the order of unity after ~ 50 orbits, in agreement with our analytic predictions.

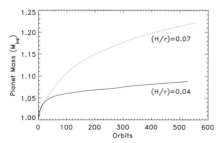
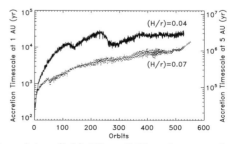

Figure 1. The total mass of the planet as a function of time (left). The solid line shows results for $\left(\frac{H}{a}\right) = 0.04$, while the dotted line shows $\left(\frac{H}{a}\right) = 0.07$. The right-hand panel shows the accretion timescale for both simulations. The left-hand ordinate shows the timescale for a planet at 1 AU, while the right-hand ordinate shows the timescale for a planet located at 5 AU.

4. Conclusion

In this paper, we present evidence to suggest that the asymptotic mass of protoplanets is determined by the structure of their nascent disk. Although the tidal interaction between the disk and massive protoplanet leads to the formation of a gap, disk gas may continue to diffuse into the gap. The main physical process highlighted here is that the tidal potential of the central star provides a barrier to the gas flow in the vicinity of the protoplanet, even if gas is allowed to diffuse into the gap. Only gas that is able to overcome this barrier is accreted. In the limit that the protoplanets' mass is small and its Roche radius is smaller than the disk thickness, the tidal potential barrier is shallow and modest surface density variation would provide adequate pressure gradient to overcome the tidal potential barrier while preserving vortensity along the stream line. However, in cold disks with thicknesses less than the protoplanets' Roche radius, a very large surface density gradient is required to create the critical pressure gradient force needed to overcome the tidal barrier along the stream line. Consequently, mass supply from the feeding zone into the vicinity of the protoplanet is quenched.

Based on this consideration, we suggest that the asymptotic mass of the protoplanet is determined by the condition $R_R \sim H$. This condition is similar to the thermal criterion for gap formation (Lin & Papaloizou 1986). The co-existence of multiple gas-giant planets around the same host star also suggests that their asymptotic mass is related to the local tidal truncation (through gap formation) rather than the global depletion of the disk. The asymptotic mass distribution within such systems is determined by both the planets' formation location and epoch. In a steady-state disk, the value of the aspect ratio H/a determined from the disk mid-plane structure generally increases with a (Garaud & Lin 2007). In a viscous evolving disk, Σ in the outer regions of the disk decreases with a much more rapidly than in a steady disk. In general, H/a attains a maximum at a radial location $a_{\max}$ which is an increasing function of the the accretion rate through the disk. During the depletion of the disk, $a_{\max}$ decreases. Outside $a_{\max}$, the disk is shielded from the stellar irradiation and the density scale decline rapidly with radius. This process may determine the mass distribution within multiple gas giant planet systems. It may also limit the domain of gas-giant formation.

References

D'Angelo, G., Henning, T., & Kley, W. 2002, A&A, 385, 647
Garaud, P. & Lin, D. N. C. 2007, ApJ, 654, 606
Goldreich, P. & Tremaine, S. 1979, ApJ, 233, 857
—. 1980, ApJ, 241, 425
Haisch, Jr., K. E., Lada, E. A., & Lada, C. J. 2001, ApJL, 553, L153
Ida, S. & Lin, D. N. C. 2004, ApJ, 604, 388
Kley, W., D'Angelo, G., & Henning, T. 2001, ApJ, 547, 457
Korycansky, D. G. & Papaloizou, J. C. B. 1996, ApJS, 105, 181
Lin, D. N. C. & Papaloizou, J. 1979, MNRAS, 186, 799
Lin, D. N. C. & Papaloizou, J. 1985, in Protostars and Planets II, ed. D. C. Black & M. S. Matthews, 981–1072
—. 1986, ApJ, 309, 846
Lubow, S. H., Seibert, M., & Artymowicz, P. 1999, ApJ, 526, 1001
Marcy, G., Butler, R. P., Fischer, D., Vogt, S., Wright, J. T., Tinney, C. G., & Jones, H. R. A. 2005, Progress of Theoretical Physics Supplement, 158, 24
Papaloizou, J. C. B. & Terquem, C. 2006, Reports of Progress in Physics, 69, 119
Tanigawa, T. & Watanabe, S.-i. 2002, ApJ, 580, 506

Formation of heavy element rich giant planets by giant impacts

H. Genda[1], M. Ikoma[2], T. Guillot[3] and S. Ida[2]

[1]Research Center for the Evolving Earth and Planets, Tokyo Institute of Technology,
2-12-1 Ookayama, Meguro-ku, Tokyo, Japan
email: genda@geo.titech.ac.jp

[2]Department of Earth and Planetary Sciences, Tokyo Institute of Technology,
2-12-1 Ookayama, Meguro-ku, Tokyo, Japan

[3]Observatoire de la Côte d'Azur,
CNRS UMR 6202, 06304 Nice Cedex 4, France

Abstract. We have performed the smoothed particle hydrodynamic (SPH) simulations of collisions between two gas giant planets. Changes in masses of the ice/rock core and the H/He envelope due to the collisions are investigated. The main aim of this study is to constrain the origin and probability of a class of extrasolar hot Jupiters that have much larger cores and/or higher core/envelope mass ratios than those predicted by theories of accretion of gas giant planets. A typical example is HD 149026b. Theoretical models of the interior of HD 149026b (Sato *et al.* 2005; Fortney *et al.* 2006; Ikoma *et al.* 2006) predict that the planet contains a huge core of 50-80 Earth masses relative to the total mass of 110 Earth masses. Our SPH simulations demonstrate that such a gas giant is produced by a collision with an impact velocity of typically more than 2.5 times escape velocity and an impact angle of typically less than 10 degrees, which results in an enormous loss of the envelope gas and complete accretion of both cores.

Keywords. planetary systems, formation, hydrodynamics

1. Introduction

More than twenty extrasolar planets are known to transit their star. From planetary radius observed by the transit method and planetary mass observed by the radial velocity method, one can determine the densities of the extrasolar planets. The density of the planet informs us about the planetary interior. According to calculation of the interior structure of gas giant planet by Guillot *et al.* (2006), the amount of the heavy elements (ice/rock core in their model) in nine transiting extrasolar planets have been estimated (see Fig. 1). It is noted that the core mass (M_{core}) and core mass ratio (x_{core}) in the planets increases with the metallicity of their star ([Fe/H]). This dependency seems to be reasonable, because the star with higher [Fe/H] had the protoplanetary disk with enough solid materials to form more heavy element-rich planet. However, as mentioned below, the simple formation theory of gas giant planets cannot fully explain this dependency.

According to the core accretion model for gas giant planet formation, when a solid core grows to a certain critical mass (called the critical core mass), the gaseous envelope around the core begins to contract and gas giant planet forms. Typical critical core mass is estimated to be $\sim 10\ M_E$ (Mizuno 1980), and upper limit is $\sim 30\ M_E$ (Ikoma *et al.* 2006), where M_E is the Earth's mass. Assuming that the composition of the gaseous envelope is the same as the composition of its star, we can estimate the total amount of the heavy elements in the planet (i.e., core + heavy element in envelope). The gray-colored region in Fig. 1 shows the estimated range of total heavy elements in the gas

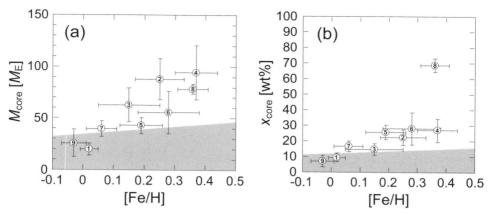

Figure 1. Core mass and core mass ratio in the planets as a function of the metal content of the parent star (figures modified from Guillot *et al.* 2006). The number in the circle corresponds to each planet; (1) HD209458b, (2) OGLE-TR-56b, (3) OGLE-TR-113, (4) OGLE-TR-132, (5) OGLE-TR-111, (6) OGLE-TR-10, (7) TrES-1, (8) HD149026b, and (9) HD189733b. Horizontal error bars correspond to the 1σ errors on the determination of the stellar metallicity ([Fe/H]). Vertical error bars are a consequence of the uncertainties on the measured planetary radii, masses and ages. (a) The core mass (M_{core}), which is normalized by the Earth's mass (M_E), increases with [Fe/H]. (b) The core mass ratio of planet (x_{core}) also increases with [Fe/H]. The gray-colored regions show the theoretically estimated range of total heavy elements in the gas giant planet with a Jupiter mass.

giant planet with a Jupiter mass. The transiting planets around the star with relatively high [Fe/H] have too much heavy elements.

Here we examine the collision between two gas giant planets as one of the possible mechanism to make a gas giant planet rich in heavy elements. We show that the collision increases both the core mass and core mass ratio of a planet owing to merging of two cores and erosion of envelopes. We discuss the dependence of collision probability of gas giants on [Fe/H] of their star.

2. Numerical Results

To perform the simulations of a collision between two gas giant planets, we use the smoothed particle hydrodynamic (SPH) code (e.g., Lucy 1977; Canup 2004). A collision is represented by 20,000 SPH particles, and the mutual gravity between the SPH particles is calculated using the special-purpose computers named micro-GRAPE (Fukushige *et al.* 2005). The equations of state we used are the Tillotson EOS (Tillotson 1962) with the parameters of ice for the core, and ideal gas with ratio of specific heat 2 or SCVH EOS (Saumon *et al.* 1995) for the envelope.

We consider the gas giant planet with 30 M_E of the core and 270 M_E of the envelope, and perform the simulations of the collision between two same-sized gas giant planets. We show two examples of time series of our simulations in Fig. 2. Panels (a)-1 to (a)-4 show the time sequence of collision with $v_{imp} = 1.5\, v_{esc}$ and $\theta = 5°$, and panels (b)-1 to (b)-4 show that of collision with $v_{imp} = 3.0\, v_{esc}$ and $\theta = 0°$, where v_{imp} is the impact velocity, v_{esc} is the two-body escape velocity, and θ is the impact angle (The collisions for $\theta = 0°$ and $90°$ correspond to head-on and grazing collisions, respectively). In both cases, the cores in two gas giant planets (grey dots) merge. The envelopes (white dots)

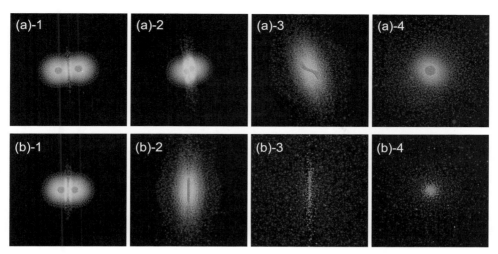

Figure 2. Snapshots of simulations for collisions between two gas giant planets. Panels from (a)-1 to (a)-4 show time sequence of a relatively low velocity collision, and panels from (b)-1 to (b)-4 show that of a relatively high velocity collision. White dots correspond to the H/He envelope, and gray dots, icy core.

are more drastically eroded in the case (b) than (a), because the impact velocity in the case (b) is higher than (a). The masses of envelope and core of the largest gravitationally bound aggregate at the end of the simulations ($\sim 10^5$ seconds) are 470 M_E and 60M_E for case (a), respectively, and 80 M_E and 57 M_E for case (b). The core mass ratios are (a) 11.3 wt% and (b) 41.6 wt%. Since 10 wt% of the initial core mass ratio is considered, the collision makes the core mass ratio increase.

Fig. 3 shows the final mass (M_1) and core mass ratio of the largest gravitationally bound aggregate at the end of the simulations for various impact velocities and impact angles. Collision with higher impact velocity results in higher core mass ratio. To produce a class of heavy element-rich planet such as HD 149026b by a single collision event, the collision with high impact velocity and low impact angle is required.

3. Discussion

As can seen from Fig. 1, the planet around the star with higher [Fe/H] has higher core mass and higher core mass ratio. To make this dependency by collision between two gas giant planets, the planetary system with higher [Fe/H] needs to have higher probability of collisions between gas giant planets. Here, we discuss the collision probability of the planetary system.

It is known that the planetary system with more than three gas giant planets is orbitally unstable. In the case for the planetary system with three gas giant planets around 5 AU, one planet is ejected on an hyperbolic trajectory for most cases, and a collision between two gas giant planets occurs for $\sim$ 5 % of the cases (Marzari & Weidenschilling 2002). If gas giant planets are located close to the star due to a certain mechanism of inward migration of the planet such as Type II migration, the collision probability rises. This is because the hill radius of the planet decreases with the semi-major axis but physical radius is constant. The collision probability also may increase with the number of gas giants. The extrasolar planets observed by the transit method so far are located close

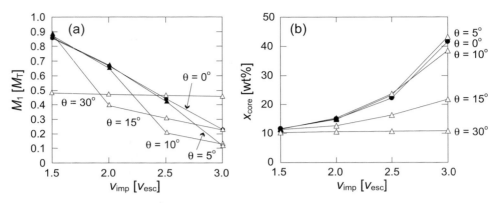

Figure 3. Collisional output for various impact velocities (v_{imp}) and impact angles (θ). (a) Mass of the largest gravitationally bound aggregate (M_1) normalized by total mass of initial two gas giant as a function of v_{imp} normalized by the two-body escape velocity (v_{esc}). (b) Core mass ratio of the largest gravitationally bound aggregate (x_{core}) as a function of v_{imp}. Two gas giant planets do not always merge. Filled symbols correspond to merging events, while open symbols hit-and-run events.

to their star. Therefore, the collision probability of the planetary system with transiting planet would be higher than 5 %, if more than three gas giant planets initially formed.

According to core accretion model for gas giant planet formation, gas giant planet forms from solid core. Therefore, we can speculate that the star with higher [Fe/H] used to have more solid material in protoplanetary disk and more gas giant planets can form. Observation also seems to be consistent with the above speculation, because 3 planetary systems out of 62 ($\sim$ 5 %) are multiple planetary systems in the case for [Fe/H] < 0, while 18 out of 138 ($\sim$ 13 %) for [Fe/H] > 0. Therefore, the collision probability qualitatively increases with [Fe/H] of the star. To make the dependency of core mass and ratio of planet on [Fe/H] precisely, we need more investigation about the number of gas giant formed in protoplanetary disk and collision probability for wide range of [Fe/H].

References

Canup, R. M. 2004, *Icarus*, 168, 433
Fortney, J. J., Saumon, D., Marley, M. S. Lodders, K., & Freedman, R. S. 2006, *ApJ*, 642, 495
Fukushige, T., Makino, J., & Kawai, A. 2005, *PASJ*, 57, 1009
Guillot, T., Santos, N. C., Pont, F., Iro, N., Melo, C., & Ribas, I. 2006, *A&A*, 453, L21
Ikoma, M., Guillot, T., Genda, H., Tanigawa, T., & Ida, S. 2006, *ApJ*, 650, 1150
Lucy, L. B. 1977, *AJ*, 82, 1013
Marzari, F. & Weidenschilling, S. J. 2002, *Icarus*, 156, 570
Mizuno, H. 1980, *Prog. Theor. Phys.*, 64, 544
Sato, B., et al. 2005, *ApJ*, 633, 465
Saumon, D., Chabrier, G., & Van Horn, H. M. 1995, *ApJS*, 99, 713
Tillotson, J. H. 1962, *Report No. GA-3216, General Atomic, San Diego, California*

SPH simulations of star/planet formation triggered by cloud-cloud collisions

Spyridon Kitsionas[1], Anthony P. Whitworth[2] and Ralf S. Klessen[3]

[1] Astrophysikalisches Institut Potsdam, An der Sternwarte 16, D-14482 Potsdam, Germany
email: skitsionas@aip.de

[2] School of Physics & Astronomy, Cardiff University, P.O. Box 913, CF24 3AA Cardiff, U.K.

[3] Institut für Theoretische Astrophysik, Universität Heidelberg,
Albert-Ueberle-Str. 2, D-69120 Heidelberg, Germany

Abstract. We present results of hydrodynamic simulations of star formation triggered by cloud-cloud collisions. During the early stages of star formation, low-mass objects form by gravitational instabilities in protostellar discs. A number of these low-mass objects are in the sub-stellar mass range, including a few objects of planetary mass. The disc instabilities that lead to the formation of low-mass objects in our simulations are the product of disc-disc interactions and/or interactions between the discs and their surrounding gas.

Keywords. hydrodynamics, methods: numerical, stars: formation, planetary systems: protoplanetary disks, planetary systems: formation

1. Introduction

In the classical picture (Shu, Adams & Lizano 1987) stars form from the collapse of a single gaseous core in isolation: all stages of the star formation process (i.e. objects of Class 0, I, II, III) as well as planet formation are terminated before the newly formed star/planetary system interacts with its environment (e.g. other stars/planetary systems, and/or its surrounding gas). In the more dynamic star formation paradigm of gravoturbulent fragmentation (Mac Low & Klessen 2004) interactions during the formation process are not only a common phenomenon but also a sufficient condition for the formation of binary and higher-order multiple systems, as well as for explaining the shape of the Initial Mass Function (Jappsen *et al.* 2005; see also Bate, Bonnell & Bromm 2003, Bate & Bonnell 2005), the evolution of the angular momentum distribution of protostars (Jappsen & Klessen 2004), their mass accretion rates (Schmeja & Klessen 2004), the spatial distribution of young stellar objects (Schmeja & Klessen 2006), etc. The competing paradigm of a more isolated star formation process advocated by Krumholz & McKee (2005), uses large-scale turbulent support against collapse in a rather isotropic way, giving rise to supercritical cores with a variety of masses (see e.g. Krumholz, McKee & Klein 2005 *vs.* Bonnel & Bate 2006). In contrast, gravoturbulent star formation is based on the small-scale density fluctuations produced by supersonic turbulence, with gravity then promoting the strongest of these fluctuations, determining the onset of collapse.

In the context of gravoturbulent fragmentation, Chapman *et al.* (1992), Turner *et al.* (1995), Whitworth *et al.* (1995), and Bhattal *et al.* (1998) have studied star formation triggered by cloud-cloud collisions. In essence, the cloud-cloud collision mechanism describes the small-scale evolution of the interaction between colliding flows (Vázquez-Semadeni *et al.* 2006, 2007). We note that several authors have studied the interaction between large-scale colliding flows in non-self-gravitating media, discussing the importance of such interactions for the formation of molecular-cloud structure and turbulence (e.g. Hunter et al. 1986, Walder & Folini 1998, Heitsch et al. 2005, Hennebelle & Audit 2007). These authors discuss the formation at best of low-mass structures such as dense cloud cores, whereas in our self-gravitating simulations we use such dense cores as the initial conditions in order to follow star formation.

In this paper we present the results of recent high-resolution hydrodynamic simulations of collisions between low-mass clumps (Kitsionas & Whitworth 2007). Due to the high numerical resolution of our simulations, we can resolve, for the first time in simulations of cloud-cloud collisions, the formation of sub-stellar objects, some of which have a mass that lies close to the boundary between brown dwarves and planetary-mass objects, i.e. $\sim 15 M_J$. These planetary-mass objects form by gravitational instabilities in the disc of their parent protostar, i.e. by the same mechanism that is responsible for the formation of low-mass stellar and/or brown dwarf companions to the central protostar of such discs. Moreover, the disc instabilities that lead to the formation of low-mass objects in our simulations are the product of disc-disc interactions and/or interactions between the discs and their surrounding gas. Furthermore, we deal here with disc instabilities of young, massive discs still in the process of formation, i.e. at the early stages of the disc life. We therefore caution the reader not to confuse the instabilities reported here with the extensive work on gravitational instabilities and related phenomena in thin, or in general evolved discs that have been recently reviewed by Durisen et al. (2007). As for the core accretion model of planet formation, we refer the reader to a number of papers in this volume dealing with the different stages of solid core formation, gas accretion and migration in protoplanetary discs.

In section 2, we give a brief description of the model and the numerical method we use. In section 3, we present our results. In section 4, we discuss our findings with respect to models of gas giant planet formation due to gravitational instabilities in protostellar discs.

2. Our model and numerical method

In Kitsionas & Whitworth (2007), we investigate, by means of numerical simulations, the phenomenology of star formation triggered by low-velocity collisions between low-mass molecular clumps. The simulations are performed using a smoothed particle hydrodynamics (SPH) code which satisfies the Jeans condition by invoking on-the-fly Particle Splitting (Kitsionas & Whitworth 2002).

Clumps are modelled as stable truncated (non-singular) isothermal, i.e. Bonnor-Ebert, spheres. Collisions are characterised by M_0 (clump mass), b (offset parameter, i.e. ratio of impact parameter to clump radius), and $\mathcal{M}$ (Mach Number, i.e. ratio of collision velocity to effective post-shock sound speed). The gas follows a barotropic equation of state, which is intended to capture (i) the scaling of pre-collision internal velocity dispersion with clump mass (Larson 1981), (ii) post-shock radiative cooling, and (iii) adiabatic heating in optically thick protostellar fragments (with exponent $\gamma \simeq 5/3$). The equation of state we use is given by Eq. 1 of Kitsionas & Whitworth (2007) and is graphically illustrated in the temperature-density plane by their Fig. 1. In short, the initial effective isothermal sound speed of the gas, i.e. when non-thermal pressure due to turbulence is included, follows the Larson (1981) relations in order to model the observed variation of the internal velocity dispersion with clump mass. As long as the collisions begin, the effective sound speed is reduced according to a $P \propto \rho^{1/3}$ relation until it reaches a value equivalent to the temperature of 10 K. Then the gas remains isothermal with increasing density up to the point where it becomes opaque to its own cooling radiation and heats up as $P \propto \rho^{5/3}$.

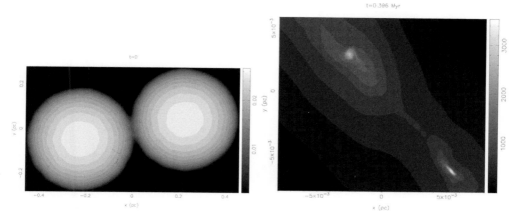

Figure 1. Column density plots for the collision with $M_0 = 10 M_\odot$, $b = 0.2$, $\mathcal{M} = 10$. *Left Panel.* Initial conditions viewed along the z-axis; $\Delta x = 0.92$ pc, $\Delta y = 0.56$ pc; the grey-scale is logarithmic, in units of g cm^{-2}, with sixteen equal intervals from 1.00×10^{-3} g cm^{-2} to 2.69×10^{-2} g cm^{-2}. *Right Panel.* $t = 0.396$ Myr; viewed along the z-axis; $\Delta x = 0.016$ pc, $\Delta y = 0.014$ pc; sixteen-interval logarithmic grey-scale, in units of g cm^{-2}, from 2.14×10^{-1} g cm^{-2} to 3.55×10^{3} g cm^{-2}.

From the variety of models we have investigated, we present here the collision between two $M_0 = 10 M_\odot$ clumps, each moving with velocity (along the x-axis) $v_{\text{clump}} = 1$ km s^{-1} (corresponding to a Mach number $\mathcal{M} = 10$, i.e. assuming a post-shock sound speed of 0.2 km s^{-1}), and with an offset parameter $b = 0.2$ taken along the y-axis. The initial conditions are shown in the left panel of Fig. 1.

We evolve the simulations using an SPH code. This is a Lagrangian method for 3D hydrodynamics. In SPH the sampling points (particles) are able to move with the fluid and their properties are distributed in space through a smoothing function that allows all hydrodynamic quantities to be continuous (Monaghan 1992). Our self-gravitating SPH code (Turner et al. 1995) uses tree-code gravity (Barnes & Hut 1986) to reduce the computational cost for the calculation of gravitational forces, and a second order Range-Kutta time integrator with multiple particle timesteps. It also uses Particle Splitting (Kitsionas & Whitworth 2002), which allows the on-the-fly increase of numerical resolution (in terms of splitting the SPH particles) when and where this becomes necessary, i.e. every time that violation of the Jeans condition (Bate & Burkert 1997, Truelove et al. 1997) becomes imminent. The benefit of the use of Particle Splitting is twofold. Firstly, gas fragmentation remains resolved at all times. Secondly, the simulations evolve with minimum computational cost, as high numerical resolution is employed only when required. Our Particle Splitting method assigns velocities to new particles by interpolating over the velocity field of their parent particle population, thus achieving conservation of energy and momenta with very high accuracy.

Our code includes sink particles (Bate, Bonnell & Price 1995). These are collisionless (star) particles that replace dense, collapsed gaseous regions in order to allow the simulations to evolve further in time. With sink particles, the simulations avoid using tiny timesteps to follow the evolution internal to the sink but at the same time all information on scales smaller than the size of the sink particle becomes invisible to us. In this work we use sink particles with radii ~ 20 AU. A summary of all features of our code can be found in Kitsionas & Whitworth (2007).

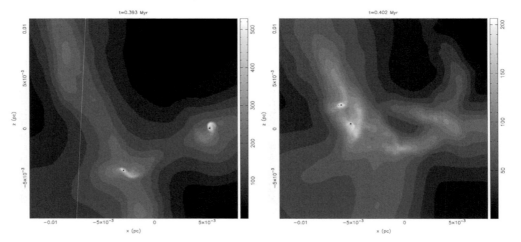

Figure 2. Column density plots for the collision with $M_0 = 10 M_\odot$, $b = 0.2$, $\mathcal{M} = 10$, with sink particles overlayed as solid circles (note that the symbol used for the sinks is larger than the actual sink radius at the scale of these plots). View along the y-axis; $\Delta x = \Delta z = 0.02\,\mathrm{pc}$ in both panels. *Left Panel.* $t \sim 0.393\,\mathrm{Myr}$, i.e. very close to the time of the right panel of Fig. 1; sixteen-interval logarithmic grey-scale, in units of $\mathrm{g\,cm^{-2}}$, from $3.31 \times 10^{-1}\,\mathrm{g\,cm^{-2}}$ to $5.25 \times 10^2\,\mathrm{g\,cm^{-2}}$. *Right Panel.* $t \sim 0.402\,\mathrm{Myr}$; sixteen-interval logarithmic grey-scale, in units of $\mathrm{g\,cm^{-2}}$, from $2.88 \times 10^{-1}\,\mathrm{g\,cm^{-2}}$ to $2.04 \times 10^2\,\mathrm{g\,cm^{-2}}$.

3. Results

A shock-compressed layer forms at the collision interface as the collision proceeds. The right panel of Fig. 1 offers a zoom-in view of this layer at $t \sim 0.396\,\mathrm{Myr}$. The layer extends along the z-axis. The x-z projection of the shock-compressed layer at this time is shown in the left panel of Fig. 2. The layer has broken up into filaments and the filaments have fragmented into protostars surrounded by circumstellar discs. Sink particles have replaced the central protostars of these discs. A second protostar forms on the upper part of the filament that extends almost vertically in this panel. At $t \sim 0.402\,\mathrm{Myr}$ (right panel of Fig. 2) the two protostars in this filament (represented by sink particles surrounded by circumstellar discs) have moved along the filament towards each other. Due to the tumbling nature of the filament, they have a close encounter in the course of which they capture each other into a binary. During this interaction, their discs merge to form a circumbinary disc (left panel of Fig. 3). This disc-disc interaction leads also to the fragmentation of the circumbinary disc (see the additional sink particles in this panel). At later times, the circumbinary disc fragments *only* when lumps of material from the filament fall on to the disc (e.g. see the lumps on the right panel of Fig. 3).

Each of the low-mass fragments, which form during the disc-disc interaction and/or during the episodes of lumpy accretion from the filament, gets ejected from the system after a close encounter with the binary. The final state of the system (Fig. 4) includes a large circumbinary disc hosting a binary, with components having mass $\sim 1.5 M_\odot$ (each) and separation $\lesssim 40\,\mathrm{AU}$, as well as a number of low-mass objects that have been ejected from the disc with moderate velocities of order a few $\mathrm{km\,s^{-1}}$. Some of these low-mass objects are sub-stellar or even of planetary mass. After their ejection from the circumbinary disc they have stopped accreting more mass.

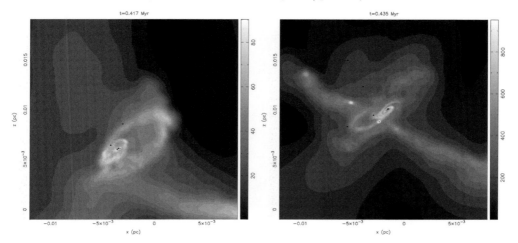

Figure 3. Column density plots for the collision with $M_0 = 10 M_\odot$, $b = 0.2$, $\mathcal{M} = 10$, with sink particles overlayed as solid circles (note that the symbol used for the sinks is larger than the actual sink radius at the scale of these plots). View along the y-axis; $\Delta x = \Delta z = 0.02 \,\mathrm{pc}$ in both panels (note that these panels have been shifted by 0.008 pc along the z-axis with respect to the panels of Fig. 2). *Left Panel.* $t \sim 0.417 \,\mathrm{Myr}$; sixteen-interval logarithmic grey-scale, in units of $\mathrm{g\,cm^{-2}}$, from $2.45 \times 10^{-1} \,\mathrm{g\,cm^{-2}}$ to $9.12 \times 10^{1} \,\mathrm{g\,cm^{-2}}$. *Right Panel.* $t \sim 0.435 \,\mathrm{Myr}$; sixteen-interval logarithmic grey-scale, in units of $\mathrm{g\,cm^{-2}}$, from $3.09 \times 10^{-1} \,\mathrm{g\,cm^{-2}}$ to $9.55 \times 10^{2} \,\mathrm{g\,cm^{-2}}$.

4. Discussion

Based on the results presented here, we can claim with confidence that dynamical interactions during the early stages of star formation provide a mechanism for the formation of low-mass objects. Disc-disc interactions and/or interactions between the discs and their gaseous environment (e.g. episodes of lumpy accretion from a filament) can lead to gravitational instabilities in protostellar discs and the formation of low-mass objects. Some of these objects may subsequently be ejected from the disc they formed in. As a result of this ejection, their accretion rates will drop significantly and they will end-up in the sub-stellar or even planetary-mass regime.

It is important to note that the gas in our discs is rather warm, as no cooling mechanism for discs is included in our code and adiabatic gas heating has already switched on at the gas densities of our discs (see section 2 for a brief description of our equation of state). Therefore, the gravitational instabilities we report here are due to an increase in the disc surface density (caused by a dynamical interaction) and not because of the decrease of the gas sound speed in the disc.

We note that the discs in our simulations contain only a few thousand SPH particles and evolve for a small number of orbital periods. Thus, due to excess numerical dissipation and limited orbital evolution, we are not yet in a position to investigate in detail the angular momentum evolution of gas and fragments in our discs. Nevertheless, fragmentation here is mostly externally triggered and therefore we are confident that it is not due to spurious angular momentum transport. Moreover, since the discs form self-consistently in a cluster environment, it is inevitable that during the initial stages of their formation they will contain a small but increasing number of particles. This makes impossible to comment on the effect of numerical dissipation on early disc evolution. Convergence studies with respect to the disc numerical dissipation in our simulations are currently underway.

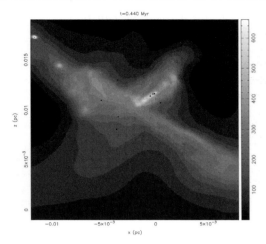

Figure 4. Column density plots for the collision with $M_0 = 10 M_\odot$, $b = 0.2$, $\mathcal{M} = 10$, with sink particles overlayed as solid circles (note that the symbol used for the sinks is larger than the actual sink radius at the scale of these plots). The final state of the simulation ($t \sim 0.440$ Myr); view along the y-axis; $\Delta x = \Delta z = 0.02$ pc; sixteen-interval logarithmic grey-scale, in units of g cm^{-2}, from 3.39×10^{-1} g cm^{-2} to 6.61×10^2 g cm^{-2}.

Gas cooling is an important ingredient in simulations of isolated protostellar discs (e.g. see Rice et al. 2003) even in the case that this cooling is not sufficient to produce gravitational instabilities and fragmentation of the disc (see the discussion in Durisen et al. 2007).

Stamatellos, Hubber & Whitworth (2007) have recently reported gravitational instabilities that lead to the formation of low-mass objects in an isolated disc using an SPH code with on-line (approximate) radiation transport. Their radiation hydrodynamics method is effective in calculating disc cooling close to the disc midplane but not further out in the disc "atmosphere". We are currently developing an improved algorithm that will be able to identify the scale-height at which each gas particle lies and thus assign the cooling efficiency of the particle accordingly.

Generalising our finding on the importance of dynamical interactions during the early stages of star formation, we assert that such interactions may be important for the formation of low-mass objects also at later stages of protostellar evolution, as long as a gas disc is still present. The outcome of such interactions need not be free-floating objects in all cases, as in less "chaotic" situations (e.g. in circumstellar discs around single stars) the low-mass objects formed may remain bound to the system they formed in. Moreover, interactions of already formed planetary systems with incoming stars or lower mass free-floating objects may disturb the planetary systems, but this is beyond the scope of this paper (see several papers in this volume on planetary system dynamics).

Taking our results at face value, one can claim that the majority of the low-mass objects forming in such simulations will end up unbound from the system of their birth, i.e. a number of free-floating objects are expected to form as a by-product of star formation.

The formation of free-floating planetary-mass objects is of particular interest. Such objects can be observed mainly through their gravitational signature. In the field, this signature could be detected only through microlensing observations. Zinnecker (2001) has calculated that a lensing planet can produce a microlensing event with duration of a few days. He has also estimated that one such event is expected out of 3×10^8 observed lightcurves. He has further advocated that campaigns for the observation of free-floating planetary-mass objects through microlensing can become possible when the VST and VISTA facilities start their operation as well as with the JWST.

The event probability and duration that Zinnecker (2001) has estimated are based on simple calculations involving the known frequency of binary stars with members hosting planetary systems and the probability for dynamical interactions between the members of such complicated systems. It would be interesting to calculate the corresponding numbers based on our models. However, we need conduct larger simulations of gravoturbulent star formation (see e.g. Klessen & Burkert 2000, 2001, Klessen 2001a,b, Jappsen et al. 2005) in order to be able to derive more reliable statistics on the number of free-floating planetary-mass objects as well as the overall mass distribution of objects forming in the calculations. Such simulations are currently underway. Moreover, by using multiple levels of Particle Splitting we are now able to employ even higher numerical resolution for our discs as well as use smaller sink particles. We are thus in a position to investigate the evolution of individual discs within a realistic cluster environment at a resolution comparable to that of simulations of isolated discs. This way we can now address issues such as the amount of numerical dissipation and its effect on the angular momentum evolution of our discs.

Acknowledgments. The authors would like to thank Hans Zinnecker for many interesting discussions. SK kindly acknowledges support by an EU Commission "Marie Curie Intra-European (Individual) Fellowship" of the 6th Framework Programme. SK also acknowledges financial support by the German Research Foundation (DFG) for attending this symposium through travel grant KON 1577/2007.

References

Barnes, J. & Hut, P. 1986, *Nature*, 324, 446
Bate, M. R. & Bonnell, I. A. 2005, *MNRAS*, 356, 1201
Bate, M. R. & Burkert, A. 1997, *MNRAS*, 288, 1060
Bate, M. R., Bonnell, I. A., & Bromm, V. 2003, *MNRAS*, 339, 577
Bate, M. R., Bonnell, I. A., & Price, N. M. 1995, *MNRAS*, 277, 362
Bhattal, A. S., Francis, N., Watkins, S. J., & Whitworth, A. P. 1998, *MNRAS*, 297, 435
Bonnell, I. A. & Bate, M. R. 2006, *MNRAS*, 370, 488
Chapman, S. J., Pongracic, H., Disney, M. J., Neslon, A. H., Turner, J. A., & Whitworth, A. P. 1992, *Nature*, 359, 207
Durisen, R. H., Boss, A. P., Mayer, L., Nelson, A. F., Quinn, T., & Rice, W. K. M. 2007, in: B. Reipurth, D. Jewitt, & K. Keil (eds.), *Protostars and Planets V* (Tucson: University of Arizona Press), p. 607
Heitsch, F., Burkert, A., Hartmann, L. W., Slyz, A. D., & Devriendt, J. E. G. 2005, *ApJ*, 633, L113
Hennebelle, P. & Audit, E. 2007, *A&A*, 465, 431
Hunter, J. H. Jr., Sandford, M. T. II, Whitaker, R. W., & Klein, R. I. 1986, *ApJ*, 305, 309
Jappsen, A.-K. & Klessen, R. S. 2004, *A&A*, 423, 1
Jappsen, A.-K., Klessen, R. S., Larson, R. B., Li, Y., & Mac Low, M.-M. 2005, *A&A*, 435, 611
Kitsionas, S. & Whitworth, A. P. 2002, *MMNRAS*, 330, 129
Kitsionas, S. & Whitworth, A. P. 2007, *MMNRAS*, 378, 507
Klessen, R. S. 2001a, *ApJ*, 550, L77
Klessen, R. S. 2001b, *ApJ*, 556, 837
Klessen, R. S. & Burkert, A. 2000, *ApJS*, 128, 287
Klessen, R. S. & Burkert, A. 2001, *ApJ*, 549, 386
Krumholz, M. R. & McKee, C. F. 2005, *ApJ*, 630, 250
Krumholz, M. R., McKee, C. F., & Klein, R. I. 2005, *Nature*, 438, 332 [astro-ph/0510412]

Larson, R. B. 1981, *MNRAS*, 194, 809
Mac Low, M.-M. & Klessen, R. S. 2004, *Rev. Mod. Phys.*, 76, 125
Rice, W. K. M., Armitage, P. J., Bate, M. R., & Bonnell, I. A. 2003, *MNRAS*, 338, 227
Schmeja, S. & Klessen, R. S. 2004, *A&A*, 419, 405
Schmeja, S. & Klessen, R. S. 2006, *A&A*, 449, 151
Shu, F. H., Adams, F. C., & Lizano, S. 1987, *ARAA*, 25, 23
Stamatellos, D., Hubber, D. A., & Whitworth, A. P. 2007, *MNRAS*, 382, L30
Truelove, K. J., Klein, R. I., McKee, C. F., Holliman, J. H., Howell, L. H., & Greenough, J. A. 1997, *ApJ*, 489, L179
Turner, J. A., Chapman, S. J., Bhattal, A. S., Disney, M. J., Pongracic, H., & Whitworth, A. P. 1995, *MNARS*, 277, 705
Vázquez-Semadeni, E., Ryu, D., Passot, T., González, R. F., & Gazol, A. 2006, *ApJ*, 643, 245
Vázquez-Semadeni, E., Gómez, G. C., Jappsen, A. -K., Ballesteros-Paredes, J., González, R. F., & Klessen, R. S. 2007, *ApJ*, 657, 870
Walder, R. & Folini, D. 1998, *A&A*, 330, L21
Whitworth, A. P., Chapman, S. J., Bhattal, A. S., Disney, M. J., Pongracic, H., & Turner, J. A. 1995, *MNARS*, 277, 727
Zinnecker, H. 2001, in: J. W. Menzies, & P. D. Sackett (eds.), *Microlensing 2000: A New Era of Microlensing Astrophysics*, ASP Conference Proceedings, Vol. 239 (San Francisco: Astronomical Society of the Pacific), p. 223

The formation of close-in planets by the slingshot model

Makiko Nagasawa[1], Shigeru Ida[2] and Taisuke Bessho[2]

[1]Global Edge Institute, Tokyo Institute of Technology,
2-12-1 Ookayama, Meguro-ku, Tokyo 152-8550, Japan
email: nagasawa.m.ad@m.titech.ac.jp

[2]Dept. of Earth & Planetary Sciences, Tokyo Institute of Technology
2-12-1 Ookayama, Meguro-ku, Tokyo 152-8551, Japan

Abstract. We investigated the efficiency of planet scatterings in producing close-in planets by a direct inclusion of the dynamical tide effect into the simulations. We considered a system consists of three Jovian planets. Through a planet-planet scattering, one of the planets is sent into shorter orbit. If the eccentricity of the scattered planet is enough high, the tidal dissipation from the star makes the planetary orbit circular. We found that the short-period planets are formed at about 30% cases in our simulation and that Kozai mechanism plays an important role. In the Kozai mechanism, the high inclination obtained by planet-planet scattering is transformed to the eccentricity. It leads the pericenter of the innermost planet to approach the star close enough for tidal circularization. The formed close-in planets by this process have a widely spread inclination distribution. The degree of contribution of the process for the formation of close-in planets will be revealed by more observations of Rossiter-McLaughlin effects for transiting planets.

Keywords. planetary systems: formation, methods: numerical

1. Introduction

More than 260 planets have been detected around both solar and non-solar type stars. In Figure 1, we display the semi-major axis versus the eccentricity of 234 planets † which have been discovered by the radial velocity techniques around solar type stars. The dotted line shows pericenter distance $q = 0.05\,\mathrm{AU}$. Approximately 17% of planets have been discovered inside of $q < 0.05\,\mathrm{AU}$. Many of these planets have circular orbits compared with other extrasolar planets and this property is accounted for by the circularization due to the tidal dissipation of energy within the planetary envelopes (Rasio & Ford 1996).

Close-in planets are considered to be formed beyond ice line and to be moved to shorter-period orbits later. One promising mechanism for migration is the tidal interaction with protoplanetary gas disk (e.g., Ida & Lin 2004a, Ida & Lin 2004b). However, while several close-in planets without nearby secondary planets have moderate eccentricities (~ 0.2), it is not clear that this mechanism alone can excite eccentricity up to that level. Alternative idea is the "Slingshot model". This is a coupled process of planet-planet interaction ("Jumping Jupiters model" (Rasio & Ford 1996, Weidenshiling & Marzari 1996, Lin & Ida 1997)) and tidal circularization. Recent studies of the planet-planet scattering show good coincidence with the observational eccentricity distribution (Marzari & Weidenschilling 2002, Chatterjee Ford, & Rasio 2007, Ford & Rasio 2007).

A formation probability of the close-in planet (i.e., the pericenter distance becomes small enough and the tidal circularization takes place) is the problem of the slingshot model. In this paper, we investigate the efficiency of the slingshot process in producing

† http://exoplanet.eu/catalog-RV.php

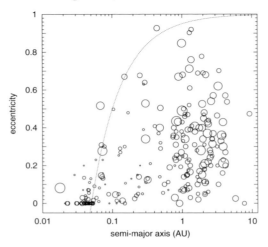

Figure 1. Distribution of orbital eccentricity and semimajor axis of observed extrasolar planets. Circle sizes are proportional to $(m \sin i)^{1/3}$. Dotted line shows pericenter distance $q = 0.05$ AU.

close-in planets by including a dynamical tide into the simulations. Following to the explanation of numerical models (§2), we present the results (§3) and give conclution (§4).

2. Methods

Following to the simulations of the Jumping Jupiter model done by Marzari & Weidenschilling 2002, we consider a system of three planets with equal mass ($m = 1/1000 m_\odot$) orbiting around a solar mass star in circular orbits. Their initial semi-major axes are $a_1 = 5.00$, $a_2 = 7.25$, and $a_3 = 9.50$ AU and their inclinations are $i_1 = 0.5°$, $i_2 = 1.0°$, and $i_3 = 1.5°$. We use a formula of the change of energy and angular momentum caused by the dynamic tide obtained by Ivanov & Papaloizou 2004. We change the orbit of planet at the time of its pericenter passage (an impulse approximation) in a same orbital plane (see Nagasawa, Ida, & Bessho 2008 for details). The tide is a function of mass and radius of planets. We performed 4 set of simulations. The radius of planet are $R = 1$ (Cases 2 and 4) and $2R_J$ (Cases 1 and 3). The planet does not rotate in Case 1 and Case 2. The change of angular momentum by tidal force is set to zero in Case 3 and Case 4.

Because of the chaotic nature of the scattering processes, we integrate 100 different systems with different orbital angles in each set. Basically, we integrate the orbit for $10^7 - 10^8$ yr. We stop the calculation when the change of energy and angular momentum by tidal force overcome the magnitude of orbital energy and angular momentum that the planet has. We also stop the calculation when a planet hits the surface of the host star of 1 Solar Radius.

3. Results

A typical evolution of semi-major axes, pericenters, and apocenters of three-planet system in the case that a close-in planet is formed is shown in Figure 2 (Case 4). After several encounter events, an initially outermost planet is scattered into 3 AU. At 2.16×10^6 yr, the planet obtains a large inclination (~ 1.4 radian) and eccentricity (~ 0.83) as a result of an encounter. The inclination is exchanged for the eccentricity by Kozai mechanism.

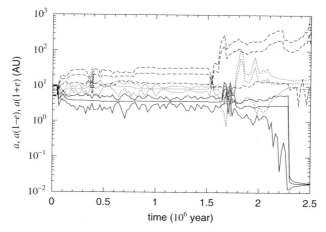

Figure 2. Typical evolution of the semi-major axes of three planets. Thin lines show evolution of pericenters and apocenters.

The pericenter of the planet approaches the star. The semi-major axis decreases immediately in 10^4 years. The damping of the eccentricity takes slightly longer time, but the planet is circularized in $\sim 2 \times 10^5$ yr.

A compiled result is shown in Table 1. There is no significant difference between the models. In total, we find the close-in planet in 33% cases. On average, a close-in planet is produced during three-planet interaction in 26% cases. In about 7% cases, a close-in planet is formed after an ejection of one of the planets. The ratio of the later case is consistent with the result of Marzari & Weidensilling 2002 that takes into account an evolution of longer period. This shows that about 3/4 close-in planets are formed during three-planet interaction stage and not after the stabilization of the system. After one of the planet is scattered inward and detached from other planets, Kozai evolution timescale for the planet can be shorter than the timescale of the planet-planet scattering. The eccentricity and inclination of the innermost planet is changed whenever one of the outer planets approaches. When the planet enters in a good Kozai-state, the pericenter approaches to its host star and the planet becomes the close-in planet.

The final eccentricity and semi-major axis of the formed planet is shown in Figure 3. When the pericenter of the planet does not approach to its host star enough for the tidal circularization, the orbital energy of planet is conserved. In this case, the minimum semi-major axis achievable ($a_{\min}$) is determined by the energy conservation, $1/a_{\min} = 1/a_1 + 1/a_2 + 1/a_3$. In our initial setting, $a_{\min} = 2.25$ AU. Therefore, the inner planets that

Table 1. Results of numerical simulations

Set	Planet radius [R_J]	Rotations	HJ_3[1]	HJ_2[2]	N_{star}[3]
Case 1	2	No	31	7	1
Case 2	1	No	21	8	1
Case 3	2	Yes	25	8	2
Case 4	1	Yes	26	6	2

Notes:
[1] Number of close-in planets produced during a three-planet interaction
[2] Number of close-in planets produced after an ejection of a planet
[3] Number of planets collided to their host star

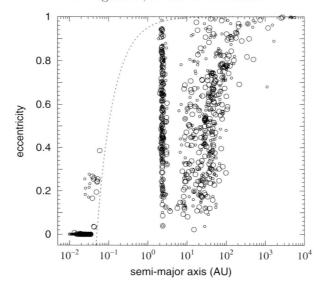

Figure 3. Eccentricities and semi-major axes of formed planets in all simulations at 10^8 year. Planets formed in Case 1 & Case 3 are shown by larger circles. Dotted line shows pericenter distance $q = 0.05$ AU.

are not tidally circularized are distributed around 2.5 AU. On the other hand, semi-major axes of outer planets are broadly spread, since a slight change of the orbital energy causes a large difference in the semi-major axis in the outer region. A planet-planet scattering spreads the eccentricities. Although a large number of the formed close-in planet are circularized in 10^8 years, some of the planets are not fully circularized in 10^8 years, and keep moderate eccentricities. This happens in a case that the circularization occurs after an ejection of a planet.

A inclination distribution of planets that are scattered inward in all simulations is shown in Figure 4. Most planets that are not tidally circularized keep relatively small inclinations. On the other hand, the close-in planet (solid line) tends to have a widely spread inclination distribution. Non-negligible fraction of close-in planets has a retrograde rotation ($i > \pi/2$) compared to an original orbital plane that is nearly perpendicular to the stellar spin axis. Basically, the Kozai mechanism is effective for planets injected to highly inclined orbits. The tidal circularization starts when the eccentricity becomes high. Therefore, although the close-in planet is formed through Kozai mechanism and has a large inclination at first, the inclination is near the minimum of allowed range when the circularization is started, and the final inclination distribution tends to be spread.

4. Conclusions

We investigated the efficiency of the slingshot process—a coupled process of a planet-planet scattering and a tidal circularization—in producing close-in planets. Starting from a system that three Jovian planets are orbiting in unstable circular orbits at $a_1 = 5.00$, $a_2 = 7.25$, and $a_3 = 9.50$ AU with small relative inclinations, we calculated their orbital evolution including the effect of dynamical tide. We found that the close-in planets are formed in about 1/3 cases. This high percentage is because that Kozai mechanism works

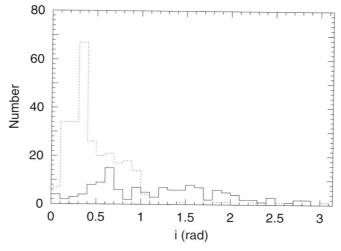

Figure 4. Histogram of inclination of planet scattered inward in all simulations. Solid line shows a number of planets that becomes a close-in planet. Dotted line shows planets that are scattered inward but that are not tidally circularized.

on a inwardly scattered planet and the pericenter of planet approaches the host star enough for tidal circularization.

The inclination of the formed close-in planet is widely spread. Although their tidal lifetime can be short, about a half of the formed close-in planets has retrograde rotations. Observations of Rossiter-McLaughlin effect of 5 transiting close-in planets show that their orbital planes are almost aligned with the stellar spin axis (Narita et al. 2007, Winn et al. 2005, Winn et al. 2006, Winn et al. 2007, Wolf et al. 2007). This may suggest that the slingshot mechanism is not a main process for the formation of close-in planets. The contribution of the slingshot process to the formation of hot planet relative to the Type-II migration will be revealed from more observation of the Rossiter-McLaughlin effect.

Acknowledgments

This work is supported by MEXT KAKENHI(18740281) Grant-in-Aid for Young Scientists (B) and MEXT's program "Promotion of Environmental Improvement for Independence of Young Researchers" under the Special Coordination Funds for Promoting Science and Technology.

References

Chatterjee, S., Ford, E. B., & Rasio, F. A. 2007, *ApJ*, submitted
Ford, E. B. & Rasio, F. A. 2007, *ApJ*, submitted
Ida, S. & Lin, D. N. C. 2004a, *ApJ*, 604, 388
Ida, S. & Lin, D. N. C. 2004b, *ApJ*, 616, 567
Ivanov, P. B. & Papaloizou J. C. B. 2004b, *MNRAS*, 347, 437
Lin, D. N. C. & Ida, S.1997, *Ap. Lett.*, 477, 781
Marzari, F. & Weidenschilling, S. J.2002, *Icarus*, 156, 570
Nagasawa, M., Ida, S., & Bessho, T. 2008, *ApJ*, in press
Narita, N., et al. 2007, *PASJ*, 59, 763
Rasio, F. & Ford, E. 1996, *Science*, 274, 954

Weidenschilling, S. J. & Marzari, F. 1996, *Nature*, 384, 619
Winn, J. N., *et al.* 2005, *ApJ*, 631, 1215
Winn, J. N., *et al.* 2006, *Ap. Lett.*, 653, 69
Winn, J. N., *et al.* 2007, *Ap. Lett.*, 665, L167
Wolf, A. S., Laughlin, G., Henry, G. W., Fischer, D. A. Marcy, G., Butler, G., & Vogt, S. 2007, *ApJ*, 667, 549

Migration and Final Location of Hot Super Earths in the Presence of Gas Giants

Ji-Lin Zhou[1] and Douglas N. C. Lin[2,3]

[1] Department of Astronomy, Nanjing University, Nanjing 210093, China
email: zhoujl@nju.edu.cn
[2] UCO/Lick Observatory, University of California, Santa Cruz, CA 95064, USA
email: lin@ucolick.org
[3] Kavli Institute of Astronomy and Astrophysics, Peking University, Beijing 100871, China

Abstract. Based on the conventional sequential-accretion paradigm, we have proposed that, during the migration of first-born gas giants outside the orbits of planetary embryos, super Earth planets will form inside the 2:1 resonance location by sweeping of mean motion resonances (Zhou et al. 2005). In this paper, we study the subsequent evolution of a super Earth (m_1) under the effects of tidal dissipation and perturbation from a first-born gas giant (m_2) in an outside orbit. Secular perturbation and mean motion resonances (especially 2 : 1 and 5 : 2 resonances) between m_1 and m_2 excite the eccentricity of m_1, which causes the migration of m_1 and results in a hot super Earth. The calculated final location of the hot super Earth is independent of the tidal energy dissipation factor Q'. The study of migration history of a Hot Super Earth is useful to reveal it's Q' value and to predict its final location in the presence of one or more hot gas giants. When this investigation is applied to the GJ876 system, it correctly reproduces the observed location of GJ876d around 0.02 AU.

Keywords. Planets and Satellites: Formation; Celestial Mechanics

1. Introduction

The search for habitable planets is an essential step in the quests to unravel the origin of the Solar System and find life elsewhere. To date, more than 250 exoplanets are detected mainly by radial velocity survey of nearby solar-type stars†. In the database, there are 17 planets with mass less than 25 Earth mass ($M_\oplus$), and among them 8 planets have orbits with period < 10 days. They are GJ876d, HD69830b, GJ674b, HD160691d, 55Cnc e, Gl581b, HD219828b and GJ436b. We call them 'hot super Earths'.

According to the conventional core-accretion scenario of planet formation, planets form in a protoplanetary disk around the host protostar. Through the sedimentation of dust, cohesive collisions of planetesimals, many embryos will form by accreting and clearing the planetesimals in their feed zone (a band centered on the embryo with a width of ~ 10 Hill radius) and result in dynamically isolated bodies. In a disk with several (f_d) times of minimum mass solar nebular, the isolation mass is (Zhou et al. 2007),

$$M_{\rm iso} = 0.51 \times 10^{-2} M_\oplus \eta k_{\rm iso}^{3/2}, \qquad (1.1)$$

where

$$\begin{aligned}
&\eta = (f_{\rm d} f_{\rm ice})^{3/2} (\tfrac{a}{1{\rm AU}})^{3/4} (\tfrac{M_*}{M_\odot})^{-3/2}.\\
&\log(k_{\rm iso}) = \sqrt{b^2 + 0.61c} - b,\\
&b = 2.8 + 0.33 \log \eta,\\
&c = 3.6 + 0.67 \log \eta + \log T_{\rm dep},
\end{aligned} \qquad (1.2)$$

† http://vo.obspm.fr/exoplanetes/, http://exoplanets.org/.

$f_{\rm ice} = 1$ for embryos inside the ice line and $f_{\rm ice} = 4.2$ outside that, $T_{\rm dep}$ is the timescale of depletion of gas disk.

According to equation (1.1), the isolation mass inside the ice line ($a \approx 2.7\,{\rm AU}$ in solar system) is to small to become a super Earth. Unless their nascent protostellar disks are highly compact, the observed super Earths are unlikely formed in situ. Some extra mechanisms are required to account for the excitation of eccentricity and the merge of isolated embryos into super Earths. In Zhou et al. (2005), we have proposed two mechanisms that may lead to the excitation of eccentricities of embryos: (1) During the type-II migration of a first-born gas giant planet outside the orbits of embryos, the locations of its mean motion resonances (mainly 2:1 resonance) sweep through the embryos region; (2) During the dispersal of the gas disk, the location of secular resonance between the gas giant and embryos sweeps through the inner orbits. Additional mechanisms have also been discussed by Raymond et al. (2007).

In this paper, we suppose a super Earth has formed through one of the above mechanisms, and study the subsequent evolution after the gas disk was depleted and the gas giant has stop its migration. First we briefly review the secular evolution of two planets under tidal dissipation. Then we show some numerical results in section 3. Conclusions are presented in section 4.

2. Secular dynamics under tidal dissipation

2.1. Tidal perturbation timescale

We adopt a two-planet system as a model. Suppose two planets with mass $m_i (i = 1, 2)$ (in the order from inner to outer) moving around a star with mass m_* in the same orbital plane. Let m_1 be an Earth-like planet, and m_2 a gas giant, S_i, Ω_i, a_i, r_i are the radius, spin rate (with spin axis perpendicular to the orbital plane), semi major axis, distance from the star of planet i ($i = 1, 2$), respectively. The acceleration to the relative motion of m_i caused by the tidal interaction between the star and planet m_i has the form of (Mignard 1979, Mardling & Lin 2002)

$$\mathbf{F}_{i,\rm tid} = -(1 + \lambda^{-1}) \frac{9 n_i}{2 Q'_i} \left(\frac{m_*}{m_i}\right) \left(\frac{S_i}{a_i}\right)^5 \left(\frac{a_i}{r_i}\right)^8 [3 v_{ir} \hat{r} + (v_{i\phi} - r_i \Omega_i) \hat{\phi}], \qquad (2.1)$$

where $\hat{r}, \hat{\phi}$ are the unit vector of radial and transversal direction of the orbital plane, $\mathbf{V}_i = v_{ir} \hat{r} + v_{i\phi} \hat{\phi}$ and n_i are the Kepler velocity and mean motion of planet i(i=1,2), respectively, Q'_* and Q'_i are the effective tidal dissipation factor of the star and planet i defined as $Q' = 3Q/(2k_L)$, where $Q^{-1} = \tan(2\epsilon)$ is the effective dissipation function, ϵ is the tidal lag angle (Goldreich & Soter 1966), k_L is the Love number or twice the apsidal constance for gaseous planets, and

$$\lambda = \left(\frac{Q'_*}{Q'_i}\right) \left(\frac{m_*}{m_i}\right)^2 \left(\frac{S_i}{S_*}\right)^5 \qquad (2.2)$$

is the ratio of tidal dissipation in the planet to that in the star. If $\lambda \gg 1$, tidal dissipation in the planet dominates the evolution.

The values Q'_* inferred form the observation of circularization period in various stellar clusters are $\sim 1.5 \times 10^5$ for young stars with age less than 0.1 Gyr, and $\sim 10^6$ for mature stars (Terquem et al. 1998, Dobbs-Dixon et al. 2004). The Q' value for Jupiter inferred form Io's orbit evolution ranges from 5×10^4 to 2×10^6 (Yoder & Peale 1981). And for Earth, $Q'_E \approx 60$ (Yoder 1995). Thus for a gas giant planet with Jupiter mass, suppose $Q'_* \approx Q'_J = 10^5$, from Eq. (2.2), $\lambda \sim 10$, while for a terrestrial planet with Earth mass, $\lambda \sim 10^4$. So in the case of tidal interaction between an Earth-like planet and a star, tidal dissipation in the planet dominates the evolution, thus we neglect the contribution of tide in star in the following study.

Take m_1 as an example. Under the perturbation of tidal effect, the averaged equations (over a period of orbital motion) governing the evolution of planet m_1 are,

$$<\dot{a}_1>_{\text{tide}} = -2a_1 \tau_{\text{tide}}^{-1} \left[f_1(e_1) - \left(\frac{\Omega_1}{n_1}\right) f_2(e_1) \right],$$
$$<\dot{e}_1>_{\text{tide}} = -9 e_1 \tau_{\text{tide}}^{-1} \left[f_3(e_1) - \frac{11}{18}\left(\frac{\Omega_1}{n_1}\right) f_4(e_1) \right], \quad (2.3)$$
$$<\dot{\varpi}_1>_{\text{tide}} = <\dot{\lambda}_1>_{\text{tide}} = 0.$$

where ϖ_1, λ_1 are the longitude of perihelion and mean longitude of the orbit of m_1 (with volume density ρ_1), respectively, and

$$\tau_{\text{tide}} = \frac{4 Q_1'}{63 n_1}\left(\frac{m_1}{m_*}\right)\left(\frac{a}{S_1}\right)^5$$
$$= 2.4 \times 10^7 Q_1' \left(\frac{a_1}{0.1\text{AU}}\right)^{13/2} \left(\frac{m_*}{m_\odot}\right)^{-3/2} \left(\frac{m_1}{m_\oplus}\right)^{-2/3} \left(\frac{\rho_1}{3\text{g cm}^{-3}}\right)^{5/3} \text{ yr.} \quad (2.4)$$

Functions used are:

$$f_1(e) = (1 + \tfrac{31}{2}e^2 + \tfrac{255}{8}e^4 + \tfrac{185}{16}e^6 + \tfrac{25}{64}e^8)/(1-e^2)^{15/2},$$
$$f_2(e) = (1 + \tfrac{15}{2}e^2 + \tfrac{45}{8}e^4 + \tfrac{5}{16}e^6)/(1-e^2)^6,$$
$$f_3(e) = (1 + \tfrac{15}{4}e^2 + \tfrac{15}{8}e^4 + \tfrac{5}{64}e^6)/(1-e^2)^{13/2},$$
$$f_4(e) = (1 + \tfrac{3}{2}e^2 + \tfrac{1}{8}e^4)/(1-e^2)^5, \quad (2.5)$$
$$f_5(e) = (1 + 3e^2 + \tfrac{3}{8}e^4)/(1-e^2)^{9/2},$$
$$f_6(e) = (1 + \tfrac{15}{7}e^2 + \tfrac{67}{14}e^4 + \tfrac{85}{32}e^6 + \tfrac{255}{448}e^8 + \tfrac{25}{1792}e^{10})/(1 + 3e^2 + \tfrac{3}{8}e^4),$$
$$f_7(e) = (1 + \tfrac{45}{14}e^2 + 8e^4 + \tfrac{685}{224}e^6 + \tfrac{255}{448}e^8 + \tfrac{25}{1792}e^{10})/(1 + 3e^2 + \tfrac{3}{8}e^4).$$

The evolution of spin rate Ω_1 is subjected to,

$$I_1 \dot{\Omega}_1 = -\frac{m_* m_1}{m_* + m_1} \mathbf{r}_1 \times \mathbf{F}_{1,\text{tide}} \quad (2.6)$$

where $I_1 \approx \frac{2}{5} m_1 S_1^2$ is the inertial momentum of m_1. The averaged change rate is

$$<\dot{\Omega}_1>_{\text{tide}} = \frac{5}{2} \tau_{\text{tide}}^{-1} \left(\frac{a_1}{S_1}\right)^2 \left[f_2(e_1) - \left(\frac{\Omega_1}{n_1}\right) f_5(e_1) \right]. \quad (2.7)$$

A stable equilibrium configuration occurs at

$$\Omega_{1,eq} = \frac{f_2(e_1)}{f_5(e_1)} n_1. \quad (2.8)$$

Since the timescale to reach the equilibrium state ($\sim \tau_{\text{tide}}(S_1/a_1)^2$) is several orders less than the tidal circularization timescale, we suppose such a state is reached. Substitute Eq. (2.8) into (2.3), we derive the timescales of tidal evolution of m_1,

$$\tau_{a-\text{tide}} \equiv \frac{a_1}{\dot{a}_1} = -\frac{(1-e_1^2)^{15/2}}{2 e_1^2 f_6(e_1)} \tau_{\text{tide}}, \quad \tau_{e-\text{tide}} \equiv \frac{e_1}{\dot{e}_1} = -\frac{(1-e_1^2)^{13/2}}{f_7(e_1)} \tau_{\text{tide}}. \quad (2.9)$$

Note that, $\tau_{a-\text{tide}} \gg \tau_{e-\text{tide}}$ when $e_1 \approx 0$. However, when $e_1 \approx 1$, $\tau_{a-\text{tide}}$ and $\tau_{e-\text{tide}}$ could be very small, and $\tau_{a-\text{tide}} < \tau_{e-\text{tide}}$ as long as $e > 0.63425\ldots$.

Due to the huge difference of Q' between the Earth-like planet m_1 and the gas giant m_2, for our later investigation of tidal evolution with $a_1 < 0.63 a_2$, we neglect the tidal effect in planet m_2.

2.2. Secular evolution in the case of $e_1 \ll e_2$

When $e_1 \ll e_2$, the secular evolution of m_1, m_2 under tidal dissipation and general relativity effect can be approximated by the following equations (Mardling 2006):

$$\begin{aligned}
\dot{e}_1 &= -W_o e_2 \sin\eta - W_T e_1, \\
\dot{e}_2 &= W_c e_1 \sin\eta, \\
\dot{\eta} &= W_q - W_o \left(\frac{e_2}{e_1}\right) \cos\eta,
\end{aligned} \quad (2.10)$$

where $\eta = \varpi_1 - \varpi_2, \alpha = a_1/a_2, \beta = \sqrt{1-e_2^2}$, and

$$\begin{aligned}
W_o &= \tfrac{15}{16} n_1 (\tfrac{m_2}{m_*}) \alpha^4 \beta^{-5}, \\
W_T &= \tau_{e-tide}^{-1}, \quad W_c = \tfrac{15}{16} n_2 (\tfrac{m_1}{m_*}) \alpha^3 \beta^{-4}, \\
W_q &= \tfrac{3}{4} n_1 (\tfrac{m_2}{m_*}) \alpha^3 \beta^{-3} [1 - \sqrt{\alpha}(\tfrac{m_1}{m_2})\beta^{-1} + \gamma\beta^3],
\end{aligned} \quad (2.11)$$

with $\gamma = 4(n_1 a_1/c)^2 (m_*/m_2)\alpha^3$, the ratio of general relativity to quadruple contribution of $\dot{\eta}$. According to these equations, the secular evolution of e_1 and e_2 mainly passes three stages:

(1) After a short time oscillation, the evolution of e_1 and η reaches a state of librating around a quasi-equilibrium configuration with $e_1 = e_1^{eq}$ and $\eta = 2n\pi$ or $(2n+1)\pi$, where (Mardling 2006)

$$e_1^{eq} = e_2 \frac{W_0}{|W_q|} = \frac{5/4\alpha e_2}{\beta^2 |1 - \sqrt{\alpha}(m_1/m_2)\beta^{-1} + \gamma\beta^3|}. \quad (2.12)$$

(2) As η librates and e_1 evolves to $e_1 = 0$ gradually, a_1 is damped according to Eq. (2.9), thus m_1 migrates inward efficiently.

(3) Finally e_2 is damped on a timescale $\tau_c \gg \tau_{e-tide}$. During this timescale, the orbit of m_1 is almost circularized, and migration of m_1 is effectively stop at a location a_{1f}.

The location of a_{1f} is what we want to find. However, due to the presence of resonance motion, the evolution of the two-planet system in real situation is more complicate, as we will show below.

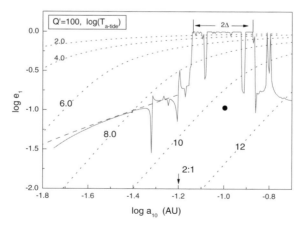

Figure 1. Maximum eccentricity (red solid line) of m_1 in an initial circular orbit of semi-major axis a_{10} excited by m_2 (black circle, with $a_{20} = 0.1$ AU, $e_{20} = 0.1$). The black dotted lines with labels $2.0, 4.0, \ldots$ denote the timescale ($\log(T_{a-tide}/\text{years})$) of m_1 from Eq. (2.9) at the specific location of (a_{10}, e_1) with $Q'_1 = 100$. The blue dashed line is obtained by two times of the equilibrium values defined by Eq. (2.12).

3. Numerical simulations

We study the migration of an Earth-like planet under the tidal and mutual planetary perturbations with general three-body model. The system consists of a solar mass host star ($m_* = 1 M_\odot$), an Earth-like planet with mass ($m_1 = 5 M_\oplus$), and a Jupiter mass gas giant ($m_2 = M_J$, where M_J is Jupiter mass). Let m_1 be initially in a nearly circular orbit ($e_{10} = 10^{-3}$), and m_2 with initial elements $a_{20} = 0.1$ AU, $e_{20} = 0.1$. To shorten the integration time, we let $Q' = 0.02$, as the migration timescale is proportional to Q'.

By integration of the full equations of the general three-body system without tidal dissipation, we plot the maximum eccentricity ($e_{1\max}$) of m_1 excited by m_2 in Figure 1. The corresponding tidal-damping timescale obtained from equation (2.9) with (a_{10}, e_1) is also shown in the background of Fig. 1.

As we can see from Figure 1, the orbits of m_1 at most locations with $a_{10} < 0.16$ AU have $T_{a-\text{tide}} < 10$ Gyr. However, most of those orbits in the Hill unstable region around m_2 with half width $\Delta = (e_{20} + 2\sqrt{3}h)a_2 \approx 0.034$ AU (where $h = [m_2/(3m_*)]^{1/3}$) will be scattered to far away in our coplanar model. According to Zhou et al. 2005, embryos formed inside the location of the 2 : 1 resonance ($a_{2:1} \approx 0.063$ AU) with m_2 are dynamically stable, so we focus on the evolution of orbits with initial semi major axis $a_{10} \leqslant 0.063$ AU.

If planet m_1 is initially located in lower order mean motion resonances with m_2, its eccentricity will be excited, thus a fast inward migration of m_1 is induced, according to Eq. (2.9). Fig. 2 shows the evolution of two orbits either from 2 : 1 resonance location ($a_{10} = 0.063$ AU) or from non-resonance location ($a_{10} = 0.040$ AU). During the subsequent passage through 5 : 2 resonance, the amplitude of eccentricity excitation is relative large. Recall that, according to Eq. (2.9), the a-damping timescale is much smaller than that of e-damping at high eccentricity. Thus a fast migration occurs until e_1 decrease to a small value ~ 0.01 (Novak et al. 2003). Then a slow migration linked with the secular dynamics occurs, with $\eta = \varpi_1 - \varpi_2$ librating along an equilibrium value (see Fig. 2b).

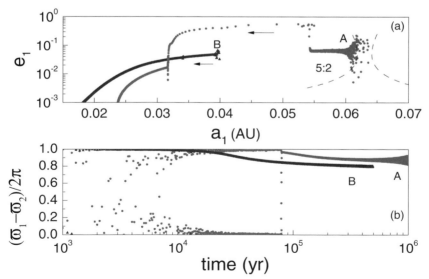

Figure 2. Evolution of orbits with $a_{10} = 0.063$ (orbit A) and $a_{10} = 0.040$ (orbit B). (a) Evolution track in $a_1 - e_1$ plane. The dashed line shows the width of 2 : 1 resonance obtained from the circular restricted-three-body problem. The 5 : 2 indicates the 5 : 2 resonance location at the place that the eccentricity of orbits A jumps up. The arrows indicate the evolution directions. (b) Evolution of ($\varpi_1 - \varpi_2$) (in radian).

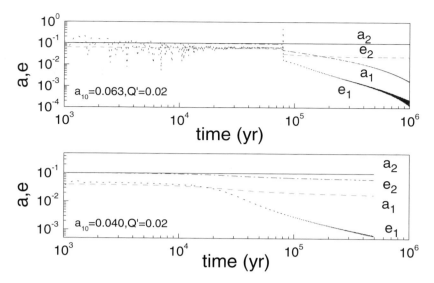

Figure 3. Evolution of orbital elements for the two orbits shown in Fig. 2 with initial elements $a_{10} = 0.040, 0.063$, respectively. The evolution of orbit with $a_{10} = 0.040$ fits with the secular evolution described in section 2, but for the orbit with $a_{10} = 0.063$, the presence of the resonance leads to a dramatic increase (decrease) of $e_1(e_2)$ at time $t \approx 8 \times 10^4$ year, when the orbit crosses the 5 : 2 resonance in Fig. 2.

The migration induced by resonant eccentricity-excitation is different from that excited by mutual secular perturbation. When we check the evolution of $a_i, e_i, (i = 1, 2)$ during the passage of 5 : 2 resonance, we find that a_1, e_1, e_2 have dramatic decrease after the crossing the 5 : 2 resonance (Fig. 3). The decrease of e_2 causes the different final states (i.e., the final state of orbit A and B) in (a_1, e_1) plane of Fig. 2. The final location is around $0.018 \sim 0.025$ AU, depending on the different evolutionary routines.

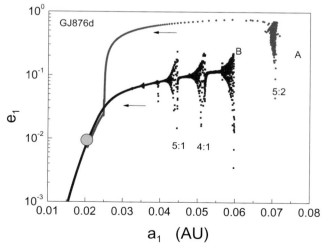

Figure 4. Evolution track of orbits with $a_{10} = 0.07074$ (the exact location of 2 : 1 resonance, orbit A) and $a_{10} = 0.060$ (orbit B). The arrows indicate the evolution directions.

In order to show that the above track correctly reproduces the observed location of extrasolar planets, we applied this study to the GJ876 system. GJ876 is a M dwarf star located 4.72 pc away from us in the solar neighborhood. To date, two gas giant planets, GJ876b and GJ876c, were observed to be located on orbits with period around 30 days and 60 days, an example of 2:1 mean motion resonance, and a hot planet GJ876d with mass around $5.7 M_\oplus$ in an orbit with period 1.94 days($a = 0.0208$ AU, e = 0). We numerical simulate the evolution of an Earth-like planet inside a gas giant located in the present orbit ($a_{20} = 0.13$ AU, $e_{20} = 0.2243$). Figure 4 shows the evolution track of m_1. The final location is around 0.02 AU according to the simulation, which is almost independent of the initial location of m_1.

4. Summary and discussions

Many super Earths are observed to be located inside the orbits of gas giants. These super Earth and gas giant pairs may be a natural consequence of planet formation and migration. Embryos formed prior and interior to the gas giants are induced to migrate, collide, and evolve into close-in Super Earths (Zhou et al. 2005). In this report, we have shown that, the migration of super Earths under tidal dissipation and the perturbation from gas giants is mainly along the secular evolution paths. Although resonances between super Earths and gas giants may excite the eccentricity and speed the migration timescale, the final evolution path can be well determined.

According to the investigation of this paper, we find that the study of the evolution path provides useful information in the following ways: (i) the migration path shows the evolution history, especially the evolution of e_1, e_2, a_1 (Figs. 2, 4). (ii) Comparing the observed location of the planet in the path, we can deduce the range of Q' values for the hot Super Earths. We will investigate in more details on these topics in the future.

Acknowledgements

This work is supported by NSFC(10778603,10233020), National Basic Research Program of China(2007CB814800), NASA (NAGS5-11779, NNG04G-191G, NNG06-GH45G), JPL (1270927), NSF(AST-0507424, PHY99-0794).

References

Dobbs-Dixon, I., Lin, D. N. C., & Mardling, R. 2004, *ApJ*, 610, 464
Goldreich, P. & Soter, S. 1966, *Icarus*, 5,375
Ida, S. & Lin, D. N. C. 2004, *ApJ*, 604, 388
Mardling, R. 2007, *MNRAS*(submitted), arXiv:0706.0224.
Mardling, R. & Lin, D. N. C. 2004, *ApJ*, 573, 829
Mignard, F. 1979, *The Moon and the Planets*, 20, 301
Novak, G. S., Lai, D., & Lin, D. N. C. 2003, in *Scientific Frontiers in Research on Extrasolar Planets*, eds D. Deming & S. Seager, (San Francisco:ASP), 177
Raymond S., Barnes, R., & Mandell A. M. 2007, arXiv:0711.2015.
Terquem, C. Papaloizou, J. C. B.,Nelson, R. P., & Lin, D. N. C. et al. 1998, *ApJ*,502,788.
Yoder C. F. 1995 in *Global Earth Physics, A Handbook of Physical Constants*, ed. T. Ahrens (American Geophysical Union ,Washington)
Zhou, J. L., Aarseth, S. J., Lin, D. N. C., & Nagasawa, M. 2005, *ApJ*, 631, L85
Zhou, J. L., Lin, D. N. C., & Sun Y. S. 2007, *ApJ*, 666, 423

Exoplanets: Detection, Formation and Dynamics
Proceedings IAU Symposium No. 249, 2007
Y.-S. Sun, S. Ferraz-Mello and J.-L. Zhou, eds.

© 2008 International Astronomical Union
doi:10.1017/S1743921308016724

Planet formation around intermediate mass stars

Katherine A. Kretke[1], D. N. C. Lin[1,2] and Neal J. Turner[3]

[1]Department of Astronomy and Astrophysics, University of California, Santa Cruz, CA
[2]Kavli Institute of Astronomy and Astrophysics, Peking University, Beijing, China
[3]Jet Propulsion Lab, Pasadena, CA

Abstract. We present a mechanism by which gas giants form efficiently around intermediate mass stars. MRI-driven turbulence effectively drives angular momentum transport in regions of the disk with sufficiently high ionization fraction. In the inner regions of the disk, where the midplane temperature is above $\sim 1000\,\mathrm{K}$, thermal ionization effectively couples the disk to the magnetic field, providing a relatively large viscosity. A pressure maximum will develop outside of this region as the gaseous disk approaches a steady-state surface density profile, trapping migrating solid material. This rocky material will coagulate into planetesimals which grow rapidly until they reach isolation mass. Around intermediate mass stars, viscous heating will push the critical radius for thermal ionization of the midplane out to around 1 AU. This will increase the isolation mass for solid cores. Planets formed here may migrate inwards due to type II migration, but they will induce the formation of subsequent giant planets at the outer edge of the gap they have opened. In this manner, gas giants can form around intermediate mass stars at a few AU.

Keywords. planetary systems: formation, protoplanetary disks

1. Introduction

Most surveys to detect extrasolar planets using radial velocity (RV) techniques have focused on solar type stars due to their favorable spectral characteristics. While on the main sequence, intermediate mass stars (stars with $1.5 M_\odot \lesssim M_* \lesssim 3 M_\odot$) make poor RV candidates as they only have a small number of spectral lines, and those tend to be relatively broad. However, once these stars evolve off the main sequence their cooler, slower rotating outer layers make them more suitable targets. Recent RV surveys targeting evolved intermediate mass stars suggest that planets around these stars have interesting distinctions from those around solar type stars (Lovis & Mayor 2007). The frequency of giant planets may be higher around intermediate mass stars. Also, they apparently lack closer in planets, even when taking into account planets whose orbits have been disrupted by stellar evolution. (Johnson 2007).

In order to understand these differences we look at the structure of the protoplanetary disk from which the planets must have formed. In the core accretion gas-capture model of planet formation (eg. Bodenheimer & Pollack 1986) gas giant formation requires a large $\sim 10 M_\oplus$ solid core to form in the presence of the gaseous disk. Therefore the we must understand how solids migrate and are retained in gaseous disks in order to understand the distribution of gas giant planets.

In most regions of the disk, the midplane gas pressure, P_{mid}, decreases with r so that the azimuthal velocity of the gas, V_ϕ, is sub-Keplerian. Grains larger than a few cm are decoupled from the gas and move at Keplerian speeds so they experience head winds and undergo orbital decay (Weidenschilling 1977). However if there are irregularities in the disk then the particles will instead drift towards local pressure maxima (Haghighipour & Boss 2003).

The inner regions of typical protostellar disks are thermally ionized (Umebayashi 1983). This ionization couples the disk gas to any existing magnetic field and, in a differentially rotating

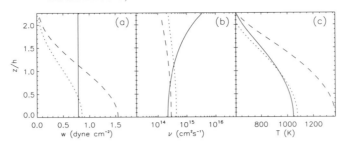

Figure 1. A comparison of this new more realistic parametrization of MRI vertical viscosity (solid line) to a disk with constant viscosity as a function of height with two different "αs (dashed and dotted lines). See text for details. The panel (a) shows the accretion stress, panel (b) the viscosity, and panel (c) the temperature

disk, the magneto-rotational instability (MRI; Balbus & Hawley 1991) generates turbulence. This turbulence provides a source of viscosity which transports angular momentum in the disk. At greater distances from the star, thermal ionization is insufficient for MRI turbulence to develop. Here stellar x-rays and diffuse cosmic rays may ionize the surface layers of the disk, resulting in a viscously active turbulent surface sandwiching an inactive "dead zone" (Gammie 1996).

In section 2 we discuss the vertical structure of an MRI active disk using a more realistic approximation of the disk viscosity as a function of height and use this parametrization to calculate the location of the inner edge of the deadzone as a function of stellar mass and mass accretion rate. In section 3 we estimate the amount of solid material which can accumulate in this region and discuss the implications for the formation of gas giant cores.

2. Thermal Structure of MRI Active Disks

The standard Shakura & Syunyaev 1973 parametrization of the viscosity in geometrically thin accretion disks ($\nu = \alpha c_s^2 \Omega_K$) is sometimes extended into two dimensions by using the local temperature and maintaining a constant α with height, which leads to a constant viscosity with height in an isothermal disk (sometimes known as the "αP-formalism"; e.g. Cannizzo 1992). However, MHD simulations demonstrate that in MRI active regions the viscosity is not constant with height. A better approximation is that the accretion stress $w = (3/2)\nu\rho\Omega_K$ is constant with height for 2 disk scale heights, above which w decreases significantly (Miller & Stone 2000). Therefore we parametrize the viscosity in an α-type manner as

$$\nu(z) = \frac{2}{3} \frac{\alpha P_{\rm mid}}{\rho(z)\Omega_K}. \quad (2.1)$$

using the midplane density instead of the more common 2D parametrization using the local pressure and density. Figure 1 compares the usual assumption that viscosity is proportional to the local sound-speed with the assumption that the accretion stress is constant. The solid curves show the disk structure in the case of a constant w, while the dashed and dotted curves are for $\nu \propto c_{s,local}$, but for two different values for the constant of proportionality α. The dashed curve shows the structure for a disk with the same α as the constant w case, such that if the two disks had the same midplane temperature then the viscosity at the midplane would be the same. The dotted curve is scaled such that the mass accretion rate through the annulus is the same as the constant w case.

In a disk with a steady-state mass accretion rate, the mass accretion rate through each radius is

$$\dot{M} = 2\pi \int_{-\infty}^{\infty} \rho\nu dz. \quad (2.2)$$

If one uses the same α parameter in the constant w and standard local description of viscosity

(Figure 1 dashed line), then the midplane temperature in the later case is significantly warmer than in the more realistic approximation, in agreement with Hirose et al. 2006. However, we can define a α_eff based on the midplane temperature similar to the original Shakura & Syunyaev model, then this can be related to the mass accretion rate as given by Pringle 1981

$$\alpha_\text{eff} = \frac{\dot{M}\Omega_K}{3\pi c_{s,mid}^2 \Sigma}. \tag{2.3}$$

Adjusting α so that the two different parameterizations of viscosity have the same α_eff yields a cooler midplane as the surface density is smaller for the same mass accretion rate (Figure 1 dotted line). The temperature differences are not as dramatic in this case, but the disk is still less isothermal than the constant w situation.

2.1. Vertical Structure

We solve the disk structure of an optically thick, viscously heated disk with radiative energy transport. Along with equations for continuity and hydrostatic equilibrium the disk structure is describe by the following equations:

$$\frac{dF_E}{d\sigma} = \frac{9}{8}\nu\Omega_K^2 \tag{2.4}$$

$$\frac{dT}{d\sigma} = -\frac{F_E}{2\rho k} \tag{2.5}$$

$$\frac{dz}{d\sigma} = \frac{1}{2\rho} \tag{2.6}$$

$$\frac{dP}{d\sigma} = -\frac{1}{2}\Omega_K^2 z \tag{2.7}$$

where F_E is the radiative flux, $\sigma(z)$ is the column depth of material between height z and $-z$, and

$$k = k_\text{rad} = \frac{4ac}{3}\frac{T^3}{\kappa\rho}. \tag{2.8}$$

In the inner regions of a rapidly accreting disk the surface density and optical depth of the disk is large enough that solving the disk structure below the photosphere (z_e) is sufficient for studying most of the disk mass. We can then use photospheric boundary conditions $T(z_e) = T_{irr} + (F_E/\sigma_{sb})^4$ and $P(z_e) = (2/3)\Omega_K z_e/\kappa$. We use a constant opacity $\kappa = 1\,\text{cm}^2\,\text{g}^{-1}$ consistent with opacities from Ferguson et al. 2005.

For high mass accretion rates, the midplane temperatures at this region is dominated by viscous heating. Therefore we use a simple prescription for stellar irradiation, simply assuming a flat disk (see Chiang & Goldreich 1997)

$$T_e = \left(\frac{2}{3\pi}\right)^{1/4}\left(\frac{R_*}{a}\right)^{3/4}T_*. \tag{2.9}$$

For the stellar parameters we use 1 Myr old stars with solar metallicity from D'Antona & Mazzitelli 1997. The 2D structure of the disk may have interesting effects on the amount of stellar light intercepted, for example a "wall" of material at the inner edge of the deadzone. But as the 1+1D approximation cannot explicitly address this issue we leave this problem for future studies.

2.2. Location of the inner edge of the Deadzone

Solving the equations for vertical structure we can calculate the location at which the midplane just satisfies the ionization criteria of $x_e \geqslant 10^{-12}$ (which corresponds to $T \sim 1000\,\text{K}$) (see Klahr et al. 2006) for different surface densities. We calculate the mass accretion rate through that annulus using equation 2.2 with $\nu(z)$ from equation 2.1. In Figure 2 the curves show the location of the inner edge of the deadzone as a function of stellar mass and accretion rate assuming that

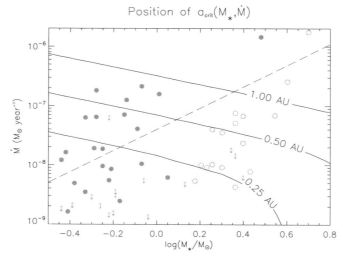

Figure 2. The curves show the position of the inner edge of the deadzone as a function of stellar mass and mass accretion rates. Symbols represent observations of mass accretion rates for stars of various masses (see text for details). The dashed line shows the best-fit for the ρ-Oph cluster from Natta *et al.* 2006.

in the active region $\alpha = 10^{-2}$. For most values of M_* and $\dot{M}$ these theoretical curves can be approximated by

$$a_{\rm crit} = 0.22 AU \left(\frac{\dot{M}}{10^{-8} M_\odot year^{-1}} \right)^{0.42} \left(\frac{M_*}{M_\odot} \right)^{0.32}. \qquad (2.10)$$

The empirical relation between stellar mass and mass accretion rate is still rather uncertain (and shows significant scatter), so for reference the symbols in Figure 2 indicate measurements of mass accretion rates onto young stars. The solid points are from Natta *et al.* 2006 observations of ρ-Oph and the dashed line shows the best fit to their data which is $\dot{M} \propto M_*^{1.8}$. As the mass accretion rate for the higher mass stars is dominated by a single object, we have also plotted as open points from the older, heterogeneously distributed intermediate mass stars (Garcia Lopez *et al.* 2006). The estimated ages on these stars range from 1-10 Myr, so they are systematically older than ρ-Oph, and when they were younger they may have had higher mass accretion rates consistent with the extrapolation from ρ-Oph (Garcia Lopez *et al.* 2006). Using the best fit relationship from ρ-Oph implies that $a_{\rm crit} \propto M_*^{1.1}$. Clarke & Pringle 2006 suggest that the correlation between stellar mass and mass accretion rate may not be this steep due to observational biases. If instead we use their estimation that $\dot{M} \propto M_*$, $a_{\rm crit} \propto M_*^{0.74}$ becomes slightly less sensitive to M_*.

3. Planet Formation at the Inner Edge of the Deadzone

Few mechanisms for angular momentum transport in protoplanetary disks are as efficient as MRI, therefore we expect the disk viscosity to decrease in the deadzone, which will cause a corresponding increase in the surface density of the disk if the disk can reach a steady-state surface density profile (Kretke & Lin 2007). This change in the surface density will create a local pressure maximum capable of accumulating solids. The $a_{\rm crit}$ calculated in the previous section corresponds to the outermost thermally ionized location. The location of the pressure maxima will be larger than $a_{\rm crit}$ by at least one scale height and perhaps more depending on how the details of the 2D structure which is neglected here.

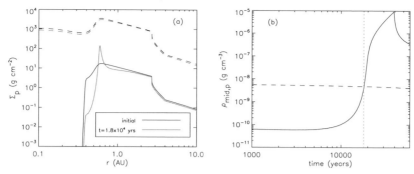

Figure 3. Panel (a) shows the of the solid surface density (solid curve) and the gas surface density(dashed curve) at two different times. Panel (b) shows the evolution of the midplane density of the two fluids at the pressure maximum. The vertical dotted line marks the time when the midplane gas and solid densities are equal.

3.1. Accumulation of Solids

In order to estimate the amount of material which will accumulate at the inner edge of the deadzone we use Garaud 2007 model (modified to have an $\alpha_{\rm eff}$ which varies with radius) to estimate the migration of material in a gaseous disk. This model approximates the evolution of solids in a disk by following the growth and migration of solids through an evolving gaseous disk. Particles undergo radial migration due to gas drag, and the radial velocity of the solids with respect to the gas is given by

$$v_p = -\frac{h^2}{a^2} v_K \frac{\partial \ln(P)}{\partial \ln(a)} \frac{2\pi St(s)}{[4\pi^2 St^2(s) + 1]} \quad (3.1)$$

where v_K is the Keplerian velocity and the Stokes number $St(s)$ is the ratio of the local stopping time to the local orbital time for particles of radius s (See Garaud 2007 for more details). For illustrative purposes, Figure 3 demonstrates the buildup of material in less than 10^5 years from an initially well-mixed disk of micron sized grains. The initial gas profile is for a steady-state mass accretion rate of $\dot{M} = 4 \times 10^{-8} M_\odot year^{-1}$ with $\alpha_{\rm eff} = 10^{-2}$ in the active regions and $\alpha_{\rm eff} = 10^{-3}$ in the MRI dead regions. At the pressure maximum at inner edge of the dead zone material accumulates rapidly once the particles grow to meter-sized, mobile objects. Particles will continue to accumulate until the midplane particle density ($\rho_{\rm mid,p}$) becomes comparable to the midplane gas surface density at the inner edge of the deadzone. At this point the solid material will force a larger region of the gas to rotate at Keplerian velocity increasing the radial extent of the gas moving at Keplerian speed and spreading out the location where particles accumulate. This makes the accumulation of particles a self-limiting process, but only once the solid surface density increases by an order of magnitude. This feedback of the particles on the gas is neglected in this contribution but will be included in future work. For these calculations we simply assume that the order-of-magnitude increase of solid surface density which should accumulate without problem is a reasonable approximation for the amount of solid material present at the inner edge of the deadzone.

3.2. Core Formation

In order to estimate the size of cores formed at the inner edge of the deadzone we use the expression from Ida & Lin 2005 for core isolation mass.

$$M_{c,iso} \simeq 0.16 \left(\frac{\Sigma_d}{10\,{\rm g\,cm^{-2}}}\right)^{3/2} \left(\frac{a}{1{\rm AU}}\right)^3 \left(\frac{M_*}{M_\odot}\right)^{-0.5} M_\oplus \quad (3.2)$$

This expression assumes that the planetesimal surface density (Σ_d) is uniform enough throughout the embryo's feeding zone to be approximated by a single value. By assuming the solids scale

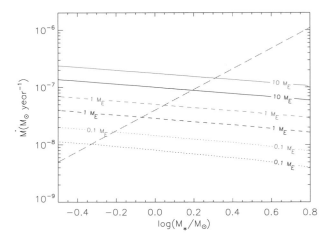

Figure 4. The core isolation mass at the inner edge of the deadzone assuming $\alpha_{\rm eff} = 10^{-4}$ (in the deadzone) and $f = 0.05$ (black curves). The blue curves show how the core isolation mass would change if either f is decreased or $\alpha_{\rm eff}$ increased by a factor of two. The long-dashed line is the same as in Figure 2 for reference.

with a steady-state gas surface density in the deadzone

$$\Sigma_d = f\Sigma_g = f\frac{\dot{M}\Omega_K}{3\pi\alpha_{\rm eff} c_s^2}. \quad (3.3)$$

Here $\alpha_{\rm eff}$ includes a large contribution from the deadzone, so it will be significantly smaller than in the thermally ionized regions. If a sufficient column of the surface layers of the disk is active to sustain MRI turbulence then the deadzone will also sustain a small degree of accretion stress even though it is laminar (Fleming & Stone 2003). We note that the total surface density at this region is sensitive to the viscous parameters, therefore the normalization may be scaled, but the trend in stellar mass should remain.

The time to reach this isolation mass is also given by Ida & Lin 2005 and is approximately

$$t_{\rm core} = 0.26 \left(\frac{\Sigma_d}{10\,{\rm g\,cm^{-2}}}\right)^{-1/2} \left(\frac{\Sigma_g}{3.4\times 10^3\,{\rm g\,cm^{-2}}}\right)^{-1/2} \left(\frac{a}{1{\rm AU}}\right)^{8/5} \left(\frac{M_*}{M_\odot}\right)^{1/3}\,{\rm Myr}. \quad (3.4)$$

This time is typically less than the viscous evolution time of the disk so cores should reach isolation mass. By assuming that $\alpha_{\rm eff}$ is independent of stellar mass, the viscosity is the same at $a_{\rm crit}$ for all stellar masses. This implies that a higher $\dot{M}$ is due to a higher surface density. The fact that typical disk mass accretion rates are higher for more massive stars and that the location of the pressure inversion is farther out combine to make the core size rise steeply with stellar mass. The isolation mass of the core at $a_{\rm crit}$ goes as $M_{c,iso} \propto \dot{M}^{1.8} M_*^{0.5}$. Using the best fit relationship for the relationship between stellar mass and mass accretion rate derived from ρ-Oph fit, $M_{c,iso} \propto M_*^{3.7}$.

Using Lodders 2003 solar composition, most of the silicates and metals will still be in solid phase at 1000 K. Therefore the mass of solids should be a little less than the 0.5% of gas at this region. This implies that the base dust-to-gas ratio would be $f = 0.005$. However, if the solid material accumulates in accordance with the results in Figure 3, f will increase by an order of magnitude. Figure 4 shows the core isolation mass for $f = 0.05$ and $\alpha_{\rm eff} = 10^{-4}$, for comparison, the blue lines show the isolation mass for the equivalent situation of f decreased or $\alpha_{\rm eff}$ begin increased by a factor of two.

3.3. Gas Giant Formation

With this estimate of core isolation mass we can continue using the approximations of Ida & Lin 2005 to look at further potential growth. Gas capture becomes important once the planet has reached core isolation mass. The planet growth due to gas capture is

$$\frac{dM_p}{dt} \approx \frac{M_p}{\tau} \tag{3.5}$$

and

$$\tau \approx \tau_0 \left(\frac{M_p}{M_\oplus}\right)^{-3} \tag{3.6}$$

where $\tau_0 \approx 10^{10}$ years. As long at the final mass of the planet is much greater than the original core mass, the timescale for planet formation is

$$t_{\text{giant}} \approx \frac{\tau_0}{3} \left(\frac{M_c}{M_\oplus}\right)^{-3}. \tag{3.7}$$

This effectively means that in order for a gas giant to form within the lifetime of the evolving disk, the core isolation mass must be on the order of 10 $M_\oplus$. Using the relationship derived for the core mass this implies that $t_{\text{giant}} \propto \dot{M}^{-5.4} M_*^{-1.5}$. This implies $t_{\text{giant}} \propto M_*^{-11}$ using the ρ-Oph data. This very steep function demonstrates that giant planets will not form effectively at the inner edge of the deadzone around less massive stars, but may form around higher mass stars with higher accretion rates.

4. Summary and Discussion

Observations indicate that planets may form systematically more efficiently around intermediate mass stars and their semi-major axis appear to be systematically larger than those around solar type stars to a degree which may not be completely explained by post-main sequence evolution. In this paper we demonstrate a mechanism to form planets preferentially around intermediate mass stars. Solid material can collect at the pressure inversion at the inner edge of the deadzone in protoplanetary disks. In order for the material accumulated to form gas giant cores, the inner edge of the deadzone must be sufficiently far from the host stars. This condition is only likely to be met around intermediate mass stars with high accretion rates.

This model does assume that the location of the inner edge of the deadzone is relatively constant with time, and least until the solids grow in size large enough to be effectively decoupled from the gas. Wünsch et al. 2006 use different parameterizations of vertical viscosity and find that the inner edge of the deadzone may be unstable to oscillations under some conditions. This may help explain why excess solids are not found at the inner edge of the deadzone in all disks. Once the first planet has formed and grown large enough to open a gap, then the outer edge of the gap provides another pressure maximum which may trap solids and encourage the growth of the next planet (Bryden et al. 2000). Therefore some of the observed planets may be subsequent planets formed in the systems, so they may be located farther out than the location of the original pressure maximum at the inner edge of the deadzone.

References

Balbus, S. A. & Hawley, J. F. 1991, ApJ, 376, 214
Bodenheimer, P. & Pollack, J. B. 1986, Icarus, 67, 391
Bryden, G., Różyczka, M., Lin, D. N. C., & Bodenheimer, P. 2000, ApJ, 540, 1091
Cannizzo, J. K. 1992, ApJ, 385, 94
Chiang, E. I. & Goldreich, P. 1997, ApJ, 490, 368
Clarke, C. J. & Pringle, J. E. 2006, MNRAS, 370, L10
D'Antona, F. & Mazzitelli, I. 1997, Memorie della Societa Astronomica Italiana, 68, 807
Ferguson, J. W., Alexander, D. R., Allard, F., Barman, T., Bodnarik, J. G., Hauschildt, P. H., Heffner-Wong, A., & Tamanai, A. 2005, ApJ, 623, 585

Fleming, T. & Stone, J. M. 2003, ApJ, 585, 908
Gammie, C. F. 1996, ApJ, 457, 355
Garaud, P. 2007, ArXiv e-prints, 705
Garcia Lopez, R., Natta, A., Testi, L., & Habart, E. 2006, A&A, 459, 837
Haghighipour, N. & Boss, A. P. 2003, ApJ, 598, 1301
Hirose, S., Krolik, J. H., & Stone, J. M. 2006, ApJ, 640, 901
Ida, S. & Lin, D. N. C. 2005, ApJ, 626, 1045
Johnson, J. A. 2007, ArXiv e-prints, 710
Klahr, H., Różyczka, M., Dziourkevitch, N., Wünsch, R., & Johansen, A. 2006, Turbulence in protoplanetary accretion disks: driving mechanisms and role in planet formation. (Planet Formation), 42
Kretke, K. A. & Lin, D. N. C. 2007, ApJL, 664, L55
Lodders, K. 2003, ApJ, 591, 1220
Lovis, C. & Mayor, M. 2007, A&A, 472, 657
Miller, K. A. & Stone, J. M. 2000, ApJ, 534, 398
Natta, A., Testi, L., & Randich, S. 2006, A&A, 452, 245
Pringle, J. E. 1981, ARA&A, 19, 137
Shakura, N. I. & Syunyaev, R. A. 1973, A&A, 24, 337
Umebayashi, T. 1983, Progress of Theoretical Physics, 69, 480
Weidenschilling, S. J. 1977, MNRAS, 180, 57
Wünsch, R., Gawryszczak, A., Klahr, H., & Różyczka, M. 2006, MNRAS, 367, 773

Giant impact, planetary merger, and diversity of planetary-core mass

S.-L. Li[1,2], C. Agnor[3] and D. N. C. Lin[2,4]

[1]Dept. of Astronomy, [4]Kavli Institute of Astronomy & Astrophysics,
Peking University, Beijing 100871, P. R. China
email: lisl@vega.pku.edu.cn

[2]Dept. of Astronomy & Astrophysics, [3]Dept. of Earth & Planetary Science,
University of California, Santa Cruz, CA 95064, USA

Abstract. Transit observations indicate a large dispersion in the internal structure among the known gas giants. This is a big challenge to the conventional sequential planetary formation scenario because the diversity is inconsistent with the expectation of some well defined critical condition for the onset of gas accretion in this scenario. We suggest that giant impacts may lead to the merger of planets or the accretion of planetary embryos and cause the diversity of the core mass. By using an SPH scheme, we show that direct parabolic collisions generally lead to the total coalescence of impinging gas giants whereas, during glancing collisions, the efficiency of core retention is much larger than that of the envelope. We also examine the adjustment of the gaseous envelope with a 1D Lagrangian hydrodynamic scheme. In the proximity of their host stars, the expansion of the planets' envelopes, shortly after sufficiently catastrophic impacts, can lead to a substantial loss of gas through Roche-lobe overflow. We are going to examine the possibility that the accretion of several Earth-mass objects can significantly enlarge the planets' photosphere and elevate the tidal dissipation rate over the time scale of 100 Myr.

1. Numerical Methods

We use two methods to simulate giant impacts between gas giants and planetesimals and mergers between two planets. In the first method, we model collisions using smooth particle hydrodynamics (SPH) (see e.g. the reviews by Monaghan (1992)). Its principle virtue for this work is its ability to simulate highly deformed flows and shocks evolving in 3-D. Thus, we are able to calculate the retention efficiency of both protoplanets' core and envelope. However, this method is limited by a small dynamical range in density and short evolution time. One important result from SPH simulations show that spherical symmetry is quickly restored after the giant impact. This allows us to use a 1-D Lagrangian hydrodynamic scheme (OLH) to study the response and evolution of the gaseous envelope after the giant impact. In the 1-D hydrodynamic scheme, we use the EOS from Saumon et al. (1995), and adopt the opacity from Alexander & Ferguson (1994). We take into account both flux- limited radiation and convective energy transfer in the code. Adaptive mesh is also included.

2. Results from SPH simulations

In table 1, we list the parameters and results for the models calculated using SPH scheme. The columns indicate the number of the model, initial total planetary mass of the target, mass of condensed material of the target, mass of the impactor, mass of

Table 1. Parameters and impact results from SPH simulations

Model	M_T ($M_\oplus$)	$M_{T,c}$ ($M_\oplus$)	M_I ($M_\oplus$)	$M_{I,c}$ ($M_\oplus$)	ξ (°)	v_{imp}/v_{esc}	$M_{f,c}$ ($M_\oplus$)	$M_{f,g}$ ($M_\oplus$)
1	100	50 fe	100	50 fe	0°	1.0	99.7	97.3
2	100	50 fe	100	50 fe	45°	1.0	100	100
3	100	50 fe	100	50 fe	0°	1.4	98.6	60.4
4	100	50 fe	100	50 fe	45°	1.4	49.4	35.6
5	100	50 fe	25	25 b	0°	1.0	75.0	50.0
6	100	50 fe	25	25 b	30°	1.0	74.9	49.1
7	100	50 fe	25	25 b	0°	1.4	74.8	46.2
8	100	50 fe	25	25 b	30°	1.4	52.4	47.6
9	100	10 fe	10	10 b	30°	1.4	10	95

condensed material of the impactor, impacting angle, the ratio of impacting velocity to escape velocity, final mass of condensed material and gaseous envelope mass of the planet after the impact. For the composition of condensed material of all the models, 'fe' denotes iron and 'b' denotes basalt. The first 4 models are simulating the impact between two giant planets. While model 5–9 are simulating the collisions between giant planets and planetesimals. We show that low-velocity collisions result in the merger of the giant planets and the impactors. Also, small impact angles generally lead to the total coalescence of impinging gas giants. While during high-velocity glancing collisions, there is high probability of core erosion and gaseous envelope loss, and the efficiency of core retention is much larger than that of the envelope.

3. Results from OLH calculations

We construct the impacted models based on static initial models with various core and envelope masses. Since we are interested in close-in planets, the tidal effect of their host stars are taken into account. We listed the model parameters and results in Table 2. Columns in Table 2 have the similar meanings as in Table 1. We also include, in the last column, the photospheric radii of the planets immediately after they have reestablished a state of hydrostatic equilibrium following the post-impact envelope expansion. This results show the extent of expansion of the gaseous envelope due to the giant impact. Two series of initial models are considered: Saturn-like models (10–13) with $M_T = 100 M_\oplus$ and $M_{T,c} = 10 M_\oplus$ and HD149026b-like models (14–17) with $M_T = 110 M_\oplus$ and $M_{T,c} = 73 M_\oplus$. The initial radii of the models with masses of $100 M_\oplus$ and $110 M_\oplus$ are around $1 R_J$ and $0.75 R_J$ respectively. With adequate impact energy, we find, in Saturn-like model 13, that giant impacts can significantly enlarge the planets' photosphere. During catastrophic impact (HD149026b-like model 17), the gaseous envelope can be substantially lost as the expansion of the planets' photosphere has largely exceeded its Hill Radius. Even though some of the Saturn-like models are collided by impactor with masses several times larger than that in the catastrophic HD149026b-like model 17, there is no significant mass loss in these models. This dichotomy may be accounted for by the following reasons. In the calculations, we assume that all the impactors can reach the core, with ablated mass and gravitational energy deposited in a region around the core. Thus, the deposited energy by the impactor is not only function of the mass of the impactor, but also related to the depositing location and the planetary mass inside the location. For the same amount of material deposited at the same location, the deposited energy would be larger in the HD149026b-like models than that in the Saturn-like models, because the former have much larger solid core than the latter.

Table 2. Parameters and impact results from OLH simulations

Model	M_T ($M_\oplus$)	$M_{T,c}$ ($M_\oplus$)	M_I ($M_\oplus$)	$M_{I,c}$ ($M_\oplus$)	v_{imp}/v_{esc}	$M_{f,c}$ ($M_\oplus$)	$M_{f,g}$ ($M_\oplus$)	R_τ/R_J
10	100	10	10	10	1.0	10	100	1.13
11	100	10	50	50	1.0	10	140	1.40
12	100	10	50	50	1.0	60	90	1.50
13	100	10	100	10	1.0	20	180	1.62
14	110	73	15	15	1.0	73	52	1.06
15	110	73	15	15	1.0	88	37	1.03
16	110	73	7.5	7.5	1.4	73	44.5	1.03
17	110	73	37	37	1.0	73	-	-

And a side effect of the massive impactors is that the Hill radius will increase with the increasing mass of impactor, which will make it more difficult for the envelope to expand out of its Hill radius. Furthermore, the Saturn-like models have very massive gaseous envelope compared with the HD149026b-like models. They obviously require much larger energy deposition to overcome the gravitational binding for the envelope to escape out of the Hill radius. In the future work, we want to examine the long-term evolution of the impacted models, in order to explore the possibility that the expansion of the planets after the impact can last for over an observable timescale.

References

Saumon, D., Chabrier, G., & van Horn, H. M. 1995, *ApJS*, 99, 713
Alexander, D. R. & Ferguson, J. W. 1994, *ApJ*, 437, 879
Monaghan, J. J. 1992 *ARAA*, 30, 543

Formation of terrestrial planets from planetesimals around M dwarfs

Masahiro Ogihara and Shigeru Ida

Department of Earth and Planetary Sciences, Tokyo Institute of Technology,
2-12-1-I2-10 Ookayama, Meguro-ku, Tokyo 152-8551, Japan
email: ogihara@geo.titech.ac.jp, ida@geo.titech.ac.jp

Abstract. We have investigated accretion of terrestrial planets from planetesimals around M dwarfs through N-body simulations including the effect of tidal interaction with disk gas. Because of low luminosity of M dwarfs, habitable zones around them are located near the disk inner edge. Planetary embryos undergo type-I migration and pile up near the disk inner edge. We found that after repeated close scatterings and occasional collisions, three or four planets eventually remain in stable orbits in their mean motion resonances. Furthermore, large amount of water-rich planetesimals rapidly migrate to the terrestrial planet regions from outside of the snow line, so that formed planets in these regions have much more water contents than those around solar-type stars.

Keywords. accretion, accretion disks, methods: n-body simulations, planetary systems: formation

1. Introduction

In the past about 10 years, over 250 extrasolar planets have been discovered mainly with radial velocity surveys. Although M dwarfs account for about 70–80% of stars in the galactic disk, their low luminosity is disadvantageous for spectroscopic observation, so that radial velocity surveys have not discovered large number of planets around M dwarfs. However, as improvement of spectroscopic observations and development of gravitational microlensing survey (its detection efficiency is independent of stellar luminosity), planetary systems around M stars are being revealed. Around M dwarfs, Jupiter-mass gas giants are generally rare (e.g., Johnson *et al.* 2007), but Neptune-mass planets are rather abundant (e.g., Beaulieu *et al.* 2006), compared with solar-type stars. Lower disk mass and lower disk temperature may account for these properties (Ida & Lin 2005). Due to the low luminosity of M dwarfs, habitable zones are close to the host stars. It allows for detection of habitable planets by radial velocity observation. In fact, the first detected potentially habitable planet is orbiting an M dwarf (Udry *et al.* 2007). On the other hand, the first detected transiting Neptune-sized planet is GJ 436b that orbits an M dwarf (e.g., Gillon *et al.* 2007). The evaluated internal density suggests that the planet is composed mainly of ice, in spite of its proximity to the host star.

Ida & Lin (2005) theoretically studied planetary systems around M dwarfs, extending their model based on solar-type stars. Although their prediction on formation and retention rates of gas giants and Neptune-mass planets is consistent with observational data, their model may not provide detailed information on formation of planets in habitable zones. Since habitable zones are much closer to host stars for M dwarfs than for solar-type stars, the formation process of habitable planets may not be directly extrapolated from that around solar-type stars. Raymond *et al.* (2007) performed N-body simulations of terrestrial planet formation from planetary embryos around low-mass stars and found that terrestrial planets in the habitable zones of most M dwarfs are likely to be small and dry. However, they neglected the effects of protoplanetary disk gas. In such inner regions, gas density is so high that both gas drag and tidal interaction are very effective, and accretion timescale of terrestrial planets would be much shorter than disk lifetime. Thereby, we carry out N-body simulations including the tidal damping of orbital eccentricity, inclination and semimajor axis due to disk gas.

When we include the effect of type-I migration, which is caused by imbalance between tidal interaction with inner disk gas and that with outer one (e.g., Goldreich & Tremaine 1980, Ward 1986), the proximity of habitable zones to inner disk edge would play an important role in the formation of habitable planets around M dwarfs, because type-I migration stops at the edge and planets would accumulate there. Terquem & Papaloizou (2007) performed N-body simulations of planetary embryos undergoing type-I migration. They found that the migrating embryos originally formed at ~ 1 AU are trapped one after another in mean motion resonances by the preceding embryos near the disk inner edge to form two to five resonant close-in planets. In the case of M dwarfs, the original formation regions of embryos are very close to the accumulation regions, so that the trapped planets would be perturbed by the embryos that emerge nearby.

2. Model

We consider a host star with $0.2\,M_\odot$ and set the locations of disk inner edge and snow line at 0.05 AU and 0.3 AU, respectively. We adopt the minimum mass solar nebula model (Hayashi 1981) for the surface density of the disks. Initially, 5,000 planetesimals with mass 3.0×10^{23} g (inside of the snow line) and 1.7×10^{24} g (outside of snow line) are placed between 0.05 AU and 0.4 AU.

The equations of motions of particle k at $\bm{r}_k$ in heliocentric coordinates are

$$\frac{d^2 \bm{r}_k}{dt^2} = -GM_* \frac{\bm{r}_k}{|\bm{r}_k|^3} - \sum_{j \neq i} GM_j \frac{\bm{r}_k - \bm{r}_j}{|\bm{r}_k - \bm{r}_j|^3} + \bm{F}_{\rm grav} + \bm{F}_{\rm gas} \qquad (2.1)$$

where $k, j = 1, 2, \ldots$, the first term is gravitational force of the central star and the second term is mutual gravity between the bodies. The self-gravity, which is the most expensive part in calculation, is calculated with GRAPE-6 $\bm{F}_{\rm grav}$ is specific gravitational drag force due to tidal interaction with disk gas and $\bm{F}_{\rm gas}$ is specific aerodynamical gas drag force. These formulas are derived by Tanaka & Ward (2004) through linear analysis. The physical radius of a body is determined by its mass and internal density. We adopt realistic value $3\,{\rm g\,cm}^{-3}$ for internal density. When physical sizes of bodies overlap, perfect accretion is assumed. After the collision, a new body is created, conserving total mass and momentum of the two colliding body.

3. Results

A typical result is shown in Figure 1. In the early stage of planetary accretion, larger planetesimals grow faster than smaller ones, resulting in runaway growth of the planetesimals. As the runaway bodies (embryos) grow, orbital decay due to tidal interaction with disk gas (type I migration) starts. At the disk inner edge, the migrations cease and they interact with preceding planets. Since embryos emerge one after another, the planets repeat close scatterings with them and occasionally merge with them. However, they end up in stable orbits captured in their mean motion resonances. In this case, three planets in mean motion resonances remain. The final configuration of the three planets is shown in Figure 2. The pair of the innermost planet and the second one has a 3 : 2

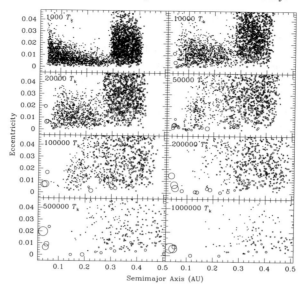

Figure 1. Time evolution of planetesimals. a and e are semimajor axis and eccentricity. Size of circles are proportional to physical radii of the bodies. Snow line and disk inner edge are located at 0.3 AU and 0.05 AU, respectively. T_k is the kepler time at 0.1 AU around a $0.2\,M_\odot$ star.

orbital period commensurability (resonance) and that of the second and third ones is in a 7 : 6 resonance. The most massive planet is the innermost one, the mass of which is about $4M_\oplus$.

In other runs, we find similar final results: three or four planets captured in mutual resonances in the disk inner edge. Even for different disk models (gas and planetesimals), this result does not change, although final planets are generally larger for more massive planetesimal disks. We take out the final planets and integrate their orbits until 10^6 years with gas surface density decaying exponentially and found that their orbital configuration is stable after the decay of disk gas.

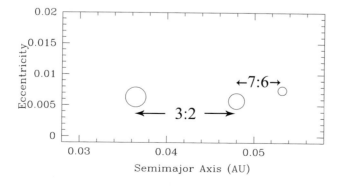

Figure 2. Final configuration of the system near the disk inner edge. The circles represent the planets. Commensurate relationships between orbital period are also shown.

4. Discussion

We have investigated accretion of terrestrial planets from planetesimals around M dwarfs through N-body simulations. We incorporated the effect of gas disk into N-body simulation. We found that three or four resonant planets are formed near the disk edge after successive close scatterings and merging. The result is similar to the case of solar type stars examined by Terquem & Papaloizou (2007). The improved radial velocity observation from ground or transit observation from space such as Corot or Kepler will test this prediction.

In our simulation, we also found that the close-in resonant planets are composed mostly of planetesimals that have migrated from the regions beyond the snow line. This result is in contrast to that obtained by Raymond *et al.* (2007) that neglected type-I migration. In comparison with solar-type stars, habitable planets around M dwarfs would contain much water. This is because growth timescale of planetary embryos and their migration timescale are much shorter than disk lifetime even outside the snow line for the case of M dwarfs. The low probability of formation of gas giants around M dwarfs allows for retention of icy planetesimals and it also facilitates supply of icy planetesimals to inner regions.

References

Beaulieu, J.-P., Bennett, D. P., Fouqué, P., Williams, A., Dominik, M., Jørgensen, U. G., Kubas, D., Cassan, A., Coutures, C., Greenhill, J., Hill, K., Menzies, J., Sackett, P. D., Albrow, M., Brillant, S., Caldwell, J. A. R., Calitz, J. J., Cook, K. H., Corrales, E., Desort, M., Dieters, S., Dominis, D., Donatowicz, J., Hoffman, M., Kana, S., Marquette, J. -B., Martin, R., Meintjes, P., Pollard, K., Sahu, K., Vinter, C., Wambsganss, J., Woller, K., Horne, K., Steele, I., Bramich, D. M., Burgdorf, M., Snodgrass, C., Bode, M., Udalski, A., Szymański, M. K., Kubiak, M., Wieckowski, T., Pietrzyński, G., Soszyński, I., Szewczyk, O., Wyrzykowski, Ł., Paczyński, B., Abe, F., Bond, I. A., Britton, T. R., Gilmore, A. C., Hearnshaw, J. B., Itow, Y., Kamiya, K., Kilmartin, P. M., Korpela, A. V., Masuda, K., Matsubara, Y., Motomura, M., Muraki, Y., Nakamura, S., Okada, C., Ohnishi, K., Rattenbury, N. J., Sako, T., Sato, S., Sasaki, M., Sekiguchi, T., Sullivan, D. J., Tristram, P. J., Yock, P. C. M., & Yoshioka, T. 2006, *Nature*, 439, 437

Gillon, M., Pont, F., Demory, B.O., Mallmann, F., Mayor, M., Mazeh, T., Queloz, D., Shporer, A., Udry, S., & Vuissoz, C. 2007, *A&A*, 472, L13

Goldreich, P. & Tremaine, S. 1980, *ApJ*, 241, 425

Hayashi, C. 1981, *Prog. Theor. Phys. Suppl.*, 70, 35

Ida, S. & Lin, D.N.C. 2005, *ApJ*, 626, 1045

Johnson, J. A., Butler, R. P., Marcy, G. W., Fischer, D. A., Vogt, S. S., Wright, J. T., & Peek, K. M. G. 2007, *ApJ*, 670, 833

Raymond, S. N., Scalo, J., & Meadows, V. S. 2007, *ApJ*, 669, 606

Tanaka, H. & Ward, W. R. 2004, *ApJ*, 602, 388

Terquem, C. & Papaloizou, J. C. B. 2007, *ApJ*, 654, 1110

Udry, S., Bonfils, X., Delfosse, X., Forveille, T., Mayor, M., Perrier, C., Bouchy, F., Lovis, C., Pepe, F., Queloz, D., & Bertaux, J. L. 2007, *A&A*, 469, L43

Ward, W. R. 1986, *Icarus*, 67, 164

Retention of protoplanetary cores near the snowline

Xiaojia Zhang[1], Katherine Kretke[3] and D. N. C. Lin[2,3]

[1]Department of Astronomy, PKU, China,
[2]KIAA, PKU, China,
[3]Department of Astronomy and Astrophysics, UCSC, CA

Abstract. Suppression of type I migration is essential for the retention of protoplanetary cores which are sufficiently massive to accrete gas in their nascent disks and evolve into gas giant planets. We explore here the possibility that special disk properties at the snow line may be the dominant process which stalled the type I migration. We simply use a 1-D model to calculate the torque with linear formula and find that, if the surface density jump near snowline is great enough, the migration can be efficiently slowed down or even halted. This mechanism offers an explanation to the observed peak, at 2–3 AU, in the extra solar planets' semi major axis distribution.

Keywords. Protoplanetary cores, Type I migration, Snowline

1. Introduction

Existence of hot Jupiters tell us that planetary migration due to the interaction between disk and planet is important for planetary formation (Lin *et al.* 1996). For a low mass planet which can not strongly perturb the disk density profile, the linear density waves excited by it will exerts torques on the planet(Goldreich & Tremaine 1979). The total torque is negative causing an inward orbital migration of the planet, which is called Type I migration(Ward 1997). According to the result of Tanaka *et al.* (2002), consider an Earth-mass planet around a sun-mass object and a low-mass protoplanetary disk model, the Type I migration time is about $10^5 - 10^6$ yr, which is shorter than the typical lifetime of the disk. Since terrestrial planets could form after gas was depleted, the type I migration is not a problem for them. However, it is a challenge for the gas giants because their formation must be preceded by the emergence of several Earth mass cores (Pollack *et al.* 1996) in a gaseous medium. With type I migration, these cores would not survive. But, the observation results challenge the planetary migration theory (Figure 1). There is such a significant fraction of solar type stars have gas giants and there is an enhancement in the period distribution around Snowline at 2–3 AU.

2. Method

The special profile near the Snowline (Figure 2) is due to the suddenly viscosity change. Given that the turbulence viscosity comes from MHI (Magneto-Rotation Instability), The depth to which MRI effectively generates turbulence is a strong function of grain size and abundance. A sharp increase in the grain-to-gas density ratio across the snow line reduces the depth of the active layer, and then reduce the viscosity sharply resulting in a local maximum value of gas surface density (Kretke & Lin 2007). Across the snow line, the change in the latent heat reduces the local temperature gradient such that the local surface density gradient also corresponds to a local pressure maximum. We use a 1-D isothermal disk model with the surface density profile showed in Figure 2, and the temperature $T = 166K$. In a centrifugal balance, gas flow is super Keplerian in regions with a positive radial pressure gradient (slightly interior to the snow line) and sub Keplerian in regions with a negative radial pressure gradient (all other locations in the disk).

We consider an Earth-mass core around a sun-mass object and take protoplanetary core's orbit as a circle. Since the core has not sufficient mass to open a gap, we neglect the core's disturbance on the disk structure and use alpha expression for the viscosity as $\nu = \alpha c_s H$, where c_s is sound speed and H is the height of disk. The total torque exerted on a planetary core is the sum of the torques arising from the inner and outer Lindblad resonances and the corotation resonance:

$$\Gamma_{tot} = \sum_m T_m + \Gamma_C \qquad (2.1)$$

where

$$T_m = -\pi^2 m \Sigma \frac{\Psi_m^2}{R\left(\frac{dD_*}{dR}\right)} \qquad (2.2)$$

is the torque from an m-order Lindblad resonance, which is calculated by Ward (1997). Σ is the surface density of disk at R and Ψ_m is the forcing function(Goldreich & Tremaine 1979). $D_* = 0$ gives the location of an m-order Lindblad resonance. We sum the torques to $m = 40$ since the torque arising form very near the protoplanetary core has some uncertainty. And

$$\Gamma_C \propto \Sigma \frac{d\log(\Sigma/B)}{d\log R} \sim h^{-2}\left(\frac{M_P}{M_*}\right)^2 R^4 \Omega_P^2 \Sigma \frac{d\log(\Sigma/B)}{d\log R} \qquad (2.3)$$

is the corotation torque calculated by Masset *et al.* (2006). Where $h = H/R$, M_p and M_* are the mass of the core and center object respectively, Ω_P is the angular velocity of the core. For a disk with surface density proportional to r^p where $p = -1.5$, Γ_C is almost zero, but in the situation near the snowline it is significantly non zero. With the total torque one can calculate the Type I migration rate as

$$\tau_{mig} = \frac{M_P R^2 \Omega_P}{2\Gamma_{tot}} \qquad (2.4)$$

Set the initial location of the core outside the snowline, we can find that the core is

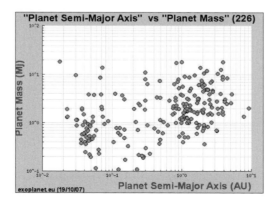

Figure 1. Mass-Semimajor axis distribution of observed exoplanets(exoplanet.edu)

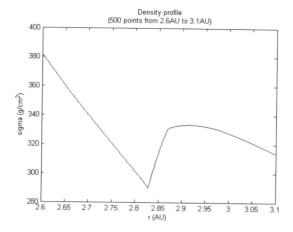

Figure 2. Density profile near snowline generated by Kretke

trapped at the snowline and will not migrate through it (Figure 3). The oscillation is because that the total torque exerted on the protoplanet reverses from negative to positive when get near to the snow line. Therefore the protoplanets will migrate outward and when they get a little further the total torque again becomes negative. So we can see the protoplanet's orbital radius oscillates around the outer edge of snowline. Tanaka et al. (2002) has already given the migration rate as a function of the radial gradient of the disk gas surface density But the expression of Type I migration in the papers of Tanaka et al. (2002) is available for the density profile with a certain gradient. When near the snow line, for a fixed location of protoplanets, different orders of Lindblad torque correspond to different density gradient. We want to accurately calculate the total torque although we can get an estimation from the result of Tanaka et al. (2002). We also use this method to calculate the migration rate for a normal disk model, by comparing the result with that of Tanaka et al. (2002) we find the method is reasonable.

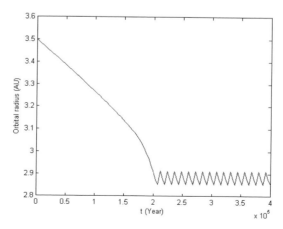

Figure 3. Orbital evolution due to Type I migration of an Earth-mass protoplanetary core.

3. Summary

Protoplanetary cores would migrate inward with a normal disk profile, but when they get close to Snowline, the torque exerted on them may change from negative to positive. Then the cores will be trapped around the Snowline. Comparing with observation result of exoplanets' semi major axis distribution, we can roughly find two groups: Protoplanetary cores formed and retained near the snowline can be explain by this result. And the smaller group may have undergone orbital decay due to Type II migration(Ida & Lin 2004).

References

Goldreich, P. & Tremaine, S. 1979, *ApJ*, 233, 857
Ida, S. & Lin, D. N. C. 2004, *ApJ*, 604, 3881
Ida, S. & Lin, D. N. C. 2004, *ApJ*, 616, 5671
Kretke, K. A. & Lin, D. N. C. 2007, *ApJ*, 664L, 55K
Lin, D. N. C., Bodenheimer, P., & Richardson, D. C. 1996, *Natur*, 380, 606L
Masset, F., Morbidelli, A., Crida, A., & Ferreira, J. 2006, *ApJ*, 642, 478M
Pollack, James B., Hubickyj, Olenka, Bodenheimer, Peter, Lissauer, Jack J., Podolak, Morris, & Greenzweig, Yuval 1996, *Icar*,124. 62P
Tanaka, H., Takeuchi, T., & Ward, W. R. 2002, *ApJ*, 565, 1257
Ward, W. R. 1997, *Icar*, 126, 261W

Formation and detectability of Earth-like planets around Alpha-Centauri B[†]

Erica Davis[1]

[1]Department of Astronomy and Astrophysics, University of California, Santa Cruz
1156 High Street, Santa Cruz, California 95064
email: edavis@ucolick.org

Abstract. We simulate the late stages of planet formation around Alpha Centauri B and analyze the detectability of the resulting terrestrial planet systems. The N-body accretionary evolution of a $\Sigma \propto r^{-1}$ disk populated with 400–900 lunar-mass oligarchs is followed for 200 Myr for each simulation. All of eight runs result in the formation of multiple-planet systems with at least one planet in the 1–2 $M_\oplus$ mass range at 0.5–1.5 AU. We examine the detectability of our simulated planetary systems by generating synthetic radial velocity observations including noise based on the radial velocity residuals to the recently published three planet fit to the nearby K0V star HD 69830. Using these synthetic observations, we find that we can reliably detect a 1.8 $M_\oplus$ planet in the habitable zone of α Centauri B after only three years of high cadence observations. We also find that the planet is detectable even if the radial velocity precision is 3 $\mathrm{m\,s^{-1}}$, as long as the noise spectrum is white.

Keywords. binaries: general, planetary systems: formation, stars: individual (α Centauri B)

1. Introduction

In the past decade, over 250 extrasolar planets have been discovered in a plethora of diverse environments. Earth-like planets in habitable-zone orbits, however, remain well below the threshold of detection. A good representation of the Doppler velocity state-of-the-art is presented by the triple planet system orbiting HD 69830. This system has been shown to contain three Neptune-mass planets, including one on a 197-day orbit, all revealed after only 74 radial velocity observations with residual noise of 0.6 $\mathrm{m\,s^{-1}}$ (Lovis *et al.* 2006). The detection of the HD 69830 system suggests that focused efforts on selected stars may be able to probe down to the characterization of planets with radial velocity half-amplitudes considerably below 1 $\mathrm{m\,s^{-1}}$. The α Centauri system provides a unique opportunity to push the limits of Doppler detection in the search for terrestrial planets in habitable orbits.

Studies of α Cen A and B show that terrestrial planet formation is possible around both stars despite their strong binary interaction (Quintana *et al.* 2002, 2006, 2007). Results to date indicate that planetary systems with one or more Earth-mass planets can form within 2.5 AU from the host stars and remain stable for gigayear scales (Quintana *et al.* 2007). Numerical simulations and stability analyses of planetesimal disks indicate that material is stable within 3 AU of α Cen B as long as the inclination of the disk with respect to the binary is $\leqslant 60°$ (Quintana *et al.* 2002; Wiegert & Holman 1997).

We assess the detectability of terrestrial planets around α Cen B. Eight simulations of the late stage of planet formation are carried out, and each resulting planetary system is tested for detectability using a Monte Carlo method for generating synthetic radial

[†] More complete results have been submitted for publication: Guedes *et al.* 2007, *ApJ*.

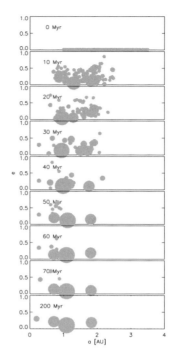

Figure 1. Evolution of a circumstellar disk initially populated by moon-mass oligarchs in circular orbits around α Centauri B. The radius of each circle is proportional to the size of the object.

velocity observations. Based on our results, we are able to accurately evaluate the detectability of planetary systems around the star.

2. Terrestrial Planet Formation

The initial conditions of the circumstellar disk in our simulations mimic conditions at the onset of the chaotic growth phase of terrestrial planet formation (Kokubo & Iba 1998, Kenyon & Bromley 2006) in which collisions of isolated oligarchs, protoplanets of approximately lunar mass, dominate the evolution of the disk. At the start of this phase, several hundred oligarchs orbit the star on nearly circular orbits. Gravitational interactions among oligarchs work to form the final planetary system around the star and clear out remaining material from the disk.

We model the α Centauri B circumstellar disk with a $\Sigma = \Sigma_0 (a/1AU)^{-1}$ surface density profile where $\Sigma_0 = 8.4 - 18.8$ g cm^{-2} as calculated from the total mass $M = Nm$, where N is the number of oligarchs in the disk and m is the oligarch's mass. The disk extends from $1 < a < 3.5$ AU and it is coplanar with the binary orbit. For each run, we populate the disk with $N = 400$ to $N = 900$ oligarchs of lunar mass ($0.0123\ M_\oplus$) with semi-major axes chosen via a rejection method in a to obtain a $\Sigma \propto r^{-1}$ density profile.

Initial orbital elements of each oligarch are randomly generated with mean anomalies, arguments of pericenter, and longitudes of ascending node extending from 0° to 360°, eccentricities in the range $0 < e < 0.001$, and inclinations in the range $0° < i < 1°$ with respect to the plane of the binary.

Each integration was run for 200 Myr using a specialized version of the symplectic hybrid integrator in the MERCURY integration package (Chambers 1999). This N-body code is designed to study planet growth in the presence of a binary companion (Chambers et al. 2002). Bodies grow via accretion through perfectly inelastic embryo-embryo collisions, and, therefore, close encounters are integrated directly rather than symplecticaly.

Figure 1 shows the late evolutionary stage of a protoplanetary disk initially containing 600 moon-mass embryos (run03). Bodies in the outer parts of the disk ($a > 3$ AU) are immediately launched into highly eccentric orbits and either migrate inward to be accreted by inner bodies, collide with the central star, or are ejected from the system ($a_{ej} = 100$ AU). In this simulation, $\sim 65\%$ of the total initial mass is cleared within the first 70 Myr. By the end of simulation run03, four planets have formed. One planet has approximately the mass of Mercury and is located at $a = 0.2$ AU, two 0.6 $M_\oplus$ planets form at $a = 0.7$ and $a = 1.8$ AU, and a 1.8 $M_\oplus$ planet forms at $a = 1.09$ AU.

Table 1 shows the orbital elements of the final systems that emerge from the calculations. All of our simulations result in the formation of 1–4 planets with semi-major axes in the range $0.7 < a < 1.9$ AU, in agreement with Quintana et al. (2002). We find that 42 % of all planets formed with masses in the range 1–2 $M_\oplus$ reside in the star's habitable zone (Fig. 2), which is taken to be 0.5 AU $< a_{hab} < 0.9$ AU (Kasting et al. 1993).

3. Detectability

We took the orbital elements of the systems emerging from our simulations and generated model radial velocities. We developed a code which effectively simulates observing conditions for any specific location on Earth. Given the latitude and longitude of an observatory and the RA and DEC of an object, the code determines when an object is observable. Two additional inputs concern the beginning and end of an observing night: the angle of the sun below the local horizon and the maximum airmass of the object, beyond which observing should not continue. We assume 25% of the observing nights are lost. This accounts for adverse weather conditions and other effects which could result in missed observations. We assumed access to a dedicated telescope at Las Campanas Observatory. At this location, α Centauri is observable for about 10 months out of the

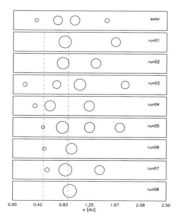

Figure 2. Results from runs 1 through 8. Each panel shows the resulting planetary system and the habitable zone (green dashed lines) for the K1 star α Cen B.

Table 1. Simulation Results

Run	N_0^a	planet	M [$M_\oplus$]	Period [yr]	a [AU]	e	I [°]
01	700	a	0.897	2.262	1.669	0.198	4.965
		b	2.165	0.812	0.843	0.142	4.516
02	700	a	1.820	0.767	0.811	0.016	1.846
		b	1.107	1.640	1.346	0.032	3.064
03	600	a	0.565	2.585	1.831	0.181	3.979
		b	0.578	0.628	0.710	0.242	6.827
		c	0.073	0.091	0.196	0.286	7.590
		d	1.771	1.189	1.086	0.031	3.124
04	800	a	0.086	0.227	0.361	0.244	19.135
		b	1.316	0.495	0.606	0.105	1.639
		c	1.279	1.453	1.242	0.168	2.042
05	900	a	2.054	0.760	0.806	0.052	1.585
		b	0.922	2.412	1.734	0.051	5.784
		c	0.036	0.361	0.491	0.094	18.108
		d	1.291	1.464	1.248	0.145	5.391
06	400	a	1.549	0.981	0.956	0.095	4.777
		b	0.049	0.388	0.515	0.345	15.378
07	800	a	0.996	1.769	1.419	0.169	6.034
		b	0.098	0.441	0.663	0.325	8.259
		c	2.435	0.835	0.858	0.024	3.759
08	700	a	2.755	0.944	0.931	0.217	4.391

Notes:
[a] Initial number of oligarchs.

year. We assumed an observing cadence of one exposure every 200 seconds, corresponding to the read out time of the detector (this is a conservative estimate, since in practice we would expect a considerably higher duty cycle). Finally, we added various values of Gaussian white noise to the model radial velocities.

Figure 3 shows a naïve result for the observed synthetic radial velocities for the system that arose from run03. In this case, we assumed Gaussian white noise with amplitude

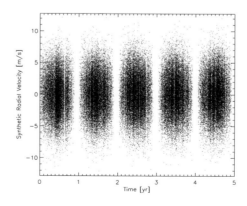

Figure 3. Synthetic radial velocities over a five year period. The 60-day gaps in the data account for the time period when the system is below the horizon. Other gaps in the data emulate nights of missed observations due to bad weather and other adverse events.

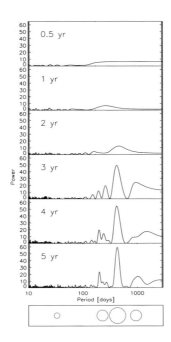

Figure 4. Evolution of the periodogram for run03 over 5 years as 97,260 synthetic radial velocity observations are made, assuming Gaussian white noise with amplitude 3 m s^{-1}. The 1.7 $M_\oplus$ planet (P = 1.2 yr) could be confidently detected in 3 years.

3 m s^{-1}. In keeping with the fact that the star is a virtual twin of HD 69830 (Lovis et al. 2006), α Cen B is expected to show very little radial velocity noise or jitter. The detection of three Neptune mass planets around HD 69830 was facilitated by the small instrumental uncertainties for its radial velocity observations, with a median value of $\approx$0.7 m s^{-1}, along with the assumption of a very small stellar jitter. It is important to note that the model radial velocity for α Cen B due to the four terrestrial planets in this system has an amplitude of 23 cm s^{-1}, or a factor of 13 below the noise.

Over the five year span, 97,260 measurements can be obtained. For this example, we find that we can confidently detect 2 or 3 planets in this time span. We can confidently detect at least one of the planets after the first three years of observations. Figure 4 shows the periodogram of the synthetic radial velocities after 0.5, 1, 2, 3, 4, and 5 years of observations for this system with 3 m s^{-1} white noise. These findings, along with the results from our eight terrestrial planet formation simulations and those of Quintana et al. (2007) suggest that we may detect a habitable terrestrial planet around one of our nearest stellar neighbors in less than 3 years of observation.

4. Discussion

The possibility that detectable terrestrial planets are orbiting α Cen B does much to fire the imagination, and, indeed, a positive identification of such a planet would be a truly landmark discovery. α Cen's proximity allows one to envision space-based follow-up efforts (astrometric, coronographic, interferometric) to characterize the planets that would be far more difficult to carry out for planets orbiting less luminous and more distant stars.

To phrase the situation another way, our current understanding of the process of terrestrial planet formation suggests that *both* components of the α Cen system may have terrestrial planets, provided that disks formed around the stars and that the system has not undergone strong interactions with other stars over the lifetime of the binary. Furthermore, assuming a negligible degree of red noise on the ultra-low frequency range for the host star, successful detection of such terrestrial planets orbiting α Cen B can be made within a few years and with the modest investment of resources required to mount a dedicated radial-velocity campaign with a 1-meter class telescope and high-resolution spectrograph. A lack of planets orbiting these stars would, thus, either provide insight into the history and physical parameters of the α Cen system or provide a critical hint that there is a significant qualitative gap in our understanding of terrestrial planet formation.

References

Chambers, J. E. 1999, *MNRAS*, 304, 793
Chambers, J. E., Quintana, E. V., Duncan, M. J., & Lissauer, J. J. 2002, *AJ*, 123, 2884
Kasting, J. F., Whitmire, D. P., & Reynolds, R. T. 1993, *Icarus*, 101, 108
Kenyon, S. J. & Bromley, B. C. 2006, *AJ*, 131, 1837
Kokubo, E. & Ida, S. 1998, *Icarus*, 131, 171
Lovis, C. *et al.*, 2006, *Nature*, 441, 305
Quintana, E. V., Lissauer, J. J., Chambers, J. E., & Duncan, M. J. 2002, *ApJ*, 576, 982
Quintana, E. V. & Lissauer, J. J. 2006, *Icarus*, 185, 1
Quintana, E. V., Adams, F. C., Lissauer, J. J., & Chambers, J. E. 2007, *ApJ*, 660, 807
Wiegert, P. A. & Holman, M. J. 1997, *AJ*, 113, 1445

Exoplanets: Detection, Formation and Dynamics
Proceedings IAU Symposium No. 249, 2007
Y.-S. Sun, S. Ferraz-Mello and J.-L. Zhou, eds.

© 2008 International Astronomical Union
doi:10.1017/S1743921308016773

Habitable planet formation in extreme planetary systems: systems with multiple stars and/or multiple planets

Nader Haghighipour

Institute for Astronomy and NASA Astrobiology Institute,
University of Hawaii-Manoa, Honolulu, HI

Abstract. Understanding the formation and dynamical evolution of habitable planets in extrasolar planetary systems is a challenging task. In this respect, systems with multiple giant planets and/or multiple stars present special complications. The formation of habitable planets in these environments is strongly affected by the dynamics of their giant planets and/or their stellar companions. These objects have profound effects on the structure of the disk of planetesimals and protoplanetary objects in which terrestrial-class planets are formed. To what extent the current theories of planet formation can be applied to such *"extreme"* planetary systems depends on the dynamical characteristics of their planets and/or their binary stars. In this paper, I present the results of a study of the possibility of the existence of Earth-like objects in systems with multiple giant planets (namely υ Andromedae, 47 UMa, GJ 876, and 55 Cnc) and discuss the dynamics of the newly discovered Neptune-sized object in 55 Cnc system. I will also review habitable planet formation in binary systems and present the results of a systematic search of the parameter-space for which Earth-like objects can form and maintain long-term stable orbits in the habitable zones of binary stars.

Keywords. (stars:) planetary systems: formation, celestial mechanics, methods: numerical

1. Introduction

The discovery of extrasolar planets during the past decade has confronted astronomers with many new challenges. The diverse and surprising dynamical characteristics of many of these objects have made scientists wonder to what extent the current theories of planet formation can be applied to other planetary systems. A major challenge of planetary science is now to explain how such planets were formed, how they acquired their unfamiliar dynamical state, and whether they can be habitable.

Among the unfamiliar characteristics of the currently known extrasolar planetary systems, the existence of systems with multiple planets in which Jovian-type bodies are in eccentric and close-in orbits, and the existence of Jupiter-like planets in multi-star systems are particularly interesting. As shown in figure 1, at the present, 26 extrasolar planetary systems contain more than one giant planet. Also, as shown in Table 1, more than 20% of planet-hosting stars are members of binary systems. The formation of terrestrial-class objects in such planetary systems, and the possibility of their long-term stability in the habitable zones of their host stars are strongly affected by the dynamical perturbations of the giant planets, and in systems of binaries with separations smaller than 100 AU, by the perturbation of the stellar companion. Whether such *"extreme"* planetary environments can be potential hosts to habitable planets is the subject of this paper. I will review the possibility of the long-term stability of terrestrial-class objects in some of multi-planet systems, and review the current status of research on planet formation in dual-star environments.

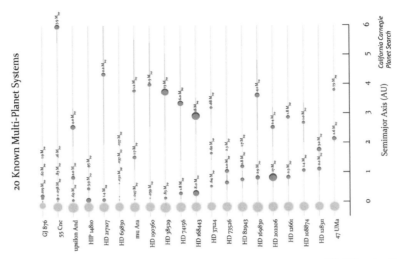

Figure 1. Currently known multi-planet extrasolar planetary systems (California-Carnegie Planet Search).

2. Habitability of Extrasolar Multi-planet Systems

In order for a planetary system to be habitable, an Earth-like planet has to maintain its orbit in the habitable zone (HZ) of the system's central star for a long time. This condition requires that the orbital eccentricity of a habitable planet to be close to zero and its interactions with other bodies of the system do not disturb its long-term stability. In a multi-planet system, these conditions may not be easily satisfied. The dynamics of an object in such systems is strongly affected by other planets and the habitability of a terrestrial planet may be influenced by the perturbations from giant bodies. The latter is more significant in systems where the orbits of giant planets are close to the habitable zone. The planetary systems of υ Andromedae (HZ = 1.68–2 AU), 47 UMa (HZ = 1.16–1.41 AU), GJ 876 (HZ = 0.1–0.13 AU), and 55 Cancri (HZ = 0.72–0.87 AU) are of this kind. In a recent article (Rivera & Haghighipour, 2007), we studied the stability of terrestrial-class objects in these systems by numerically integrating the orbits of several hundred test particles, uniformly distributed along the x-axis, in initial circular orbits. Figure 2 shows the graphs of the lifetimes of these particles for 10 Myr. As shown here, unlike the stable orbit of the newly discovered Earth-like planet of GJ 876 (Rivera et al., 2005), the orbit of the small close-in planet of 55 Cnc, as reported by McArthur et al. (2004) is unstable. Our results also indicate that it is unlikely that υ Andromedae and GJ 876 harbor habitable planets. This has also been confirmed by the direct integration of the orbit of an Earth-sized object in the habitable zone of the system by Dove & Haghighipour (2006). The two systems of 47 UMa and 55 Cnc, however, have stable habitable zones, although direct integrations of actual Earth-like objects in these systems are necessary to confirm their habitability. The results of our test particle simulations also indicated the capability of 55 Cnc system in harboring stable planet(s) in the region between 0.7 AU and 2.2 AU. As shown by Fischer et al. (2007) and as in figure 2, the newly discovered Neptune-sized planet of this system is located in this region.

Table 1. Extrasolar Planet-Hosting Stars in Binary Systems (Haghighipour 2006)

Star	Star	Star	Star
HD142 (GJ 9002)	HD3651	HD9826 (v And)	HD13445 (GJ 86)
HD19994	HD22049 (ϵ Eri)	HD27442	HD40979
HD41004	HD75732 (55 Cnc)	HD80606	HD89744
HD114762	HD117176 (70 Vir)	HD120136 (τ Boo)	HD121504
HD137759	HD143761 (ρ Crb)	HD178911	HD186472 (16 Cyg)
HD190360 (GJ 777A)	HD192263	HD195019	HD213240
HD217107	HD219449	HD219542	HD222404 (γ Cephei)
HD178911	HD202206	PSR B1257-20	PSR B1620-26

3. Habitability of Multiple Star Systems

As shown in Table 1, more than 20% of currently known planet-hosting stars are members of binary systems (Haghighipour 2006). Many of these systems are wide with separations ranging from 200 AU to 6000 AU. In such systems, the perturbative effect of the stellar companion is negligible and planet formation around the other star may proceed in the similar fashion as around a single star. There are, however, three binary systems, namely, GL 86 (Els et al. 2001), γ Cephei (Hatzes et al. 2003), and HD 41004 (Zucker et al. 2004, Raghavan et al. 2006), in which the primary star is host to a Jovian-type planet and the binary separation is smaller than 20 AU. How these planets were formed, and whether such *binary-planetary* systems can be habitable are now among major theoretical challenges of planetary dynamics.

Planet formation in close binary systems is strongly affected by the perturbation of the binary companion. This star may remove planet-forming material by truncating the primary's circumstellar disk (Artymowicz & Lubow 1994) and destabilizing the regions where planetesimals and protoplanets may undergo collisional growth (Thébault et al. 2004). In binary systems where the primary hosts a giant planet, the perturbative effect of the planetary companion will also affect the growth of protoplanetary objects. However, as shown by numerical integrations of the orbits of Earth-sized planets in γ Cephei system (Haghighipour 2006), it is possible for a terrestrial-class body to maintain a long-term stable orbit at distances close to the primary star and outside the giant planet's influence zone. Figure 3 shows the graph of the lifetime of an Earth-sized object in the system of γ Cephei. As shown here, the HZ of the system is unstable. However, an Earth-like planet can main a stable orbit close to the primary star.

Based on the results of the simulations shown in figure 3, we recently studied habitable planet formation in moderately closed binary star systems that host giant planets (Haghighipour & Raymond 2007). We simulated the late stage of terrestrial planet formation for different values of the semimajor axis and orbital eccentricity of the binary, as well as different binary mass-ratios. Our system consisted of a Sun-like star as the primary, a disk of protoplanetary bodies with 120 Moon- to Mars-sized objects distributed randomly between 0.5 AU and 4 AU, and a Jupiter-sized planet at 5 AU. To study the effect of the orbital dynamics of the secondary star on the formation and water contents of planets in the HZ of the primary, we considered the orbit of the giant planet to be circular. We also assumed that, following the model of Morbidelli et al. (2000), in which the main source of Earth's water is the water-rich bodies in the solar system's asteroid belt, the water in the habitable planet of a binary system is originated from the asteroid region of its primary and is delivered to it during the collisional growth of the system's protoplanetary objects. We considered an initial gradient in the water contents of em-

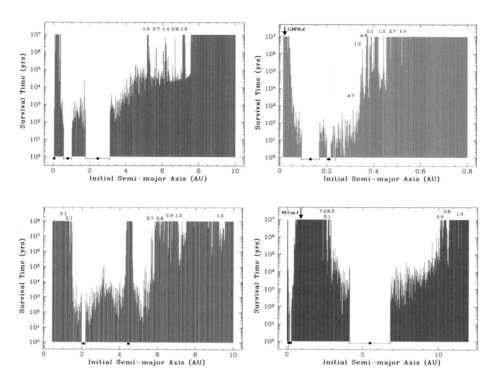

Figure 2. Graphs of the lifetimes of test particles in υ Andromedae (top left, HZ = 1.68–2 AU), GJ 876 (top right, HZ = 0.1–0.3 AU), 47 UMa (bottom left, HZ = 1.16–1.41), and 55 Cnc (bottom right, 0.72–0.87 AU). The graphs and habitable zones are from Rivera & Haghighipour (2007). The islands of stability and instability, with their corresponding mean-motion resonances with the inner and/or outer planet are also shown. As shown here, the habitable zones of υ Andromedae and GJ 876 are unstable implying that the planetary systems of these stars will not be habitable. The habitable zones of 47 UMa and 55 Cnc, on the other hand, are stable. Also as shown here, the recently detected Earth-like planet of GJ 876 (Rivera *et al.* 2005), and the newly discovered fifth planet of 55 Cnc (Fischer *et al.* 2007) are in stable orbits.

bryos similar to the distribution of water in the primitive asteroids of the asteroid belt (Abe *et al.* 2000). That is, embryos inside 2 AU were taken to be dry, the ones between 2 to 2.5 AU were considered to contain 1% water, and those beyond 2.5 AU were assumed to have a water to mass ratio of 5%. Figure 4 shows some of the results for binary mass-ratios $\mu_b = 0.5, 1.5$. As shown here, it is possible to form Earth-like objects with substantial amount of water in the HZ of the primary star. The sizes of these planets and their water contents vary with the semimajor axis and eccentricity of the stellar companion. In binaries where the secondary star has a small periastron, the interaction between this object and the giant planet of the system, which transfers angular momentum to the disk of planetary embryos, causes many of these bodies to be ejected from the system. As a result, in closer and eccentric binaries, the final planets are smaller and contain less or no water. Figure 5 shows the relation between the periastron of the binary (q_b) and the semimajor axis of the outermost terrestrial planet (a_{out}). As shown in the left graph of figure 5, similar to Quintana *et al.* (2007), simulations with no giant planets favor

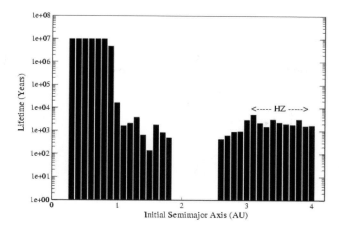

Figure 3. Lifetime of an Earth-sized planet in γ Cephei system. The giant planet of the system (1.67 Jupiter-mass) is at 2.13 AU with an eccentricity of 0.12. As shown here, the HZ of the system is unstable. However, a terrestrial-class object can maintain a long-term orbit at close distances to the primary star (Haghighipour 2006).

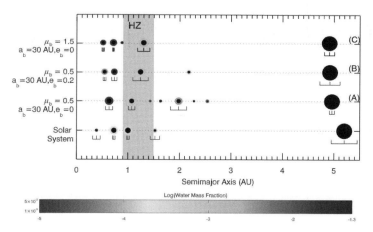

Figure 4. Habitable planet formation in binary-planetary systems with 0.5 and 1.5 stellar mass-ratios. As shown here, binaries with moderate periastron distances are more favorable for the formation of terrestrial-class planets with considerable amounts of water (Haghighipour & Raymond 2007).

regions interior to $0.19q_b$ for the formation of terrestrial objects. That means, around a Sun-like star, where the inner edge of the habitable zone is at ∼ 0.9 AU, a stellar companion with a perihelion distance smaller than $0.9/0.19 = 4.7$ AU would not allow habitable planet formation. In simulations with giant planets, on the other hand, figure 5 shows that terrestrial planets form closer-in. The ratio a_{out}/q_b in these systems is between 0.06 and 0.13. A detailed analysis of our simulations also indicate that the systems, in which habitable planets were formed, have large periastra. The right graph of figure 5 shows this for simulations in a binary with equal-mass Sun-like stars. The circles in this figure represent systems with habitable planets. The numbers on the top of the circles show the mean eccentricity of the giant planet. For comparison, systems with unstable giant

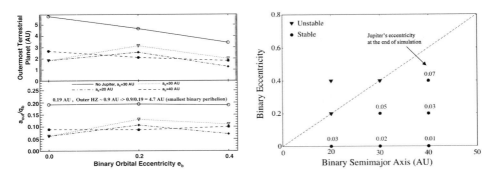

Figure 5. The graph on the left shows the relation between the periastron of an equal-mass binary and the location of its outermost terrestrial planet. The graph on the right shows the region of the (e_b, a_b) space for a habitable binary-planetary system (Haghighipour & Raymond 2007).

planets have also been marked. Since at the beginning of each simulation, the orbit of the giant planet was considered to be circular, a non-zero eccentricity is indicative of the interaction of this body with the secondary star. As shown here, Earth-like objects are formed in systems where the interaction between the giant planet and the secondary star is weak and the average eccentricity of the giant planet is small. That implies, habitable planet formation is more favorable in binaries with moderate to large perihelia, and with giant planets on low eccentricity orbits.

Acknowledgements

Support by the NASA Astrobiology Institute under Cooperative Agreement NNA04CC08A with the Institute for Astronomy at the University of Hawaii-Manoa is acknowledged.

References

Abe, Y., Ohtani, E., Okuchi, T., Righter, K., & Drake, M. 2000, in *Origin of the Earth and the Moon*, ed. K. Righter & R. Canup (Tucson: Univ. Arizona Press), 413
Artymowicz, P. & Lubow, S. H. 1994, *ApJ*, 421, 651
Dove, A. & Haghighipour, N. 2006, *BAAS*, 37, 1284
Els, S. G., Sterzik, M. F., Marchis, F., Pantin, E., Endl, M., & Kruster, M. 2001, *A&A*, 370, L1
Fischer, D. A., Marcy, G. W., Butler, R. P., Vogt, S. S., Laughlin, G., Henry, G. W., Abouav, D., Peek, K. M. G., Wright, J. T., Johnson, J. A., McCarthy, C., & Isaacson, H. 2008, *ApJ*, 675, 790.
Haghighipour, N. 2006, *ApJ*, 644, 543
Haghighipour, N. & Raymond, S., N. 2007, *ApJ*, 666, 436
McArthur, B. E., Endl, M., Cochran, W. D., Benedict, G. F., Fischer, D. A., Marcy, G. W., Butler, R. P., Naef, D., Mayor, M., Queloz, D., Udry, S., & Harrison, T. E. 2004, *ApJ*, 614, L81
Morbidelli, A., Chambers, J., Lunine, J, I., Petit, J. M., Robert, F., Valsecchi, G., B., & Cyr, K., E. 2000, *Meteorit. Planet. Sci.*, 35, 1309
Quintana, E. V., Adams, F. C., Lissauer, J. J., & Chambers, J. E. 2007, *ApJ*, 660, 807
Raghavan, D., Henry, T. J., Mason, B. D., Subasavage, J. P., Jao, W. C., Beaulieu, T. D., & Hambly, N. C. 2006, *ApJ*, 646, 523
Rivera, E. J., Lissauer, J. J., Butler, R. P., Marcy, G. W., Vogt, S. S., Fischer, D. A., Brown, T. M., Laughlin, G., & Henry, G. W. 2005, *ApJ*, 634, 625
Rivera, E. & Haghighipour, N. 2007, *MNRAS*, 374, 599
Thébault, P., Marzari, F., Scholl, H., Turrini, D., & Barbieri, M. 2004, *A&A*, 427, 1097
Zucker, S., Mazeh, T., Santos, N. C., Udry, S., & Mayor, M. 2004, *A&A*, 426, 695

On the formation age of the first planetary system

T. Hara, S. Kunitomo, M. Shigeyasu and D. Kajiura

Department of Physics, Kyoto Sangyo University, Kyoto 603-8555, Japan
email: hara@cc.kyoto-su.ac.jp

Abstract. Recently, it has been observed the extreme metal-poor stars in the Galactic halo, which must be formed just after Pop III objects. On the other hand, the first gas clouds of mass $\sim 10^6 M_\odot$ are supposed to be formed at $z \sim 10$, 20, and 30 for the 1σ, 2σ and 3σ, where the density perturbations are assumed of the standard ΛCDM cosmology. Usually it is approximated that the distribution of the density perturbation amplitudes is gaussian where σ means the standard deviation. If we could apply this gaussian distribution to the extreme small probability, the gas clouds would be formed at $z \sim 40$, 60, and 80 for the 4σ, 6σ, and 8σ where the probabilities are approximately 3×10^{-5}, 10^{-9}, and 10^{-15}. Within our universe, there are almost $\sim 10^{16}$ ($\sim 10^{22} M_\odot / 10^6 M_\odot$) clouds of mass $10^6 M_\odot$. Then the first gas clouds must be formed around $z \sim 80$, where the time is ~ 20 Myr ($\sim 13.7/(1+z)^{3/2}$ Gyr). Even within our galaxy, there are $\sim 10^5$ ($\sim 10^{11} M_\odot / 10^6 M_\odot$) clouds, then the first gas clouds within our galaxy must be formed around $z \sim 40$, where the time is ~ 54 Myr ($\sim 13.7/(1+z)^{3/2}$ Gyr).

The evolution time for massive star ($\sim 10^2 M_\odot$) is ~ 3 Myr and the explosion of the massive supernova distributes the metal within a cloud. The damping time of the supernova shock wave in the adiabatic and isothermal era is several Myr and stars of the second generation (Pop II) are formed within a free fall time ~ 20 Myr. Even if the gas cloud is metal poor, there is a lot of possibility to form the planets around such stars. The first planetary systems could be formed within $\sim 6 \times 10^7$ years after the Big Bang in the universe. Even in our galaxies, the first planetary systems could be formed within $\sim 1.7 \times 10^8$ years. If the abundance of heavy elements such as Fe is small compared to the elements of C, N, O, the planets must be the one where the rock fraction is small. It is interesting to wait the observations of planets around metal-poor stars. For the panspermia theory, the origin of life could be expected in such systems.

Keywords. Formation, First Planetary system, Gaussian distribution, Panspermia.

1. Introduction

Recently it has been reported that the detection of the metal poor stars: HE0107-5240 is a giant star ($36,000$ lyr, $M \simeq 0.8 M_\odot$), having just $1/200,000$ of the solar metal abundance (Chrtlieb *et al.* 2002): HE1327-2326, discovered in 2005, is a star ($\geqslant 1,500$ lyr, $M \simeq 0.7 M_\odot$), having just $1/300,000$ of the solar metal abundance (the lowest known iron abundance to date) (Frebel *et al.* 2005). It has been speculated that these stars are the second generation, born out of the gas clouds which were polluted by the primordial Population III stars.

In the following we estimate the formation age of the first stars and consider the formation of the first planetary systems. If we consider the panspermia theory seriously, it could be speculated that first life began in such primordial planetary systems.

2. Gaussian Distribution

It is now understood that the first luminous stars are formed through the contraction of gas cloud due to cooling into the dark matter potential where the amplitudes of density perturbation become order one. Without the density perturbation of dark matter, baryonic gas could not contract due to background homogeneity. The formation of primordial gas clouds has been investigated (Matsuda, Sato and Takeda 1969, Palla, Salpeter, and Stahler 1983) and recently numerically elaborated (Bromm, Coppi and Larson 2002, Yoshida et al. 2003).

The first gas clouds of mass $\sim 10^6 M_\odot$ are supposed to be formed at $z \sim 10$, 20, and 30 for the 1σ, 2σ and 3σ, where the density perturbations are assumed of the standard ΛCDM cosmology (Nishi and Susa 1999). Usually the distribution of the perturbation amplitudes is approximated as gaussian where σ means the standard deviation.

If we could apply this gaussian distribution to the extreme small probability, the gas clouds would be formed at $z \sim 40$, 60, and 80 for the 4σ, 6σ, and 8σ where the probabilities are approximately 3×10^{-5}, 10^{-9}, and 6×10^{-16}. Within our universe, there are almost $\sim 10^{16}$ ($\sim 10^{22} M_\odot/10^6 M_\odot$) clouds of mass $10^6 M_\odot$. Then the first gas clouds must be formed around $z \sim 80$, where the time is ~ 20 Myr ($\sim 13.7/(1+z)^{3/2}$ Gyr). Even within our galaxy, there are $\sim 10^5$ ($\sim 10^{11} M_\odot/10^6 M_\odot$) clouds, then the first gas clouds within our galaxy must be formed around $z \sim 40$, where the time is ~ 54 Myr ($\sim 13.7/(1+z)^{3/2}$ Gyr). The probability for each sigma is displayed in the table.

σ	probability	σ	probability
1	0.158	6	9.90×10^{-10}
2	2.27×10^{-2}	7	1.28×10^{-12}
3	1.35×10^{-3}	8	6.25×10^{-16}
4	3.17×10^{-5}	9	1.13×10^{-19}
5	2.87×10^{-7}	10	7.66×10^{-24}

Although the probability will increase if we take into account the non-gaussianity of the density perturbations (Bartolo et al. 2004), we did not consider the effect.

3. Formation of Population III stars

It is now understood that the first luminous stars are formed through the following stages such as Jean's instability, free fall and Kelvin contraction. Comparing the free fall time and cooling time (mainly H_2 cooling), the gas clouds of $\sim 10^6 M_\odot$ become Jean's instability and contracts into the dark matter potential as stated before.

In the following, we take $z \sim 80$ as the representative age for the first gas cloud formation, where the time includes the contraction of the gas cloud (Susa, 2002). The age at $z \sim 80$ is

$$t_z \sim \frac{13.7 \text{Gyr}}{(1+z)^{3/2}} \simeq 1.91 \times 10^7 \left(\frac{z}{80}\right)^{-3/2} \text{yr.} \qquad (3.1)$$

During the contraction of gas cloud, the central part contracts almost free fall and becomes opaque (Omukai and Nishi 1998). The massive star of $\sim 10^2 M_\odot$ is formed there. In the opaque stage where the configuration becomes stable, radiation radiates the gravitational potential energy. The time scale of Kelvin contraction is

$$t_K \sim \frac{GM^2}{2R}/L \simeq 2 \times 10^7 \left(\frac{M}{100 M_\odot}\right)^2 \left(\frac{R}{10 R_\odot}\right)^{-1} \left(\frac{L}{L_\odot}\right)^{-1} \text{ yr.} \quad (3.2)$$

Although the formation age is a little bit different, the detailed contraction process of primordial gas clouds are numerically investigated and the results are not so much different (Omukai and Nishi 1998, Bromm, Coppi and Larson 2002, Yoshida et al. 2003).

The evolution time for massive star ($\sim 10^2 M_\odot$) is ~ 3 Myr and the explosion of the massive supernova distributes the metal within a cloud.

4. Formation of Planetary System

The damping time of the supernova shock wave in the adiabatic and isothermal era is several Myr. Here we assume it takes the time scale of the following free fall time.

The contraction and the free fall time of the cloud around $z \sim 80$ is

$$t_{ff} \sim \frac{1}{\sqrt{\pi G \rho}} \simeq 2 \times 10^7 \left(\frac{z}{80}\right)^{-3/2} \text{ yr.} \quad (4.1)$$

(the gas cloud is assumed to collapse at $z \sim 80$ so the mean density at turn around is two times of the background density at $z \sim 1.6 \times 80$ (Susa, 2002)). If the mass of gas cloud is smaller than $\sim 10^6 M_\odot$, the explosion energy of SN is greater than the gravitational binding energy. So the mass of the gas cloud must be greater than $\sim 10^6 M_\odot$. Then the increase of metal abundance δZ in the cloud is order of $\delta Z \sim 10^{-5} \sim 10 M_\odot/10^6 M_\odot$.

After the contraction of metal contaminated gas cloud, the stars of the second generation (Pop II) are formed within a free fall time ~ 20 Myr. Then the first planetary systems could be supposed to be formed within $\sim 6 \times 10^7$ years after the Big Bang.

The same arguments could be applied to the case of $z \sim 40$ for 4σ where the collapsed time is ~ 54 Myr. After the explosion of the gas cloud, the expected elapsed time for the adiabatic, isothermal expansion and the free fall time is $t_{ff} \sim 6 \times 10^7 (z/40)^{-3/2}$ yr. So the first planetary systems for this case could be supposed to be formed within $\sim 1.7 \times 10^8$ years after the Big Bang.

The Population II stars reside in Halos and globular clusters. Even stars in globular clusters could take planetary systems for enough time (several Gyr), it is expected the life could start there.

The fragmentation of the supernova shocked shell is investigated by Salvaterra, Ferrara and Schneider (2004) and they estimated the instability of the shell sets in $0.2 \sim 50$ Myr after the explosion, depending on the explosion energy and the density of the surrounding medium. If we adopt their estimation of the short time scale ~ 0.2 Myr, the formation age of the first planetary system will become as fast as ~ 43 Myr after the Big Bang. They also discussed the metallicity of the fragments through the mixing of the heavy elements into the swept up matter in the shell. Although there are many uncertainty mechanism, they estimated the mean metallicity of the fragments $10^{-3.5} Z_\odot \leqslant Z \leqslant 10^{-2.6} Z_\odot$. If we include their estimation, the metallicity of the secondary stars will much increase and the formation of the planetary systems around the secondary stars will be much expected in the earlier stage of the cosmography.

5. Conclusions and Discussion

The formation of the first planetary system is related to gaussian distribution of dark matter perturbation. Within our universe, there are almost $\sim 10^{16}$ ($\sim 10^{22} M_\odot/10^6 M_\odot$)

clouds of mass $10^6 M_\odot$. Then the first gas clouds must be formed around $z \sim 80$, where the time is ~ 20 Myr ($\sim 13.7/(1+z)^{3/2}$ Gyr). Considering the evolution, SN explosion, expansion, and contraction, the first planetary systems with metals could be supposed to be formed within $\sim 6 \times 10^7$ years after the Big Bang.

Even within our galaxy, there are $\sim 10^5$ ($\sim 10^{11} M_\odot / 10^6 M_\odot$) clouds. As stated before the first gas clouds within our galaxy must be formed around $z \sim 40$, where the time is ~ 54 Myr ($\sim 13.7/(1+z)^{3/2}$ Gyr). The first planetary systems for this case could be supposed to be formed within $\sim 1.7 \times 10^8$ years after the Big Bang.

It is interesting to wait the observations of planets around metal-poor stars in the halos and globular clusters. For the panspermia theory, the origin of life could be expected in such systems. The ejected rocks from such planets could transfer micro-organism within galaxies (Wallis and Wickramasinghe 2004). Even in our solar system, there must be many types of life in Mars, Europa, Ganymede, Io, Uranus, Neptune, Pluto, and Titan. It is interesting to wait the investigation of these planets and satellites.

References

Chrtlieb, N. *et al.* 2002, *Nature*, 419, 904
Frebel, A. *et al.* 2005, *Nature*, 434, 871
Matsuda, T., Sato, H., & Takeda, H. 1969, *Prog. Theor. Phys.*, 42, 219
Palla, F., Salpeter, S., & Stahler, S. 1983, *ApJ*, 271, 632
Nishi, R. & Susa, H. 1999, *ApJ*, 523, L103
Omukai, K. & Nishi, R. 1998, *ApJ*, 508, 141
Bromm, V., Coppi, P., & Larson, R. 2002, *ApJ*, 564, 23
Fuller, T. & Couchman, H. 2000, *ApJ*, 544, 6
Yoshida, N., Abel, T., Hernquist, L., & Sugiyama, N. 2003, *ApJ*, 544, 6
Bartolo, N., Komatsu, E., Matarrese S., & Riotto, A. 2004, *Phys. Rept.*, 402, 103
Susa, H. 2002, *Prog. Theo. Phys. Sup.*, 147, 11
Salvaterra, R., Ferrara, A., & Schneider, R. 2004, *New Astron.*, 10, 113
Wallis, M. & Wickramasinghe, N. 2004 *MNRAS*, 348, 52.

Part 4

Protoplanet Disks and Migration

Planetary migration in gaseous protoplanetary disks

Frédéric S. Masset

Laboratoire AIM, CEA/DSM - CNRS - Université Paris Diderot, DAPNIA/Service d'Astrophysique, CEA/Saclay, 91191 Gif/Yvette Cedex, France, and IA-UNAM, Ciudad Universitaria, Apartado Postal, 70-264, Mexico, D.F. 04510, Mexico

Abstract. Tides come from the fact that different parts of a system do not fall in exactly the same way in a non-uniform gravity field. In the case of a protoplanetary disk perturbed by an orbiting, prograde protoplanet, the protoplanet tides raise a wake in the disk which causes the orbital elements of the planet to change over time. The most spectacular result of this process is a change in the protoplanet's semi-major axis, which can decrease by orders of magnitude on timescales shorter than the disk lifetime. This drift in the semi-major axis is called planetary migration. In a first part, we describe how the planet and disk exchange angular momentum and energy at the Lindblad and corotation resonances. Next we review the various types of planetary migration that have so far been contemplated: type I migration, which corresponds to low-mass planets (less than a few Earth masses) triggering a linear disk response; type II migration, which corresponds to massive planets (typically at least one Jupiter mass) that open up a gap in the disk; "runaway" or type III migration, which corresponds to sub-giant planets that orbit in massive disks; and stochastic or diffusive migration, which is the migration mode of low- or intermediate-mass planets embedded in turbulent disks. Lastly, we present some recent results in the field of planetary migration.

Keywords. planets and satellites: formation, planetary systems: formation, planetary systems: protoplanetary disks

1. Introduction

The importance of the tidal interaction between a protoplanetary disk and a forming planet was first recognized long before the discovery of the first extrasolar planet in 1995. Goldreich & Tremaine (1980) discussed the case of Jupiter in a conservative protoplanetary nebula, and found that its semi-major axis should evolve as a result of the gravitational interaction between the planet and the nebula (although they could not determine whether it should increase or decrease).

When the first extrasolar planet was discovered orbiting 51 Peg with a period of 4.23 days (Mayor *et al.* 1995) at a distance of only 0.052 AU from the central star, theories of orbital migration received renewed attention. None of the reasonable planetary formation scenarios was able to account for the formation of a planetary core that close to the star. It therefore appeared likely that this planet had formed farther out in the protoplanetary disk and then migrated towards the star, along the lines of predictions made by the theoretical work of the eighties (Lin *et al.* 1996).

Had anyone doubted that significant planetary migration is common in forming planetary systems, additional clues were provided by the discovery of planetary systems exhibiting low-order mean motion resonance. Under the effect of differential migration (i.e., the outer planet migrates inwards faster than the inner one), two planets can converge and be captured in a low-order mean motion resonance.

The present contribution is organized as follows: (i) in section 2 we define the notation; (ii) in section 3, we present the torque expressions at the Lindblad and corotation resonances; (iii) in section 4, we present the different migration modes that have been envisaged so far; (iv) finally, in section 5 we present a list of recent results that numerical simulations have recently brought to our knowledge of planet–disk interactions.

2. Notation

We consider a Keplerian gaseous disk with vertical scaleheight $H(r)$, surface density $\Sigma(r)$, kinematic viscosity $\nu(r)$, and orbital frequency $\Omega(r)$, where r is the distance to the central object. We consider a planet with a prograde orbit coplanar to the disk, of mass M_p and orbital frequency Ω_p. Whenever we consider a single azimuthal Fourier component of a given quantity, we denote by m its azimuthal wavenumber.

3. Disk torque at an isolated resonance

The problem of determining the torque between any perturbing potential and the disk, in the linear regime, amounts to determining the torque exerted on the disk by the Fourier components of the potential. Goldreich & Tremaine (1979) have shown that angular momentum exchange between the perturbing potential and the disk occurs only at the Lindblad and corotation resonances. Lindblad resonances correspond to locations in the disk where the perturbing potential's frequency in the matter frame ($\tilde{\omega}(r) = m[\Omega_p - \Omega(r)]$) matches $\pm\kappa(r)$ (the epicyclic frequency). The corotation resonance occurs where the perturbing potential's frequency is zero in the matter frame, that is to say at a radius where the disk material rotates along with the perturbing potential.

3.1. Torque at a Lindblad resonance

3.1.1. Torque expression

The torque expression at a Lindblad resonance by a single Fourier component of the potential with m-fold symmetry is (Goldreich & Tremaine 1979, Meyer-Vernet & Sicardy 1987, Artymowicz 1993)

$$\Gamma_m = -\frac{m\pi^2\Sigma}{rdD/dr}\left(r\frac{d\Phi_m}{dr} + \frac{2\Omega}{\Omega - \Omega_p}\Phi_m\right)^2, \qquad (3.1)$$

where Γ_m is the torque exerted on the disk material by the perturbing potential, $D = \kappa(r)^2 - m^2[\Omega(r) - \Omega_p]^2$ represents a distance to the resonance and $\Phi_m(r)$ is the amplitude of the potential component. In Eq. (3.1), the term in brackets and rdD/dr are both to be evaluated at the resonance location. In a Keplerian disk, rdD/dr is positive at the ILR (Inner Lindblad Resonance, where $\tilde{\omega} = -\kappa$) and negative at the OLR (Outer Lindblad Resonance, where $\tilde{\omega} = +\kappa$). The perturbing potential therefore exerts a negative torque on the disk at the ILR, and a positive torque at the OLR. Newton's third law thus implies that the disk exerts a positive (negative) torque on the perturber at the ILR (OLR).

3.1.2. Lindblad resonance location

In order to evaluate the torques given by Eq. (3.1), one has to know the location of the Lindblad resonances. As stated previously, a Lindblad resonance is found where $\tilde{\omega} = \pm\kappa$ (the upper sign stands for the OLR, while the lower sign stands for the ILR). Using the fact that $\kappa = \Omega$ in a Keplerian disk, we obtain: $\Omega(r_{\mathrm{LR}}) = \frac{m}{m\pm 1}\Omega_p$. Note that owing to pressure effects, the waves launched by the potential components are slightly offset from the resonance locations. In particular, as $m \to \infty$ the turning point locations tend to pile up at a radius given by: $r = r_c \pm \frac{\Omega}{2A} H$ where $A = (1/2) r \Omega \, d\Omega/dr$ is Oort's first constant. These points of accumulation correspond to the radius at which the flow becomes supersonic in the corotating frame (Goodman & Rafikov 2001). In the case of a Keplerian disk, these points are located $\pm (2/3) H$ away from the corotation radius.

3.2. Torque at a corotation resonance

The angular momentum exchange at a corotation resonance and a Lindblad resonance are due to different physical processes. In the latter case the perturbing potential tends to excite epicyclic motion, and the angular momentum deposited is evacuated through pressure-supported waves. On the other hand, these waves are evanescent in the corotation region and therefore unable to remove the angular momentum brought there by the perturber (Goldreich & Tremaine 1979). In the linear regime, the corotation torque exerted by a perturbing potential with m-fold symmetry on the disk is

$$\Gamma_C = \frac{\pi^2 m}{2} \left[\frac{\Phi_m^2}{d\Omega/dr} \frac{d}{dr} \left(\frac{\Sigma}{B} \right) \right]_{r_c}, \tag{3.2}$$

where the term in brackets is to be evaluated at the corotation radius. The corotation torque is thus proportional to the gradient of Σ/B, evaluated at the corotation radius, where B is equal to half the flow vorticity. The corotation torque is therefore proportional to the gradient of the vortensity (ratio of the vorticity to the surface density). The corotation torque is therefore zero in a disk with $\Sigma \propto r^{-3/2}$, such as the minimum mass solar nebula (MMSN).

The physical picture of the flow at a corotation resonance with azimuthal wavenumber m is characterized by a set of m eye-shaped libration islands in which fluid elements move along closed streamlines. The corotation torque is prone to saturation, which can be described as follows: when the disk viscosity is close to zero, the vortensity is conserved along a fluid element's path. The libration of fluid elements redistributes the vortensity within the libration islands. Once the vortensity has been sufficiently stirred up, even an infinitesimally small amount of viscosity suffices to render the vortensity uniform over the whole libration island. The corotation torque then goes to zero (i.e., saturates), because it scales with the vortensity gradient.

In order to avoid saturation, the viscosity must be high enough to prevent the vortensity from becoming uniform over the libration islands. This is possible if the viscous timescale across these islands is smaller than the libration timescale, as shown by Ogilvie & Lubow (2003). In this case, viscous diffusion across the libration islands permanently imposes the large-scale vortensity gradient over the libration islands. Finally, it should be noted that saturation properties cannot be captured by a linear analysis, since saturation requires a finite libration time, and thus a finite resonance width.

4. Planetary migration

4.1. Type I migration

We consider the case of a low-mass planet, so that the overall disk response can be treated as a linear superposition of its responses to individual Fourier components of the potential. Each component torques the disk at its Lindblad and corotation resonances. We denote by Γ_{ILR}^m the torque of the mth potential component at its ILR, and adopt similar notation for the torques at the Outer Lindblad Resonance (Γ_{OLR}^m) and corotation resonance (Γ_{CR}^m). The total tidal torque exerted by the disk on the planet, which is equal

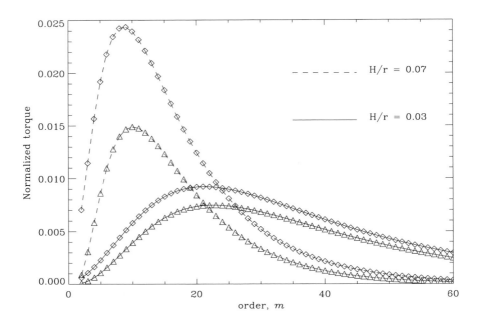

Figure 1. The absolute value of individual inner (triangle) and outer (diamond) torques as a function of m. The torques are normalized to the value $\Gamma_0 = \pi q^2 \Sigma a^4 \Omega_p^2 h^{-3}$, where q is the planet mass to star mass ratio. The one-sided Lindblad torques scale as $(H/r)^{-3}$, hence the total areas under each curve are of the same order of magnitude. The differential Lindblad torque scales as $(H/r)^{-2}$.

and opposite to the torque exerted by the planet on the disk, can therefore be written as

$$\Gamma = \sum_{m>0} \Gamma_{ILR}^m + \sum_{m>0} \Gamma_{OLR}^m + \sum_{m>0} \Gamma_{CR}^m. \qquad (4.1)$$

The first series in this sum is the total Inner Lindblad torque, and the second is the total Outer Lindblad torque. The absolute value of either term is also called the one-sided Lindblad torque. The last term is called the coorbital corotation torque. The sum of the two Lindblad torques is referred to as the differential Lindblad torque.

4.1.1. *Differential Lindblad torque*

The acoustic shift of the effective Lindblad resonances mentioned at section 3 has an important consequence: there is a sharp cut-off in the high-m torque components (for $m \gg r/H$) as shown by Artymowicz (1993), since the high-m potential components become localized in increasingly narrow annuli around the perturber orbit. The value of a potential component at the accumulation point (where the torque is exerted) therefore tends to zero as m tends to infinity.

Fig. 1 illustrates the behavior the one-sided Lindblad torques. In particular, one can see that the cutoff occurs at larger m in a thinner disk. Also, for both disk aspect ratios there is a very apparent mismatch between the inner and the outer torques; the former is systematically smaller than the later. If we consider the torque of the disk acting on the planet, then the outer torque is negative and the inner torque is positive; the total torque on the planet is therefore negative. As a consequence, migration is directed inwards and the orbit decays towards the central object (Ward 1986).

4.1.2. Pressure buffer

One remarkable feature of the differential Lindblad torque is its weak dependence on the slope of the surface density function. This is not what one would naively expect, since as one increases the slope the surface density increases at the Inner Lindblad Resonances and decreases at the Outer Lindblad Resonances. As one increases the surface density gradient, however, one simultaneously increases the radial pressure gradient. This makes the disk more and more sub-Keplerian. As a consequence, the Outer Lindblad Resonances approach the planet's orbit while the Inner Lindblad Resonances recede from it. This process plays against the more obvious effect of the surface density. This effect is known as the pressure buffer (Ward 1997, Tanaka *et al.* 2002), and frustrates any reasonable attempt to revert the differential Lindblad torque by tuning the power law indexes of the surface density and temperature profiles. This makes inward type I migration inevitable, at least in disks where the surface density and temperature are power laws of the radius.

4.1.3. Type I migration timescale

The most up-to-date estimate of the total (*i.e.* Lindblad plus corotation) tidal torque in the linear regime between a three dimensional disk and a low mass planet is the estimate by Tanaka (2002). It yields a migration timescale of 8×10^5 yrs for an Earth-mass object embedded at 5 AU in the MMSN. This is much shorter than the disk lifetime.

4.2. Type II migration

4.2.1. Shock appearance and horseshoe asymmetry

The wake excited by a planet eventually turns into a shock. The location at which profile steepening produces a shock depends on the planet mass; the larger the mass, the closer the shock will be to the orbit. For planets above some critical mass, the wake becomes a shock within the excitation region. Under these circumstances, the fluid elements circulating just outside the co-orbital region receive a kick of angular momentum every time they cross the wake. This is represented in Fig. 2. As a consequence horseshoe U-turns are not symmetric. A fluid element initially located inside the libration region thus progressively recedes from the orbit as it performs a sequence of horseshoe U-turns, until it ends up in the inner disk or the outer disk (Lubow *et al.* 1999). The co-orbital region is thereby emptied, and an annular gap eventually appears around the orbit. The timescale for emptying the co-orbital region can readily be estimated from Fig. 2. After each horseshoe U-turn, the distance of a fluid element from the orbit increases by an amount between 10 and 20 %. The characteristic emptying time of the horseshoe region is therefore between 5 and 10 times half the libration time, which is given by $\tau_{\mathrm{lib}}/2 = 2\pi a/(3/2)\Omega_p x_s = (2/3)T_o(a/x_s)$. Here we can estimate from the figure that $x_s \approx 0.16$, so $\tau_{\mathrm{lib}}/2 \approx 4\,T_0$. In this particular example, the co-orbital region is therefore emptied after about 20 to 40 orbits. This simple estimate also shows that the smaller the planet mass, the longer the gap clearance timescale. Indeed, as the planet mass decreases, the horseshoe region becomes more and more symmetric so that more libration times are needed to get rid of the co-orbital material, while the libration time itself increases. For a $1\,M_J$ planet orbiting in a disk with $H/r = 0.05$, the clearance timescale of the gap is about 100 orbits.

4.2.2. Accretion

A planet engaged in type II migration has a mass much larger than the critical mass for runaway gas accretion (Pollack *et al.* 1996). It therefore accretes gas from the nebula at the same time as it migrates. Kley (1999) has devised a scheme to simulate the

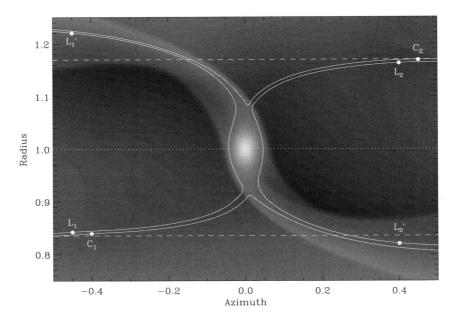

Figure 2. Asymmetry of the horseshoe region. The circulating fluid elements C_2 (moving towards the left) and C_1 (moving towards the right) recede from the orbit after crossing the shock excited by the planet. Similarly, the librating fluid elements recede from the orbit after executing their horseshoe U-turns. This particular example shows streamlines of the flow in the corotating frame of a 2 M_J planet in a disk with $h = 0.05$. The planet is on a fixed circular orbit, and this snapshot was taken after 22.5 orbits.

accretion in the planet's vicinity, and found that a giant planet can still accrete significant amounts of gas despite the presence of the gap. Nelson et al. (2000) have found that the final mass of a protogiant of initially one Jupiter mass could be of several Jupiter masses. The ability of a giant planet to accrete the surrounding gas depends on the equation of state of the gas and its ability to get rid of the gravitational energy released by the accretion. D'Angelo et al. (2003) and Klahr and Kley (2006) have performed high resolution hydrodynamics calculations taking into account radiative transfer in order to assess the dynamics of the inner Roche lobe and its impact on accretion. They still find significant accretion but the geometry of the flow in the Roche lobe is that of a bubble rather than a thin accretion disk.

4.2.3. Gap opening criteria

Classically, the gap opening conditions once consisted of two independent criteria (Lin & Papaloizou 1979, Lin & Papaloizou 1993, Bryden et al. 1999) that needed to be simultaneously fulfilled. The first, referred to as the thermal criterion (since it imposes a limit on the disk thickness, and hence on the disk temperature) requires that the wake becomes a shock just as it is excited. The flow must therefore be strongly non-linear in the planet's vicinity, and the parameter R_H/H must be larger than some critical value, where R_H is the planetary Hill radius. This critical value is ~ 1, although its precise value can be slightly different. The second criterion is that the viscosity is sufficiently low, so that the surface density jump across the edges of the excavated region is a sizable fraction of the unperturbed surface density. This condition, which is known as the viscous criterion, is expressed as $q > \frac{40}{\mathcal{R}}$ where $\mathcal{R} = a^2\Omega_p/\nu$ is the Reynolds number.

Crida et al. (2006) have used another condition, namely that the circulating streamline just outside the separatrix should be closed, to derive the gap surface density profile semi-analytically. They require that the integral of the viscous, gravitational, and pressure torques cancels out over one synodic period of a given fluid element. They provide an ansatz expression for the pressure torque that is approximately valid for a reasonable range of planetary masses and disk thicknesses, and derive the following unique criterion for gap opening: $\frac{3}{4}\frac{H}{R_H} + \frac{50}{q\mathcal{R}} < 1$. Broadly speaking, this criterion is approximately equivalent to the previous two except in the case where both are only marginally fulfilled.

4.2.4. *Migration of planets that open a gap*

A "clean" gap (i.e., a gap with little residual surface density) splits the disk material into an outer disk and an inner disk. Therefore, the planet must drift inwards at the same rate that the outer disk spreads inwards. In other words, the migration rate of a giant planet that has opened a gap in the disk is the same as the viscous drift rate of the disk (Lin & Papaloizou 1986). This type of migration is referred to as type II migration (Nelson et al. 2000 and refs. therein). It is usually said that in this regime, the planet's orbit is locked to the disk's viscous evolution. The migration drift rate of the planet is therefore

$$\frac{da}{dt} \sim -\frac{\nu}{a}. \qquad (4.2)$$

For a $M_p = 1\ M_J$ planet that undergoes type II migration in a disk with $H/r = 0.04$ and $\alpha = 6 \cdot 10^{-3}$, the migration time starting from $a = 5$ AU is about $1.6 \cdot 10^4$ orbits. This corresponds to $\sim 1.6 \cdot 10^5$ years, if the central object has one solar mass.

Using two-dimensional numerical simulations, Nelson et al. (2000) have shown that the migration of giant planets (with masses greater than or equal to one Jupiter mass) in a viscous disk obeys the scenario outlined above, at least broadly speaking. In particular, they found that the timescale of variation in the planet's semi-major axis is similar to the viscous timescale of the disk. These results have been obtained by assuming that the effective viscosity of the disk is adequately modeled by the Navier–Stokes equation. In this approach the kinematic viscosity is chosen to account for the accretion rates inferred from observations of T Tauri objects. Nelson & Papaloizou (2003) and Papaloizou et al. (2004) have performed much more numerically demanding calculations; instead of resorting to the purely hydrodynamical scheme including an *ad hoc* kinematic viscosity, their model describes the self-sustained magnetohydrodynamic (MHD) turbulence arising from the magnetorotational instability (MRI).

They find that a giant protoplanet still opens a gap in the disk, in much the same manner as in a disk modeled by the Navier–Stokes equations. Surprisingly, the gap in a turbulent disk tends to be larger and deeper than in a laminar disk (Papaloizou et al. 2007). The mass accretion rate tends to be larger in the MHD turbulent case, most likely because of magnetic breaking of the circumplanetary disk (Papaloizou et al. 2004).

4.2.5. *Type II migration of several planets*

The migration properties of two or more giant planets is a topic that has received a lot of attention, primarily because we detect extrasolar giant planets that are in mean motion resonance, which is a natural outcome of convergent type II migration (Snellgrove et al. 2001, Lee & Peale 2002, Kley et al. 2004, Kley et al. 2005), and secondly because under some circumstances (if the outer planet is sufficiently lightweight and barely opens a gap), the migration of the whole system may be reversed and be directed outwards (Masset & Snellgrove 2001, Morbidelli & Crida 2007, Zhang & Zhou 2008). Finally, as was first noted by Kley (2000) the distance between the giant planets brought to close orbits by convergent migration can be sufficiently short to render the system unstable after gas clearance, which may account for the eccentric orbits of some extrasolar planets.

4.3. Type III migration

Type III migration refers to a mode of migration for which the major driver is material flowing through the coorbital region. In the previous sections, the torque acting on a migrating planet was considered independent of its migration rate. However, the corotation torque implies material that crosses the planet orbit on the U-turn of the horseshoe streamlines. In a non-migrating case, only the material trapped in the horseshoe region participates in these U-turns, but in the case of an inward (or outward) migrating planet, material of the inner disk (outer disk) has to flow across the co-orbital region and executes one horseshoe U-turn to do so. By doing this, it exerts a corotation torque on the planet that scales with the drift rate. We call x_s the half radial width of the horseshoe region. The amount of specific angular momentum that a fluid element near the separatrix takes from the planet when it crosses the planet orbit and goes from the orbital radius $a - x_s$ to the orbital radius $a + x_s$ is $\Omega_p a x_s$. The corresponding torque exerted on the planet in steady migration is therefore, to lowest order in x_s/a:

$$\Gamma_2 = (2\pi a \Sigma_s \dot{a}) \cdot (\Omega_p a x_s), \qquad (4.3)$$

where we keep the same notation as in Masset & Papaloizou (2003), hereafter MP03, and where Σ_s is the surface density at the upstream separatrix. As the system of interest, we take the system composed of the planet and all fluid elements trapped in libration in its co-orbital region, namely the whole horseshoe region (with mass M_{HS}) and the Roche lobe content (with mass M_R), because all of these parts perform a simultaneous migration. The drift rate of this system is then given by :

$$(M_p + M_{HS} + M_R) \cdot (a\dot{a}\Omega_p/2) = (4\pi a x_s \Sigma_s) \cdot (a\dot{a}\Omega_p/2) + \Gamma_{LR} \qquad (4.4)$$

which can be rewritten as:

$$m_p \cdot (a\dot{a}\Omega_p/2) = (4\pi a \Sigma_s x_s - M_{HS}) \cdot (a\dot{a}\Omega_p/2) + \Gamma_{LR} \qquad (4.5)$$

where $m_p = M_p + M_R$ is all the mass content within the Roche lobe, which for now on for convenience we refer to as the planet mass. The first term of the first bracket of the r.h.s. corresponds to the horseshoe region surface multiplied by the upstream separatrix surface density, hence it is the mass that the horseshoe region would have if it had a uniform surface density equal to the upstream surface density. The second term is the actual mass of the horseshoe region. The difference between these two terms is called in MP03 the coorbital mass deficit and denoted δm. Eq (4.5) yields a drift rate :

$$\dot{a} = \frac{\Gamma_{LR}}{2Ba(m_p - \delta m)} \qquad (4.6)$$

This drift rate is faster than the standard estimate in which one neglects δm. This comes from the fact that the coorbital dynamics alleviates the task of the differential Lindblad torque by advecting fluid elements from the upstream to the downstream separatrix. The angular momentum they extract from the planet by doing so favors its migration. As δm tends to m_p, most of the angular momentum lost by the planet and its coorbital region is gained by the orbit crossing circulating material, making migration increasingly cost effective. When $\delta m \geqslant m_p$, the above analysis, assuming a steady migration ($\dot{a}$ constant), is no longer valid. Migration undergoes a runaway, and has a strongly time varying migration rate, that increases exponentially over the first libration

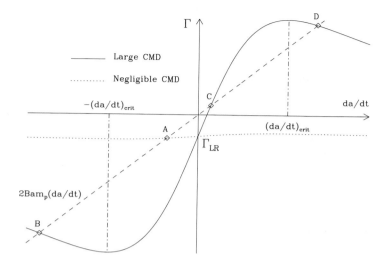

Figure 3. The solid curve shows the total torque on the planet in a massive disk (hence with a large coorbital mass deficit) as a function of the drift rate. For $|\dot{a}| \ll \dot{a}_{\rm crit}$ the torque exhibits a linear dependence in $\dot{a}$. The dotted line shows the torque in a low mass disk (i.e. with a negligible coorbital mass deficit), in which case the torque is almost independent of the migration rate and is always close to the differential Lindblad torque Γ_{LR}. The dashed line represents the planet angular momentum gain rate as a function of $\dot{a}$, assuming a circular orbit. For a given situation, the migration rate achieved by a steadily migrating planet is given by the intersection of the dashed line with the torque curve. In the low mass disk case, the intersection point, A, is unique, and stable. It yields a negative drift rate controlled by the differential Lindblad torque. In the high mass disk case (type III case), there are 3 points of intersection (B, C and D). The central point (C) is unstable, while the extreme ones (B and D) are stable and correspond to the maximum drift attained by the planet, either inwards (point B) or outwards (point D).

times. Runaway (also said type III) migration is therefore a mode of migration of planets that deplete their coorbital region and embedded in sufficiently massive disks, so that the above criterion be satisfied. An analysis similar to the above calculation may be performed, in which the corotation torque depends on the migration rate, except that one now has to introduce a delay τ between the mass inflow at the upstream separatrix and the corotation torque. Fluid elements passing through the upstream separatrix need indeed on average a fraction of a libration timescale to reach the planet and execute a horseshoe U-turn. This delay represents the latency of the feedback loop.

$$\Gamma_{CR}(t) = 2Ba\delta m \dot{a}(t-\tau) \qquad (4.7)$$

A Taylor expansion in time of $\dot{a}(t-\tau)$ yields a first order differential equation for $\dot{a}$ (see MP03 for details). The linear dependence of the corotation on the drift rate remains valid as long as the semi-major axis variation over a horseshoe libration time is smaller than the horseshoe zone width, i.e.:

$$|\dot{a}| < \dot{a}_{\rm crit} = \frac{Ax_s^2}{2\pi a} \qquad (4.8)$$

The corotation torque then reaches a maximum and slowly decays for larger values of $\dot{a}$ (see fig. 3).

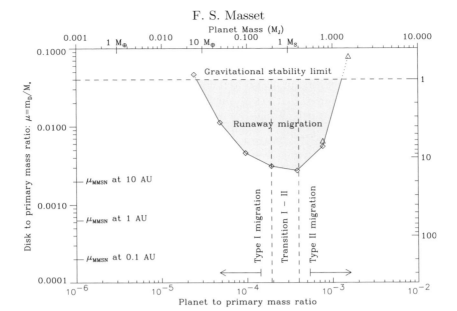

Figure 4. Runaway (or type III) limit domain for a $H/r = 0.04$ and $\nu = 10^{-5}$ disk, with a surface density profile $\Sigma \propto r^{-3/2}$. The variable $m_D = \pi\Sigma r^2$ features on the y axis. It is meant to represent the local disk mass, and it therefore depends on the radius. Type III is most likely for Saturn mass planets. These would undergo type III migration in disks no more massive than a few times the MMSN.

The terminal drift rate of a type III steadily migrating planet can be estimated by a standard bifurcation analysis as illustrated in fig. 3. The transition from one case to the other (one intersection point to three intersection points) occurs when the line showing the rate of change of the angular momentum (which has slope $am_p\Omega_p/2$) and the torque curve near the origin are parallel. Since this latter has a slope $a\delta m\Omega_p/2$ near the origin, the transition occurs near $m_p = \delta m$. The disk critical mass above which a planet of given mass undergoes a runaway depends on the disk parameters (aspect ratio and effective viscosity). The limit has been worked out by MP03 for different disk aspect ratios and a kinematic viscosity $\nu = 10^{-5}$. We reproduce in figure 4 the type III migration domain for a disk with $H/r = 0.04$.

A number of comments can be made from figure 4:

• The MMSN was barely massive enough to yield type III migration of Saturn. This suggests that in many protoplanetary disks, inferred to be several times more massive than the MMSN, type III migration is very likely for Saturn mass protoplanets.

• Type III is impossible for massive planets ($m_p > 1\ M_J$) as the horseshoe separatrices sample the gap edges in regions significantly depleted, yielding a small coorbital mass deficit.

• The sharp limit on the high mass side of the runaway domain might be related to the fact that most of the extrasolar planets known as "hot Jupiters", with a semi-major axis $a < 02$. AU, happen to have sub-Jovian masses. A forming protoplanet, as it passes through the runaway domain, would migrate very fast towards the central object in a series of type III episodes, and at the same time it would accrete gas from the nebula. If the protoplanet happens to get out of the runaway domain before it reaches the central regions of the disk, it enters the slow, type II migration regime, having at least about a

Jupiter mass. Otherwise, it may reach the central regions through type III migration (if the surface density profile is steep enough), still as a sub-Jovian object.

Type III migration, for the same disk profile and planet mass, can be directed either outwards or inwards, depending on the initial conditions. This is related to the fact that the differential equation obeyed by the semi-major axis, in the type III regime, is a second order differential equation in which one can specify independently a and $\dot{a}$. The type III migration of a planet therefore depends on its migration history, the "memory" of this history being stored in the way the horseshoe region is populated, i.e. in the repartition of the coorbital mass deficit. Note that owing to the strong variation of the drift rate in a type III episode, the horseshoe streamlines are not exactly closed, so that the coorbital mass deficit can be lost and type III migration can stall. This has been observed in numerical simulations, which show that the semi-major axis is varied by a factor of a few at most during a single type III episode.

4.4. Stochastic migration

Oftentimes, the protoplanetary disk is considered laminar while the accretion of material onto the central object is ensured by an *ad hoc* kinematic viscosity chosen to account for the mass accretion rate measured for T Tauri stars. The molecular viscosity of protoplanetary disks appears to be insufficient by many orders of magnitude, however, to reproduce the accretion rates typically measured. The source of the high effective viscosity in these disks is thought to be turbulence. The MRI (see section 4.2) has been identified as a powerful source of MHD turbulence in magnetized disks (Balbus & Hawley 1991, Hawley & Balbus 1991, Hawley & Balbus 1992), and this section will exclusively focus on the impact of this kind of turbulence on planetary migration.

MRI can develop only in regions of the disk where the matter and magnetic field are coupled, which requires a sufficiently high (albeit weak) ionization rate. In the planet-forming region $(1-10\mathrm{~AU})$, it is thought that only the upper layers of the disk are ionized by X-rays from the central star or cosmic rays (Gammie 1996, Fromang *et al.* 2002). The bulk of the disk, however, should be ionized outside this region. This has led Gammie (1996) to the concept of layered accretion: the upper layers of the region between 1 and 10 AU participate in accretion onto the central star, whereas its magnetically inactive equatorial parts, usually called the *dead zone*, do not participate in the inwards flow of disk material.

There already exist a large number of works describing numerical simulations that self-consistently describe an MHD turbulent disk with embedded planets (Nelson & Papaloizou 2004, Papaloizou *et al.* 2004, Nelson & Papaloizou 2003, Nelson 2005, Winters *et al.* 2003). They exclusively consider a fully magnetized disk (hence with no dead zone), however, without any vertical stratification for reasons of computational cost.

Not surprisingly, the torque felt by a planet in a turbulent disk displays large temporal fluctuations. One can assign an order of magnitude to their amplitude by considering an overdense region of size H, located at a distance H from the planet such that the perturbed density in this region is of the same order as the unperturbed density. This yields an order of magnitude for the torque fluctuations of $G\Sigma a$ (Nelson & Papaloizou 2004, Nelson 2005).

Nelson and Papaloizou (2004) and Nelson (2005) have investigated the migration of low and intermediate mass planets embedded in turbulent disks. Laughlin *et al.* (2004) have also investigated this problem, but rather than tackling it through self-consistent numerical simulations they performed a two-dimensional calculation which mimicked the effects of turbulence using a time-varying, non-axisymmetric potential acting on the gas disk, rather than directly on the planet. The migration of low-mass planets embedded in turbulent disks is significantly different from the type I migration expected for laminar disks. The large torque variations due to turbulence induce the planet's semi-major axis to evolve on a random walk rather than systematically decay.

One question that is still open is whether the total torque felt by a planet in a turbulent disk can be decomposed into a laminar torque and the effect of fluctuations arising from the turbulence. We call the latter component the stochastic torque. One might expect that the time average of the stochastic torques is negligible compared to the total mean torque (which might be the same as the laminar torque, but this is still unknown), provided that this average is performed over a time interval that is much longer than the turbulence recurrence time. Under this assumption, the behavior of the planet should exhibit a systematic trend reminiscent of type I migration. Nelson (2005) has investigated the statistical properties of these torque fluctuations, finding significant power at low frequencies, corresponding to timescales comparable to the simulation time. As a consequence, in many of his calculations no systematic trend is observed; stochastic migration dominates type I migration over the entire run time of his calculations, or about 150 orbits. The reason for such significant power at very low frequencies is still unknown.

The amplitude of the specific stochastic torque is independent of the planet mass, whereas the specific wake torque scales with the planet mass. Nelson (2005) found that for planets up to $\sim 10\ M_\oplus$ the stochastic migration overcomes the systematic trend (over a simulation run time of 150 orbits), whereas systematic effects are dominant for larger masses. We mention however the recent work by Fromang & Nelson (2006), who argue that density fluctuations are smaller in a stratified, turbulent disk than in the unstratified models currently used to assess stochastic torques. This argument suggests that systematic effects could be dominant at masses even lower than $10\ M_\oplus$.

As pointed out by Johnson et al. (2006), if the turbulence has a finite correlation time then the stochastic (or diffusive) migration of low-mass planets can be reduced to an advection-diffusion equation. They show that diffusion always reduces the mean migration time of the planets, although a fraction of them still "survive" an extended period of migration.

5. Recent results on planetary migration

In the past few years, a number of new results have been obtained in the field of planetary migration either by the inclusion of new physical ingredients to the customary picture of planetary migration in a locally isothermal power law disk, or by intensive numerical modeling, or both. We draw hereafter a non comprehensive list of such results.

5.1. *Planetary migration and magnetic field*

Notwithstanding the issue of MRI and its non-linear outcome as MHD turbulence, the role of a toroidal or poloidal magnetic field on type I migration in a laminar disk as been contemplated by several authors. Terquem (2003) considers a disk threaded by a toroidal magnetic field, and shows that when the magnetic field as a function of radius decreases sufficiently fast, the total torque felt by the planet is positive, hence the planet migrates outwards. Fromang et al. (2005) have performed two-dimensional numerical simulations which essentially confirmed the analytic predictions of Terquem (2003). More recently, Muto et al. (2007, 2008) worked out the analytic torque expression both for a disk threaded by a poloidal magnetic field and a disk with a toroidal magnetic field, which enables them to make a variety of predictions about type I migration in magnetized disks.

5.2. Inclusion of the disk's self gravity

The protoplanetary disk's self-gravity, usually neglected on the grounds of its large Toomre parameter, has recently been contemplated by a number of authors, which were led to contradictory statements (Nelson & Benz 2003, Pierens & Huré 2005). Baruteau & Masset (2008a, 2008b) have revisited previous works on this subject and found that the inclusion of self-gravity slightly speeds up migration with respect to analytical drift rate estimates. They also exhibited a strong bias that systematically affects numerical calculations in which a planet is released and freely migrates in a non self-gravitating disk.

5.3. Role of the corotation torque at a cavity edge

Eq. (3.2) shows that the corotation torque can be a large, positive quantity at a surface density jump such that the surface density is larger on the outside, and that it may possibly overcome the differential Lindblad torque. This happens at a cavity edge, even if the cavity is shallow. This has led Masset et al. (2006) to the concept of planetary trap: type I migrating embryos are stopped whenever they reach a relatively abrupt drop of surface density, such as could be found at the inner edge of a dead zone, at the inner edge of a tidally truncated disk (Pierens & Nelson (2007)) or at the snow line (Zhang et al. 2008, Kretke et al. 2008).

5.4. Planetary migration and radiative transfer

Radiative transfer plays a very important role for planetary migration scenarios, for many different reasons. Menou and Goodman (2004) exploit the differential Lindblad torque's extreme sensitivity to the location of the Lindblad resonances. They consider realistic models of T Tauri α-disks instead of the customary power law models, and show that type I migration can be significantly slowed at opacity transitions. Jang-Condell and Sasselov (2005) argue that taking into account the temperature perturbations due to shadowing and irradiation of the disk photosphere could significantly reduce the type I migration rate. More recently, Paardekooper and Mellema (2006), hereafter PM06, consider a low-mass planet embedded in a disk with inefficient radiative cooling. A complex temperature structure develops in the vicinity of the planet which gives an underdense region behind the planet. As a consequence, the disk ultimately exerts a positive torque on the planet. This result clearly indicates that radiative transfer effects may prove crucial in resolving the problem of type I drift.

Baruteau and Masset (2008a, 2008b) have undertaken a follow up study of the results of PM06 in order to understand their physical origin. For this purpose, they considered the limiting case of a two dimensional, adiabatic flow. They firstly consider the linear case for an isolated resonance, for which they find an expression of the corotation torque which reduces to the usual dependence on the vortensity gradient in the limit of a cold disk. In the general case, they find an additional dependence on the entropy gradient at corotation. This dependence is associated to the advection of entropy perturbations. Secondly, they consider the case of a planet embedded in a Keplerian disk. They find, in the same manner, that the horseshoe drag contains a term that scales with the entropy gradient, and which may be strong enough to overcome the differential Lindblad torque, thus yielding a migration reversal. Although at the early stages of a calculation, the horseshoe drag's excess corresponds exactly to the series of torque excesses on individual corotation resonances, Paardekooper and Papaloizou (2007) argue that non-linear effects arise even at very low planetary mass and boost this excess with respect to its linear estimate. They further argue that the linearly estimated excess would be by itself not sufficient to halt migration. In any case, the self-consistent three dimensional calculation of PM06 shows that the migration of low-mass planets is reversed in the radiatively inefficient inner parts of protoplanetary disks. It will therefore be halted on the outer edge of this inner region. Paardekooper and Mellema (2007) estimate this location to be at $r \sim 10 - 15$ AU. Although the saturation properties of the corotation torque in a radiatively inefficient disk are still to be properly investigated, it is reasonable to expect that it remains unsaturated if the timescale for the relaxation of the temperature perturbations fulfills the following two requirements: (i) it is shorter than the horseshoe libration time (so that a nearly unperturbed entropy distribution reaches the horseshoe U-turns) (ii) it is longer than the horseshoe U-turn time, so that the flow can be considered as adiabatic over this timescale. Should these expectations be confirmed, this would provide a solution to the long-standing problem of type-I migration being too fast and flushing embryos onto the central object before they can accrete gas and become giant protoplanets.

6. Summary and discussion

Significant progress has been recently accomplished in the theories of planet-disk tidal interactions. Most of the new results have primarily been brought by large-scale calculations using modern supercomputer resources. In particular, the problem of type I migration is on the way of being solved, by relaxing the customary barotropic assumption, thereby enabling a new kind of perturbation (the so-called contact discontinuity familiar to the Riemann solvers community) to arise in the co-orbital region in the presence of an entropy gradient.

In the parts of a disk that are magnetically active, stochastic migration changes dramatically the migration properties of low- and intermediate-mass objects. Many questions remain open in this field which requires considerable computing resources.

The description of type II and type III migration also requires considerable computational resources. Primarily captured by simple two dimensional calculations in laminar disks, type II migration is now described by three dimensional hydrodynamics or magnetohydrodynamics calculations. An interesting problem, which has not yet been tackled, is the impact of gap edge irradiation on type II migration. In the same respect, while the physical picture of type III migration had been initially illustrated by two dimensional laminar calculations, an accurate description of its properties must be investigated by means of high resolution, three dimensional calculations. A first step forward has been made by Pepliński et al. (2007a, 2007b) who performed high resolution AMR two-dimensional calculations. A challenge for type III migration, whose onset depends on a subtle balance of gravitational and inertial effects, is to get rid of any possible artifact that may alter the effective inertial or gravitational mass of the migrating system (namely the planet and any fluid element trapped in libration with it, be it horseshoe or circumplanetary).

The actual trend among numericists performing calculations of planet-disk interactions is to include more and more physics relevant to planetary migration in their schemes. These efforts render scenarios of planetary migration progressively more quantitative and predictive, and they should in the future eventually bridge the gap between the properties of the protoplanetary disk, and the structure of the planetary systems that may emerge from it.

References

Artymowicz, P., 1993, *ApJ*, 419, 155
Balbus, S. A. & Hawley, J. F., 1991, *ApJ*, 376, 214

Baruteau, C. & Masset, F., 2008, *ApJ*, 672, 1054

Baruteau, C. & Masset, F., 2008, in: Y.-S. Sun, S. Ferraz-Mello & J.-L. Zhou, (eds.), *Exoplanets: Detection, Formation and Dynamics*, Proc. IAU Symposium No. 249 (Suzhou,China), p. 391

Bryden, G., Chen, X., Lin, D. N. C., Nelson, R. P., & Papaloizou, J. C. B., 1999, *ApJ*, 514, 344

Crida, A., Morbidelli, A., & Masset, F., 2006, *Icarus*, 181, 587

Fromang, S. & Nelson, R. P., 2006, *A&A*, 457, 343

Fromang, S., Terquem, C., & Balbus, S. A., 2002, *MNRAS*, 329, 18

Fromang, S., Terquem, C., & Nelson, R. P., 2005, *MNRAS*, 363, 943

Gammie, C. F., 1996, *ApJ*, 457, 355

Goldreich, P. & Tremaine, S., 1979, *ApJ*, 233, 857

Goldreich, P. & Tremaine, S., 1980, *ApJ*, 241, 425

Goodman, J. & Rafikov, R. R., 2001, *ApJ*, 552, 793

Hawley, J. F. & Balbus, S. A., 1991, *ApJ*, 376, 223

Hawley, J. F. & Balbus, S. A., 1992, *Bulletin of the American Astronomical Society* 24, 1234

Jang-Condell, H. & Sasselov, D. D., 2005, *ApJ*, 619, 1123

Johnson, E. T., Goodman, J., & Menou, K., 2006, *ApJ*, 647, 1413

Kley, W., 1999, *MNRAS*, 303, 696

Kley, W., 2000, *MNRAS*, 313, L47

Kley, W., Peitz, J., & Bryden, G., 2004, *A&A*, 414, 735

Kley, W., Lee, M. H., Murray, N., & Peale, S. J., 2005, *A&A*, 437, 727

Kretke, K. A. Lin, D. N. C., & Turner, N. J. 2008, in: Y.-S. Sun, S. Ferraz-Mello & J.-L. Zhou, (eds.), *Exoplanets: Detection, Formation and Dynamics*, Proc. IAU Symposium No. 249 (Suzhou,China), p. 293

Laughlin, G., Steinacker, A., & Adams, F. C., 2004, *ApJ*, 608, 489

Lee, M. H., & Peale, S. J., 2002, *ApJ*, 567, 596

Lin, D. N. C., Bodenheimer, P., & Richardson, D. C., 1996, *Nature* 380, 606

Lin, D. N. C. & Papaloizou, J., 1979, *MNRAS*, 186, 799

Lin, D. N. C. & Papaloizou, J., 1986, *ApJ*, 309, 846

Lin, D. N. C. & Papaloizou, J. C. B., 1993, in E. H. Levy & J. I. Lunine (eds.), *Protostars and Planets III*, pp 749–835

Lubow, S. H., Seibert, M., & Artymowicz, P., 1999, *ApJ*, 526, 1001

Masset, F., & Snellgrove, M., 2001, *MNRAS*, 320, L55

Masset, F. S. & Papaloizou, J. C. B., 2003, *ApJ*, 588, 494

Mayor, M., Queloz, D., Marcy, G., Butler, P., Noyes, R., Korzennik, S., Krockenberger, M., Nisenson, P., Brown, T., Kennelly, T., Rowland, C., Horner, S., Burki, G., Burnet, M., & Kunzli, M.: 1995, *IAU Circ.* 6251, 1

Menou, K. & Goodman, J., 2004, *ApJ*, 606, 520

Meyer-Vernet, N. & Sicardy, B., 1987, *Icarus*, 69, 157

Morbidelli, A., & Crida, A., 2007, *Icarus*, 191, 158

Muto T., Machida M., & Inutsuka, S., 2007, *ArXiv Astrophysics e-prints 0712.1060*

Muto, T., Machida, M. N., & Inutsuka, S.-I., 2008, in: Y.-S. Sun, S. Ferraz-Mello, & J.-L. Zhou, (eds.), *Exoplanets: Detection, Formation and Dynamics*, Proc. IAU Symposium No. 249 (Suzhou,China), p. 399

Nelson, A. F., & Benz, W., 2003, *ApJ*, 589, 556

Nelson, R. P., 2005, *A&A* 443, 1067

Nelson, R. P. & Papaloizou, J. C. B., 2003, *MNRAS*, 339, 993

Nelson, R. P. & Papaloizou, J. C. B., 2004, *MNRAS*, 350, 849

Nelson, R. P., Papaloizou, J. C. B., Masset, F. S., & Kley, W., 2000, *MNRAS*, 318, 18

Ogilvie, G. I. & Lubow, S. H., 2003, *ApJ*, 587, 398

Paardekooper, S. & Mellema, G., 2006, *A&A*, 459, 17

Paardekooper, S. & Mellema, G., 2007, *ArXiv Astrophysics e-prints 0711.3601, accepted by A&A*

Paardekooper, S. & Papaloizou, J., 2007, *submitted to A&A*

Papaloizou, J. C. B., Nelson, R. P., Kley, W., Masset, F. S., and Artymowicz, P., 2007, *Protostars and Planets V*, 655

Papaloizou, J. C. B., Nelson, R. P., & Snellgrove, M. D., 2004a, *MNRAS*, 350, 829

Papaloizou, J. C. B., Nelson, R. P., & Snellgrove, M. D., 2004b, *MNRAS*, 350, 829
Pepliński, A., Artymowicz, P., & Mellema, G., 2007a, *ArXiv Astrophysics e-prints 0709.3622*, submitted to MNRAS
Pepliński, A., Artymowicz, P., & Mellema, G., 2007b, *ArXiv Astrophysics e-prints 0709.3754*, submitted to MNRAS
Pierens, A., & Huré, J.-M., 2005, *A&A*, 433, L37
Pierens, A., & Nelson, R. P., 2007, *A&A*, 472, 993
Pollack, J. B., Hubickyj, O., Bodenheimer, P., Lissauer, J. J., Podolak, M., & Greenzweig, Y., 1996, *Icarus*, 124, 62
Snellgrove, M. D., Papaloizou, J. C. B., & Nelson, R. P., 2001, *A&A*, 374, 1092
Tanaka, H., Takeuchi, T., & Ward, W. R., 2002, *ApJ*, 565, 1257
Terquem, C. E. J. M. L. J., 2003, *MNRAS*, 341, 1157
Ward, W. R., 1986, *Icarus*, 67, 164
Ward, W. R., 1997, *Icarus*, 126, 261
Winters, W. F., Balbus, S. A., & Hawley, J. F., 2003, *ApJ*, 589, 543
Zhang, X. J., Kretke, K. A. Lin, & D. N. C. 2008, in: Y.-S. Sun, S. Ferraz-Mello & J.-L. Zhou, (eds.), *Exoplanets: Detection, Formation and Dynamics*, Proc. IAU Symposium No. 249 (Suzhou,China), p. 309
Zhang, H., & Zhou, J.-L. 2008, in: Y.-S. Sun, S. Ferraz-Mello & J.-L. Zhou (eds.), Exoplanets: Detection, Formation and Dynamics, Proc. IAU Symposium No. 249 (Suzhou, China), p. 413

On the solar system–debris disk connection

Amaya Moro-Martín

Department of Astrophysical Sciences, Princeton University,
Princeton, NJ 08544, USA
e-mail: amaya@astro.princeton.edu

Abstract. This paper emphasizes the connection between solar and extra-solar debris disks: how models and observations of the Solar System are helping us understand the debris disk phenomenon, and vice versa, how debris disks are helping us place our Solar System into context.

Keywords. asteroids – cirumstellar matter – infrared: stars – Kuiper Belt – planetary systems – Solar System.

1. Introduction

Debris disks are disks of dust 10s–100s AU in size that surround main sequence stars of a wide range of stellar types (A to M) and ages (0.01–10 Gyr). In general, debris disks are not spatially resolved and are identified in the infrared from the dust thermal emission that results in an excess over the expected stellar values. Debris disks surveys carried out with *Spitzer* indicate that they contain a few lunar masses of dust and negligible quantities of gas, and that they are present around >33% of A-type stars (Su *et al.* 2006) and 10–15% of solar-type FGK stars (Bryden *et al.* 2006; Beichman *et al.* 2006; Trilling *et al.* 2008; Hillenbrand *et al.* 2008; Carpenter *et al.* in preparation). However, these results are calibration limited because the disks can only be detected at a certain level above the stellar photosphere due to uncertainties in the stellar flux. Figure 1 shows examples of some nearby spatially resolved debris disks.

The term *debris* refers to the fact that the dust cannot be primordial, because the expected lifetime of the dust grains due to Poynting-Robertson drag ($t_{PR} = 710(\frac{b}{\mu m})(\frac{\rho}{g/cm^3})$ $(\frac{R}{AU})^2(\frac{L_\odot}{L_{star}})\frac{1}{1+albedo}$ yr, where R, b and ρ are the grain location, radius and density, respectively – Burns, Lamy and Soter, 1979 and Backman and Paresce, 1993) and mutual grain collisions ($t_{col} = 1.26 \times 10^4 (\frac{R}{AU})^{3/2}(\frac{M_\odot}{M_*})^{1/2}(\frac{10^{-5}}{L_{dust}/L_*})$yr – Backman and Paresce, 1993) is much shorter than the age of the star, which means that the dust is likely being regenerated by planetesimals like the asteroids, Kuiper Belt objects (KBOs) and comets in our Solar System.

Indeed, the Solar System is filled in with dust. The sources of dust are the asteroids and comets in the inner region and the KBOs and interstellar dust in the outer region. The dust produced in the inner region can be seen in scattered light with our naked eyes, either in the zodiacal light on in the coma of comets, and has extensively been observed in thermal emission by space-based observatories (*IRAS* and *COBE*). Evidence of the presence of dust originated in the Kuiper Belt (KB) comes from dust collision events detected by Pioneer 10 and 11 beyond the orbit of Saturn (Landgraf *et al.*, 2002). Figure 2 shows the location of the planetesimals in the outer Solar System (left) and the expected spatial distribution of the dust generated in that region (right).

It is important to study the connection between the Solar System debris disks and the much brighter extra-solar debris disks because models and observations of the Solar

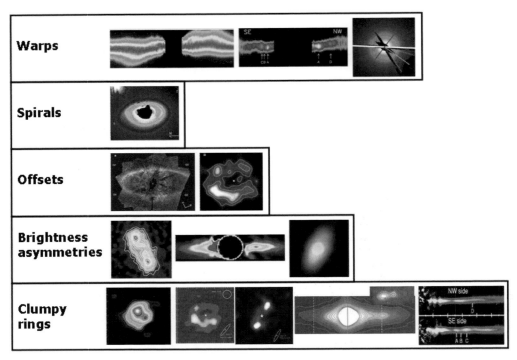

Figure 1. Spatially resolved images of nearby debris disks showing a wide diversity of debris disk structure. From left to right the images correspond to: (*1st row*) β-Pic (0.2–1 μm; Heap et al., 2000), AU-Mic (1.63 μm; Liu, 2004) and TW Hydra (0.2–1 μm; Roberge, Weinberger and Malumuth, 2005); (*2nd row*) HD 141569 (0.46–0.72 μm; Clampin et al., 2003); (*3rd row*) Fomalhaut (0.69–0.97 μm; Kalas et al., 2005) and ϵ-Eri (850 μm; Greaves et al., 2005); (*4th row*) HR 4796 (18.2 μm; Wyatt et al., 1999), HD 32297 (1.1 μm; Schneider, Silverstone and Hines, 2005) and Fomalhaut (24 and 70 μm; Stapelfeldt et al., 2004); (*5th row*) Vega (850 μm; Holland et al., 1998), ϵ-Eri (850 μm; Greaves et al., 1998), Fomalhaut (450 μm; Holland et al., 2003), β-Pic (12.3 μm; Telesco et al., 2005) and Au-Mic (0.46–0.72 μm; Krist et al., 2005). All images show emission from 10s to 100s of AU.

System can help us understand the debris disk phenomenon, and vice versa, models and observations of extra-solar debris disks can help us place our Solar System into context.

2. Debris Disk Evolution

2.1. *Steady collisional evolution*

It is thought that the Solar System was significantly more dusty in the past because both the Asteroid Belt (AB) and the Kuiper Belt (KB) were more densely populated. The system then became progressively less dusty as the planetesimal belts eroded away by mutual planetesimal collisions. Evidence of collisional evolution comes from the modeling and observation of the asteroid and KBO size distributions. In the AB, Bottke et al. (2005) showed that the initial size distribution progressively changes from a power-law to the observed wavy distribution, with peaks at $D \sim 120$ km (leftover from the accretion process) and $D \sim 200$ m (marking the transition at which the energy required to catastrophically destroy a particle is determined by self-gravity rather than strength forces). In the KB, Bernstein et al. (2004) found that its current size distribution shows a strong

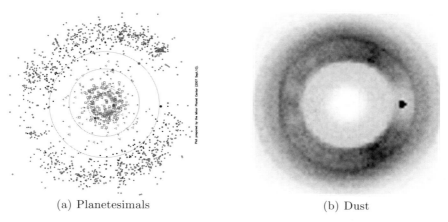

(a) Planetesimals (b) Dust

Figure 2. (*Left*) Distribution of planetesimals in the outer Solar System (courtesy of G. Williams at the Minor Planet Center). The outer circle is the orbit of Neptune. (*Right*) Distribution of dust in the outer Solar System resulting from dynamical simulations of dust particles originated in the Kuiper Belt (from Moro-Martín and Malhotra, 2002). The scale is the same as in the previous panel, with the black dot representing the location of Neptune. The structure is the result of gravitational perturbations of the giant planets on the orbit of the dust particles (see Sec. 3).

break to a shallower distribution at $D < 100$ km (when particles become more susceptible to collisional destruction).

Models show that this collisional evolution likely resulted in the production of large quantities of dust, as it can be seen in Figure 3 (from Kenyon and Bromley, 2005): in a planetesimal belt, Pluto-sized bodies ($D \sim 1000$ km) excite the eccentricities of the smaller and more abundant 1–10 km sized planetesimals, triggering collisions and starting a collisional cascade that produces dust and changes the planetesimal size distribution. Because the dust production rate is proportional to the number of collisions, and this is proportional to the square of the number of planetesimals, as the planetesimals erode and grind down to dust, the dust production rate decreases and the expected thermal emission from the dust slowly decays with time as $1/t$. This decay is punctuated by large spikes that are due to particularly large planetesimal collisions happening stochastically. Examples of stochastic events in the recent history of the Solar System are the fragmentation of the asteroids giving rise to the Hirayama and Veritas asteroid families (the latter happening 8.3 Myr ago and accounting for 25% of the present zodiacal thermal emission; Dermott *et al.*, 2002) and the dust bands observed by *IRAS* (Sykes and Greenberg, 1986).

Recent surveys carried out by *Spitzer/MIPS* have enabled the detection of debris disks around hundreds of A-type and solar-type stars with a wide range of ages, showing that the dust emission follows a $1/t$ decay and there is a large variability likely due to individual collisions (see Figure 4), in broad agreement with the results from collisional cascade models (Su *et al.*, 2006; Siegler *et al.*, 2007). Because solar and extra-solar planetary systems seem to follow similar evolutions, the imaging of debris disks at different evolutionary stages could be equivalent to a Solar System "time machine".

2.2. *Stochastic non-collisional evolution*

As discussed above, there is observational and theoretical evidence that collisional evolution played a role in the evolution of solar and extra-solar debris disks. However, there is also evidence that additional non-collisional processes, likely related to the dynamical

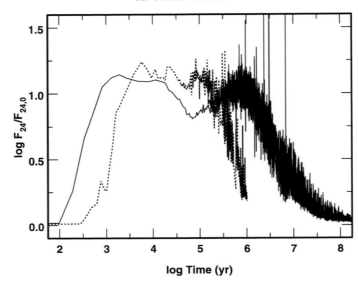

Figure 3. Evolution with time of the 24 μm dust thermal emission expected from the collisional evolution of two planetesimal belts extending from 0.68–1.32 AU (dashed line) and 0.4–2 AU (solid line) around a solar type star (Kenyon and Bromley, 2005).

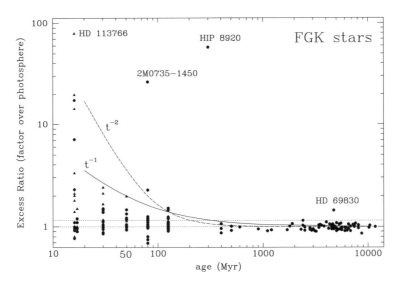

Figure 4. Ratio of the dust emission to the expected stellar emission at 24 μm for a survey of solar-type (FGK) stars. The stars aligned vertically belong to clusters or associations, therefore sharing the same age. The main features are the $1/t$ decay and the large variability found for a given stellar age. A few particularly massive debris disks are labeled. Figure from Siegler *et al.* (2007).

depletion of planetesimals that can result from gravitational interactions with massive planets, have also played a major role is disk evolution.

In the Solar System, evidence comes from the Late Heavy Bombardment (LHB, or Lunar Cataclysm), a period of time in the Solar System past during which a large number of impact craters in the Moon and the terrestrial planets were created (with an impact rate at Earth of ∼20000× the current value). This event, dated from lunar samples of impact melt rocks, happened during a very narrow interval of time – 3.8 to 4.1 Gyr ago (∼600 Myr after the formation of the terrestrial planets). Thereafter, the impact rate decreased exponentially with a time constant ranging from 10–100 Myr (Chyba, 1990). Strom et al. (2005) compared the impact cratering record and inferred crater size distribution on the Moon, Mars, Venus and Mercury, to the size distribution of different asteroidal populations, showing that the LHB lasted ∼20–200 Myr, the source of the impactors was the main AB, and the mechanism was size independent. The most likely scenario is that the orbital migration of the giant planets caused a resonance sweeping of the AB and as a result many of the asteroidal orbits became unstable, causing a large scale ejection of bodies into planet-crossing orbits (explaining the observed cratering record), and an increased rate of asteroidal collisions that would have been accompanied by the production of large quantities of dust. Under this scenario, the LHB was a single event in the history of the Solar System (Strom et al., 2005).

A handful of extra-solar debris disks observed with *Spitzer* also show evidence of non-collisional evolution (e.g. BD+20307, HD 72905, eta-Corvi and HD 69830; Wyatt et al., 2007). A particularly interesting case is that of HD 69830, a system that harbors three Neptune-like planets inside 0.63 AU(Lovis et al. 2006), shows a strong 24 μm dust emission indicative of large quantities of warm grains, no 70 μm dust emission indicative of a lack of cold dust (Beichman et al. 2005), and a dust emission spectra thought to arise from highly processed material similar to that of a disrupted P- or D-type asteroid plus small icy grains, likely located outside the outermost planet (Lisse et al. 2007). Wyatt et al. (2007) showed that its 24 μm emission, seen as an outlier in Figure 4, implies a very high dust production rate that could not possibly have been sustained for the entire lifetime of the star and must therefore be a transient event rather than the results of steady collisional evolution.

The models and observations described above for both solar and extra-solar systems indicate that in a planetesimal swarm there is collisional evolution that produces dust, triggered by the largest (Pluto-sized) planetesimals in the swarm, and on top of that, depending on the planetary configuration, there may be drastic dynamical events that produce very significant depletion of planetesimals and an increased rate of planetesimal collisions and dust production. The next Section discusses how the presence of planets not only can affect the production of debris dust, but can also sculpt the debris disk by creating a rich diversity of spatial structure.

3. Debris Disk Structure

Even though the great majority of debris disks observations are spatially unresolved, their structure can be studied in some detail through the spectral energy distribution (SED) of the disk because different wavelengths in the SED trace different distances to the star, so that an SED with sufficiently high spectral resolution can be used to constrain roughly the radial distribution of dust. Recent *Spitzer* debris disks surveys suggest that debris disks commonly show evidence of the presence of inner cavities, as most systems show 70 μm dust emission (from cold dust), but no emission at $\lambda \leqslant 24\,\mu$m (i.e., no warm dust; see e.g. Meyer et al., 2004; Beichman et al., 2005b; Bryden et al., 2006; Kim et al., 2005; Moro-Martín, Wolf and Malhotra, 2005; Moro-Martín et al., 2007a; Hillenbrand et al., 2008). High resolution spatially resolved observations have been obtained for a handful of nearby debris disks and indeed these images show the presence of inner cavities together with more complex morphology, like warps, spirals, offsets, brightness asymmetries and clumpy rings (see Figure 1).

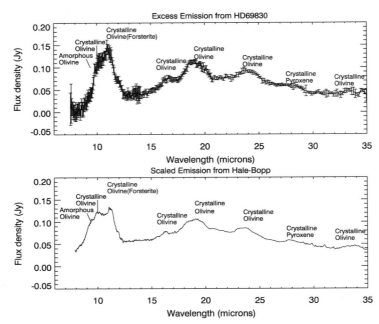

Figure 5. Spectrum of the dust emission around HD 69830 (*top*) compared to the spectrum of the comet Hale-Bopp normalized to a blackbody temperature of 400 K (*bottom*). Figure from Beichman *et al.* (2005a).

Dynamical simulations of the orbits of dust particles and their parent planetesimal in systems where massive planets are present suggest that this complex morphology could be the result from gravitational perturbations by planets (e.g. Roques *et al.*, 1994; Mouillet *et al.*, 1997; Wyatt *et al.*, 1999; Wyatt, 2005, 2006; Liou and Zook, 1999; Moro-Martín and Malhotra, 2002, 2003, 2005; Moro-Martín, Wolf and Malhotra, 2005; Kuchner and Holman, 2003; see Moro-Martín *et al.*, 2007b for a review). The basic mechanisms by which the planets can affect the debris disks structure are the following:

• *Ejection by gravitational scattering:* This process can affect dust particles as they spiral inward under P-R drag, and dust-producing planetesimals, in the case when the planet migrates outwards, resulting in a depletion of dust inside the orbit of the planet (an inner cavity). Dynamical simulations show that this process can be very efficient, ejecting >90% of the particles in the case of a 3–10 M_{Jup} planet located between 1–30 AU around a solar-type star.

• *Trapping in mean motion resonances (MMR) with the planet:* In a system where the dust producing planetesimals are located outside the orbit of the planet, as the dust particle drift inward due to P-R drag they can get trapped in MMRs with the planet. The MMRs are located where the orbital period of the planet is $(p + q)/p$ times that of the particle, where p and q are integers, $p > 0$ and $p + q \geqslant 1$. At these locations the particle receives energy from the perturbing planet that can balance the energy loss due to P-R drag, halting the inward motion of the particle and giving rise to planetary resonant rings. Due to the geometry of the resonance, the spatial distribution of material in resonance is asymmetric with respect to the planet, and this can explain the clumpy structure observed in some disks (Figure 1). An example of MMR trapping of KB in the Solar System can be seen in Figure 2, where the ring-like structure, the asymmetric clumps along the orbit of Neptune, and the clearing of dust at Neptune's location are all due to the trapping of particles in MMRs with the planet, while the dust depleted region inside 10 AU is due to gravitational scattering by Saturn and Jupiter. MMRs can also affect the location of the planetesimals and the dust when the planets migrate outward.

• *Effects of secular perturbations:* These are the long-term average of the perturbing forces and act on timescales >0.1 Myr (see review by Wyatt *et al.*, 1999). If the planet and the planetesimal disk are not coplanar, the secular perturbations tend to align the orbits and in the process they will create a warp in the disk. If the planet is in an eccentric orbit, the secular perturbations will force an eccentricity on the dust particles, creating an offset in the disk center with respect to the star that can result in a brightness asymmetry. Other effects of secular perturbations are spirals and inner gaps.

Finally, it is important to point out that because the debris disk structure is sensitive to the presence of planets located far from the star, the study of the structure could be used as a potential planet detection method that would be complementary to the well-established radial velocity and transit techniques (sensitive to close-in planets).

4. Concluding Remarks

Large surveys of debris disks over a wide range of evolutionary states, enabled by high sensitivity spaced-based IR telescopes like *Spitzer*, are starting to provide a "movie" of how planetary systems evolve with time. In this regard, debris disks help us place our Solar system into a broader context and vice versa, the study of the Solar System, in particular its dynamical history and the characterization of its small body population, sheds light on the physical processes giving rise to the debris disk phenomenon. Debris disks surveys, together with the results from planet searches, can help us understand the frequency of planetesimal and planet formation and the diversity of planetary systems, which ultimately addresses one of the most fundamental questions: is the Solar System common or rare?

Acknowledgments

A.M.M. warmly thanks the organizers of the IAU Symposium No. 249 for their kind invitation and generous financial support. A.M.M. is under contract with the Jet Propulsion Laboratory (JPL) funded by NASA through the Michelson Fellowship Program. JPL is managed for NASA by the California Institute of Technology. A.M.M. is also supported by the Lyman Spitzer Fellowship at Princeton University.

References

Backman, D. E. & Paresce, F., in *Protostars and Planets III* (E. H. Levy and J. I. Lunine, eds.) (Univ. of Arizona, Tucson), 1253 (1993).
Beichman, C. A., Bryden, G., Rieke, G. H. *et al.*, *ApJ* **622**, 1160 (2005b).
Beichman, C. A., Bryden, G., Gautier, T. N. *et al.*, *ApJ* **626**, 1061 (2005a).
Beichman, C. A., Bryden, G., Stapelfeldt, K. R. *et al.*, *ApJ* **652**, 1674 (2006).
Bernstein, G. M., Trilling, D. E., Allen, R. L. *et al.*, *AJ* **128**, 1364 (2004).
Bottke, W. F., Durda, D. D., Nesvornry, D. *et al.*, *Icarus* **175**, 111 (2005).
Bryden, G., Beichman, C. A., Trilling, D. E. *et al.*, *ApJ* **636**, 1098 (2006).
Burns, J. A., Lamy, P. L., & Soter, S., *Icarus* **40**, 1 (1979).
Chyba, C. F., *Nature* **343**, 129 (1990).
Clampin, M., Krist, J. E., Ardila, D. R. *et al.*, *AJ* **126**, 385 (2003).

Dermott, S. F., Kehoe, T. J. J., Durda, D. D. et al., in *Proceedings of Asteroids, Comets, Meteors - ACM 2002* (B. Warmbein, ed.) (ESA Publications Division, Noordwijk, Netherlands), 319 (2002).
Greaves, J. S., Holland, W. S., Moriarty-Schieven, G. et al., *ApJ* **506**, L133 (1998).
Greaves, J. S., Holland, W. S., Wyatt, M. C. et al., *ApJ* **619**, L187 (2005).
Heap, S. R., Lindler, D. J., Lanz, T. M. et al., *ApJ* **539**, 435 (2000).
Hillenbrand, L. A. et al., *ApJ*, in press (2008) (arXiv:0801.0163).
Holland, W. S., Greaves, J. S., Zuckerman, B. et al., *Nature* **392**, 788 (1998).
Holland, W. S., Greaves, J. S., Dent, W. R. F. et al., *ApJ* **582**, 1141 (2003).
Kalas, P., Graham, J. R. & Clampin, M., *Nature* **435**, 1067 (2005).
Kenyon, S. J. & Bromley, B. C., *AJ* **130**, 269 (2005).
Kim, J. S., Hines, D. C., Backman, D. E. et al., *ApJ* *632*, 659 (2005).
Krist, J. E., Ardila, D. R., Golimowski, D. A. et al., *AJ* **129**, 1008 (2005).
Kuchner, M. J. & Holman, M. J., *ApJ* **588**, 1110 (2003).
Landgraf, M., Liou, J.-C., Zook, H. A., & Grün, E., *AJ* **123**, 2857 (2002).
Liou, J.-C. & Zook, H. A., *AJ* **118**, 580 (1999).
Lisse, C. M., Beichman, C. A., Bryden, G., & Wyatt, M. C., *ApJ*, **658**, 584 (2007).
Liu, M. C., *Science* **305**, 1442 (2004).
Lovis, C., et al., Nature, **441**, 305 (2006).
Meyer, M. R., Hillenbrand, L. A., Backman, D. E. et al., *ApJ Supp.* **154**, 422 (2004).
Moro-Martín, A. & Malhotra, R., *AJ* **124**, 2305 (2002).
Moro-Martín, A. & Malhotra, R., *AJ* **125**, 2255 (2003).
Moro-Martín, A. & Malhotra, R., *ApJ* **633**, 1150 (2005).
Moro-Martín, A., Wolf, S., & Malhotra, R., *ApJ* **621**, 1079 (2005).
Moro-Martín, A., Malhotra, R., Carpenter, J. M. et al., *ApJ*, **668**, 1165 (2007a).
Moro-Martín, A., Wyatt, M. C., Malhotra, R., Trilling, D., in *Kuiper Belt* (A. Barucci, H. Boehnhardt, D. Cruikshank, A. Morbidelli, eds.) (Univ. of Arizona, Tucson), in press (2007b).
Mouillet, D., Larwood, J. D., Papaloizou, J. C. B., & Lagrange, A. M., *Mon. Not. R. Astron. Soc.* **292**, 896 (1997).
Roberge, A., Weinberger, A. J., & Malumuth, E. M., *ApJ* **622**, 1171 (2005).
Roques, F., Scholl, H., Sicardy, B., & Smith, B. A., *Icarus* **108**, 37 (1994).
Schneider, G., Silverstone, M. D., & Hines, D. C., *ApJ* **629**, L117 (2005).
Siegler, N., Muzerolle, J., Young, E. T. et al., *ApJ* **654**, 580 (2007).
Stapelfeldt, K. R., Holmes, E. K., Chen, C. et al., *ApJ Supp.* **154**, 458 (2004).
Strom, R. G., Malhotra, R., Ito, T. et al., *Science* **309**, 1847 (2005).
Su, K. Y. L., Rieke, G. H., Stansberry, J. A. et al., *ApJ* **653**, 675 (2006).
Sykes, M. V. & Greenberg, R., *Icarus* **65**, 51 (1986).
Telesco, C. M., Fisher, R. S., Wyatt, M. C. et al., *Nature* **433**, 133 (2005).
Trilling, D. E., Bryden, G., Beichman, C. A. et al., *ApJ* **674**, 1086 (2008).
Wyatt, M. C., *A&A* **433**, 1007 (2005).
Wyatt, M. C., *ApJ* **639**, 1153 (2006).
Wyatt, M. C., Dermott, S. F., Telesco, C. M. et al., *ApJ* **527**, 918 (1999).
Wyatt, M. C., Smith, R., Greaves, J. S. et al., *ApJ* **658** 569 (2007).

Disc signatures in a new population of low mass YSOs in ρ Ophiuchi

Catarina Alves de Oliveira[1] and Mark Casali[1]

[1]ESO, Karl-Schwarzschild-Str. 2,
85748 Garching bei München, Germany
email: coliveir@eso.org

Abstract. The deepest near-IR variability survey of the ρ Ophiuchi cluster with the WF-CAM/UKIRT has been used to uncover a new population of young low mass objects. Evidence for the existence of discs around the new objects has been found when combining data with IRAC/Spitzer observations. A new insight on the frequency and dynamics of discs around low mass YSOs is given, essential for understanding the origin and evolution of circumstellar discs and ultimately the environment for planet formation.

Keywords. Stars: pre-main-sequence,activity

1. Introduction

Variability is a common characteristic of pre-main-sequence stars (Joy 1942) and a useful tool for uncovering new cluster members in star formation regions. Young stellar objects (YSOs) are known to be variable in the near-IR, especially suitable for studying the circumstellar environment (e.g, Carpenter *et al.* 2001). With a large field of view (~ 0.8 deg^2 in four exposures), the WFCAM IR imager (Casali *et al.* 2007) on the UK Infrared Telescope has made possible deep variability studies of nearby star forming regions which extend over degree scales. The results of such a study are presented, with multi-epoch H and K observations of the ρ Ophiuchi cluster used to search for variability and uncover new pre-main-sequence stars (Alves de Oliveira & Casali 2008). Combining recently released IRAC/Spitzer data from the c2d point-source catalogues of the third data delivery (Evans *et al.* 2005), the list of variable objects is searched for IR-excesses characteristic of stars surrounded by an accretion disc.

2. Variable stars in ρ Ophiuchi

The reduced chi-square χ^2 (χ_ν^2) of the magnitudes together with cross-correlation indexes are used to detect variability; 182 variable stars are found, 53 are previously known *members* of the cluster and the others are named *candidate members*. The colours and magnitudes of the variable stars are an important clue in investigating the youth and masses of these objects. The colour-magnitude diagram, K vs. H-K, shown in Fig. 1, displays the variable members of ρ Ophiuchi (open triangles) and the candidate members (filled circles), overlaid on the distribution of all other detected sources, expected to be in its majority field stars. The models indicate that the brighter candidates are above the hydrogen burning limit, but the faint sources would extend to very low masses.

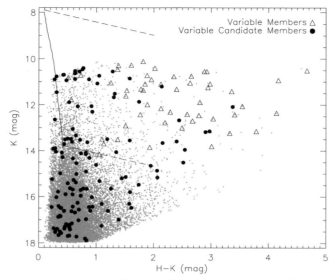

Figure 1. Colour-magnitude diagram for variable members and variable candidate members of ρ Ophiuchi, overlaid on the distribution of all other detected sources. The locus for the 5 Myr pre-main sequence down to $20 M_{Jup}$ (Baraffe et al. 1998) is shown along with reddening vectors (Rieke & Lebofsky 1985).

3. Disc signatures

Young stars show infrared emission which originates from dusty envelopes and circumstellar discs surrounding the central object. Lada & Wilking (1984), based on the level of long wavelength excess with respect to a stellar photosphere emission, identified three different classes which define an IR or SED classification scheme: Class I, low mass protostars surrounded by an infalling envelope with large IR excess; Class II, young stars with accretion discs and a moderate IR excess (as classical T Tauri stars, CTTSs); and Class III, stars which no longer accrete matter from a circumstellar disc and show no IR excess (as weak-line T Tauri stars, WTTSs). The IRAC data from Spitzer allows the study of these objects in the mid-IR, where the excess contribution from discs and envelopes is predominant. The IRAC colour-colour diagram ([3.6]−[4.5] vs. [5.8]−[8.0]) was presented as a tool to separate young stars of different classes (Allen et al. 2004).

The Ophiuchus molecular cloud has been mapped with IRAC in the 3.6, 4.5, 5.8 and 8.0 μm bands, over a region of 8.0 deg^2, which encompasses the WFCAM field. Figure 2 shows the IRAC colour-colour diagram for all variable objects with detections in the four bands, where the top diagram displays ρ Ophiuchi members and the bottom diagram the candidate members. The objects are displayed accordingly to their class, i.e., Class I (open squares), Class II (filled circles), and Class III (open triangles). The extinction vector is from Flaherty et al. (2007). For comparison, a sample of objects has been chosen from a region of the sky away from the central cloud with little extinction which should mainly contain field stars. The objects tend to cluster around three main regions of the diagram. Centred in the origin, [3.6]−[4.5],[5.8]−[8.0] = (0,0), are sources which have colours consistent with stellar photospheres and have no intrinsic IR excess. These can be foreground and background stars, but also Class III stars which do not have significant circumstellar dust. Another preferred region for objects in the diagram is located within the box defined by Allen et al. (2004) which represents the colours expected from models of discs around young, low-mass stars. In fact, many of the Class II objects classified by

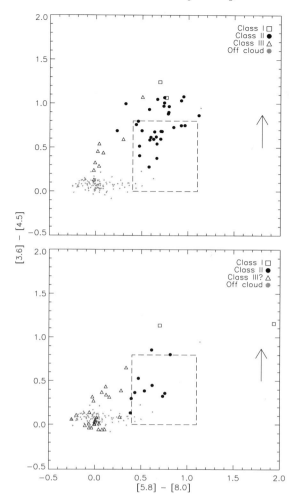

Figure 2. IRAC colour-colour diagram for variable members (top panel) and candidate members (bottom panel). Arrow represents the $A_k = 4$ extinction vector for ρ Oph (Flaherty et al. 2007).

Bontemps et al. (2001) lie within that range. However, some sources previously identified as Class II do not fall within the predicted limits. These objects have colours inconsistent with Class II sources (higher [3.6]–[4.5] colours than Class II but lower [5.8]–[8.0] colours than Class I) and were previously classified as candidate flat spectrum objects. Their location in this diagram does not confirm them as transition objects between Class I and II (Bontemps et al. 2001) since they can be explained as reddened Class II sources. Finally, from models of infalling envelops, Allen et al. (2004) predict the colours of Class I sources to have ([3.6]–[4.5]) > 0.8 and/or ([5.8]–[8.0]) > 1.1, which agrees well with the two Class I objects identified with ISOCAM (Bontemps et al. 2001). The same criteria can be applied to classify the candidate member variables on the bottom panel of the diagram. Combining WFCAM near-IR observations with IRAC/Spitzer mid-IR data, it is possible to classify 12 of the candidate members of ρ Ophiuchi as Class I/II, and 25 into possible Class III. Furthermore, the SED classes from the literature for the variable members are confirmed, with the exception of few objects.

4. Conclusions

Using near-IR variability, 129 new candidate members of Ophiuchus are uncovered. Combined with IRAC/Spitzer data, a subsample of 10 candidate members is classified as Class II objects, believed to be YSOs with accretion discs. This study is an important contribution for obtaining a complete census of stars surrounded by circumstellar disks in nearby star forming regions. A new coeval sample of YSOs spanning a wide range of masses is presented, ideal to study the early onset of planet formation.

Acknowledgements

We thank the UKIRT observatory staff and WFCAM Science Archive team for acquiring and pipeline processing the observations used for this project. C. A. O. acknowledges partial support from a Marie Curie Fellowship for Early Research Training. This work is based in part on observations made with the Spitzer Space Telescope, which is operated by the Jet Propulsion Laboratory, California Institute of Technology under a contract with NASA.

References

Allen, L. E., Calvet, N., D'Alessio, P., Merin, B., Hartmann, L., Megeath, S. T., Gutermuth, R. A., Muzerolle, J., Pipher, J. L., Myers, P. C., & Fazio, G. G. 2004, *ApJS*, 154, 363
Alves de Oliveira, C. & Casali, M. 2008, *submitted to A&A*
Baraffe, I., Chabrier, G., Allard, F., & Hauschildt, P. H. 1998, *A&A*, 337, 403
Bontemps, S., André, P., Kaas, A. A., Nordh, L., Olofsson, G., Huldtgren, M., Abergel, A., Blommaert, J., Boulanger, F., Burgdorf, M., Cesarsky, C. J., Cesarsky, D., Copet, E., Davies, J., Falgarone, E., Lagache, G., Montmerle, T., Pérault, M., Persi, P., Prusti, T., Puget, J. L., & Sibille, F. 2001, *A&A*, 372, 173
Carpenter, J. M., Hillenbrand, L. A., & Skrutskie, M. F. 2001, *ApJ*, 121, 3160
Casali, M., Adamson, A.,Alves de Oliveira, C. Almaini, O., Burch, K., Chuter, T., Elliot, J., Folger, M., Foucaud, S., Hambly, N., Hastie, M. Henry, D., Hirst, P., Irwin, M., Ives, D., Lawrence, A., Laidlaw, K., Lee, D., Lewis, J., Lunney, D., McLay, S., Montgomery, D., Pickup, A., Read, M., Rees, N., Robson, I., Sekiguchi, K., Vick, A., Warren, S., & Woodward, B. 2007, *A&A*, 467, 777
Evans, II, N. J., Allen, L. E., Blake, G. A., Boogert, A. C. A., Bourke, T., Harvey, P. M., Kessler, J. E., Koerner, D. W., Lee, C. W., Mundy, L. G., Myers, P. C., Padgett, D. L., Pontoppidan, K., Sargent, A. I., Stapelfeldt, K. R., van Dishoeck, E. F., Young, C. H., & Young, K. E. 2005, *Third Delivery of Data from the c2d Legacy Project: IRAC and MIPS(Pasadena, SSC)*
Flaherty, K. M.,Pipher, J. L.,Megeath, S. T.,Winston, E. M., Gutermuth, R. A., Muzerolle, J., Allen, L. E., & Fazio, G. G. 2007, *ApJ*, 663, 1069
Joy, A. H. 1942, *PASP*, 54, 35
Lada, C. J. & Wilking, B. A. 1984, *ApJ*, 287, 610
Rieke, G. H. & Lebofsky, M. J. 1985, *ApJ*, 288, 618

Searching for H_2 emission from protoplanetary disks using near- and mid-infrared high-resolution spectroscopy

A. Carmona[1,2,3]†, M. E. van den Ancker[3], Th. Henning[2], Ya. Pavlyuchenkov[2], C. P. Dullemond[2], M. Goto[2], D. Fedele[2,3,4], B. Stecklum[5], W. F-.Thi[6], J. Bouwman[2] and L. B. F. M. Waters[7,8]

[1] ISDC & Geneva Observatory, University of Geneva, chemin d'Ecogia 16, CH-1290 Versoix. Switzerland
email:Andres.Carmona@obs.unige.ch

[2] Max Planck Institute for Astronomy, Königstuhl 17, 69117 Heidelberg, Germany

[3] European Southern Observatory, Karl Schwarzschild Strasse 2 , 85748 Garching bei München, Germany

[4] Dipartimento di Astronomia, Università di Padova, Vicolo dell'Osservatorio 2, 35122 Padova, Italy

[5] Thüringer Landessternwarte Tautenburg, Sternwarte 5, 07778 Tautenburg, Germany

[6] Royal Observatory Edinburgh, Blackford Hill, Edinburgh, EH9 3HJ, UK

[7] Astronomical Institute, University of Amsterdam, Kruislaan 403, NL-1098 SJ Amsterdam, The Netherlands,

[8] Instituut voor Sterrenkunde, Katholieke Universiteit Leuven, Celestijnenlaan 200B, B-3030 Heverlee, Belgium

Abstract. The mass and dynamics of protoplanetary disks are dominated by molecular hydrogen (H_2). However, observationally very little is known about the H_2. In this paper, we discuss two projects aimed to constrain the properties of H_2 in the disk's planet forming region (R<50AU). First, we present a sensitive survey for pure-rotational H_2 emission at 12.278 and 17.035 μm in a sample of nearby Herbig Ae/Be and T Tauri stars using VISIR, ESO's VLT high-resolution mid-infrared spectrograph. Second, we report on a search for H_2 ro-vibrational emission at 2.1228, 2.2233 and 2.2477 μm in the classical T Tauri star LkHα 264 and the debris disk 49 Cet employing CRIRES, ESO's VLT high-resolution near-infrared spectrograph.

VISIR project: none of the sources show H_2 mid-IR emission. The observed disks contain less than a few tenths of $M_{Jupiter}$ of optically thin H_2 at 150 K, and less than a few M_{Earth} at $T > 300$ K. Our non-detections are consistent with the low flux levels expected from the small amount of H_2 gas in the surface layer of a Chiang and Goldreich (1997) Herbig Ae two-layer disk model. In our sources the H_2 and dust in the surface layer have not significantly departed from thermal coupling ($T_{gas}/T_{dust} < 2$) and the gas-to-dust ratio in the surface layer is very likely < 1000.

CRIRES project: The H_2 lines at 2.1218 μm and 2.2233 μm are detected in LkHα 264. An upper limit on the 2.2477 μm H_2 line flux in LkHα 264 is derived. 49 Cet does not exhibit H_2 emission in any of observed lines. There are a few M_{Moon} of optically thin hot H_2 in the inner disk ($\sim$0.1 AU) of LkHα 264, and less than a tenth of a M_{Moon} of hot H_2 in the inner disk of 49 Cet. The shape of the 1–0 S(0) line indicates that LkHα disk is close to face-on ($i < 35°$). The measured 1–0 S(0)/1–0 S(1) and 2–1 S(1)/1–0 S(1) line ratios in LkHα 264 indicate that the H_2 is thermally excited at $T < 1500$ K. The lack of H_2 emission in the NIR spectra of 49 Cet and the absence of Hα emission suggest that the gas in the inner disk of 49 Cet has dissipated.

† This paper is based on materials of our A&A papers Carmona *et al.* 2008 & 2007.

Keywords. stars: emission-line – stars: pre-main sequence – planetary systems:protoplanetary disks

1. Introduction

Circumstellar disks surrounding low- and intermediate- mass stars in their pre-main sequence phase are the locations where planets presumably form. Such protoplanetary disks are composed of gas and dust. Their mass and dynamics are dominated by gas (99%), specifically by molecular hydrogen (H_2). Observationally, very little is known about the gas compared to the dust. However, if we want to answer major questions in planet formation such as: how massive are the disks?, how extended are the disks?, and how long protoplanetary disks last ? we require information about the gaseous component of the disk. In particular, we would like to characterize warm gas in the inner disk (R < 50 AU), the region where planets form (see reviews by Najita et al. 2007; Carr 2005).

H_2 is by far the most abundant molecular species in protoplanetary disks. Unfortunately, H_2 is one of the most challenging molecules to detect. H_2 is a homonuclear molecule that lacks a permanent dipole moment, therefore, H_2 transitions are electric quadrupole in nature, thus, very weak (i.e., small Einstein coefficients). In addition, in the case of protoplanetary disks, the H_2 lines are not sensitive to the gas in the optically thick regions where the dust and gas are at equal temperature. Practical observational challenges also have to be faced. The H_2 lines from the disk needs to be detected on the top of a strong infrared continuum. From the ground, the infrared windows (specially the mid-infrared) are strongly affected by sky and instrument background emission, and the H_2 transitions lie close to atmospheric absorption lines highly dependent on atmospheric conditions. The advent of high spectral resolution infrared spectrographs mounted on large aperture telescopes, allows for the first time the study of H_2 emission from the ground.

In this paper, we discuss two projects aimed to constrain the properties of H_2 gas in the planet forming region of circumstellar disks. In Sect. 1 we present the results of a large, sensitive survey for pure-rotational molecular hydrogen emission at 12.278 and 17.035 microns in a sample of nearby Herbig Ae/Be and T Tauri stars using the high-resolution mid-infrared spectrograph VISIR at ESO-VLT (Carmona et al. 2008). In Sect. 2. we report on the results of a search for H_2 ro-vibrational emission at 2.1228, 2.2233 and 2.2477 μm in the classical T Tauri star LkHα 264 and the debris disk 49 Cet using CRIRES, the new ESO's VLT Adaptive Optics high resolution near-infrared spectrograph (Carmona et al. 2007).

2. Searching for mid-IR H_2 emission from protoplanetary disks with VISIR

2.1. Motivation.

To probe the gas in the giant planet forming region of the disk. H_2 mid-IR lines probe warm gas at $T \sim 150$–1000 K. This gas is located from a few AU up to 50 AU.

2.2. Previous work.

H_2 mid-IR emission from protoplanetary disks has been reported from ISO observations (Thi et al. 2001). However, subsequent ground-based efforts (Richter et al. 2002; Sheret et al. 2003; Sako et al. 2005) did not confirm the ISO detections. H_2 emission in the mid-IR has been searched towards debris disks using Spitzer (Hollenbach et al. 2005, Pascucci et al. 2006, Chen et al. 2006) with no detection reported. Most recently, Bitner et al. (2007) and Martin-Zaïdi et al. (2007) reported the detection of mid-IR H_2 emission in two Herbig Ae/Be stars (AB Aur and HD 97048) from the ground, and Lahuis et al. (2007) reported the detection of mid-IR H_2 emission in 6 T Tauri stars with *Spitzer*.

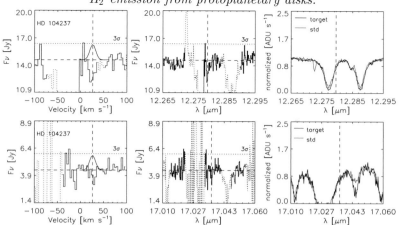

Figure 1. Example of the VISIR spectra obtained in the case of HD 104237. The upper panels show the spectra for the H_2 0–0 S(2) line at 12.278 μm. The lower panels display the spectra for the H_2 0–0 S(1) line at 17.035 μm. The left panels show a zoom to the -100 to 100 km s^{-1} interval of the atmospheric corrected spectra. A Gaussian of $FWHM = 15$ km s^{-1} and integrated line flux equal to the line-flux upper limits obtained is overplotted at the expected velocity shifted location (vertical dashed lines). The central panels show the full corrected spectra. Dotted lines show spectral regions strongly affected by telluric or standard star absorption features. The right panels show the continuum normalized spectra of the standard star and the target before telluric correction. The spectra are not corrected for the radial velocity of the targets (Carmona et al. 2008).

2.3. Observations & Data Reduction.

We observed the Herbig Ae/Be stars UX Ori, HD 34282, HD 100453, HD 101412, HD 104237 and HD 142666, and the T Tauri star HD 319139 in the first semester of 2006 and 2007 with VISIR, a combined imager and spectrograph designed for observations in the N ($\approx 8-13$ μm) and Q bands ($\approx 16.5-24.5$ μm) (Lagage et al. 2004), mounted at the ESO-VLT Melipal telescope in Cerro Paranal, Chile. We selected a sample of well known nearby Herbig Ae/Be and T Tauri stars based on evidence of large disk reservoirs. The targets have either reported detections of cold CO gas at (sub)-mm wavelengths, or dust continuum emission at mm wavelengths. We chose stars with 12 μm continuum fluxes >0.5 Jy (otherwise too faint for acquisition with VISIR) and < 25 Jy (hard to detect weak lines on top of a strong continuum). The H_2 $v = 0-0$ $S(1)$ line at 17.035 μm was observed in the *high-resolution long-slit mode* with a 0.4 arcsec slit, giving a spectral resolution R ≈ 21000, or 14 km s^{-1}. The H_2 $v = 0-0$ $S(2)$ line at 12.278 μm was observed in the *high-resolution echelle mode* with a 0.4 arcsec slit, giving a spectral resolution R ≈ 20000, or 15 km s^{-1}. The total integration time in each line was 1 h. The slit was oriented in the North-South direction. Sky background was subtracted by chopping the telescope by $\sim 8''$ in the direction of the slit. Asymmetrical thermal background of the telescope was subtracted by nodding the telescope by $\sim 8''$ in the direction of the slit.

For correcting the spectra for telluric absorption and obtaining the absolute flux calibration, spectroscopic standard stars at close airmasses to that of the science targets were observed immediately preceding and following the 12 μm exposure, and preceding or following the 17 μm exposure. After assuring that all the half-chop cycles in the VISIR data cubes had the same wavelength in the same row of pixels, the data-cubes were processed with the VISIR pipeline (Lundin 2006). The science spectrum was extracted by summing the number of counts inside the PSF in the dispersion direction in the 2D spectrum. To correct for telluric absorption and flux-calibrate the science spectrum, the one-dimensional extracted science spectrum was divided by the one-dimensional extracted

Table 1. H_2 mid-IR emission line flux upper limits and optically thin H_2 gas mass limits.

Star	H_2 line	λ [μm]	continuum [Jy]	line flux [a] [$\times 10^{-14}$ ergs s^{-1} cm^{-2}]	H_2 mass limits in M_J $T=150$ K	$T=300$K	$T=1000$K
UX Ori	0–0 S(2)	12.278	1.9 (1.1)	<1.4	27.9	1.9×10^{-1}	1.2×10^{-2}
	0–0 S(1)	17.035	2.0 (1.5)	<1.3	1.0	6.8×10^{-2}	2.0×10^{-2}
HD 34282	0–0 S(2)	12.278	0.3 (0.4)	<0.5	13.8	9.7×10^{-2}	6.0×10^{-3}
	0–0 S(1)	17.035	...	...	...	...	...
HD 100453	0–0 S(2)	12.278	9.0 (0.7)	<0.9	1.9	1.4×10^{-2}	0.9×10^{-3}
	0–0 S(1)	17.035	14.8 (1.7)	<1.5	0.1	0.9×10^{-2}	2.5×10^{-3}
HD 101412	0–0 S(2)	12.278	3.5 (1.0)	<1.2	5.3	3.7×10^{-2}	2.3×10^{-3}
	0–0 S(1)	17.035	1.6 (1.6)	<1.4	0.2	1.6×10^{-2}	4.8×10^{-3}
HD 104237	0–0 S(2)	12.278	14.6 (1.8)	<2.2	5.0	3.6×10^{-2}	2.2×10^{-3}
	0–0 S(1)	17.035	4.4 (1.5)	<1.3	0.1	0.8×10^{-2}	2.3×10^{-3}
HD 142666	0–0 S(2)	12.278	...	...	...	...	...
	0–0 S(1)	17.035	8.0 (1.2)	<1.1	0.2	1.0×10^{-2}	3.1×10^{-3}
HD 319139	0–0 S(2)	12.278	...	...	...	...	...
	0–0 S(1)	17.035	2.5 (1.5)	<1.3	0.2	1.2×10^{-2}	3.7×10^{-3}

Notes: [a] Upper limits calculated using a line $FWHM$ of 15 km s^{-1}.
spectrum of the standard star (for further details on the data reduction procedure, telluric correction and flux calibration see Sect. 3 of Carmona *et al.* 2008).

2.4. Results.

None of the observed sources show evidence for H_2 emission at 12 or 17 μm (see Fig. 1 for an example of the spectra obtained in the case of HD 104237, for all the spectra see Carmona *et al.* 2008). Assuming that the H_2 lines are unresolved, we calculated 3σ upper limits to the integrated H_2 line fluxes by multiplying the 3σ continuum flux noise with the instrument resolution line width ($\sim$15 km s^{-1}). Our results are summarized in Table 1. The typical sensitivity limit of our observations is a line flux of 10^{-14} ergs s^{-1} cm^{-2}. Our flux limits are of the order of magnitude of the H_2 0–0 S(1) and 0–0 S(2) lines fluxes (1.1 and 0.53 $\times 10^{-14}$ ergs s^{-1} cm^{-2}, respectively) reported by Bitner *et al.* (2007) for AB Aur† and the H_2 line fluxes of 0.33 to 1.70 $\times 10^{-14}$ ergs s^{-1} cm^{-2} reported for the H_2 0–0 S(2) line in the observations by Lahuis *et al.* (2007).

2.5. Discussion.

Under the assumption that the H_2 emission and the emission of the accompanying dust are optically thin, that the emitting H_2 is in local thermodynamical equilibrium (LTE), and that the source size is equal or smaller than VISIR's beam size, we derived upper limits to the H_2 mass as a function of the temperature employing (Thi *et al.* 2001)

$$M_{\text{gas}} = f \times 1.76 \times 10^{-20} \frac{4\pi d^2 \, F_{ul}}{E_{ul} \, A_{ul} \, x_u(T)} \; M_\odot, \qquad (2.1)$$

where F_{ul} is the upper limit to the integrated line flux, d is the distance in pc to the star, E_{ul} is the energy of the transition, A_{ul} is the Einstein coefficient of the $J = u - l$ transition and x_u is the population of the level u at the excitation temperature T in LTE; f is the conversion factor required for deriving the total gas mass from the H_2-ortho or H_2-para mass determined. In Table 1, we present our results. The disks contain less than a few tenths of Jupiter mass of optically thin H_2 at 150 K, and less than a few Earth masses of optically thin H_2 at 300 K and higher temperatures.

Using a two-layer Chiang and Goldreich (1997, CG 97) disk model implementation (CGplus, Dullemond *et al.* 2001) with physical parameters aimed to fit the spectral

---

† Note that the observations of AB Aur were performed with TEXES, which provides a spectral resolution of 100000, increasing the line-to-continuum contrast compared with our observations.

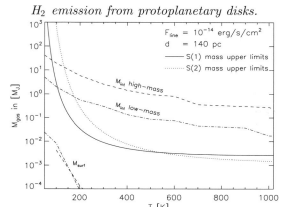

Figure 2. Mass limits of optically thin H_2 derived from H_2 0–0 S(1) (solid line) and H_2 0–0 S(2) (dotted line) as a function of the temperature for a line flux limit of 10^{-14} ergs s^{-1} cm^{-2} for a source at a distance of 140 pc. Dashed and dot-dashed lines show the gas mass as function of the temperature for a Chiang and Goldreich (1997) optically thick two-layer model for a low-mass ($M_{\rm DISK} = 0.02 M_\odot$) and a high-mass ($M_{\rm DISK} = 0.11 M_\odot$) disk assuming a gas-to-dust ratio of 100. $M_{\rm int}$ is the mass of the interior layer. $M_{\rm surf}$ is the mass of the surface layer. H_2 emission arises only from the optically thin molecular gas in surface layer of the disk (Carmona et al. 2008).

energy distribution (SED) of prototypical Herbig Ae/Be stars (Chiang et al. 2001), we computed the expected amount of gas in the interior and surface layer as function of the temperature for disks of mass 0.02 $M_\odot$ and 0.11 $M_\odot$ (see Fig. 2). The mass limits derived from the H_2 0–0 S(1) and 0–0 S(2) line observations are smaller than the amount of warm gas in the interior layer, but much larger than the amount of molecular gas in the surface layer. Fig. 2 shows that the amount of gas in the surface layer is very small ($<10^{-2}$ $M_J \sim 3 M_\oplus$) and almost independent of the total mass of the disk. *If the two-layer model is an adequate representation of the structure of the disk, the thermal flux levels of H_2 mid-infrared emission are below the detection limit of the observations, because the mass of H_2 in the surface layer is very small.*

In Carmona et al. (2008), based on a CG97 two-layer model, assuming LTE thermal emission, we calculated the H_2 mid-IR fluxes expected from optically thick Herbig Ae disks (see Sect. 5.2 of Carmona et al. 2008 for a detailed description of the calculation). We found that for a source distant 140 pc, the expected fluxes are of the order of 10^{-16} erg s^{-1} cm^{-2} for the 0–0 S(1) line and 10^{-17} erg s^{-1} cm^{-2} for the 0–0 S(2) line, and line-to-continuum flux ratios of $<10^{-3}$. *These line flux levels are two orders of magnitude below the sensitivity limits of our observations* (5×10^{-15} erg s^{-1} cm^{-2}). If the two-layer approximation to the structure of the disk is correct, we are essentially "blind" to most of the warm H_2 in the disk because it is located in the optically thick interior layer of the disk.

Nevertheless, some detections of mid-IR H_2 emission from disks have been reported in the Herbig Ae/Be stars AB Aur (Bitner et al. 2007) and HD 97048 (Martin-Zaïdi et al. 2007), and in the T Tauri stars Sz 102, EC 74, EC 82, Ced 110 IRS6, EC 92, and ISO-Cha 237 (Lahuis et al. 2007). An interesting question to address is the reason for the high H_2 fluxes observed in these sources. A possibility is that an additional mechanism (X-rays, UV heating, e.g., Glassgold et al. 2007; Nomura et al. 2007) heats the molecular gas in the surface layer making $T_{\rm gas\,surf} > T_{\rm dust\,surf}$. An additional scenario is to invoke a gas-to-dust ratio much larger than the canonical value of 100 in the surface layer of the disk (e.g., as a result of dust coagulation and sedimentation). In order to explore the influence of a change in the gas-to-dust ratio and the thermal decoupling of gas and dust in the surface layer, in Carmona et al. (2008) we calculated the expected 0–0 S(0) to 0–0 S(4) H_2 line fluxes from a M_{DISK} 0.01 $M_\odot$ disk model, as a function of the surface gas-to-dust ratio (ranging from 100 to 14000) for $T_{gas\,surf}/T_{dust\,surf}$ ranging from 1.0 to 2.0. We obtained that detectable 0–0 S(1) and 0–0 S(2) H_2 line flux levels can be achieved if $T_{gas\,surf}/T_{dust\,surf} > 2$ and if the gas-to-dust ratio in the surface layer is greater than 1000. *H_2 emission levels are very sensitive to departures from the thermal coupling between the molecular gas and dust in the surface layer.* Our results suggest that in the observed sources the molecular gas and the dust in the surface layer have not significantly departed from thermal coupling and that the gas-to-dust ratio in the surface layer is very likely lower than 1000. A definitive interpretation of our results awaits the development of future, more sophisticated models.

3. Searching for near-IR H_2 emission from protoplanetary disks with CRIRES

3.1. Motivation.

To probe the gas in the terrestrial planet forming region of the disk. H_2 near-IR lines probe hot gas at $T \sim 1000$–3000 K. This gas is located from a tenth of AU up to a few AU.

3.2. Previous work.

The $v = 1$-0 S(1) H_2 line at 2.2218 μm has been detected in few classical T Tauri stars (CTTS): TW Hya, GG Tau A, LkCa 15 (Weintraub et al. 2000, Bary et al. 2002, 2003), AA Tau, CW Tau, UY Aur, GM Tau (Shukla et al. 2003), CS Cha (Weintraub et al. 2005), ECHAJ0843.3-7905 (Howat & Greaves 2007) and LkHα 264 (Itoh et al. 2003, Carmona et al. 2007), and in four weak-line T Tauri stars (WTTS): DoAr 21 (Bary et al. 2003), V773 Tau (Shukla et al. 2003), Sz33 and Sz 41 (Weintraub et al. 2005).

3.3. Observations & Data Reduction.

We observed the classical T Tauri star LkHα 264 and the debris disk 49 Cet, with the ESO-VLT cryogenic high-resolution ($R \sim 45000$, 6.6 km s^{-1}) infrared echelle spectrograph CRIRES (Käufl et al. 2004), mounted on ESO UT1 "Antu" 8-m telescope atop Cerro Paranal Chile, during the CRIRES science-verification phase (November 8 - 9, 2006). We employed the wave-ID 27/1/n and the wave-ID 25/-1/n, providing a spectral coverage from 2.0871 to 2.1339 μm and from 2.2002 to 2.2552 μm respectively. The observations were performed using a 46" long, 0.4" wide, north-south oriented slit, nodding the telescope 10" along the slit. A random jitter smaller than 2" was added to the telescope in addition to the nodding offset at each nodding position to correct for bad pixels and decrease systematics due to the detector. The total integration time was of 720s for LkHα 264 and of 240s for 49 Cet. Spectrophotometric standard stars at similar airmass to the science target were observed immediately following the science observations for performing the telluric correction. We searched for $v = 1 - 0$ S(1) H_2 emission at 2.1218 μm, $v = 1 - 0$ S(0) H_2 emission at 2.2233 μm and $v = 2 - 1$ S(1) H_2 emission at 2.2477 μm. The data was reduced using the CRIRES pipeline and the ESO/CPL recipes. To correct for telluric absorption and flux-calibrate the science spectrum, the one-dimensional science spectrum obtained was divided by the one-dimensional spectrum of the standard star (for further details on the data reduction procedure, telluric correction and flux calibration see Sect. 2.1 of Carmona et al. 2007)

3.4. Results.

Our observations confirm the previous detections of the H_2 1–0 S(1) line reported by Itoh et al. (2003). However, in contrast to Itoh et al. (2003), the H_2 1–0 S(0) line is detected in our CRIRES spectra of LkHα 264. Our CRIRES observation show, for the first time, the simultaneous detection of the 1–0 S(1) and 1–0 S(0) H_2 line from a protoplanetary disk.

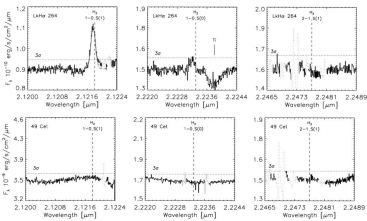

Figure 3. CRIRES spectra of LkHα 264 (upper panels) and 49 Cet (lower panels) in the regions of the H$_2$ v=1-0 S(1), H$_2$ v=1-0 S(0) and H$_2$ v=2-1 S(1) emission lines. The H$_2$ v=1-0 S(1) and the H$_2$ v=1-0 S(0) lines are detected in LkHα 264. Photospheric Ti features at 2.2217 (not shown) and 2.2238 μm (central upper panel) are observed in LkHα 264. The Gaussian fits to the detected lines are illustrated in dash-dot lines. The H$_2$ v=2-1 S(1) line is not present in LkHα 264. In the case of 49 Cet none of the three H$_2$ features are present in emission or absorption. Horizontal dotted lines show the 3σ continuum flux limits. The spectra are not corrected for V$_{LSR}$ of the star. Regions of poor telluric correction are in gray-dotted lines in the spectra (Carmona et al. 2007).

Table 2. H$_2$ near-IR emission line fluxes and upper limits

Star	H$_2$ line	λ [μm]	continuum [× 10^{-10} ergs s^{-1} cm^{-2} μm^{-1}]	line flux^a [× 10^{-14} ergs s^{-1} cm^{-2}]
LkHα 264	1–0 S(1)	2.1218	0.9 (0.04)	0.30
	1–0 S(0)	2.2233	1.5 (0.08)	0.10
	2–1 S(1)	2.2477	1.6 (0.10)	<0.05
49 Cet	1–0 S(1)	2.1218	35 (1.0)	<0.5
	1–0 S(0)	2.2233	17 (1.8)	<0.9
	2–1 S(1)	2.2477	15 (3.1)	<1.6

Notes: ^{a} Upper limits calculated using a line $FWHM$ of 6.6 km s^{-1}.

The H$_2$ 2–1 S(1) line is not seen in LkHα 264. The central wavelength of the 1–0 S(1) H$_2$ emission in LkHα 264 was measured to be 2.121757 ± 0.000005 μm. This corresponds to a velocity shift of −5.6 ± 1.0 km s^{-1} (taking into account Earth's velocity at the time of observation), a velocity coincident with the rest velocity of the star (−5.9 ± 1.2 km s^{-1}, Itoh et al. 2003). The H$_2$ 1–0 S(0) feature at 2.2233 μm is detected with a 3σ level confidence. Employing a Gaussian fit, the central wavelength of the line found is 2.22321 ± 0.00005 μm. This corresponds to a velocity shift of −12 ± 7 km s^{-1} which is in agreement with the velocity shift found in the 1–0 S(1) line. The $FWHM$ of the lines are 20.6 ± 1 km s^{-1} for the H$_2$ 1–0 S(1) line and 19.8 ± 1 km s^{-1} for the H$_2$ 1–0 S(0) line. The measured line fluxes are 3.0 ×10^{-15} and 1.0 ×10^{-15} erg s^{-1} cm^{-2} for the H$_2$ 1–0 S(1) and H$_2$ 1–0 S(0) respectively. The 2–1 S(1) H$_2$ line is not observed in LkHα 264. Assuming a $FWHM$ of 6.6 km s^{-1} (CRIRES resolution), a 3σ flux upper limit of 5.3 ×10^{-16} ergs s^{-1} cm^{-2} is derived for the line. Assuming an error of 20% in the flux calibration of the spectra, the 1–0 S(0)/1–0 S(1) line ratio in LkHα 264 is 0.33 ± 0.1 and the 2–1 S(1)/1–0 S(1) line ratio is <0.2. These line ratios are consistent with the line ratios of a gas at LTE at a temperature cooler than 1500 K (Mouri 1994).

In the case of 49 Cet none of the three H$_2$ features are present in emission or absorption. A summary of the 3σ flux upper limits is presented in Table 2.

3.5. Discussion.

H$_2$ emission in LkHα 264: a disk or an outflow?. The first question is to determine whether the H$_2$ emission observed in LkHα 264 originates in an outflow (shock excited emission) or in a disk. The small velocity shift, the line shape (well reproduced by a disk model), and the fact that the emission is spatially unresolved are not in favor of shock excited H$_2$. An additional strong argument against shock excitation of H$_2$ is that LkHa 264 does not exhibit [OI] forbidden emission at 6300 Å (Cohen and Kuhi 1979); a classical signature of outflows in T Tauri stars. The lack of this line indicates that in LkHα 264 the outflow is not present or at least that it is very weak. We conclude that the H$_2$ emission observed in LkHα 264 originates very likely in a disk.

Excitation mechanism of the H$_2$ emission in LkHα 264. Thermal and non-thermal excitation mechanisms are distinguishable on the basis of line ratios (Mouri 1994 and references there in). With Figure 3b of Mouri (1994), we find that the measured 1–0 S(0)/1–0 S(1) (0.33 ± 0.1) and the 2–1 S(1)/1–0 S(1) (<0.2) line ratios in LkHα 264 are consistent with thermal emission of a gas cooler than 1500 K thermally excited by UV photons (i.e., UV photons NOT in the Lyman-Werner band at 912–1108 Å). Assuming a gaussian error distribution, given the line 1–0 S(0)/1–0 S(1) ratio measured, the probability that the heating mechanism is X-ray excitation is less than 1%.

H$_2$ emitting region and inclination of the LkHα 264 disk. The spectral resolution of CRIRES ($\approx$ 6.6 km s^{-1}) and the thermal width of a 1500 K line ($\approx$2.4 km s^{-1}) are significantly smaller than the $FWHM$ of 20 km s^{-1} of the H$_2$ lines observed in LkHα 264. Therefore, the line width must be linked to the dynamics of the gas in the region that is emitting the line. Implementing the two-layer Chiang and Goldreich (1997) disk model code CG plus (Dullemond *et al.* 2001), we modeled the disk around LkHα 264, and found that the regions of the disk with a surface layer at $T_s < 1500$ K are located at $R > 0.1$ AU. Modeling of the single peaked 1–0 S(1) line shape (for the details of the model see Carmona *et al.* 2007) indicates that the disk is close to face-on ($i < 35°$). The best model fit suggests that the disk of LkHα 264 is inclined 20° for a H$_2$ emitting region extending from 0.1 to 10 AU with a power law relation of the intensity as a function of radius with exponent $\alpha = -2$. If the 1–0 H$_2$ S(1) line intensity decreases with an exponent $\alpha = -2$ as a function of radius, then 50% of the line flux is produced within 0.1 AU and 1 AU of the LkHα 264 disk, 40% of the line flux is emitted within 1 and 7 AU and the rest of the flux at larger radii.

Mass of optically thin hot H$_2$ in LkHa 264 and 49 Cet. Using a similar equation to Eq. 2.1 from the 1–0 S(1) H$_2$ line flux in LkHα 264 and the upper limit to the flux of the same line in 49 Cet, assuming optically thin gas in LTE, we calculated the mass H$_2$ gas at 1500 K in both sources. We found that there are a few lunar masses of hot H$_2$ gas in the inner disk of LkHα 264 and less than a tenth of a lunar mass of hot H$_2$ in the inner disk of 49 Cet. The lack of H$_2$ ro-vibrational emission in the spectra of 49 Cet, combined with non detection of pure rotational lines of H$_2$ (Chen *et al.* 2006) and the absence of Hα emission suggest that the gas in the inner disk of 49 Cet has dissipated. These results together with the previous detection of ^{12}CO emission at sub-mm wavelengths (Zuckerman *et al.* 1995; Dent *et al.* 2005) point out that the disk of 49 Cet should have a large inner hole, and it is strongly suggestive of theoretical scenarios in which the disk disappears inside-out. This could be due to inside-out photoevaporation, or to the presence of an unseen low-mass companion(s).

Detections and non detections of near-IR H$_2$ emission from T Tauri stars. In Carmona *et al.* (2007) we present a detailed comparative analysis of the physical properties of classical T Tauri stars in which the H$_2$ 1-0 S(1) line has been detected versus non-detected. We found that there is a higher chance of observing the H$_2$ near-infrared lines in CTTS with a high $U - V$ excess and a strong Hα line. This result suggests that there is a higher probability of detecting the H$_2$ 1–0 S(1) line in systems with high accretion. In contrast to weak-lined T Tauri stars, there is no apparent correlation between the X-ray luminosity and the detectability of the H$_2$ 1–0 S(1) line in classical T Tauri stars. Taken as a group, LkHα 264 and the CTTS in which near-IR H$_2$ emission has been detected exhibit typical properties of classical T Tauri stars. Therefore, we expect near-IR ro-vibrational H$_2$ lines from T Tauri disks to be detected on a routine basis in the near future.

4. Conclusions

The advent of infrared high-resolution spectrographs mounted on large aperture telescopes allow for the first time the study of the gaseous component of the disk in the planet forming region (R < 50 AU). Here, we presented two efforts for constraining the properties of the molecular hydrogen in a sample of Herbig Ae/Be, T Tauri stars and a debris disk.

In the first project, we searched for fundamental emission of H$_2$ at 17 and 12 micron employing VISIR, ESO's mid-infrared high resolution spectrograph. Our aim was to probe the warm gas ($T \sim 150 - 1000$ K) in the giant planet forming region of the disk. None of the sources show fundamental H$_2$ emission. From the upper limits on the line fluxes derived, we estimated that there is less than a tenth of Jupiter mass of optically thin H$_2$ at $T = 150$ K and less than a few Earth mass at higher temperatures. These results are consistent with the two-layer disk model (CG 97). In this model only the molecular gas in the optically thin surface layer of the disk emits the H$_2$ lines. Typical Herbig Ae disks have a very small mass of molecular gas ($<10^{-2}$ M$_J$) in the surface layer. We calculated the expected H$_2$ fluxes assuming a gas in LTE and a gas-to-dust ratio of 100. We found that the line fluxes are of the order of 10^{-16} to 10^{-17} erg s^{-1} cm^{-2}, two orders of magnitude fainter than our detection limits. If the two-layer model is correct we are "blind" to the H$_2$ gas in the interior layer. This result explains the numerous non-detections of mid-IR H$_2$ emission towards several pre-main sequence stars with disks. The few detections reported can be explained as the result of a thermal decoupling of the gas and the dust (gas hotter than the dust) or a change in gas-to-dust ratio in the surface layer. Our non-detections suggest that in our sources the H$_2$ and dust in the surface layer have not significantly departed from thermal coupling (T$_{\rm gas}$/T$_{\rm dust}$ < 2) and the gas-to-dust ratio in the surface layer is very likely < 1000.

In the second project, we searched for near-infrared ro-vibrational H$_2$ emission from LkHα 264 a classical T Tauri star and 49 Cet a debris disk with detections of CO emission in the sub-mm. Our aim was to probe the hot gas ($T \sim 1000$ K) present in the terrestrial planet region of the disks. We detected the 1–0 S(1) and 1–0 S(0) H$_2$ lines in LkHα 264 and derived an upper limit to the 2–1 S(1) H$_2$ line. Given that the velocity of the lines is coincident with the velocity of LkHα 264, that both lines have a similar *FWHM* of 20 km s^{-1} (well reproduced by a disk model) and that LkHα 264 does not have evidence for outflows, we concluded that the observed emission very likely arises from a circumstellar disk. This is the first time that the 1–0 S(1) and 1–0 S(0) H$_2$ lines are detected in a protoplanetary disk. From the line ratios, we deduced that the H$_2$ gas emitting the line is at temperature lower than 1500 K and that the gas is likely thermally excited by UV photons. Modeling of the disk around LkHα 264 and the shape of the 1–0 S(1) H$_2$ line suggests that the disk is observed close to face on ($i < 35°$) and that most of the near-IR H$_2$ emission is produced at less than 1 AU. The flux of the 1-0 S(1) H$_2$ line indicates that there is a few lunar masses of hot ($T \sim 1500$ K) optically thin H$_2$ in the inner disk

(R$\sim$0.1 AU) of LkHα 264. In the case of 49 Cet none of the three H$_2$ near-IR lines are observed in the spectra. The flux upper limits on the 1-0 S(1) H$_2$ line indicate that there is less than a fraction of lunar mass in the inner (R$<$1 AU) disk of 49 Cet. This result combined with the absence of near-IR excess in the spectral energy distribution, the lack of Hα in emission and the non-detection of mid-IR H$_2$ emission (from Spitzer), suggest that the inner disk of 49 Cet has a hole in the gas and in the dust.

These projects show how high-resolution infrared spectroscopy is a useful tool for constraining the structure of protoplanetary disks, even in the case non-detections. Future studies, in larger samples are essential to identify statistical trends between the gas and dust properties of the disk as a function of the age and multiplicity.

References

Bary, J. S., Weintraub, D. A., & Kastner, J. H. 2002, *ApJL*, 576, L73
Bary, J. S., Weintraub, D. A., & Kastner, J. H. 2003, *ApJL*, 586, 1136
Bitner, M. A., Richter, M. J., Lacy, J. H., Greathouse, T. K., Jaffe, D. T., & Blake, G. A. 2007, *ApJL*, 661, L69
Carmona, A., et al. 2008, *A&A*, 477, 839
Carmona, A., van den Ancker, M. E., Henning, T., Goto, M., Fedele, D., & Stecklum, B. 2007, *A&A*, 476, 853
Carr, J. S. 2005, in High Resolution Infrared Spectroscopy in Astronomy, Edited by H.U. Kufl, R. Siebenmorgen, and A.F.M. Moorwood. Springer-Verlag Berlin/Heidelberg, 2005, p. 203.
Chen, C. H., et al. 2006, *ApJS*, 166, 35
Chiang, E. I. & Goldreich, P. 1997, *ApJ*, 490, 368
Chiang, E. I., Joung, M. K., Creech-Eakman, M. J., Qi, C., Kessler, J. E., Blake, G. A., & van Dishoeck, E. F. 2001, *ApJ*, 547, 1077
Cohen, M., & Kuhi, L. V. 1979, *ApJS*, 41, 743
Dent, W. R. F., Greaves, J. S., & Coulson, I. M. 2005, *MNRAS*, 359, 663
Dullemond, C. P., Dominik, C., & Natta, A. 2001, *ApJ*, 560, 957
Glassgold, A. E., Najita, J. R., & Igea, J. 2007, *ApJ*, 656, 515
Hollenbach, D., et al. 2005, *ApJ*, 631, 1180
Howat, S. K. R., & Greaves, J. S. 2007, *MNRAS*, 379, 1658
Itoh, Y., Sugitani, K., Ogura, K., & Tamura, M. 2003, *PASJ*, 55, L77
Käufl, H. U. et al. 2004, *SPIE*, 5492, 1218
Lagage, P. O. et al. 2004, *The Messenger*, 117, 12.
Lahuis, F., van Dishoeck, E. F., Blake, G. A., Evans, N. J., II, Kessler-Silacci, J. E., & Pontoppidan, K. M. 2007, *ApJ*, 665, 492
Lundin, L. K. VLT VISIR Pipeline User Manual. VLT-MAN-ESO-19500-3852. 2006
Martin-Zaïdi, C., Lagage, P.-O., Pantin, E., & Habart, E. 2007, *ApJL*, 666, L117
Mouri, H. 1994, *ApJ*, 427, 777
Najita, J. R., Carr, J. S., Glassgold, A. E., & Valenti, J. A. 2007, in Protostars and Planets, V, Edited by B. Reipurth, D. Jewitt, and K. Keil, University of Arizona Press, Tucson, 2007, p. 507–522
Nomura, H., Aikawa, Y., Tsujimoto, M., Nakagawa, Y., & Millar, T. J. 2007, *ApJ*, 661, 334
Pascucci, I., et al. 2006, *ApJ*, 651, 1177
Richter, M. J., Jaffe, D. T., Blake, G. A., & Lacy, J. H. 2002, *ApJL*, 572, L161
Sako, S., Yamashita, T., Kataza, H., Miyata, T., Okamoto, Y. K., Honda, M., Fujiyoshi, T., & Onaka, T. 2005, *ApJ*, 620, 347
Sheret, I., Ramsay Howat, S. K., & Dent, W. R. F. 2003, *MNRAS*, 343, L65
Shukla, S. J., Bary, J. S., Weintraub, D. A., & Kastner, J. H. 2003, *Bulletin of the American Astronomical Society*, 35, 1209
Thi, W. F., et al. 2001, *ApJ*, 561, 1074
Weintraub, D. A., Kastner, J. H., & Bary, J. S. 2000, *ApJ*, 541, 767
Weintraub, D. A., Bary, J. S., Kastner, J. H., Shukla, S. J., & Chynoweth, K. 2005, *Bulletin of the American Astronomical Society*, 37, 1165
Zuckerman, B., Forveille, T., & Kastner, J.H. 1995, *Nature*, 373, 494

Astromineralogy of protoplanetary disks

O. Schütz[1], G. Meeus[2], M. F. Sterzik[1] and E. Peeters[3,4]

[1]European Southern Observatory, Alonso de Cordova 3107, Santiago 19, Chile
[2]Astrophysikalisches Institut Potsdam, An der Sternwarte 16, D-14482 Potsdam, Germany
[3]The University of Western Ontario, London, ON N6A 3K7, Canada
[4]SETI Institute, 515 N. Whisman Road, Mountain View, CA 94043, USA

Abstract. We review mid-infrared N-band spectra (8–13 μm) for a sample of 28 targets, obtained with the TIMMI2 camera at La Silla Observatory. The sample contains 5 FU Orionis stars, 6 Herbig Ae/Be objects, 7 T Tauri stars and 10 Vega-type main sequence objects. All targets show infrared excess, but for several the proof of circumstellar matter was lacking up to our observations. We model the N-band emission features with a mixture of silicates consisting of different grain sizes and composition, and determine the status of dust processing in these disks. While for some targets the emission spectrum resembles those of known pre-main sequence stars of evolved dust, other objects show strong isolated PAH bands but no silicate emission. For the first time we find evidence of PAH processing occurring in a T Tauri star. The Vega-type object HD 113766 exhibits highly-processed secondary generation dust, likely released by the collision of planetesimal-sized bodies. The findings of our dust analysis are set in context to previous dust studies of young stellar objects.

Keywords. Circumstellar matter, Planetary systems: protoplanetary disks, Stars: pre-main sequence, Infrared: stars

1. Introduction

Up to now a full understanding of dust evolution and grain growth in protoplanetary disks is missing. There is no global correlation of dust properties with age, not even when considering stellar spectral types (e.g. Schegerer *et al.* 2006). Therefore, the observed grain sizes in pre-main sequence disks must also depend on other factors like, e.g., turbulence and the gas content in the disk (Kessler-Silacci *et al.* 2006), while the exact dependencies thereof remain to be characterised.

This work describes mid-infrared (MIR) spectra of circumstellar disks. We selected targets with infrared excess that have not yet been discussed conclusively in the literature, i.e. objects which were suspected disk candidates, but for which no MIR spectroscopy had been obtained yet. In Schütz *et al.* (2005a), hereafter Paper I, we analysed silicate processing in a sample of pre-main sequence stars, while Schütz *et al.* (2005b), hereafter Paper II, focuses on debris disks and their collisional dust. A mixed target sample is discussed in Schütz *et al.* (2008), hereafter Paper III.

In this contribution we review our previous target samples, summarised in Table 1, and describe observed trends in the dust properties for the pre-main sequence objects. For details on all sources, as well as for a description of the observations and data reduction, we refer to Paper I–III.

2. Data analysis

The data were obtained in the years 2002–2006 with the ESO mid-infrared camera TIMMI2 at La Silla Observatory. One object, HD 34700, is also observed with the IRS on board the Spitzer Space Telescope.

Table 1. Our target sample. Spectral types are obtained from SIMBAD, or the literature if a revised classification exists. V-band magnitudes and IRAS 12 µm fluxes are from SIMBAD. Some targets are multiple systems or variable. Stellar ages and distances are from the literature.

Object	Class	Spectral Type	V [mag]	$F_{12\mu m}$ [Jy]	Age [Myr]	d [pc]
HD 3003	Vega	A0V	5.07	0.48	50	47
HD 10647	Vega	F8V	5.52	0.82	300–3500	17.4
HD 34282	HAeBe	A0e	9.85	0.70	5–10	400
HD 34700	TTS	G0IVe	9.15	0.60	–	–
HD 38678	Vega	A2Vann	3.55	2.18	230	21.5
HD 72106	HAeBe	A0IV	8.50	2.22	10	290
HD 80951	Vega	A1V	5.29	0.44	–	220
HD 98800	TTS	K5 / K7 / M1 V	9.11	1.98	8.5	47.6
HD 109085	Vega	F2V	4.31	2.18	~1000	18
HD 113766	Vega	F3 / F5 V	7.56	1.59	16	130
HD 123356	Vega	G1V	10.0 / 12.2	1.36	–	–
HD 143006	TTS	G5Ve	10.21	0.86	5	82
HD 155555	TTS	G5IV / K0IV-V	6.88	0.69	–	32
HD 172555	Vega	A5IV-V	4.78	1.47	12	29.2
HD 181296	Vega	A0Vn	5.03	0.54	12	48
HD 190073	HAeBe	A2IIIpe	7.82	7.16	1.2	> 290
HD 207129	Vega	G0V	5.58	0.81	6000	15.6
HD 319139	TTS	K5Ve / K7Ve	10.44	0.45	5.5	83
BBW 76	FUOR	G0–G2 I	~12	1.02	–	1800
CD-43 344	TTS (?)	M2	9.42	3.73	–	–
FU Ori	FUOR	G0II	8.94	5.95	0.3	450
KK Oph	HAeBe	A6Ve / G6Ve	9.4–12.9	9.87	7	160
MP Mus	TTS	K1IV	10.32	0.88	8	86
PDS 144 N	HAeBe	A2IV	14.2	–	–	~1000
PDS 144 S	HAeBe	A5V	13.1	–	–	~1000
V 346 Nor	FUOR	–	16.3	7.50	–	700
V 883 Ori	FUOR	–	~15	52.5	–	460
Z CMa	FUOR	–	9–11	126.6	0.3	930

2.1. *Silicates*

To derive the composition of the circumstellar dust, we first determine a local continuum to our TIMMI2 spectra, by fitting a blackbody to the 8–13 µm region. Subsequently, we model the spectra with a linear combination of emission features from the following dust species which are commonly found in disks of pre-main sequence stars:

- Amorphous olivine ($[Mg,Fe]_2SiO_4$)
- Amorphous pyroxene ($[Mg,Fe]SiO_3$)
- Crystalline silicates: magnesium forsterite (Mg_2SiO_4) and enstatite ($MgSiO_3$)
- Silica (SiO_2).

Van Boekel et al. (2005) showed that the size distribution of grains radiating in the 10 µm region can be represented by two grain sizes: "small" ($r_V = 0.1$ µm) and "large" ($r_V = 2.0$ µm in our model), with the given volume-equivalent radii r_V. We used absorption coefficients of van Boekel et al. (2005) and Bouwman (private communication). For

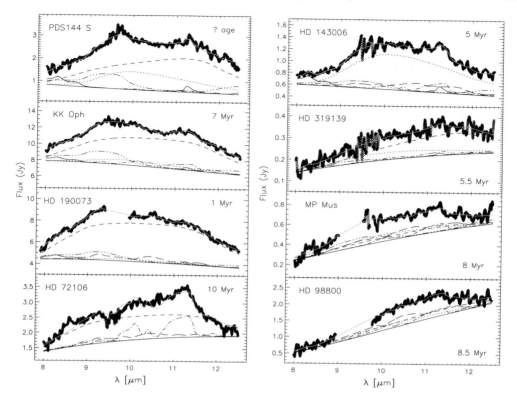

Figure 1. Decomposition of the silicate emission for the Herbig (left) and T Tauri stars (right) in our sample. The different linestyles represent small amorphous silicates (*dotted*), large amorphous silicates (*dashed*), silica (SiO_2, *dash-three-dots*), crystalline forsterite (*dash dot*) and crystalline enstatite (*long dashes*). The very thick black curve corresponds to the observed spectrum, including – for some targets – noise features. A summary of all silicate components is given by the grey curve. The targets are sorted according to the apparent degree of dust processing.

the dust decomposition we perform a χ^2-fit with 10 free parameters (the mass fractions of above 5 silicate types, with 2 grain sizes each). Resulting model spectra, together with the contribution of each dust type, are shown in Figs. 1–2. While we model the spectra with 10 (grain and size) dust types, we only plot six components for a better visibility. In particular, we added amorphous olivine and pyroxene of 0.1 μm and plot them as "small amorphous silicates". Similarly, "large amorphous silicates" contain the 2.0 μm pyroxene and 2.0 μm olivine grains. For silica, forsterite and enstatite, we each plot the sum of the 0.1 and 2.0 μm grains.

For a discussion of the FU Orionis targets (FUORs) we refer to Paper I and III. These sources show the silicate feature either in emission or absorption. Given the large number of FUORs with silicate absorption features, which cannot be explained only by geometry effects (pole-on vs. edge-on view to the disk), Quanz *et al.* (2007) defined two categories of FUORs: objects with the silicate feature in absorption are likely still embedded in a circumstellar envelope, while the silicate band in emission is thought to originate from the surface layer of their accretion disks, similar as for the Herbig and T Tauri stars.

2.2. PAHs

A key result in observational studies of PAH bands is that their profiles show clear variations in terms of peak positions, profile shape and relative intensities (Peeters et al. 2004, and references therein). Based upon their profile, PAHs have therefore been classified into categories A, B and C. These classes also depend on the object type and evolutionary status.

We find PAH bands towards the Herbig stars HD 34282, PDS 144 N and the T Tauri HD 34700 (see Fig. 3). Their appearance is rather expected for Herbig objects and commonly explained with the excitation of PAH molecules by UV photons. However, Li & Draine (2002) and Mattioda et al. (2005) have shown that PAH molecules around cool stars can also be excited.

The PAH band profiles of the T Tauri star HD 34700 are clearly unique and show that PAH processing and/or formation has occurred in this source. This is similar to PAH observations towards isolated Herbig stars and suggests that similar PAH processing occurs in protoplanetary disks around low-mass and intermediate-mass stars.

3. Trends observed with stellar, disk and silicate properties

We analyse trends in the dust properties for the pre-main sequence targets, as the dust in Vega-type objects is of collisional nature and would not correlate with the dust properties of the younger stars. The silicate mass fractions of each dust component show no clear relation with stellar parameters. Instead we use the silicate shape, i.e. the ratio of the fluxes at 11.3 and 9.8 μm in the continuum normalised spectra. The silicate shape

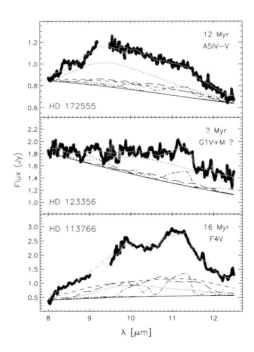

Figure 2. Decomposition of the silicate emission for the Vega-type sources, sorted according to the apparent degree of dust processing. Note the strength in the dust emission of HD 113766 and the diversity between this source and HD 172555, despite the similar age.

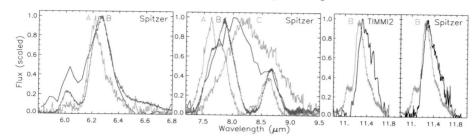

Figure 3. Comparison of the PAH profiles towards HD 34700 with the PAH classes A–C. Observations of HD 34700 are shown in thin black, and the PAH classes in grey shades. The PAH bands of this T Tauri star are clearly unique and point to ongoing PAH processing.

can be seen as an indicator for the grain size composition and the evolutionary dust status (Bouwman et al. 2001).
- Silicate shape vs. peak strength: Our sample is lacking sources of very pristine dust, but nevertheless shows the same trend as found in previous works (van Boekel et al. 2005, Kessler-Silacci et al. 2007): a decrease of the silicate peak strength is observed with higher dust processing status (i.e. with a larger 11.3/9.8 ratio).
- Kessler-Silacci et al. (2007) have shown a correlation of the silicate shape with stellar luminosity, in form of an apparent *higher* dust evolution with lower luminosity. This trend can be understood when considering that the 10 μm silicate emission region lies further inward for stars with lower luminosity, while resolved observations (e.g. van Boekel et al. 2004) have shown that the degree of dust processing increases towards the star. Our sample is much smaller, but suggests a similar trend with luminosity.
- We see no correlation of the dust and disk evolution with stellar binarity in our sample, i.e. whether the time spans of dust processing and disk lifetime in binary systems would differ from those of isolated stars with the same age. This non-correlation is also affected by a rather imprecise age determination for some targets.
- With the evolution of protoplanetary into debris disks, the IR emission from the inner disk decreases. This results in a falling 2–25 μm slope with progressing disk evolution. An eventual dependence of silicate emission with inner disk evolution should be recognisable as a correlation between silicate strength and SED slope. Sicilia-Aguilar et al. (2007) noticed a very weak to absent correlation, to which our data agree.

We refer to Paper III for more details about these trends and correlations.

4. Conclusion

By mid-infrared spectroscopy and modelling of the 10 μm emission feature we characterised circumstellar disks, detected new ones and found the first T Tauri star which shows processing of PAH – HD 34700. For our pre-main sequence targets we confirm a correlation of the spectral silicate shape with decreasing silicate peak strength and with decreasing stellar luminosity, but we find no clear dependence of dust processing with age.

References

Bouwman, J., Meeus, G., de Koter, A., et al. 2001, A&A, 375, 950
Li, A. & Draine, B. T. 2002, ApJ, 572, 232
Kessler-Silacci, J. E., Augereau, J.-C., Dullemond, C. P., et al. 2006, ApJ, 639, 275
Kessler-Silacci, J. E., Dullemond, C. P., Augereau, J.-C., et al. 2007, ApJ, 659, 680
Mattioda, A. L., Hudgins, D. M., & Allamandola, L. J. 2005, ApJ, 629, 1188

Peeters, E., Allamandola, L. J., Hudgins, D. M., Hony, S., & Tielens, A. G. G. M. 2004, Astrophysics of Dust, ASP Conference Series, Vol. 309, Edited by Adolf, N. Witt, Geoffrey, C. Clayton, and Bruce T. Draine, p. 141
Quanz, S. P., Henning, Th., Bouwman, J., et al. 2007, ApJ, 668, 359
Schegerer, A., Wolf, S., Voshchinnikov, N. V., Przygodda, F., & Kessler-Silacci, J. E. 2006, A&A, 456, 535
Schütz, O., Meeus, G., & Sterzik, M. F. 2005a, A&A, 431, 165
Schütz, O., Meeus, G., & Sterzik, M. F. 2005b, A&A, 431, 175
Schütz, O., Meeus, G., Sterzik, M. F., & Peeters, E. 2008, A&A, submitted
Sicilia-Aguilar, A., Hartmann, L. W., Watson, D., et al. 2007, ApJ, 659, 1637
van Boekel, R., Min, M., Leinert, Ch., et al. 2004, Nature, 432, 479
van Boekel, R., Min, M., Waters, L. B. F. M., et al. 2005, A&A, 437, 189

Dust evolution in protoplanetary disks

Jean-François Gonzalez[1], Laure Fouchet[2], Sarah T. Maddison[3] and Guillaume Laibe[1]

[1] Université de Lyon, Lyon, F-69003, France; Université Lyon 1, Villeurbanne, F-69622, France;
CNRS, UMR 5574, Centre de Recherche Astrophysique de Lyon,
École Normale Supérieure de Lyon, 46 allée d'Italie, F-69364 Lyon cedex 07, France
email: Jean-Francois.Gonzalez@ens-lyon.fr, Guillaume.Laibe@ens-lyon.fr

[2] Department of Physics, ETH Zurich, CH-8093 Zurich, Switzerland
email: fouchet@phys.ethz.ch

[3] Centre for Astrophysics and Supercomputing, Swinburne University of Technology,
PO Box 218, Hawthorn, VIC 3122, Australia
email: smaddison@swin.edu.au

Abstract. We investigate the behaviour of dust in protoplanetary disks under the action of gas drag using our 3D, two-fluid (gas+dust) SPH code. We present the evolution of the dust spatial distribution in global simulations of planetless disks as well as of disks containing an already formed planet. The resulting dust structures vary strongly with particle size and planetary gaps are much sharper than in the gas phase, making them easier to detect with ALMA than anticipated. We also find that there is a range of masses where a planet can open a gap in the dust layer whereas it doesn't in the gas disk. Our dust distributions are fed to the radiative transfer code MCFOST to compute synthetic images, in order to derive constraints on the settling and growth of dust grains in observed disks.

Keywords. planetary systems: protoplanetary disks – hydrodynamics – methods: numerical

---

1. Introduction

Dust grains in disks around young stars are thought to be the building blocks of planets (Dominik *et al.* 2007). They grow via sticking in low-velocity collisions from (sub)micron sizes up to decimetric sizes. The higher collision velocities of larger grains makes them shatter upon impact and prevents them from growing further. Yet we know that they must reach planetesimal sizes. How they overcome the decimetric barrier is the subject of much debate. Part of the solution to this problem could be found in the reduction of collision velocities via the increase of the dust layer density. This can be achieved by the settling of dust to the disk midplane.

Dust is also essential to the interpretation of disk observations. Indeed, the dust opacity largely dominates that of the gas, except in particular lines of molecular gas, which are more difficult to detect than dust thermal emission. Therefore, most images of disks trace the spatial distribution of dust, not that of the gas. Both components are often assumed to be well mixed, with a uniform gas-to-dust mass ratio of 100, but, as discussed in Sect. 2, this is not always the case and one has to be careful when deriving gas distributions from observations of dust in circumstellar disks.

We present here our work on dust evolution in protoplanetary disks using 3D SPH simulations. We treat vertical settling and radial migration of dust in planetless disks in Sect. 2 and present the resulting synthetic images and comparison to observations in Sect. 3. We describe the variety of structures of disks with gap-forming planets in Sect. 4 and our first treatment of grain growth in Sect. 5.

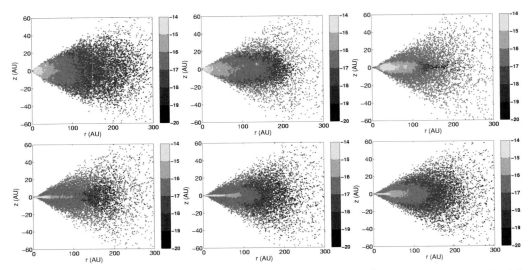

Figure 1. Gas (grey) and dust (colour represents $\log \rho_{\mathrm{dust}}$, in $\mathrm{g\,cm^{-3}}$) distributions in our CTTS disk viewed edge-on. Dust grain sizes are, from left to right and top to bottom: 10 μm, 100 μm, 1 mm, 1 cm, 10 cm, 1 m.

2. Settling and migration

In a non self-gravitating disk, gas feels the gravity of the central star and its own pressure gradient. As a result, it will orbit at sub-keplerian velocities. On the other hand, dust grains do not interact with each other and only feel the star's gravity: they have keplerian orbits. In a disk with both components, this velocity difference gives rise to aerodynamic drag, which slows down the dust and makes it migrate radially towards the central star. In the vertical direction, drag makes the dust settle towards the midplane.

Analytical studies of the radial migration of solid particles have been done by Weidenschilling (1977) and of their vertical settling by Garaud et al. (2004). In order to simultaneously study both effects, we developed a 3D, bi-fluid (gas+dust), SPH code to model vertically isothermal, non self-gravitating dusty disks (Barrière-Fouchet et al. 2005). The two inter-penetrating phases representing gas and dust interact via aerodynamic drag. Grain growth is not taken into account at this point (see Sect. 5).

We ran simulations for a 0.01 $M_\odot$ disk in orbit around a 1 $M_\odot$ star, representative of a classical T Tauri star (CTTS) disk, and composed of 99% gas and 1% dust in mass. We obtained the resulting dust spatial distributions for a set of grain sizes ranging from 1 μm to 10 m (Barrière-Fouchet et al. 2005), a selection of which is illustrated in Fig. 1.

We found that for small sizes (1–10 μm), with the strongest drag force, the dust is so strongly coupled to the gas that it follows its motion and both components are well mixed. On the other hand, the drag force for large grain sizes (1–10 m) is weak and the dust is almost insensitive to the gas, it occupies the whole disk again. It is for the median sizes (100 μm–10 cm) that all the interesting dynamics happens and the gas drag has a strong influence on the resulting distributions, showing important vertical settling in the inner regions and depletion in the outer disk due to inwards radial migration, with varying strength depending on grain size.

We would like to point out that the efficiency of both processes depends not only on grain size but also on the nebula parameters, via the densities. Therefore, the size of the fastest inwards migrating grains is not universal. It is commonly thought to always be of the order of 1 m, as found by Weidenschilling (1977) for the minimum mass solar nebula

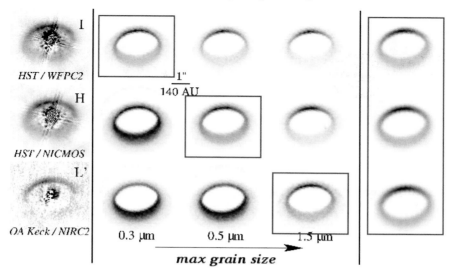

Figure 2. Observations (left panel) and synthetic images for models without (center panel) and with dust settling (right panel) of GG Tau's circumbinary ring in the I, H and L' bands.

(MMSN), whereas for our CTTS disk, we find it to be smaller, around 1 mm to 1 cm, and to depend on the position in the disk.

3. Synthetic images

Numerous scattered light images of protoplanetary disks are now available (Watson et al. 2007). Light scattering depends on grain size and, because of the wavelength dependency of opacity, observations at different wavelengths reveal different depths in the disk. As a result, multi-wavelength observations in scattered light constitute a probe of dust settling.

In order to link our hydrodynamical simulations to observations of disks, we used our resulting dust distributions as an input to MCFOST, a continuum 3D Monte-Carlo radiative transfer code (Pinte et al. 2006). It can produce synthetic scattered light images, as well as spectral energy distributions (SEDs) and polarization maps. We found that dust settling and migration cause wavelength variations of the disk apparent size and of the dark lane width, in agreement with observations. They also affect SEDs, causing a change of slope between 10 and 200 μm. However, these effects are small, and a single observation would not give strong indications on the amount of settling. It is with a combination of multi-wavelength, multi-technique observations that one can hope to efficiently constrain the dust distribution.

This is beautifully illustrated in the case of GG Tau. The left panel of Fig. 2 shows three scattered light images of its circumbinary ring observed in the I, H, and L' bands. The middle panel displays the corresponding synthetic images for three models without dust settling, therefore with gas and dust well mixed in the disk, for which the dust size distribution (with the same slope as that of the interstellar medium), is truncated at a maximum size of 0.3, 0.5 and 1.5 μm, from left to right. No single model is able to simultaneously reproduce the observations in all three bands, and in particular the contrast between the front and back sides of the ring, but each one seems appropriate for a single band. This points towards a variation of the dust size distribution with depth in the disk. The right panel of Fig. 2 shows synthetic images produced from new

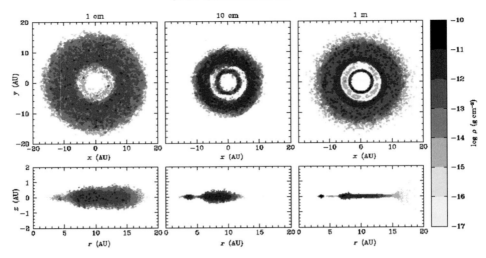

Figure 3. Gap created in the dust phase of the MMSN disk for 1-cm, 10-cm and 1-m grains, from left to right, by a 1 M_J planet at 5.2 AU. Top panel: face-on view, bottom panel: edge-on view.

hydrodynamical simulations of GG Tau's circumbinary ring including dust settling and migration. Without any fit to the data, the agreement with observations is now much better, showing that dust had indeed settled in this ring. Polarization data comforting this result, as well as more detailed azimuthal brightness profile comparisons of observations and models, are presented in Pinte et al. (2007). The remaining discrepancies are likely due to missing physics in our simulations, among which grain growth.

4. Planetary gaps

In a disk containing a planet, gravitational perturbations of the planet cause density waves, leading to an exchange of angular momentum with the disk. The planet pushes the disk exterior (resp. interior) to its orbit further out (resp. in), therefore opening a gap. The gap is sustained if there is an equilibrium between tidal torques, acting to clear the gap, and viscous torques, acting to fill it. The observation of such a gap in a disk would provide a clear indication of the presence of a planet. Simulations of gaps in 2D gas disks by Wolf & D'Angelo (2005) showed that a 1 M_J planet can open a gap that will be detectable by ALMA at sub-mm wavelengths up to a distance of 100 pc. In 2D simulations of dusty disks, Paardekooper & Mellema (2006) found that a smaller planet mass is needed to open a gap in the dust phase than in the gas.

We ran new 3D simulations of dusty disks with an embedded planet for two configurations, the CTTS disk we modeled previously and a MMSN disk, in which we vary the grain size and planet mass. We found that gap formation is much more rapid and striking in the dust layer rather than in the gas disk, and even more so than in 2D simulations (Maddison et al. 2007; Fouchet et al. 2007). Additionally, the structures caused in the dust phase depend strongly on the grain size, as illustrated in Fig. 3, due to varying drag and radial drift velocity throughout the disk.

Gaps are more prominent as the planet mass increases. Since they are also much deeper in the dust than in the gas, much lighter planets are needed to open one in the dust phase, the one visible in most disk observations, than in the gas phase. In our MMSN models, we found that a 0.2 M_J planet can open a gap in the dust layer while only a small

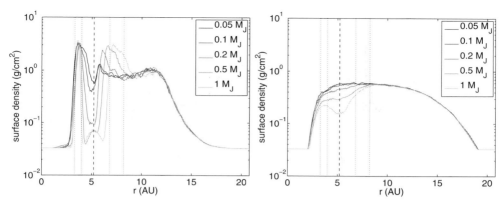

Figure 4. Azimuthally averaged surface density profile of the MMSN disk for the dust (1-m boulders, left) and the gas (right) with a planet of varying mass at 5.2 AU. The gas density is divided by 100 for direct comparison to the dust density.

surface density dip is observed in the gas (see Fig. 4). Lighter planets down to 0.05 M_J create density dips in the dust, with steep gradients on both sides that should be visible by ALMA (Varnière *et al.* 2006). This suggests that planet gaps will be much easier to detect, with a lower threshold on planet mass, than previously anticipated.

Since the influences of grain size, affecting the disk thickness and radial profile (Fig. 3) and of planet mass, affecting the gap structure (Fig. 4) are different (see also Fouchet *et al.* 2007 for more detail), future observations of planet gaps will therefore allow to constrain the dust grain size distribution in addition to the planet mass.

Gaps can even play a role in planet formation. Because of a very efficient vertical settling, the volume density of the dust reaches that of the gas in several regions of the disk, and even exceeds it at the gap inner edge (see Fig. 5). This affects the dynamics in that region and can favour planetesimal growth, ultimately leading to the formation of a second planet in the vicinity of the first one.

5. Grain growth

We recently started to add grain growth in our code by implementing a scheme able to treat the variation of grain sizes via an analytical prescription. We tested it with the simple model of Stepinski & Valageas (1997), who grow solid particles made of water ice by sticking, without taking shattering into account. We found that grain growth is very efficient, especially in the denser inner regions of the disk, but is too fast, as can be expected (Dullemond *et al.* 2005). For details, see the poster by Laibe (this volume) and references therein. More extensive tests are presented in Laibe *et al.* (2008).

6. Conclusion

Our results demonstrate that, with or without an embedded planet, gas and dust are not well mixed in protoplanetary disks. The density enhancements we find in the dust component shape their observed images, and help solid particles to aggregate.

Similarly to our work on GG Tau, we have started to compute synthetic images of disks with gaps in order to make better quantitative predictions of what ALMA will be able to detect. Gas dispersal, not taken into account here, can lower the drag force and alter the dust distribution, and should be considered when interpreting observations.

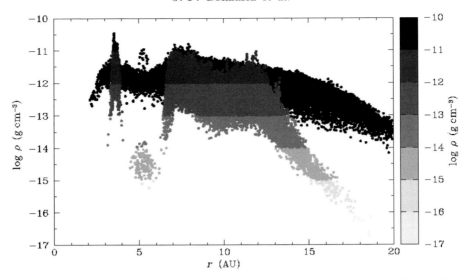

Figure 5. Volume density profile of the gas (black) and dust (colour) in the MMSN disk with a 1 M_J planet at 5.2 AU for 1-m boulders.

Now that we have implemented in our code a mechanism to treat grain growth, we are working on an improved and more realistic physical model, taking into account detailed microscopic processes, as well as shattering, which produces small grains via collisional cascade. Our goal is to be able overcome the decimetric barrier. Our approach is different, but complementary to the one of Johansen et al. (2007), who find a concentration of solid particles in transient high pressure regions.

A physically consistent model of grain growth needs a proper implementation of turbulence in the gas, because it determines the kinematics of grains. As it is unfeasible to include a full treatment of turbulence in our code, we started working on a simplified, but realistic description.

References

Barrière-Fouchet, L., Gonzalez, J.-F., Murray, J. R., Humble, R. J., & Maddison, S. T. 2005, A&A, 443, 185
Dominik, C., Blum, J., Cuzzi, J. N., & Wurm, G. 2007, in: B. Reipurth, D. Jewitt & K. Keil (eds.), *Protostars and Planets V* (Tucson: University of Arizona Press), p. 783
Dullemond, C. P. & Dominik, C. 2005, A&A, 434, 971
Fouchet, L., Maddison, S. T., Gonzalez, J.-F., & Murray, J. R. 2007, A&A, 474, 1037
Garaud, P., Barrière-Fouchet, L., & Lin, D. N. C. 2004, ApJ, 603, 292
Johansen, A., Oishi, J. S., Mac Low, M.-M., et al. 2007, Nature, 448, 1022
Laibe, G., Gonzalez, J.-F., Fouchet, L., & Maddison, S. T. 2008, in preparation
Maddison, S. T., Fouchet, L., & Gonzalez, J.-F. 2007, Ap&SS, 311, 3
Paardekooper, S.-J. & Mellema, G. 2006, A&A, 453, 1129
Pinte, C., Ménard, F., Duchêne, G., & Bastien, P. 2006, A&A, 459, 797
Pinte, C., Fouchet, L., Ménard, F., Gonzalez, J.-F., & Duchêne, G. 2007, A&A, 469, 963
Stepinski, T. F. & Valageas, P. 1997, A&A, 319, 1007
Varnière, P., Bjorkman, J. E., Frank, A., et al. 2006, ApJ, 637, L125
Watson, A. M., Stapelfeldt, K. R., Wood, K., & Ménard, F. 2007, in: B. Reipurth, D. Jewitt & K. Keil (eds.), *Protostars and Planets V* (Tucson: University of Arizona Press), p. 523
Weidenschilling, S. 1977, MNRAS, 180, 57
Wolf, S. & D'Angelo, G. 2005, ApJ, 619, 1114

Origin of the dusty disks around white dwarfs

R. B. Dong[1], Y. Wang[1], D. N. C. Lin[2,3] and X. W. Liu[1,2]

[1] Department of Astronomy, Peking University,
100871, Beijing, China,
email: rbdong@pku.edu.cn

[2] Kavli Institute of Astronomy and Astrophysics, Peking University,
100871, Beijing, China
email: liuxw@bac.pku.edu.cn

[3] Department of Astronomy and Astrophysics, University of California,
Santa Cruz, CA 95064, USA
email: lin@ucolick.org

Abstract. Some circumstantial evidence for residual planetesimals is constructed based on the recent discovery of a dusty ring around a young white dwarf at the center of the Helix nebula (Su et al. 2007). This ring extends between about 35 and 150 AU from the nebula center, and have a total mass of about 0.13 $M_\oplus$. In this paper we propose that this ring is the by-product of planets and planetesimals' orbital evolution during the epoch when the central star rapidly lost most of its mass. We examine the dynamical evolution of planetary systems similar to the solar system (i.e. with gas giant planets and residual planetesimals) as their host stars evolve off the main sequence. During the process, some planetesimals will be captured by the gas giants into mean motion resonances and their mutual collisions will form a dust ring similar to that observed at the center of the Helix nebula.

Keywords. Giant planet; planetary formation; debris disks; planet dynamics; n-body simulation

1. The dynamical model

We idealize the dynamics of the system into a series of interaction between three populations of objects: planetesimals with mass m_j, planets with mass M_j, and a central star with mass M_*.

There are two mean factors which influence the orbital change of a planet around the central star during the epoch of its mass loss: the change of the star's mass, which reduces its gravity and causes the planet's orbit to expand; and the gas drag by the outflow, which induces orbital decay. Even though the magnitude of the gas drag on a giant planet is negligible, it has a significant effect on the orbit of the planetesimals, which are also perturbed by the gravity of the planet.

But the drag effect has a significant effect on the orbit of the planetesimal. In addition, it is also perturbed by the gravity of the planet. In a non rotating frame centered on the star, the equation of motion of a planetesimal at a position $\mathbf{r}$ can be expressed as

$$\frac{d^2 \mathbf{r}_i}{dt^2} = -\frac{GM_* \mathbf{r}_i}{r_i^3} - \Sigma_j \frac{GM_j(\mathbf{r}_i - \mathbf{r}_j)}{|\mathbf{r}_i - \mathbf{r}_j|^3} - \Sigma_j \frac{GM_j \mathbf{r}_j}{r_j^3} - \frac{\mathbf{f}_i}{m} \qquad (1.1)$$

where $\mathbf{r}_i$ is the position of the planetesimal, $\mathbf{r}_j$ is the position of the planet and M_* is a function of time. The magnitude of the drag force on a planetesimal by the wind of its host star can be expressed as

$$\mathbf{f}_i = \pi \rho_g (\dot{\mathbf{r}}_\mathbf{i} - \mathbf{v}_\mathbf{g})|\dot{\mathbf{r}}_\mathbf{i} - \mathbf{v}_\mathbf{g}|(\mathbf{s}^2 + \mathbf{s}_\mathbf{c}^2). \tag{1.2}$$

where v_g is the wind velocity, $s = (3m_i/4\pi\rho_p)^{1/3}$ and $s_c = Gm_i/v_p^2$ are the physical and gravitational radius of a planetesimal with a mass m and density ρ_p. In the radial direction, this drag contribution is much smaller than the gravity from all three bodies. However, in the azimuthal direction, the drag slows down the planetesimals' orbit and induces them to migrate inward. The density of the background gas is determined by the stellar mass loss rate

$$\dot{M}_* = 4\pi R^2 \rho_g v_g \tag{1.3}$$

The equation of motion is solved with a Hermit scheme which is kindly provided by Dr Sverre Aarseth.

2. Simulations of orbital evolutions and collisions between planets and planetesimals

Simulations based on the dynamical model above show there are three types of orbital evolution of planets and planetesimals during the mass lose phase of the red giant progenitor. Regarding the initial conditions of the simulations, we mostly consider a star with a main-sequence mass in the range $2M_\odot < M_* < 4M_\odot$ (A-F stars). thought to be the progenitor of the $\sim 0.5 M_\odot$ white dwarf at the center of the Helix nebula. The analysis to be presented below is particularly relevant for the planetary nebula phase when the typical mass loss rate is $\dot{M}_* \sim 10^{-4} M_\odot$ yr^{-1}. The typical outflow speed is ~ 10 km s^{-1} and $\dot{M}_*$ may vary on the time scale of 10^4 yr.

Results of our simulations show that the orbits of the gas giant planets and super-km-size planetesimals expand adiabatically. The gas drag is insignificant in their orbital evolutions, so their semi major axis are in inverse ratio with the mass of central star. On the other hand, if a planetesimal is smaller than a critical size (about 10 m in our simulation), orbital decay will be introduced due to gas drag; this reduced the particle's angular momentum resulting in a non-adiabatic orbital evolution that makes the particle migrate toward the central star.

Massive planets affect the orbital evolution of intermediate-size planetesimals via trapping at mean motion resonances, where the period of the planetesimal and the period of the planet are in a ratio of two integer numbers. The final orbital semi-major axis of the captured particles does not depend on their initial positions. These intermediate-size planetesimals also acquire modest eccentricities around 0.2, which are determined by the mass of the perturbing planet and the stellar mass loss rate.

When the mass loss rate eventually decreases, a large number of planetesimals would be swept from various starting orbits into several mean motion resonances (usually two or three), with am modest eccentricity. Collisions between them would take place and a dust ring would form. For example, a 100 $M_\oplus$ planetesimal belt with objects 100 m in size, located between 50–150 AU around a 0.58 $M_\odot$ star, results in a collision frequency of 10^{-8} per year. During 10^5 years, the probability of collision for a given planetesimal is 10^{-3}, resulting in the formation of a dust cloud 0.1 $M_\oplus$ in mass, similar to that observed around the Helix nebula.

We thank Sverre Aarseth for provide the Hermit integrator and Y. H. Chiu for useful conversation. This work is supported by NASA (NAGS5-11779, NNG04G-191G, NNG06-GH45G), JPL (1270927), NSF(AST-0507424, PHY99-0794)

References

Adachi, I., Hayashi, C. & Nakazawa, K. 1976, *Prog. of Theo. Phys.*, 56, 1756
Artymowicz, P. 1993, *ApJ*, 419,155
Bodenheimer, P., Lin, D. N. C. & Mardling, R. A. 2001 ,*ApJ*, 548, 466
Bodenheimer, P., & Pollack, J.B. 1986, *Icarus*, 67,391
Burrows, A. *et al.* 2000, *ApJ*, 534,L97
Bryden, G. *et al.* 2000, *ApJ*, 540, 1091
Dobbs-Dixon, I., Li, S.L., Lin, D. N. C. 2007, Tidal Barrier and the Asymptotic Mass of Proto Gas-Giant Planets, arXiv:astro-ph/0701269.
Frank, J., King, A. & Raine, D. 2002, Accretion power in Astrophysics, Cambridge university press.
Franklin, F. A. *et al.* 1980, *Icarus*, 42, 271
Garaud, P. & Lin, D. N. C. 2007, *ApJ*, 654, 606
Goldreich, P., Lithwick, Y. & Sari, R. 2004, *ApJ*, 614, 497
Goldreich P.& Tremaine, S. 1978, *ApJ*, 222, 850
Goldreich P.& Tremaine, S. 1980, *ApJ*, 241, 425
Greenzweig, Y., & Lissauer, J. J. 1992, *Icarus*, 100, 440
Guillot, T., Gautier, D., & Hubbard, W. B. 1997, *Icarus*, 130, 534
Hasegawa, M., Nakazawa, K. 1990, *A&A*, 227, 619
Hayashi, C., Nakazawa, K. & Nakagawa, Y. 1985, in Protoplanets and Planets II ed. D. C. Balck & M. S. Mathew (Tucson:Univ. Arizona Press), 1100
Hayashi, C. 1981, *Prog. Theor. Phys. Suppl.*,70,35
Hayashi, C., Nakazawa, K., & Adachi, I. 1977, *Publ. Astron. Soc. Japan.* 29, 163
Hubickyj, O., Bodenheimer, P., & Lissauer, J. J. 2005, *Icarus*, 179, 415
Ida, S. & Lin, D. N. C. 1996, *AJ*, 112,1239
Ida, S. & Lin, D. N. C. 2004, *ApJ*, 604,388
Ida, S. & Lin, D. N. C. 2007, submitted to *ApJ*
Ikoma, M., Nakazawa, K., Emori, H. 2000, *ApJ* 537, 1013
Inaba, S. & Ikoma, M. 2003, *A&A*, 410, 711
Kokubo, E. & Ida, S. 1996, *Icarus*, 123, 180
Kokubo, E. & Ida, S. 2000, *Icarus*, 143, 15
Kokubo, E. & Ida, S. 2002, *ApJ*, 581, 666
Koller, J., Li, H., & Lin, D. N. C 2003,*ApJ* 596,L91
Kominami, J., & Ida, S. 2002, *Icarus* 157, 43
Laughlin, G., Steinacker, A. & Adams, R. 2004, *ApJ*, 608, 489
Levison, H. F. & Duncan, M. J. 1994, *Icarus* 108, 18
Lin, D. N. C, Bodenheimer, P. & Richardson, D. C. 1996, *Nature*, 380, 606
Lin, D. N. C., Ida, S. 1997 *ApJ* 477, 781
Lin, D.N.C, & Papaloizou, J. C. B. 1979, *MNRAS*, 186, 799
Lin, D. N. C, & Papaloizou, J. C. B. 1986, *ApJ*, 309, 846
Lin, D. N. C, & Papaloizou, J. C. B. 1993, in: E. H. Levy & J. I. Lunine (eds.), *Protostars and Planets III*, (Tucson: Unv. Arizona), p. 749
Leinhardt, Z. & Richardson, D. C. 2005, *ApJ*, 625, 427
Lissauer, J. J. 1987, *Icarus*, 69, 249
Marcy, G., *et al.* 2005, *Prog. Theor. Phys. Suppl.*, 158, 24
Mizuno, H. 1980, *Prog. Theor. Phys.*, 64, 544
Murray, C. D., & Dermott, S. F. 1999, Solar system Dynamcis, Cambridge Unversity.
Nelson, R. P. & Papaloizou, J. C. B. 2003, *MNRAS*, 339, 993
Palmer, P. L., Lin, D. N. C. & Aarseth, S. J. 1993, *ApJ*, 403, 336
Perri, F. & Cameron, A. G. W. 1974, *Icarus* 22,416
Pollack, J. B. *et al.* 1996, *Icarus*, 124, 62
Rafikov, R. R. 2001, *AJ*, 122, 2713
Rice, W. K. M. & Armitage, P. J. 2003, *ApJ*, 598 , L55

Safronov, V.S. 1969, Evolution of the Protoplanetary Cloud and Formation of the Earth and Planets (Moscow: Nauka)
Sato, B. *et al.* 2005, *ApJ*, 633, 465
Saumon, D. & Guillot, T. 2004, *ApJ*, 609, 1170
Shakura, N.I. & Sunyaev, R.A. 1973, *AAp*, 24,337
Tanaka, H. & Ida, S. 1999, *Icarus*, 139,350
Tanaka, H., Takeuchi, T. & Ward, W. R. 2002, *ApJ*565, 1257
Ward, W. R. 1986, *Icarus*, 67,164
Ward, W. R. 1989, *ApJ*, 336,526
Ward, W. R. 1997, *Icarus*, 126,261
Wetherill, G. W & Stewart, G. R. 1989, *Icarus*, 77, 330
Wuchterl, G., Guillot, T., & Lissauer, J. J. 2000, in Protostars and Planets IV eds. V. Mannings, A.P. Boss & S. S. Russell (Tucson : Univ. Arizona Press), 1081
Zhou, J. L., Lin, D. N. C.& Sun Y. S. 2007,*ApJ*, 666,423.

Exoplanets: Detection, Formation and Dynamics
Proceedings IAU Symposium No. 249, 2007
Y.-S. Sun, S. Ferraz-Mello and J.-L. Zhou, eds.

3D SPH simulations of grain growth in protoplanetary disks

Guillaume Laibe[1], Jean-François Gonzalez[1], Laure Fouchet[2] and Sarah T. Maddison[3]

[1]Université de Lyon, Lyon, F-69003, France; Université Lyon 1, Villeurbanne, F-69622, France;
CNRS, UMR 5574, Centre de Recherche Astrophysique de Lyon,
École Normale Supérieure de Lyon, 46 allée d'Italie, F-69364 Lyon cedex 07, France
email: Jean-Francois.Gonzalez@ens-lyon.fr, Guillaume.Laibe@ens-lyon.fr

[2]Department of Physics, ETH Zurich, CH-8093 Zurich, Switzerland
email: fouchet@phys.ethz.ch

[3]Centre for Astrophysics and Supercomputing, Swinburne University of Technology,
PO Box 218, Hawthorn, VIC 3122, Australia
email: smaddison@swin.edu.au

Abstract. We present the first results of the treatment of grain growth in our 3D, two-fluid (gas+dust) SPH code describing protoplanetary disks. We implement a scheme able to reproduce the variation of grain sizes caused by a variety of physical processes and test it with the analytical expression of grain growth given by Stepinski & Valageas (1997) in simulations of a typical T Tauri disk around a one solar mass star. The results are in agreement with a turbulent growing process and validate the method. We are now able to simulate the grain growth process in a protoplanetary disk given by a more realistic physical description, currently under development. We discuss the implications of the combined effect of grain growth and dust vertical settling and radial migration on subsequent planetesimal formation.

Keywords. planetary systems: protoplanetary disks — hydrodynamics — methods: numerical

1. Introduction

Collisions and aggregation govern the first steps of planet formation from micron-sized particles to decimetric pre-planetesimals. Observations of protoplanetary disks support this mechanism by showing evidence of dust grain growth (e.g. Apai *et al.* 2004). We describe here the implementation of grain growth in our simulations of protoplanetary disks. Depending on their relative velocities and material properties (Dominik & Tielens 1997), collisions between solid particles can make them stick and grow, or conversely break them into smaller pieces.

2. Grain growth in protoplanetary disks

A first description of the grain growth process is given by Stepinski & Valageas (1997). They model a turbulent, vertically isothermal protoplanetary disk, in which gas and dust are represented by two separate phases interacting via aerodynamic drag in the Epstein regime. Their solid particles are supposed to stick perfectly during collisions, and can therefore only grow. The variation of their size s is given by the following analytical expression of the growth rate:

$$\frac{ds}{dt} = \sqrt{2^{3/2}\,\text{Ro}\,\alpha}\,\frac{\rho_s}{\rho_d}\,C_s\,\frac{\sqrt{\text{Sc}-1}}{\text{Sc}}, \qquad (2.1)$$

where Ro is the Rossby number for turbulent motions, α the Shakura & Sunyaev (1973) viscosity parameter, ρ_s the density of matter concentrated into solid particles, ρ_d the intrinsic density of the grains, C_s the local gas sound speed, and Sc the Schmidt number of the flow which estimates the effect of gas turbulence on the grains. Sc is defined by

$$\text{Sc} = (1 + \Omega_k\, t_s)\sqrt{1 + \frac{\bar{v}^2}{V_t^2}}, \quad (2.2)$$

where Ω_k is the local keplerian velocity, t_s the dust stopping time, $\bar{v}$ the mean relative velocity between gas and dust, and V_t a turbulent velocity. The growth rate depends on s through the stopping time

$$t_s = \frac{\rho_d\, s}{\rho_g\, C_s}, \quad (2.3)$$

where ρ_g is the gas density.

The motion of a dust grain and the grain growth process, which both depend on the grain size s, are coupled phenomena. The understanding of the global evolution of dust in disks therefore requires a numerical treatment.

3. Grain growth with an SPH code

Our 3D, bi-fluid, Smooth Particle Hydrodynamics (SPH) code has been developed to model vertically isothermal, non self-gravitating protoplanetary disks. Gas and dust are treated as two separate phases and are coupled by aerodynamic drag. The code and its results on dust migration and settling are presented in Barrière-Fouchet et al. (2005). It has also been applied to grain stratification in GG Tau's circumbinary ring (Pinte et al. 2007) and to gaps opened by planets in the dust phase of protoplanetary disks (Maddison et al. 2007; Fouchet et al. 2007; see also talk by Gonzalez, this volume).

The assumptions we have made are very similar to those of Stepinski & Valageas (1997) mentioned in Sect. 2, their prescription for grain growth is therefore easy to implement in our code. We allow the grain size s assigned to each SPH particle, assuming it to represent the typical size of dust grains at its position, to vary with time following Eq. (2.1). We take the initial grain size distribution to be uniform.

4. Results

We model grain growth in a typical T Tauri disk of mass $M_{\text{disk}} = 0.01\ M_\odot$, with a total dust mass $M_{\text{dust}} = 0.01\ M_{\text{disk}}$, around a central star of mass $M_\star = 1\ M_\odot$. We ran simulations with 200,000 SPH particles, evolved over 50,000 years, for a series of initial dust grain sizes ranging from $s_0 = 1\ \mu$m to 1 mm. Fig. 1 and Fig. 3 show the resulting grain sizes in face-on and edge-on views of the disk, for $s_0 = 1\ \mu$m.

We find that grain growth occurs very quickly, especially in the inner disk where the density is the highest, as one would expect from Eq. (2.1). In this region, the dust grains reach centimetric size (see Fig. 2) in only a few timesteps. In the outer parts of the disk, where the density is far lower, the grains grow much more slowly and their size stays below the millimeter.

An inwards migrating grain coming into a denser region grows almost instantaneously to a size characteristic of the grains at its new position, which in turn evolves slowly over time. Hence, the global shape of the plot shown in Fig. 2 stays the same as time goes on, but the size distribution slowly progresses to larger sizes. The initial size s_0 has

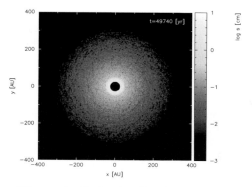

Figure 1. Grain size distribution in an equatorial plane cut of the disk.

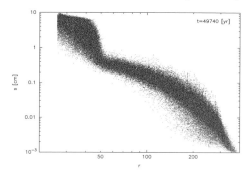

Figure 2. Grain size distribution as a function of radial distance to the star.

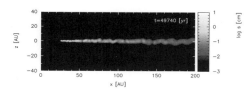

Figure 3. Grain size distribution in a meridian plane cut of the disk. Figures are made with SPLASH (Price 2007).

very little influence on the final size distribution, and in particular on the maximum size reached.

Our results are consistent with those obtained by Dullemond & Dominik (2005). With a model solving the coagulation equation in presence of turbulence, they found a very fast grain growth in T Tauri protoplanetary disks, depleting very small sizes and producing centimetric grains.

5. Conclusion

We have implemented in our 3D SPH code a mechanism able to treat grain growth and validated it through the use of the simple model of Stepinski & Valageas (1997).

In accordance with physical intuition, dust grains grow much more quickly in the denser, central regions of the disk, where centrimetric sized are reached. The growth time is smaller than the migration time and a quasi-stationnary distribution of grain size appears in the disk. The small grains are depleted too rapidly to be consistent with observations of protoplanetary disks, showing the need to take into account other processes such as microscopic interactions between the grains, kinetic energy dissipation, porosity, and re-fragmentation. This will be addressed in future work.

References

Apai, D., Pascucci, I., Sterzik, M. F., *et al.* 2004, *A&A*, 426, L53

Barrière-Fouchet, L., Gonzalez, J.-F., Murray, J. R., Humble, R. J., & Maddison, S. T. 2005, *A&A*, 443, 185

Dominik, C., Tielens, A. G. G. M. 1997, *ApJ*, 480, 647

Dullemond, C. P. & Dominik, C. 2005, *A&A*, 434, 971

Fouchet, L., Maddison, S. T., Gonzalez, J.-F., & Murray, J. R. 2007, *A&A*, in press (arXiv:0708.4110)

Maddison, S. T., Fouchet, L., & Gonzalez, J.-F. 2007, *Ap&SS*, 474, 1037

Pinte, C., Fouchet, L., Ménard, F., Gonzalez, J.-F., & Duchêne, G. 2007, *A&A*, 469, 963

Price, D. 2007, *PASA*, accepted (arXiv:0709.0832)

Shakura, N. I. & Sunyaev, R. A. 1973, *A&A*, 24, 337

Stepinski, T. F. & Valageas, P. 1997, *A&A*, 319, 1007

Origin of debris disks and the supply of metals in DZ white dwarfs

Y. Wang[1], R. B. Dong[1], D. N. C. Lin[2,3] and X. W. Liu[1,2]

[1]Department of Astronomy, Peking University,
100871, Beijing, China,
email: yanwang@pku.edu.cn

[2]Kavli Institute of Astronomy and Astrophysics, Peking University,
100871, Beijing, China
email: liuxw@bac.pku.edu.cn

[3]Department of Astronomy and Astrophysics, University of California,
Santa Cruz, CA 95064, USA
email: lin@ucolick.org

Abstract. We discuss the dynamical evolution of minor planetary bodies in the outer regions of planetary systems around the progenitors of DZ white dwarfs. We show that during the planetary-nebula phase of these stars, mass loss can lead to the expansion of all planetary bodies. The orbital eccentricity of the minor bodies, as relics of planetesimals, may be largely excited by the perturbation due to both gas drag effects and nearby gas giant planets. Some of these bodies migrate toward the host star, while others are scattered out of the planetary system. The former have modest probability of being captured by the sweeping secular resonances of giant planets, and induced to migrate toward the host star. When they venture close to their host stars, their orbits are tidally circularized so that they form compact disks where they may undergo further collisionally driven evolution. During the subsequent post main sequence evolution of their host stars, this process may provide an avenue which continually channels heavy elements onto the surface of the white dwarfs. We suggest that this scenario provides an explanation for the recently discovered Calcium line variation in G29-38.

Keywords. Giant planet, debris disks, planet dynamic, n-body simulations, white dwarfs

1. Introduction

In the recent years, more than 200 planets have been found around solar-type stars. Despite these impressive gains, observational selection effects limit the discovery space to only a small fraction of all potential orbital configurations.

An effective method to find indirect evidence for the presence of planets is to search for remnant of planetary system in post-main sequence stars. Debris disks have been detected around a few white dwarfs with the highest photospheric metal abundances, such as G29-38 (Heppel & Thompson 2007), and they are generally considered to represent the relics of planet building process. All of them are metal-rich, with substantial abundances of photospheric Calcium and Silicates and show no signs of H or He. (Koester *et al.* 1997)

A survey of DA white dwarfs reveals 25% of white dwarfs have excess metals in their atmosphere (Zuckerman *et al.* 2003). This excess is unexpected due to their short estimated gravitational settling time scales ($10^{-2} \sim 10^6$ years; Dupuis et al. 1992; comparison with the typical ages of these white dwarfs ($10^8 \sim 10^9$ years). This dichotomy suggest that these metal contents may be continually replenished. One class of possible reservoir is debris disks in the proximity of these white dwarfs. The inferred detection of a debris disk at $0.14 \sim 1\ R_\odot$, within the tidal disruption radius of the white dwarf (Jura 2003) provides support for this scenario.

During planetary nebula phase, the stellar envelopes extend to several AU's. Any planets and residual planetesimals within this region would encounter a strong hydrodynamic drag and undergo orbital decay. Outside the red giants' envelope, planetary orbits expand due to stellar mass losses. Residual planetesimals also experience hydrodynamic drag induced by the stellar wind as well as the dynamical perturbation due to any neighboring gas giant planets.

In this paper, our goal is to examine the dynamic evolution of outer planets and planetesimals, which may have once resided in regions similar to the asteroid or the Kuiper belt regions. We show that they may be scattered to the stellar proximity and form the disks around DZ white dwarfs. We find that the orbits of the residual planetesimals are destablished by the planetary perturbations during the mass loss of their host stars.

2. A working model

We assume the planetary system is composed of three populations of objects, including the host star, several gas giant planets, and a large number of planetesimals.

The model of planetary system we used is analogous to the solar system. We consider the main sequence progenitors of the DZ white dwarfs to have masses in the range of $2\,M_\odot < M_* < 4\,M_\odot$ (A-F stars). We assume the mass of the white dwarf is $\sim 0.5\,M_\odot$, which is common for the remnant white dwarf of such stars. During the mass loss, we assume the mass of star changes linearly with time. Beyond the stellar envelope during the red giant phase ($\sim$ a few AUs), planets' orbits expand with the stellar mass loss. The region within $\sim 10^2$ AU is essentially depleted (see contribution by Dong et al. in this volume).

In addition to the expansion of the planets' and planetesimals' orbits, the ratio of their mass to that of their host stars increases. Consequently their Roche lobes occupies a greater fraction of their newly acquired semi major axis. This change leads to an enhanced planetary perturbation which destabilizes the orbits of residual planetesimals.

In order to demonstrate this effect, we carry out a series of simulations with 100 planetesimals. The initial separation between these small bodies varies between $0.01 \sim 1$ AU. We choose a median semi-major axis at 350 AU. We also include two planets which have a asymptotic (at the end of stellar mass-loss epoch) semi major axis 50 and 100 AU and mass $\sim 10^{-6} M_\odot$. Such kind of planets are analogous to Earth-like planets which may have formed at a few AU and undergone adiabatic orbital expansion during the stellar mass loss. Additional and more massive planets may be present around some other planetary systems, but the physical process we intend to study is similar. The initial eccentricities and inclinations of planets and planetesimals are all set to be zero. Although the mass of the planets are relatively small, planetesimals can be efficiently scattered when their eccentricities are excited and their orbits cross those of planets.

To determine the dynamic evolution of the system, we neglected the interaction between small planetesimals. Therefore, the planetesimals generally move in a central gravitational field, with the perturbation of of the giant planets. In a non-rotating frame centered on the star, the equation of motion can be expressed as:

$$\frac{d^2 \mathbf{r}_i}{dt^2} = -\frac{GM_* \mathbf{r}_i}{r_i^3} - \Sigma_j \frac{GM_j(\mathbf{r}_i - \mathbf{r}_j)}{|\mathbf{r}_i - \mathbf{r}_j|^3} - \Sigma_j \frac{GM_j \mathbf{r}_j}{r_j^3} \qquad (2.1)$$

where M_* and M_j are the mass of the central star and that of the planets, respectively. $\mathbf{r}_i$ and $\mathbf{r}_j$ are the position of a particular planetesimal and planets. The perturbation term in equation (2.1) is normally small. However, the effect of small perturbation force can accumulate with time.

3. Scenarios and simulations

Enhancement of dynamical instability. As the semi-major axis of planets changes proportionally to their mass of host star, the mass ratios of the planets μ increases with time, and the the scaled separation k of the planetesimals decreases simultaneously. Here μ and k are defined as

$$\mu(t) = \frac{M_p}{M_*(t)} \tag{3.1a}$$

$$k(t) = \frac{a_{i+1}(t) - a_i(t)}{R_H(t)} \tag{3.1b}$$

where a_i is the separation of the planetesimals, and R_H is the Hill's radius. Using the empirical fitting formula (Zhou, Lin & Sun 2007), for the time scale for orbit crossing,

$$log\left(\frac{T_c}{yr}\right) = A + Blog\left(\frac{k_0}{2.3}\right)$$
$$A = -2 - 0.27 log(\mu) \tag{3.2}$$
$$B = 18.7 + 1.1 log(\mu)$$

we find that for planetesimals in systems with initial $k_0 = 20$, can survive the planets' perturbation during the main sequence evolution of their host stars. The magnitude of the orbit-crossing time scale T_c can decrease from 10^{10} years to 10^8 years during their host stars' mass loss. On the other hand, the period of giant planets T_p increases proportional to $M_*^{-1.5}$ due to the Keplerian motion. Both of these two factors enhances the tendency toward dynamical instability of the system. In this process, the eccentricity of planetesimals is exited, but the orbits of planets expand adiabatically, which leads to the orbit crossing of planetesimals and planets.

Migration toward host star. If two or more planets survive the mass loss scenery, they would scatter the nearby planetesimals, after orbit crossing. As the eccentricity of planetesimals can be greatly excited and their semi-major axis remain largely unaffected, a fraction of them can venture into the inner region within 100 AU from the host star. Using a Hermit integrator kindly provide by Dr. Sverre Aarseth, we simulate these scattering events. Our results show that the course of the scattering can last for a long time. If the initial separation between planetesimals is 0.5 AU and the initial eccentricity is 0.7, $\sim 25\%$ of the planetesimals would become unbounded in 5 million years. This timescale becomes smaller if the mass of the planets, the initial eccentricity of planetesimals, or the initial semi-major axis is larger.

Eccentricity of planetesimals can be excited to magnitudes close to unity, but the semimajor axis can either increase or decrease. A fraction of planetesimals migrate toward the host star, as their semi-major axis becomes smaller and the orbits of them no longer cross those of the planets, which makes their final orbits stable. The final periastron can be well within 30 AU from the host star. For other planetesimals, eccentricity may exceed unity and they are scattered out of the system. Such objects may become a source of freely floating rocky planets and gas giants. (Lucas & Roche 2000) Result of our simulation shows that only ~ 5 of the planetesimals can migrate into the inner region.

Secular resonance. During $10^8 \sim 10^9$ years, as planets are immersed in numerous, low-mass planetesimals, their orbits migrate slowly and their secular resonances sweep across a wide region, analogous to the late heavy bombardment scenario (Gomes *et al.* 2005). Duncan *et al.* (2007) point out that a planet embedded near 20 AU in an initially low eccentricity minimum-mass planetesimal disk typically migrates 1 AU in 10^4 years, and the ratio of semi-major axis of different planets changes after migration. Considering the time scale of DZ white dwarfs, the migration rate should be much slower, which means the density of planetesimals disk would be lower than typical density in Duncan's work.

When planetesimals are captured by secular resonant point of planets, their eccentricity would be exited while semi-major axis remains unchanged. This tendency brings their periastron close to their host star. As secular resonance depends mainly on the ratio of planets' semi-major axis, planetesimals in different regions migrate to the host star due to secular resonance. This process could continuously provide materials to the disks around white dwarfs. At the proximity of their host stars, intense tidal force disrupt the planetesimals and their debris form debris disks and supply heavy elements to the atmosphere of white dwarfs, such as G29-38.

We thank Sverre Aarseth for provide the Hermit integrator, and Y.H. Chiu, Ji-Lin Zhou for valuable suggestions. This work is supported by NASA (NAGS5-11779, NNG04G-191G, NNG06-GH45G), JPL (1270927), NSF (AST-0507424, PHY99-0794)

References

Aarseth, S. J., Lin, D. N. C., & Palmer, P. L. 1993, *ApJ*, 403, 351
Brasser, R. & Lehto, H. J. 2002, *MNRAS*, 334, 241
Brasser, R., Duncan, M. J., & Levison, H. F. 2007, *Icarus*, 191, 413
Chambers, J. E. 1999, A Hybrid Symplectic Integrator that Permits Close Encounters between Massive Bodies, *MNRAS*, 304, 793
Debes, J. H. & Sigurdsson, S. 2002, *ApJ*, 572, 556
Duncan, M. J., Kirsh, D., Capobianco, C., Brasser, R., & Levison, H. F. 2007, *American Astronomical Society, DPS meeting*, 39, 60.01
Duncan, M. J., Kirsh, D., Brasser, R., & Levison, H. F. 2007, *American Astronomical Society, DDA meeting*, 38, 6.05
Dupuis, J., Fontaine, G., Pelletier, C., & Wesemael, F. 1992, *ApJS*, 82, 505
Gomes, R., Levison, H. F., Tsiganis, K., & Morbidelli, A. 2005, *Nature*, 435, 466
Gu, P. G., Lin, D. N. C., & Bodenheimer, P. H. 2003, *ApJ*, 588, 509
von Hippel, T. & Thompson, S. E. 2007, *ApJ*, 661, 477
Ida, S. & Lin, D. N. C. 1996, *AJ*, 112,1239
Ida, S. & Lin, D. N. C. 2004, *ApJ*, 604,388
Ida, S. & Lin, D. N. C. 2007, *ApJ*, 626,1045
Jiang, I., Duncan, M., & Lin, D. N. C. 2006, RevMexAA(SC), 21, 217
Jura, M. 2003, *ApJ*, 584, L91
Koester, D., Provencal, J., & Shipman, H. L. 1997, *A&A*, 320, L57
Koester, D. & Wilken, D. 2006, *A&A*, 453, 1051
Kokubo, E. & Ida, S. 1996, *Icarus*, 123, 180
Kokubo, E. & Ida, S. 2000, *Icarus*, 143, 15
Kokubo, E. & Ida, S. 2002, *ApJ*, 581, 666
Lee, M. H. & Peale, S. J. 2003, *ApJ*, 592, 1201
Levison, H. F. et al. 2001, *Icarus*, 151, 286
Lin, D. N. C & Papaloizou, J. C. B. 1986, *ApJ*, 309, 846
Lucas, P. W. & Roche, P. F. 2000, *MNRAS*, 314, 858
Murray, C. D. & Dermott, S. F. 1999, Solar system Dynamcis, Cambridge Unversity.
Nagasawa, M. & Ida, S. 2000, *AJ*, 120, 3311
Nagasawa, M., Ida, S., & Tanaka, H. 2002, *Icarus*, 159, 322
Nagasawa, M., Lin, D. N. C., & Ida, S. 2003, *ApJ*, 586, 1374
Nagasawa, M., Lin, D. N. C., & Thommes, E. 2005, *ApJ*, 635, 578
Palmer, P. L., Lin, D. N. C., & Aarseth, S. J. 1993, *ApJ*, 403, 336
Zhou, J. L., Lin, D. N. C., & Sun, Y. S. 2007, *ApJ*, 666, 447
Zuckerman, B., Koester, D., Reid, I. N., & Hünsch, M. 2003, *ApJ*, 596, 477

Type I planetary migration in a self-gravitating disk

Clément Baruteau[1] and Frédéric Masset[1,2]

[1]Laboratoire AIM, CEA/DSM - CNRS - Université Paris Diderot
DAPNIA/Service d'Astrophysique, CEA/Saclay, 91191 Gif/Yvette Cedex, France
email: clement.baruteau@cea.fr

[2]IA-UNAM, Ciudad Universitaria, Apartado Postal 70-264, Mexico D.F. 04510, Mexico
email: fmasset@cea.fr

Abstract. We investigate the impact of the disk self-gravity on type I migration. We first show that considering a planet migrating in a disk without self-gravity can lead to a significant overestimate of the migration rate. Unbiased drift rates can be obtained only if the planet and the disk feel the same gravitational potential. We then confirm that the disk gravity slightly accelerates type I migration.

Keywords. accretion, accretion disks, gravitation, hydrodynamics, methods: numerical

1. Introduction

A low-mass planet embedded in a protoplanetary disk should undergo an orbital decay towards the central object. This process, known as type I migration, has been extensively studied analytically and numerically. Recent works on planet-disk interactions have challenged to include more physical ingredients and to yield precise estimates of the migration rate. Two of them investigated the impact of the disk self-gravity, but reported *a priori* contradictory statements: the *disk self-gravity* slows down migration for Nelson & Benz (2003), whereas *the disk gravity* accelerates it for Pierens & Huré (2005). This communication aims at clarifying and quantifying the impact of the disk gravity on type I migration.

2. Torque estimate in a non self-gravitating disk

Numerical setup. We performed two-dimensional calculations of a low-mass planet embedded in a non self-gravitating gaseous disk. The planet to primary mass ratio is $q = 5 \times 10^{-6}$, and the softening length of the planet potential is $\varepsilon = 0.3\,hr_p$, where h is the disk aspect ratio, taken uniform, and r_p denotes the planet's orbital radius. In its unperturbed state, the disk has axisymmetric temperature and surface density profiles which read, respectively, $T(r) = T_p(r/r_p)^{-1}$ and $\Sigma(r) = \Sigma_p(r/r_p)^{-3/2}$, where T_p and Σ_p denote the temperature and the surface density at $r = r_p$. Since Σ decreases as $r^{-3/2}$, and the disk is described by an isothermal equation of state, the corotation torque cancels out (Masset 2001, Baruteau & Masset 2008). The torque exerted by the disk on the planet therefore reduces to the differential Lindblad torque. The disk feels the gravity of the central star and of the planet, while two situations are considered for the planet:

- It is held on a *fixed* circular orbit. In this case (the fixed case), the planet only feels the star gravity. The planet and the disk therefore feel the same gravitational potential. This situation has been contemplated for analytical estimates of the migration rate (e.g. Tanaka *et al.* 2002).

- It *freely* migrates in the disk. In this case (the free case), the planet feels the star and the disk gravity. Contrary to the fixed case, this situation is not self-consistent since the planet and the disk do not orbit under the same potential. Nevertheless, this situation is of interest since it corresponds to the standard scheme of all planet-disk simulations.

An unexpected result. We display in figure 1 the torque Γ exerted on the planet when varying Σ_p (we take $h = 0.05$, T_p is therefore fixed for this series). When the planet is on a fixed orbit, Γ scales with Σ_p, as expected. However, the free case reveals two unexpected results: Γ increases faster than linearly with Σ_p, and, for a given value of Σ_p, Γ is far larger in the free case than in the fixed one.

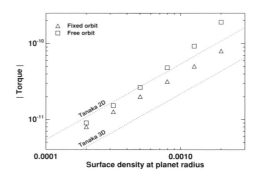

Figure 1. Torque exerted on a $M_p = 5 \times 10^{-6} M_*$ planet mass by a non self-gravitating disk. The surface density at the planet radius is varied from one to ten times that of the MMSN. Triangles refer to the fixed case whereas squares refer to the free case (see text). The two dotted lines depict the two- and three-dimensional analytical estimates of Tanaka *et al.* (2002).

Shift of Lindblad resonances. These results can be explained with the relative positions of the Lindblad Resonances (LR) in the fixed and free cases. We display in figure 2a the locations $r_{\rm ILR}$ ($r_{\rm OLR}$) of an Inner (Outer) LR, when the planet is on a fixed orbit. They are given by $r_{\rm ILR} = \Omega^{-1}(\Omega_{\rm ILR})$ and $r_{\rm OLR} = \Omega^{-1}(\Omega_{\rm OLR})$, with $\Omega(r)$ the disk's rotation profile (solid curve), and with $\Omega_{\rm ILR}$ ($\Omega_{\rm OLR}$) the frequency of the ILR (OLR), simply deduced from the planet frequency Ω_p. When the planet is on a free orbit (figure 2b), its frequency is larger than in the fixed case, depending on the mass distribution of the disk (Baruteau & Masset 2007). Thus, the frequencies of the LR are also larger in the free case, and all the resonances are shifted inwards.

Torque discrepancy. The free situation induces a spurious inward shift of all the LR, which reduces the (positive) inner torque, and increases the (negative) outer one. Thus, the (negative) total torque is artificially larger in the free case. This torque overestimate increases with Σ_p and can typically reach a factor of two (figure 1). In order to give unbiased torque estimates, the planet and the disk must feel the same potential.

3. Torque estimate in a self-gravitating disk

Impact of the disk gravity. In addition to the star gravity, the disk gravity is now felt by both the planet and the disk. The impact of the disk gravity is three-fold:

(*a*) It increases the planet frequency, as already stated in section 2. This shifts all the LR inwards (figure 2b).

(*b*) Similarly, it increases the disk frequency, which shifts all the LR outwards (see figure 3). The increase of the planet and of the disk frequencies induces two shifts that

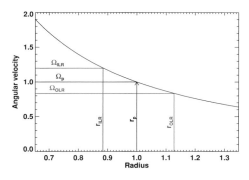 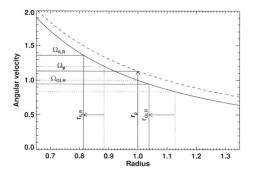

Figure 2. Location of two Lindblad resonances in the fixed case (left panel) and in the free case (right panel): the ILR of $m = 6$ ($\Omega_{\rm ILR} = 6/5\,\Omega_p$), and the OLR of $m = 5$ ($\Omega_{\rm OLR} = 5/6\,\Omega_p$). The disk's rotation profile is depicted without self-gravity (solid curve) and with self-gravity (dashed curve, right panel). The difference between both profiles has been exaggerated to improve legibility. In the right panel, arrows highlight the shift of the LR with respect to the left panel.

nearly compensate, and the net impact on the torque is negligible (Baruteau & Masset 2007).

(*c*) It yields an additional term in the dispersion relation of density waves, which shifts the LR towards the planet location. This shift prevails over the sum of the two previous shifts (Pierens & Huré 2005), and the disk gravity should enhance the torque.

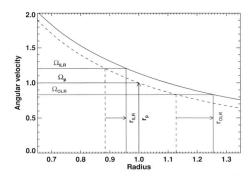

Figure 3. Same as figure 2, except that we examine the shift of the LR when the disk is self-gravitating (its rotation profile is now the solid, upper curve). The frequency of the planet, and therefore that of the LR, is the same as in figure 2a.

Solving the contradiction. The torques obtained with self-gravitating calculations are displayed in figure 4a. The results of figure 1 are overplotted for comparison:

- Comparing with the results of the self-consistent fixed case, as in Pierens & Huré (2005), confirms that the disk gravity slightly accelerates type I migration.
- Comparing with the results of the biased free case, as done by Nelson & Benz (2003), would also lead us to conclude that the disk self-gravity slows down migration significantly. However, this conclusion is misleading since it compares a self-consistent situation to a biased one.

What if one discards the disk gravity? The torque increase due to the disk gravity can be exclusively accounted for by a shift of the LR, and, under some limitations, be reproduced with an anisotropic pressure tensor, without self-gravity (Baruteau & Masset

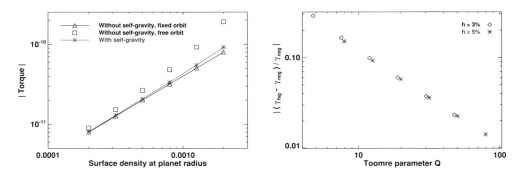

Figure 4. *Left:* same as figure 1, except that the results of the self-gravitating calculations are overplotted (stars). *Right:* relative difference of the torques obtained with the self-gravitating calculations ($\gamma_{\rm fsg}$) and the calculations without disk gravity ($\gamma_{\rm nog}$, those of the fixed case), versus the Toomre parameter at the planet location.

2007). For a given softening length, the relative difference of the torques without and with disk gravity depends only on the Toomre parameter Q at the planet location. It does not depend on the global mass distribution of the disk. Figure 4b shows that this relative difference scales with Q^{-1}, at least for high to moderate values of Q. This figure yields a quantitative estimate of the torque underestimate if one discards the disk gravity (here for $\varepsilon = 0.3\,hr_p$, see Baruteau & Masset (2007) for a study of its dependence with the softening length).

Acknowledgements

C.B. thanks Andrés González Carmona, Takayuki Muto and Hui Zhang for fruitful discussions at the symposium.

References

Baruteau, C. & Masset, F. 2007, submitted to *ApJ*
Baruteau, C. & Masset, F. 2008, *ApJ*, 672, 1054
Masset, F. S. 2001, *ApJ*, 558, 453
Nelson, A. F. & Benz, W. 2003, *ApJ*, 589, 578
Pierens, A. & Huré, J.-M. 2005, *A&A*, 433, L37
Tanaka, H., Takeuchi, T., & Ward, W. R. 2002, *ApJ*, 565, 1257

On type I planetary migration in adiabatic disks

Clément Baruteau[1] and Frédéric Masset[1,2]

[1] Laboratoire AIM, CEA/DSM - CNRS - Université Paris Diderot
DAPNIA/Service d'Astrophysique, CEA/Saclay, 91191 Gif/Yvette Cedex, France
email: clement.baruteau@cea.fr

[2] IA-UNAM, Ciudad Universitaria, Apartado Postal 70-264, Mexico D.F. 04510, Mexico
email: fmasset@cea.fr

Abstract. We investigate type I migration in a two-dimensional adiabatic disk. We find entropy perturbations that are advected in the planet's coorbital region. These entropy perturbations yield an excess of corotation torque that scales with the unperturbed entropy gradient at corotation. This torque excess can be large enough to slow down migration significantly, or even stop it.

Keywords. accretion, accretion disks, hydrodynamics, methods: numerical

1. Introduction

Recently, Paardekooper & Mellema (2006) have revisited type I migration with high-resolution 3D calculations, including radiative transfer. They find that the migration can be reversed for sufficiently large values of the disk opacity, which may solve the lingering problem of type I migration. This work is a first step to further investigate this topic. For this purpose, we consider a more restricted situation, namely 2D adiabatic flows.

2. Adiabatic vs. isothermal calculations

We performed 2D calculations of a low-mass planet embedded in an inviscid gaseous disk. In its unperturbed state, the disk has axisymmetric surface density and temperature profiles, which read, respectively, $\Sigma(r) \propto r^{-0.5}$ and $T(r) \propto r^{-0.6}$. The gas pressure p is given by $p = \Sigma T$. The disk is perturbed by a planet on a fixed circular orbit. The planet to primary mass ratio is $q = 2.2 \times 10^{-5}$, and the softening length of the planet potential is $\varepsilon = 0.03\,r_p$, where r_p denotes the planet's orbital radius.

Two series of runs were performed:
- The disk is described by a locally isothermal equation of state, as is customarily assumed in disk-planet calculations.
- The disk is described by an adiabatic energy equation. This equation corresponds to the Lagrangian conservation of the gas entropy, referred to as the quantity $S = p/\Sigma^\gamma$, where γ is the adiabatic index ($\gamma = 1.4$ in our calculations).

Figure 1 shows the relative perturbation of the gas surface density in both situations. The planet's wake is less tightly wound in the adiabatic case, since the adiabatic sound speed c_s is larger than the isothermal one, by a factor of $\sqrt{\gamma}$. The differential Lindblad torque, which scales as c_s^{-2} (Ward 1997), is therefore reduced by a factor of γ in an adiabatic disk. The same is true of the corotation torque, when there is no entropy gradient (Baruteau & Masset 2008). Furthermore, the adiabatic calculation displays additional

397

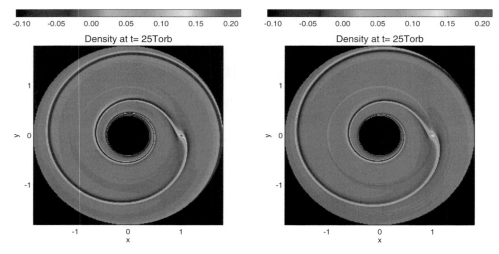

Figure 1. Relative perturbation of the gas surface density obtained with a locally isothermal equation of state (left panel), and with an adiabatic energy equation (right panel). The planet is located at $x = 1$, $y = 0$.

density perturbations, located in the planet's coorbital region. These perturbations can have a dramatic impact on the corotation torque, as we shall see.

3. Additional density perturbations in the planet's coorbital region

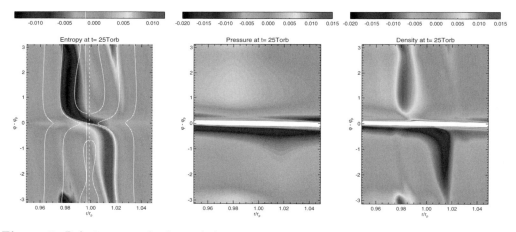

Figure 2. Relative perturbations of the gas entropy, pressure and surface density obtained with the adiabatic calculation of section 2. The planet is located at $r = r_p$, $\varphi = \varphi_p$. In the entropy panel, the vertical dashed line represents the planet's corotation radius, and streamlines are overplotted to show the extent of the planet's coorbital region. In the pressure and surface density panels, the nearly horizontal (saturated) structure at $\varphi = \varphi_p$ is the planet's wake. The planet's Hill sphere radius is $R_H \approx 0.02\, r_p$.

We display in figure 2 the relative perturbations of the gas entropy, pressure and surface density obtained with the adiabatic calculation of section 2. While the azimuthal range spans the whole $[0, 2\pi]$ interval, the radial range depicted is restricted to a band of width $2.5\, x_s$ around the corotation radius, where x_s denotes the half-width of the planet's coorbital region. Streamlines are overplotted to the entropy panel to show the extent of the coorbital region, and the vertical dashed line displays the location of the corotation radius r_c. In the pressure and surface density panels, the nearly horizontal (saturated) structure is the planet's wake. The entropy and surface density panels show the propagation of perturbations sliding along the separatrices of the coorbital region, whereas the gas pressure is weakly perturbed.

As reported by Baruteau & Masset (2008), these perturbations arise only in adiabatic disks, where a non-vanishing entropy gradient induces a discontinuity in the entropy and surface density fields at corotation, while the pressure field is continuous. In the linear limit, the perturbed fields (and their contribution to the torque) become singular, since the coorbital region becomes infinitely narrow. In our calculation, the unperturbed entropy profile is a decreasing function of the radius. The horseshoe dynamics therefore advects a negative entropy perturbation to the inner side of the coorbital region ($r < r_c$), and a positive entropy perturbation to the outer side ($r > r_c$). Since the pressure is only weakly perturbed, each entropy perturbation is related to a surface density perturbation of opposite sign and, in relative value, of same order of magnitude.

4. Excess of corotation torque and entropy gradient

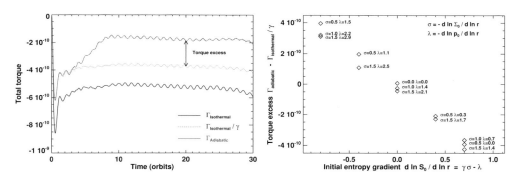

Figure 3. *Left:* torques obtained with the adiabatic and isothermal runs of section 2. The dotted curve depicts the torque expected in the adiabatic situation when discarding the surface density perturbations advected by the horseshoe dynamics. *Right:* torque excess versus the unperturbed entropy gradient at corotation $(d \ln S_0/d \ln r)_{r_c}$, for different values of the unperturbed pressure and surface density gradients.

The torque variation with time of our adiabatic calculation is depicted in figure 3a. The torque obtained with a locally isothermal equation of state is displayed for comparison. If there were no unperturbed entropy gradient, hence no entropy perturbations advected by the horseshoe dynamics, the adiabatic torque would be equal to the isothermal torque divided by the adiabatic index (dotted line, see section 2). The adiabatic run displays here a positive excess of corotation torque, which accounts for the perturbations advected by the horseshoe dynamics. The sign of this torque excess is related to the sign of the surface density perturbations inside the coorbital region, hence to the sign of the unperturbed entropy gradient at corotation.

As shown in figure 3b, the excess of corotation torque scales with the gradient of the unperturbed entropy at corotation $(d \ln S_0/d \ln r)_{r_c}$. A positive entropy gradient induces a negative torque excess, which accelerates inward migration. On the contrary, a negative entropy gradient yields a positive torque excess, which slows down type I migration. If the entropy gradient is sufficiently negative, the torque excess can be positive enough to reverse type I migration. However, the torque excess saturates as libration flattens out the entropy profile across the coorbital region. The long-term behavior of this process will depend on the irreversible processes at work in the disk (viscous heating, radiative cooling), which can maintain an unsaturated torque excess (forthcoming work).

References

Baruteau, C., & Masset, F. 2008, *ApJ*, 672, 1054
Paardekooper, S.-J., & Mellema, G. 2006, *A&A*, 459, L17
Ward, W. R. 1997, *Icarus*, 126, 261

Baroclinic generation of potential vorticity in an embedded planet-disk system

Jianghui JI[1], Shangli OU[2] and Lin LIU[3]

[1] Purple Mountain Observatory, Chinese Academy of Sciences, Nanjing 210008, China
email: jijh@pmo.ac.cn

[2] High Performance Computing, Center for Computation and Technology/Information Technology Services, Louisiana State University, Baton Rouge, LA 70803
email: ou@cct.lsu.edu

[3] Department of Astronomy, Nanjing University, Nanjing 210093, China

Abstract. We use a multi-dimensional hydrodynamics code to study the gravitational interaction between an embedded planet and a protoplanetary disk with emphasis on the generation of vortensity (Potential Vorticity or PV) through a Baroclinic Instability. We show that the generation of PV is very common and effective in non-barotropic disks through the Baroclinic Instability, especially within the coorbital region. Our results also complement previous work that non-axisymmetric Rossby-Wave Instabilities (RWIs) are likely to develop at local minima of PV distribution that are generated by the interaction between a planet and an inviscid barotropic disk. The development of RWIs results in non-axisymmetric density blobs, which exert stronger torques onto the planet when they move to the vicinity of the planet. Hence, large amplitude oscillations are introduced to the time behavior of the total torque acted on the planet by the disk. In current simulations, RWIs do not change the overall picture of inward orbital migration but cause a non-monotonic behavior to the migration speed. As a side effect, RWIs also introduce interesting structures into the disk. These structures may help the formation of Earth-like planets in the Habitable Zone or Hot Earths interior to a close-in giant planet.

Keywords. accretion, accretion disks - Baroclinic Instability - Rossby-wave instability - hydrodynamics - numerical methods - planetary systems: protoplanetary disks

---

1. Introduction

The standard core-accretion theory (Safronov 1969; Lissauer 1993) suggests that the formation of planets in circumstellar disks around T Tauri stars consists of the formation of planetesimals via collisions/coalitions of dust grains in the early stage and then gravitationally accretion after they accumulate enough mass. The lifetime of this T Tauri star phase is estimated to be short ($\lesssim 10^7$ years). In order for cores of protoplanets with Jupiter mass to accumulate enough material in the T Tauri stage, it is thought that their cores have to form outside the so-called "ice line" located far away from the central star (typically beyond ~ 4 AU, see Ida & Lin 2004) so that the temperature is low enough to allow the condensation of gas materials to solid ice. These additional solid grains helps to increase the dust coagulation speed and shorten the time needed to form the embryos of protoplanets. From the observational side, recent discoveries of extrasolar planets show that a large number of the host stars are surrounded by "hot Jupiters" and close-in super Earths. Around 80% of the extrasolar planets are in orbits with semi-major axes in the range $0.01 \lesssim a \lesssim 2.5$ AU, and $\sim 25\%$ of the total population are short-period planets with $a \lesssim 0.1$ AU. This has brought one of the most interesting puzzles to theorists: if protoplanets had formed in a disk region beyond ~ 4 AU from the central star, how did the observed extrasolar planets end up with orbits that are so close to their host stars? If the standard theory for the formation of protoplanetary cores holds (Pollack et al. 1996; Ida & Lin 2004), then the giant planets (like hot-Jupiters) must have undergone inward orbital migration to their current locations.

Goldreich & Tremaine (1978) and many other authors studied the gravitational interaction between a gaseous protoplanetary disk and an embedded planet (or satellite). The planet causes the formation of spiral waves inside the disk at the Lindblad resonances; as a result of the density asymmetry induced by spiral waves, the inner disk exerts a positive gravitational torque onto the planet and the outer disk exerts a negative gravitational torque onto the planet. According to analytical analysis, the overall torque is generally negative and, hence, forces the planet to migrate inward. Ward (1997) pointed out that depending on the mass of the protoplanet two major kinds of migration exist. In the so called Type I migration, the planet's mass is small and the response of the disk is linear; the migration speed is very fast so that the migration timescale is as short as $\sim 10^4$ yr for a planetary core of 10 $M_\oplus$ in a minimum mass solar nebula at ~ 5.2 AU with sufficient viscosity (see Figure 14 of Ward 1997). In Type II migration, the protoplanet is massive enough to open a gap inside the disk, and migrates on a much longer viscous timescale. Besides classical analytical analysis, many groups have studied the nonlinear evolution of a disk-planet system using numerical multi-dimensional hydrodynamics (Kley 1999; Nelson et al. 2000). These numerical simulations showed that the nonlinear evolution of the orbital migration of a planet inside a disk agrees with linear analysis in a qualitative manner. Masset (2001) and Masset & Papaloizou (2003) suggested that the corotation torque originated from the coorbital region may play a very different role from that of Lindblad torques; this leads to a third kind of migration referred to as Type III migration, in which the migration happens on a timescale as short as a few tens of orbits and can be directed outward in some cases.

Klahr & Bodenheimer (2003) described a Baroclinic Instability in non-barotropic disks that may contribute to vorticity and global turbulence, then argued that strong vorticities may contribute to rapid formation of Jupiter-size gas planet (Klahr & Bodenheimer 2006). Koller et al. (2003) and Li et al. (2005) showed that the so-called RWIs may develop at the local minima of PV, or vortensity (defined as the ratio between local vorticity and surface density), in an inviscid disk with initially uniform PV distribution. Li et al. (2005) showed that non-axisymmetric RWIs lead to the formation of vorticies and density blobs, which exert stronger torque onto the planet when they travel around its vicinity and bring large oscillations to the total torque acted on the planet. They further argued that this mechanism may be possible to change the direction of the migration. Non-axisymmetric RWIs are also relevant to evolution of a single disk (Li et al. 2000) and stellar models with strong differential rotation (Ou & Tohline 2006). In this work, we study the generation of PV through baroclinic effect and subsequent development of RWIs in systems consisting of a planet and a non-barotropic disk.

2. Basic Equations, Methods and Initial Setup

To investigate the interaction between a disk and an embedded planet requires coupling hydrodynamics and orbital dynamics together. Here, we follow Nelson et al. (2000) and many previous investigations to reduce the problem to a two-dimensional (2D) one since the disk thickness is of the order of or smaller than the planetary Hill radius. Three dimensional (3D) investigations will be postponed to future. The fluid motion inside the disk is described by the vertically integrated continuity equation (2.1), radial and azimuthal components of the Navier-Stokes equation (2.2) and (2.3),

$$\frac{\partial \Sigma}{\partial t} + \nabla \cdot (\Sigma \vec{v}) = 0 \tag{2.1}$$

$$\frac{\partial (\Sigma v_r)}{\partial t} + \nabla \cdot (\Sigma v_r \vec{v}) = \frac{\Sigma v_\phi^2}{r} - \frac{\partial P}{\partial r} - \Sigma \frac{\partial \Phi}{\partial r} + f_r \tag{2.2}$$

$$\frac{\partial (\Sigma v_\phi)}{\partial t} + \nabla \cdot (\Sigma v_\phi \vec{v}) = -\frac{\Sigma v_r v_\phi}{r} - \frac{1}{r}\frac{\partial P}{\partial \phi} - \frac{\Sigma}{r} \frac{\partial \Phi}{\partial \phi} + f_\phi, \tag{2.3}$$

where Σ is disk surface density, $\vec{v}$ is two fluid velocities, P is vertically integrated pressure, f_r and f_ϕ are two components of viscous forces, and Φ is the gravitational potential felt by fluid elements. Details regarding viscous terms can be found in Nelson *et al.* (2000). The EOS of the disk fluid is considered as locally isothermal (Nelson *et al.* 2000) as given by $P = c_s^2 \Sigma$, where the local isothermal sound speed is $c_s = \frac{H}{r}\sqrt{GM_*/r}$ with disk aspect ratio $H/r = 0.05$. In order to compare with globally isothermal situation, we also carried out one run with a uniform value of $c_s = \frac{H}{r}\sqrt{GM_*/r_p}$, where $r_p = 1$.

We further simplify our study to non-self-gravitational systems, in which the self-gravity of the fluid is not taken into account for the fluid motion; hence, $\Phi = \Phi_* + \Phi_p$, where Φ_* is the potential field of the central star and Φ_p is the potential field of the planet, which is given by $\Phi_p = -M_p/\sqrt{r^2 + \epsilon^2}$, where M_p is the planet mass and ϵ is taken to be 0.2 times the Roche Lobe of the planet. The initial disk model has Keplerian rotational profile and uniform density, which results in an initial radial PV profile $\xi(r)$ that is proportional to $r^{-\frac{3}{2}}$. The value of density and viscosity are chosen to follow those specified in de Val-Borro et al. (2006). To handle the hydrodynamics part, we adopted a legacy code developed by the astrophysical group at Louisiana State University to study star formation (Tohline 1980). The code is explicit and 2nd order in both space and time. It splits the source term and advection term in a manner similar to Zeus (Stone & Norman 1992). Other features implemented include Van Leer upwind scheme, artificial viscosity to handle shock, and, staggered cylindrical grids. The code is originally three-dimensional, but adapted to 2D in this work. At the boundary of our computational grids, mass is allowed to flow off the grids but no inflow is allowed. We also implemented the wave-killing boundary condition specified in de Val-Borro *et al.* (2006). The planet is put on a fixed orbit at $r = 1$ AU in most of our simulations. We also allowed the planet to move in some runs to study the effect of RWIs on its migration. The units adopted are the following: the gravitational constant $G = 1$, length unit is 1 AU, and $M_* + M_p = 1$. Some brief results follow in the next section.

3. Simulation results

We present simulation results for systems with a Neptune-mass planet embedded in either a locally isothermal disk (non-barotropic) or an isothermal disk (barotropic). Fig.1a illustrates a linear color map of surface density distribution on a polar plot for a locally isothermal disk at $t \approx 150$ orbits. Focusing on the formation of high density areas (red/brown regions), we observe not only red Lindblad spiral arms, but also, other non-axisymmetric high density structures at different locations: the inner edge of the outer disk (brown arc-shaped region centered around 7 o'clock) , the outer edge of the inner disk (red/brown arc-shaped region around its edge); interestingly, the density inside the gap is no longer axisymmetric any more, as suggested by the light blue region around 10 oclock. The radial locations of these non-axisymmetric structures match exactly with the local minima of the PV distribution (see Fig.4 of Ou *et al.* 2007), which is consistent with previous studies on RWIs (Li *et al.* 2000) that RWIs are capable of introducing

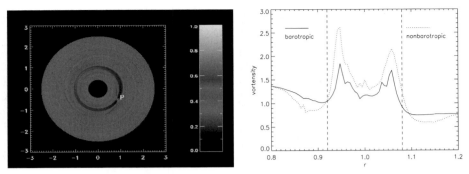

Figure 1. Simulations for an embedded Neptune-mass planet in an inviscid disk. *Left panel*: (1a) Color map of the disk surface density distribution in linear scale $t \approx 150$ orbits for a resolution run (400 x 1600). A letter "P" is labeled next to the location of the planet. The color bar represents relative rather than absolute values. *Right panel*: (1b) Comparison of azimuthally-averaged radial PV profile between two disk-planet systems with a planet embedded in a locally isothermal disk (non-barotropic) and an isothermal disk (barotropic). Vertical dashed lines illustrate edges of the horseshoe region. The planet is located at $r = 1$.

non-axisymmetric density distribution in a disk. The total torque acted on the planet is negative in average; however, as time goes on, it varies greatly in amplitude (see Ou et al. 2007 for more details). This is because those high density blobs formed through RWIs exert a stronger torque on the planet when they are traveling in the vicinity of the planet, hence, large amplitude of oscillations show up in the time evolution of the total torque. Our results on the temporal oscillating behavior of the total torque is consistent with those in Li et al. (2005).

In general, we expect vorticity (hence, PV) to be generated around the shock outside coorbital region, where density gradient and pressure gradient are not aligned to each other. However, detailed analysis shows that PV is generated both near shocks and around the planet in the coorbital region. To understand the mechanism that generates PV within the coorbital region, we note that our disk models have a non-barotropic EOS with a radial variation of sound speed c_s, which brings in misalignment of pressure gradient and density gradient wherever azimuthal density gradient appears ($\nabla c_s \times \nabla \Sigma \neq 0$). This misalignment acts as a source term for the generation of vorticity within the coorbital region, where no shock but strong azimuthal density gradient presents. Therefore, the large density depression in the same region naturally gives birth to PV increment. Fig.1b shows a comparison of azimuthally averaged radial PV profiles for two runs at $t \approx 60$ orbits: the disk in one run has an isothermal EOS (barotropic) and the other disk has a locally isothermal EOS (non-barotropic). The PV peaks in the non-barotropic disk is significantly higher than those of barotropic disk. Furthermore, a peak located around the planet ($r = 1$) shows up in the non-barotropic disk. To further examine if our results are resolution-dependent, we carried out a run with resolution 800×3200 for ~ 70 orbits. The simulation exhibits PV maxima within the coorbital region as well (see Ou et al. 2007). As a further evidence that PV is generated around where azimuthal density gradient exists, Figure 2a and 2b illustrate the distribution of PV increment and azimuthal density gradient, respectively, for the run with resolution 800×3200 at $t \sim 70$ orbits. For a better view, they are zoomed in the neighborhood of the planet. It is observed that PV is generated within the Roche lobe, where strong azimuthal density gradient, instead of shocks, exists. Such a generating mechanism of PV inside a planet's Roche lobe has the same origin as baroclinic instability discussed in Klahr & Bodenheimer (2003), which contributes to global turbulence within the disk. On the other hand, we also observe that

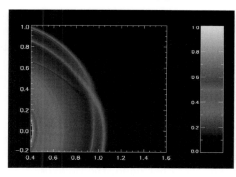 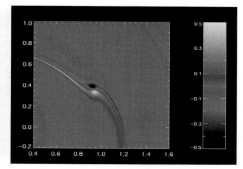

Figure 2. Simulations for the Neptune-mass planet in disk. *Left panel*: (2a) Distribution of PV around the Neptune-mass planet at $t = 70$ orbits for the run with resolution 800 x 3200. Shocks are well resolved. Vortensity is generated within the coorbital region, especially in the Roche lobe, where no shock exists. The azimuthal density gradient within the same region acts as a source term of PV generation. *Right panel*: (2b) Distribution of azimuthal density gradient at the same time. Strong density gradient is observed within the Roche lobe and coorbital region.

the oscillations of the torque acted on a Neptune-mass planet appear much earlier in our simulations compared to previous studies, which suggests that PV generation is even more effective and common in non-barotropic disk models (see also Ou *et al.* 2007).

As shown in Figure 2b, the azimuthal density gradient flips the sign across the planet; thus, we expect the PV increment also flip the sign there. However, the PV change within the Roche-lobe shown in Figure 2a is always positive, this is possibly due to PV mixture along librating stream lines in the horse region where fluid elements make the U-turn. A detailed and rigorous analysis of fluid dynamics and PV mixture around a Neptune-mass planet embedded in a non-barotropic disk will be carried out in a separate investigation.

In our simulations of a freely moving Neptune-mass planet (Ou *et al.* 2007), RWIs do not change the overall picture of inward migration of a Neptunian planet; but they have significant influence on the torque exerted on the planet and make the migration speed of the planet non-monotonic.

4. Summary and Discussion

In this work, we carried out high resolution simulations on the interaction between a protoplanetary disk and an embedded planet with emphasis on the interplay between a disk and a planet under the influence of baroclinic generation of PV. Our results are consistent with classical analysis on the interaction between a protoplanetary disk and an embedded planet through Lindblad torques. We confirmed previous outcomes that non-axisymmetric RWI is likely to develop under certain circumstances and have an important influence on the migration of a planet inside an inviscid disk. We also found that the generation of PV is more common and effective in disks with non-barotropic EOS through the baroclinic instability, further favoring the development of RWIs. As the asymmetry of the density distribution induced by RWIs becomes prominent, the resulting density blobs exert periodical and enhanced gravitational pull onto the planet as they pass by the vicinity of the planet, which causes the total torque received by the planet undergo large amplitude oscillations.

Our analysis shows that strong vorticity has been generated around the planet through baroclinic effect, this may help its core to accrete materials faster in a way suggested in Klahr & Bodenheimer (2006) and shorten the time scale needed to form a Jupiter mass planet. As a side effect of an inwardly migrating planet, RWIs introduce non-axisymmetric density blobs along the way. These enhanced density blobs with strong vortices may help rapid formation of new planet cores within them, especially in inner regions of circumstellar disks where rapid precipitation and coagulation of solid materials are likely to happen (Silverstone 2006). If these new-born cores could survive during the migration of a giant planet (Raymond 2006), they may produce Earth-like planets in the Habitable Zones (Ji et al. 2007) or Hot Earths interior to a close-in giant planet (Raymond 2006). One important issue is where this non-barotropic mechanism is expected to occur in a protoplanetary disk. Typically, at small radii ($r < 10$ AU), the disk flow can not dissipate its internal energy in a time less than the horseshoe U-turn time, the locally isothermal approximation does not hold there. Therefore, we expect this mechanism is likely to happen at larger radii. On the other hand, the breakdown of the locally isothermal approximation does not necessarily remove the temperature variation. If temperature variation preserves at smaller radii, it may still favor baroclinic effect. In realistic situations, this could be a much more complicated process than what is described here. Further investigations on these issues with full 3D simulations and disk self-gravity (see also Baruteau & Masset 2008; Zhang & Zhou 2008 in this issue) are under way.

Acknowledgements

We thank the anonymous referee for useful comments and suggestions that helped to improve the contents. J.H.J. acknowledges the financial support by the National Natural Science Foundations of China (Grants 10573040, 10673006, 10203005, 10233020) and the Foundation of Minor Planets of Purple Mountain Observatory. S.O. was partially supported by NSF grant AST-0407070.

References

Baruteau, C. & Masset, F. 2008, *IAU S249*, this issue
Goldreich, P., Tremaine, S., 1978, *ApJ*, 222, 850
Ida, S., & Lin, D.N.C. 2004, *ApJ*, 604, 388
Ji, J., Kinoshita, H., Liu, L., & Li, G.Y. 2007, *ApJ*, 657, 1092
Klahr, H.H., Bodenheimer, P., 2003, *ApJ*, 582, 869
Klahr, H.H., Bodenheimer, P., 2006, *ApJ*, 639, 432
Kley, W. 1999, *MNRAS*, 303, 696
Koller, J., Li, H., Lin, D.N.C., 2003, *ApJ*, 596, 91
Li, H., et al., 2000, *ApJ*, 533, 1023
Li, H., et al. 2005, *ApJ*, 624, 1003
Lissauer, J. J. 1993, *ARAA*, 31, 129
Masset, F.S., 2001, *ApJ*, 558, 453
Masset, F.S., Papaloizou, J.C.B., 2003, *ApJ*, 588, 494
Nelson, R.P., Papaloizou, J.C.B., Masset, F., Kley, W., 2000, *MNRAS*, 318, 18
Ou, S., Tohline, J.E., 2006, *ApJ*, 651, 1068
Pollack, J. B., et al. 1996, *Icarus*, 124, 62
Raymond, S. N., Mandell, A. M., & Sigurdsson, S. 2006, *Science*, 313, 1413
Safronov, V.S. 1969, Evolution of the Protoplanetary Cloud and Formation of the Earth and the Planets (Moscow:Nauka)
Silverstone, M.D., et al., 2006, *ApJ*, 639, 1138
Stone, J.M., Norman, M.L., 1992, *ApJS*, 80, 753
Tohline, J. E. 1980, *ApJ*, 235, 866
de Val-borro, M., et al. 2006, *MNRAS*, 370, 529
Ward, W. R. 1997, *Icarus*, 126, 261
Zhang, H., Zhou, J.-L. 2008, *IAU symposium 249: Exoplanets: Detection, Formation and Dynamics*, eds. Y.-S. Sun, S. Ferraz-Mello & J.-L. Zhou, p. 413

Runaway migration in a multiple-protoplanet system

Hui Zhang and Ji-Lin Zhou

Department of Astronomy,Nanjing University,Nanjing 210093,China

Abstract. We investigate the migration of two giant planets embedded in a proto-stellar disk. The inner planet(initially located at $R_{10} = 1$) is of 1 Jupiter mass and the outer one($R_{20} = 1.5$) is of 1 Saturn mass. We find that due to the existence of the inner massive planet, the outer planet can not open a clear gap. Instead of an inward migration and being captured by the mean motion resonance of the inner planet, the outer planet undergoes an outward runaway migration. We conclude that this runaway migration is caused by the co-rotation torque in the co-orbital region of the outer planet and sustained by the wave(flow) driven by the inner massive planet.

Keywords. Exoplanet, multiple planets, migration, runaway migration

1. Introduction

Migration of planets is one of the important processes that affect the final architecture of a planet system. Important migrations such as Type I, Type II and Type III migrations may change substantially the configuration of multiple-planet systems(See Masset 2008 for a review). A protoplanet normally undergoes Type I or Type II migration depending on its mass. These two kinds of migration had been well studied both analytically and numerically during the past years. However, the cause as well as the consequence of Type III migration remains unclear and needs further investigations especially in a multiple-planet system. Linear analysis suggests that the interaction between an embedded planet and the disk leads to a negative torque on the planet, which is called differential Lindblad torque(Goldreich & Tremaine 1980, Ward 1986). So it is believed that low mass protoplanets will undergo rapid inward migration(Type I) and be trapped in the mean motion resonances of the giant planet in a multiple-planet system(Kley 2000, Cresswell & Nelson 2006). The first numerical simulation of Jupiter and Saturn embedded in a gas disk has been done by Masset and Snellgrove(2001). They found the Saturn was captured into the 2:3 mean motion resonance with Jupiter. And then the two planets reversed their migration outward in parallel. Morbidelli and Crida (2007) extend their work by exploring a wider set of initial conditions and disk parameters. They found the two planets will eventually end up locked in a MMR. The situation maybe different if planets embedded in a disk where the disk gas is still massive. In such a situation, the co-rotation torque on the protoplanets may arise and play a significant role on the orbital evolution of planets(Masset & Papaloizou 2003).

In this report, we study the onset of type III migrations in a two-planet system. The inner planet is set to have a mass of 1 Jupiter mass and the outer one is of 1 Saturn mass. The planets are embedded in a massive disk with totally more than 20 Jupiter masses distributed in the disk uniformly. Numerical simulation indicated that the outer planet can not open a clear gap due to the presence of inner planet perturbation, and a rapid runaway outward migration occurs on the outer planet. By comparing with the single-planet case, we conclude that this outward runaway migration is probably caused by the existence of inner planet which performs strong perturbation to the outer planet's co-orbital region and results in a strong co-orbital torque on the planet. We present our numerical methods in section 2 and our results in section 3. In section 4 we summary and discuss the implication of our results.

2. Physical and Numerical Model

2.1. *Physical model*

We construct a 2D numerical hydrodynamic scheme to solve the continuity and momentum equations, neglecting the effect of any explicit viscosity. We solve the governing equations in the Cartesian coordinate.

The vertically averaged continuity equation for the disk gas is given by

$$\frac{\partial \sigma}{\partial t} + \frac{\partial (\sigma u_x)}{\partial x} + \frac{\partial (\sigma u_y)}{\partial y} = 0. \tag{2.1}$$

The equations of motion in the Cartesian coordinates are,

$$\frac{\partial (\sigma u_x)}{\partial t} + \frac{\partial (\sigma u_x^2)}{\partial x} + \frac{\partial (\sigma u_x u_y)}{\partial y} = -\frac{\partial P}{\partial x} - \sigma \frac{\partial \Phi}{\partial x}, \tag{2.2}$$

$$\frac{\partial (\sigma u_y)}{\partial t} + \frac{\partial (\sigma u_x u_y)}{\partial x} + \frac{\partial (\sigma u_y^2)}{\partial y} = -\frac{\partial P}{\partial y} - \sigma \frac{\partial \Phi}{\partial y}, \tag{2.3}$$

where P is pressure and Φ is the gravity potential of the star-planets-disk system, which includes the softened potential of central star (Φ_s), softened potential of the planets (Φ_{pi}), potential of the disk itself (Φ_d) and indirect potential (Φ_i) due to the acceleration of origin by the planets and the disk. σ is the vertical integrated surface density.

We assume the disk gas has an isothermal equation of state. The sound speed is $c_s = (H/r)v_k$ where we set $H/r = 0.04$ and $v_k = 1$(Keplerian velocity at $r = 1$). We don't add any explicit viscosity in the simulation. There is however some numerical viscosity associated with our computational scheme and grid effects. To focus on the migration and reduce variables, we don't allow the planet to accrete gas from the disk.

2.2. *Computational Units*

For numerical convenience we set gravitational constant $G = 1$, solar mass $M_\odot = 1$ and the radius of inner planet's initial orbit $R_{10} = 1$, where $R_{10} = 5.2 AU$. The unit of time is $1/2\pi$ of the inner planet's initial orbit period P_{10}. Total mass within the disk are about 20 Jupiter masses, which corresponds to a uniform surface density $\sigma = 0.0006$.

2.3. *Numerical Method*

The **Antares** code we have developed is adopted in the calculations. It is a 2-D Godunov code based on the exact Riemann solution for isothermal or polytropic gas, featured with non-reflecting boundary conditions. At the boundaries we adopt non-reflecting (absorbing) boundary condition to make sure there is no wave reflect from the boundaries. To avoid initial impact of the system we adopt a 'quiet start' initial condition. For the orbit of planets we use RK78 to integrate it. For more details of the physical and numerical methods please see Zhang *et al.* 2008.

3. Numerical Results

We performed a hydrodynamic simulation to investigate the orbit evolution of two giant planets embedded in a massive disk. The computational domain is from -3 to 3 in

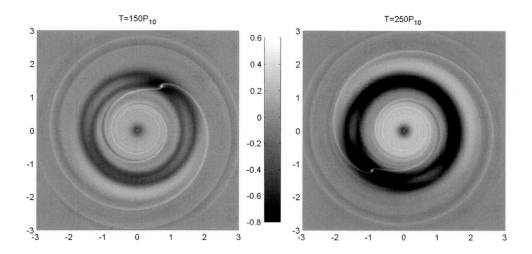

Figure 1. Density evolution of the disk contains only one planet. Without the inner massive planet, the outer planet may open a clear gap when it is of 1 Saturn mass.

x direction and from -3 to 3 in y direction. Total resolution is $N_x \times N_y = 1024 \times 1024$. At beginning the disk is uniform and have 20 Jupiter masses. The initial locations of two planets are $R_{10} = 1$(Jupiter mass) and $R_{20} = 1.5$(Saturn mass). We hold their orbits and increase their mass gently from 1×10^{-7} to their final mass. After $200P_{10}$ which is the orbit period of the inner planet, two planets had grown up and we release them at the same time. In first case, we neglect the inner planet by setting its mass equal to 0.

As we can see in Fig.1, the (outer)planet opens a gap and a regular, symmetric horseshoe zone appears within its co-orbital region. In second case,we turn on the effects of inner planet. During the first $200P_{10}$, the inner planet had already form a gap but the outer planet do not clear any gap. Since we set $H/r = 0.04$ in our simulations, it will open a clear gap when the embedded planet is above tens of Earth masses(as in the first case). However, the existence of the inner massive planet prevents the outer planet from opening a clear gap. From Fig.2 we can see the inner planet had already open a gap while the outer planet's co-orbital region is still replete of gas.

Further more the waves excited by the two giant planets perturb each other's co-orbital region significantly. During a short period after release, outer planet undergo inward Type I migration and soon be perturbed by the inner planet. Gas flows in and out from its co-orbital region rapidly and form a large vortensity gradient, then this vortensity gradient generates great co-rotation torque on the planet. As a result, the migration of outer planet become unstable. We see a outward runaway migration occurs soon after the release. We also note that during a short period the outer planet is captured into the 5:3 MMR of the inner planet. But as soon as the the outer planet's migration became unstable this MMR is broken up.In the first case when the inner planet is absent, the outer planet is embedded in the gap and undergoes Type II migration which is smooth and slow. See Fig.3-a.

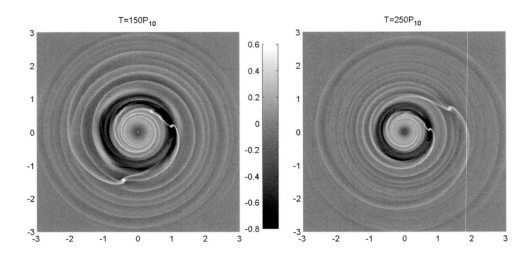

Figure 2. Density evolution of the disk contains two planets. The interaction between the two planets significantly perturbs each planet's co-orbital region, and prevent the outer less massive one from clearing gap.

4. Conclusion and Discussion

From the comparison, we can conclude that the runaway migration is triggered by the perturbation comes from the inner planet. Masset & Papaloizou (2003) have shown that a protoplanet embedded in a massive disk may undergo runaway migration when the co-orbital mass deficit δm is comparable with the planet mass M_p. They analyzed co-rotation torque and gave a relation between the migration rate and co-orbital mass deficit. In a Keplerian case it reads:

$$\frac{1}{2}a\Omega_p(M_p - \delta m)\dot{a} = \Delta\Gamma_{LR} - \frac{\pi a^2 \delta m}{3x_s}\ddot{a}, \qquad (4.1)$$

where a is the semi-major axis of the planet orbit, Ω_p is the angular velocity of the planet, $\Delta\Gamma_{LR}$ is the differential Lindblad torque and x_s is the half-width of planet horseshoe zone. The co-orbital mass deficit ($\delta m = M_{HS0} - M_{HS}$) can be roughly understood as the mass variation within the planet horseshoe zone, where M_{HS0} is the initial mass within planet horseshoe zone and M_{HS} is the mass within planet horseshoe zone which changes with time. When $\delta m \geqslant M_p$, above equation indicates that small perturbation to semi-major axis (a) leads to its exponentially growing on a timescale of few tens of orbital periods. And in specific initial condition(a and $\dot{a}$) outward runaway migration may occur naturally.

A strong perturbation to planet orbit makes gas flush into or out of its co-orbital region and therefor leads to a large δm. Then the planet will undergo runaway migration triggered by the co-rotation torque. Once the rapid migration occurs, it can self-sustained because the gas will keep flooding in its co-orbital region as it travels a large radial shift within very short time(Zhang et al. 2008).

In our simulations,the existence of the inner giant planet gives strong perturbations to the gas traveling within the outer planet horseshoe region, which leads to a large δm. Fig.3-b shows the evolution of M_{HS} of each planet. Clearly, the rapid variation of M_{HS} greatly affects the planet migration. As soon as δm becomes larger than the

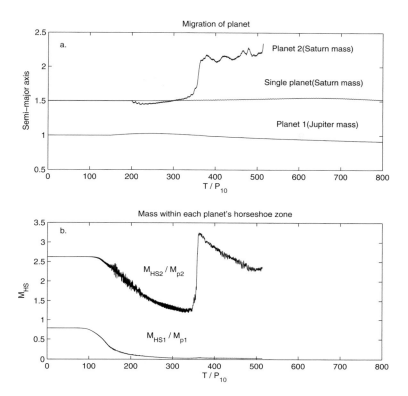

Figure 3. (a).Migration curves of multiple planets system. Two cases are shown in this figure. The first case contains only single Saturn mass planet. The second case contains two planets, one is of a Jupiter mass while the other is of a Saturn mass. (b).Evolution of M_{HS} in second case. Curves are normalized by each planet's mass. Clearly the rapid change of M_{HS} significantly affects the migration of planet 2. At around $T = 350 P_{10}$, a large δm appears and leads to the runaway migration of planet 2.

outer planet($M_{p2} = 3 \times 10^{-4}$), a runaway migration occurs. Within a few orbits after $T = 350 P_{10}$, M_{HS} increases sharply and $\delta m \sim 2 M_{p2}$. When the planet 2 is far away from planet 1, this runaway migration can be halted since the perturbation has damped. But its orbit is still unstable. The properties of this kind of runaway migration substantially depends on the initial condition of the system, which merits further studies.

5. Acknowledgement

We thank Prof. Lin, D.N.C and Dr. F.S. Masset for very useful discussions. The work is in part supported by a grant from National Basic Research Program of China (2007CB4800), Natural Science Foundation of China (10778603,10403004).

References

Cresswell, P.& Nelson, R. P. 2006, *A&A*, 450, 833
Goldreich, P. & Tremaine, S. 1980, *ApJ*, 241, 441
Kley, W. 2000,*MNRAS*, 313, 47

Masset, F. S. & Snellgrove, M. 2001, *MNRAS*, 320, L55-L59
Masset, F. S. & Papaloizou, J. C. B. 2003, *ApJ*, 588, 494
Masset, F. S. 2008, in IAU Symposium No.249. Exoplanets: Detection, Formation and Dynamics, Y.-S. Sun, S. Ferraz-Mello & J.-L. Zhou, eds.(Suzhou, China), p. 331
Morbidelli, A. & Crida, A. 2007, *Icarus*, 191, 158
Ward, W. R. 1986, *ICARUS*, 67, 164
Zhang, H., Yuan,C., Lin, D. N. C. & Yen, D. C. C. 2008, *ApJ*, (acceptted), Arxiv-astroph:0709.0338.

Effects of dissipating gas drag on planetesimal accretion in binary systems

Ji-Wei Xie and Ji-Lin Zhou

Department of Astronomy, Nanjing University, China
email: xjw0809@163.com; zhoujl@nju.edu.cn

Abstract. We numerically investigate the conditions for planetesimal accretion in the circumprimary disk under the perturbing presence of a companion star, with focus on the γ Cephei system. Gas drag is included with a dissipating time scale of 10^5 years. We show at the beginning(within $10^3 \sim 10^4$ years), gas drag damps the $\triangle V$ between planetesimals of same sizes and increases $\triangle V$ between planetesimals of different sizes. However, after increasing to high values(300~800m/s), we find the $\triangle V$ between bodies of different sizes decrease to very low values (below $10m/s$) in a few $10^5 yrs$ (depending on the gas-dissipating time scale T_{damp}, radial size R_p and semi-major axis a_p of planetesimals). Hence, the high $\triangle V$ is somewhat short-lived, and runaway accretion can be turned on later. We conclude that the conditions for planetary formation in binary systems (even close binary systems) are much better than what we expected before.

Keywords. planetary formation, accretion, planetesimals, binary stars

1. Introduction

Because of the increasing number of discovered planets in binary systems and the belief that a majority of solar type stars reside in binary or multiple systems, problem of planetary formation in binary systems becomes a crucial one. Most of the discovered binary systems are S-type systems(e.g. γ Cephei system, see Hatzes *et al.* 2003) in which planets orbit the primary star and with a companion star surrounding them on an outer orbit. According to the classical planetary formation scenario, planets form in a protoplanetary disk of gas and dust orbiting a protostar. The formation process is usually treated in three stages: [stage I] formation of kilometers-size plantesimals ($10^{18} - 10^{22}g$) from dust accretion; [stage II] accretion of plantesimals to planetary embryos ($10^{25} - 10^{26}g$, Mercury- to Mars-size) in $\sim 10^5$ years, depending on the initial disk density; [stage III] giant impacts between embryos, producing full-size $\sim 10^{26}$ to $\sim 10^{27}$ g terrestrial planets in $\sim 10^7 - 10^8$ years. Here we focus on the stage II to see the influence of the companion on the planetesimal accretion.

The companion star, especially when it is in a close orbit with a high eccentricity, may prevent planetary formation because the companion reduces the size of the accretion disk, and it excites high relative velocities between colliding planetesimals. The relative impact velocity ($\triangle V$) is a critical parameter which determines whether accretion or erosion dominates the planetesimal collisional evolution. Due to the perturbation by the companion, $\triangle V$ may exceed the planetesimal escape velocities($V_{esc} \sim 10 \times (R_p/km)ms^{-1}$), thus preventing runaway accretion, or even the threshold velocity ($V_{ero} \sim b \times V_{esc}$) for which erosion dominates accretion, where b is a dimensionless coefficient of $2 \sim 5$ roughly depending on the prescription on collision. However, the presence of gas drag does not only damp the secular perturbation, but it also forces a strong periastron alignment of planetesimal orbits. This alignment significantly reduces $\triangle V$, favoring the accretion process(Marzari & Scholl 2000). Nevertheless, the alignment forced by the gas drag induces

Table 1. initial parameters for runs

semi-major axis of companion	$a_B = 18.5 \pm 1.1 AU$
eccentricity of binary	$e_B = 0.361 \pm 0.023$
mass of primary	$M_A = 1.6 M_\odot$
mass of companion	$M_B = 0.4 M_\odot$
physical radius of planetesimals	$R_p = 2.5, 5, 15, 50 km$
initial eccentricities of planetesimals	$0 < e < 10^{-3}$
initial inclinations of planetesimals	$0 < i < 5.10^{-4}$
planetesimal density	$3 g \cdot cm^{-3}$
initial gas density at 2AU	$\rho_{g20} = 2.10^{-9} g \cdot cm^{-3}$
initial gas density radial profile	$\rho_g \propto r^{-2.75}$
gas-dissipating time scale	$T_{damp} = 10^5 years$

another problem. Since the alignment is size-dependent, it can only reduce $\triangle V$ between planetesimals of same size, and at the same time it increases $\triangle V$ between planetesimals of different sizes. Thebault *et al.* (2006) found that the differential orbital alignment was very efficient, leading to a significant $\triangle V$ increase for any departure from the exact equal-size condition($R_1 = R_2$, where R_1 and R_2 are the radial sizes of two colliding bodies). We reinvestigate the effects of gas drag on relative velocities $\triangle V$ among a swarm of planetesimals in γ Cephei system(as our model system, and the mechanism can be applied to other binary systems). We will show that the presence of a dissipating gas can induce a $\triangle V$ decrease between planetesimals of different sizes in a timescale $(2 \sim 5) \times T_{damp}$(gas-dissipating timescale). After $(2 \sim 5) \times T_{damp}$ the $\triangle V$ between planetesimals of different sizes will reach very low values, allowing runaway accretion.

2. Numerical model and method

Here we adopt the parameters listed in table 1. We concentrate on four sizes planetesimals ($R_p = 2.5, 5, 15, 50 km$). The relative velocity $\triangle V(2.5, 5)$ between bodies of $R_p = 2.5 km$ and $R_p = 5 km$ can be representative for small planetesimals, and so does $\triangle V(15, 50)$ for large planetesimals (Thebault *et al.* 2006).

Following Weidenschilling and Davis (1985), the gas drag can be expressed as:

$$\mathbf{F} = -Kv\mathbf{v} \tag{2.1}$$

where $\mathbf{F}$ is the force per unit mass, $\mathbf{v}$ the relative velocity between planetesimal and gas, v the velocity modulus, and K is the drag parameter defined as:

$$K = \frac{3\rho_{g0} C_d}{8\rho_p R_p} \exp\left(-t/T_{damp}\right) \tag{2.2}$$

where ρ_{g0} is the initial gas density, ρ_p and R_p the planetesimal density and radius, respectively. C_d is a dimensionless coefficient related to the shape of the body($\simeq 0.4$ for spherical bodies). The *exp* function is used to include the dissipation of gas with a

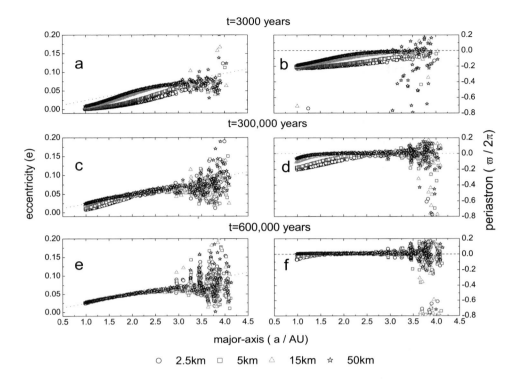

Figure 1. Distributions of planetesimal eccentricities and periastrons vs. semi-major axes at three different epochs. Bodies of different sizes are plotted in different styles. The dashed lines which are in the 3 left panels and 3 right ones denote e_f (forced eccentricity) and $\varpi = 0$ (which means the planetesimal periastrons $w_p = w_f$ the forced periastron), respectively.

damping time scale T_{damp}. As shown in Eq.(2), the gas drag force is size-dependent($\propto 1/R_p$).

It has been shown by Gladman (1993) that two isolated planetesimals will eventually collide if they are within a critical distance $2.4(\mu_1 + \mu_2)^{1/3}a$, where μ_1, μ_2 and a are the masses in solar units and average semi-major axis of two colliding planetesimals, respectively. We adopt it as the criterion of searching for impact, and calculate the $\triangle V$ for each impact. In all the runs, we used the fourth order Hermite integrator (Kokubo et al., 1998), including the gas drag force and the perturbation of companion. As gas drag also forces inward drift of planetesimals, we adopt the following boundary conditions: bodies whose major-axes are less than R_{in}(greater than R_{out}), will be reset to $R_{out}(R_{in})$, where R_{in} and R_{out} are inner and outer boundaries of planetesimal belt. In these resetting processes, only the semi-major axes of those bodies are changed, and other orbital elements are preserved.

3. Result

We first performed a simulation where 1000 planetesimals (4 equal-number groups: $R_p = 2.5, 5, 15, 50 km$, mutual interactions are neglected) were initially distributed between 1 AU and 4 AU from the primary. Fig. 1 shows the distributions of planetesimal eccentricities and periastrons vs. semi-major axes at different epochs. Beyond 3 AU, the distribution of planetesimal eccentricities and periastrons are random because the shorter period perturbations and mean motion resonances dominate. Hence, hereafter only planetesimals within 3 AU are discussed. In Fig. 1a(b), within 3 AU (hereafter), every eccentricity (periastron) reaches an equilibrium phased value at 30,000 years. This equilibrium phased value depends on the balance between the perturbation by the companion and the gas drag force. Due to the side-dependence of gas drag force, bodies of different sizes reach different equilibrium eccentricities (periastrons). In Fig. 1a(b) the four lines are associated with bodies of four kinds of sizes ($R_p = 2.5, 5, 15, 50 km$). As the gas dissipates gradually, the equilibrium phased eccentricities (periastrons) move to larger values, but at the same time the differences in eccentricity (periastron) distributions among bodies of different sizes become smaller (see Fig. 1c(d)). After a long time (600,000 years, see Fig. 1e(f)), almost all eccentricities (periastrons) fix on e_f (w_f). Hence, the differential phasing effect caused by the size-dependence of gas drag is reduced if the gas dissipation is included.

As shown in Fig. 1, dissipating gas drag reduces the effect of size-dependence of orbital alignment. Hence, we can expect a decrease in $\triangle V$ between bodies of different sizes. Fig. 2a shows the $\triangle V$ (R_1, R_2) as the function of time at 2 AU from the primary star. Fig. 2b and Fig. 2c show the average eccentricity and periastron of bodies at 2 AU as the functions of time. It is evident that a correlation(the larger differences in orbital elements, the larger value of $\triangle V$) exists between Fig. 2a and Fig. 2b(c). In this calculation, 1000 Planetesimals were initially distributed with major-axes between 1.5 and 3 AU. This ring is so wide that it includes most of the impacts that occur at 2AU (not semi-major axis of planetesimal orbit but distance from the center star). Although some bodies with

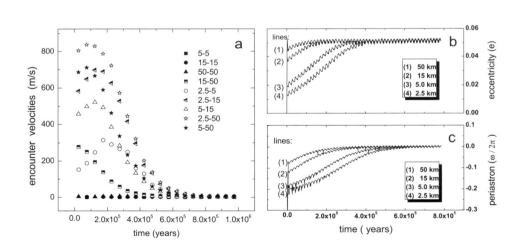

Figure 2. Correlations between $\triangle V$ and distributions of orbital elements. (a)Average encounter velocities $\triangle V$ at 2 AU from the primary star v.s. time in dissipating gas case. $\triangle V$ between bodies of different sizes are plotted in different styles. (b) and (c): Average eccentricity and periastron(ϖ) at 2 AU from the primary v.s. time in dissipating gas case. Bodies of different sizes are plotted in different styles

high eccentricities on outer orbits (a>3AU) can cross the orbits of bodies near 2 AU region, the rate of this kind impact is much lower because of the longer orbital periods and smaller number of planetesimals in outer region. Hence, a initial planetesimal belt between 1.5 and 3.0 AU is reasonable for tracing the $\triangle V$ at 2 AU.

From Fig. 2a, it appears that the $\triangle V$ between bodies of equal-side are always small because of orbital alignment. However, the $\triangle V$ between bodies of different sizes first increase to high values quickly(e.g. 300∼800 m/s). Then each of them experiences a relatively slow decrease. After about 5×10^5 years they all stay steadily at low values (below 10m/s). These low $\triangle V$ favor planetesimal accretion and even runaway growth. The time scale for $\triangle V$ decreasing depends on the gas damping time scale T_{damp}, body size R_p and semi-major axis of body a_p. From Fig. 2a, $\triangle V$ (R_1, R_2) with R_1 or R_2 below 5 km decrease to low values within $\sim 5 \times T_{damp}$, but only $\sim 2 \times T_{damp}$ is needed for $\triangle V$ (R_1, R_2) with R_1 and R_2 both greater than 15 km. Furthermore, from Fig. 1b(c), bodies at outer region converge their eccentricities (periastrons) faster than inner bodies. For this reason, we find $\triangle V$ (R_1 or $R_2 < 5km$) decreases to and stays steadily on $30m/s$ ($100m/s$) within $\sim 4(3) \times T_{damp}$ for bodies at 2.5 (3.0) AU. The larger equilibrium value of $\triangle V$ at outer region is due to the larger perturbation there.

4. Conclusion

The previous results show that $\triangle V$ depends on the delicate balancing between gas drag and secular perturbation by the companion. Bodies of different sizes reach different equilibrium values in eccentricities and periastrons. This is called size-dependence of orbital alignment. Its effect is very efficient and induces a significant $\triangle V$ increase between bodies of different sizes. Based on these facts, Thebault et al. (2006) investigated $\triangle V$ for binary systems of various orbital parameters. They found on the region at a distance of 1 AU from the primary, $\triangle V$ is always high preventing runaway accretion, except in a narrow $e_b \simeq 0$ domain. In this paper, focusing on the γ Cephei system, we adopt a fast dissipating and massive(initial density is 10 MMS) gas disk model, which is possible for some close binary systems. We also find a significant increase of $\triangle V$ between bodies of different sizes caused by size-dependence of orbital alignment, but at the same time this high $\triangle V$ decreases to a very low value in about (2∼5) $\times T_{damp}$. Notice that the $\triangle V$ will not increase again even when the gas is dissipated totally. After (2∼5) $\times T_{damp}$, the $\triangle V$ is low enough for runaway accretion, and there is enough gas ($0.1 \sim 1.0 MMS$) left for gas giant formation.

Acknowledgments

We thank W. Kley for useful discussions and suggestions. This work is supported by NSFC(10778603), National Basic Research Program of China(2007CB4800).

References

Desidera, S. & Barbieri, M. 2007, A&A, 462, 345
Eggenberger, A., Udry, S., & Mayor, M. 2004, A&A, 417, 353
Gladman, B., 1993, Icarus, 106, 247
Hatzes, A.P., Cochran, W. D., Endl, M., McArthur, B., Paulson, D., Walker, G. A. H., Campbell, B., Yang, S., 2003, ApJ, 599, 1383
Hayashi, C., 1981, PthPS, 70, 35
Kokubo, E., Yoshinaga K., & Makino, J. 1998, MNRAS, 297, 1067
Marzari, F. & Scholl, H. 2000, ApJ, 543, 328

Thebault, P., Marzari, F., Scholl, H., Turrini,D., Barbieri,M.,2004,A&A,427,1097
Thebault, P., Marzari, F., Scholl, H., 2006, Icarus, 183, 193
Tsukamoto, Y. & Makino, J. 2007, Formation of protoplanets from massive planetesimals in binary systems, eprint arXiv: astro-ph/07072928
Weidenschilling, S. J. & Davis, D. R., 1985, Icarus, 62, 16

Part 5

Dynamics of Multiple Exoplanet Systems

Orbital determination and dynamics of resonant extrasolar planetary systems

C. Beaugé[1], S. Ferraz-Mello[2], T. A. Michtchenko[2] and C. A. Giuppone[1]

[1] Observatorio Astronómico, Universidad Nacional de Córdoba, Laprida 854, (X5000BGR) Córdoba, Argentina

[2] Instituto de Astronomia, Geofísica e Ciências Atmosféricas, USP, Rua do Matão 1226, 05508-900 São Paulo, Brasil

Abstract. In this communication we review some properties and applications of mean-motion resonances in extrasolar planetary systems, with particular emphasis on the 2/1 commensurability. A first part is devoted to the dynamical structure of the 2/1 resonance, including (but not restricted to) the so-called apsidal corotations. In a second part we discuss the orbital evolution of resonant systems under the effects of non-conservative forces. Special attention is given to the use of apsidal corotations as markers of largescale orbital decay, possibly due to disk-planet interactions in primordial times. Finally, we analyze the interplay between dynamical analysis and orbital fitting. Using the HD82943 planetary system as an example, we discuss: (i) up to what point present orbital fits allow us to distinguish between different resonant configurations, and (ii) in what ways may the dynamical structure of resonances be used as a complementary part of the orbital fitting process.

Keywords. celestial mechanics, planetary systems, planets and satellites: general

1. Introduction

A mean-motion resonance (MMR) occurs when the ratio of their orbital periods lies close to the ratio of two small integers. We can write this condition in a generic form as

$$\frac{n_1}{n_2} \simeq \frac{(p+q)}{p}, \qquad (1.1)$$

where n_i are the mean motions (orbital frequencies) of the planets and p, q are integers. Following general usage, we identify the planet with smaller semimajor axis by the index 1, while 2 is reserved for its outer companion (i.e. $a_1 < a_2$). In the planar case, the dynamics of a resonant system is characterized by the evolution of two critical (or resonant) angular variables, defined as

$$\theta_i = (p+q)\lambda_2 - p\lambda_1 - q\varpi_i \qquad (i = 1, 2), \qquad (1.2)$$

where λ_i are the mean longitudes and ϖ_i the longitudes of pericenter. The value of q defines the *order* of the resonance.

If at least one of the critical angles is found to be librating, then the system is said to be *inside* the mean-motion resonance. For the present work, we define *libration* as the case in which the angle does not take all values between zero and 2π, but oscillates with a certain amplitude around an equilibrium value θ_{i_0}. We will not analyze whether this libration is dynamical (i.e. associated to a separatrix crossing) or merely kinematical, but concentrate solely on the observed behavior of the variable under question. If neither angular variable librates, the system is said to be *outside* the resonance, or in a non-resonant configuration.

Definition (1.2) implies that $\theta_1 - \theta_2 = q(\varpi_2 - \varpi_1) = q\Delta\varpi$, with $\Delta\varpi$ the difference in the longitudes of the pericenter. This a a purely secular (non-resonant) angle and its behavior defines the secular evolution of the system. Analogous to a MMR, if $\Delta\varpi$ librates, the system is said to be in an apsidal alignment. Usually this behavior is referred to as a secular resonance, although Michtchenko and Malhotra (2004) showed that true secular resonance -dynamical libration- only occur for very high eccentricities (typically of the order of 0.7). Thus, for most cases the apsidal alignment is a purely kinematical phenomenon, and no separatrix divides the librational and circulatory regions.

Since rational numbers form a dense set in the real axis, equation (1.1) does not appear to be a very strict condition. At least in principle, any two planets (with arbitrary semimajor axes) could be said to be resonant, provided the values of p, q are chosen sufficiently large. However, it can be shown (e.g. Murray and Dermott 1999) that, unless the values of p, q are small integers, the magnitude of the resonant terms in the perturbation are smaller than the short period contributions, and the corresponding resonance is not effective in dominating the dynamics of the system. Thus, only small order resonances are important: 2/1, 3/2, 3/1, 5/3, etc. Of these, the strongest is the 2/1, corresponding to the smallest integers ($p = q = 1$).

Since the important resonances only occupy a very limited region in the domain of semimajor axis, it is not expected that planetary systems should be largely associated to MMR. However, this does not appear to be the case. Among the 26 presently known multiple-planetary systems, 13 contain planets in nearby orbits in which their mutual perturbations are significant. Of these, probably 8 are in the immediate vicinity of mean-motion resonances. Of course, many of them have been detected only recently, and their orbits are still not well determined. Consequently, it is not certain whether all seemingly resonant configurations are real. Nevertheless, the proportion of resonant systems is still very significant.

Among the different resonant systems, most of them seem to be associated to the 2/1 MMR. These are GJ876, HD82943, HD73526, HD128311 and HD160691. Two planets in 55Cnc are believed to lie within the 3/1 MMR, although the most recent orbital fit by Fischer et al. (2007) seems to place these planets outside MMR. Other examples could also exist in higher-order commensurabilities, such as HD202206 and HD12661 in the 5/1. However, these resonances are very weak, and the statement of their resonant motion is more related to stability criteria than to observational evidence. Even so, it appears that approximately 8/13 ($\approx 61\%$) of the presently known planetary systems in nearby orbits contain bodies associated to MMR. Even if some are questionable and are later removed from the list, the percentage is still very significant, indicating that the frequency of resonant systems among extrasolar planets is much larger than expected from a random distribution of orbits.

In our own Solar System, a similar picture occurs in satellite systems of the outer planets, where many of the regular moons display resonant motion. Probably the best known example are the three inner Galilean satellites of Jupiter (Io, Europa and Ganymedes). Not only are these bodies locked in two successive 2/1 MMR, as the orbital configuration of all three lie in the so-called Laplace resonance where the resonant angle $\theta_L = \lambda_E - 3\lambda_G + 2\lambda_I$ librates around zero with an amplitude of less than one tenth of a degree. As far back as fifty years ago (e.g. Roy and Ovenden 1954) it was known that the frequency of resonances in the satellite systems is much higher than expected. It was later shown that these configurations could be explained by orbital evolution due to tidal effects (Goldreich 1965). These resonances are not primordial, but a consequence of *resonance trapping* due to an exterior non-conservative force.

In the past years, resonance capture has also been proposed for extrasolar planets, although in this case the driving mechanism is not tidal friction but interactions between the planets and planetesimals or gaseous disks. However, a good scenario must not only explain the presence of planets in resonance, but also the specific resonant configurations in which they presently lie. For that sake, we must understand the structure of the phase space associated to planetary mean-motion resonances, identify possible regions of motion for planetary bodies, and analyze which solutions are compatible with planetary migration.

2. Dynamics of Planetary Mean-Motion Resonances

In the vicinity of a MMR, limiting ourselves to coplanar orbits, the gravitational potential of the interaction between both planets (the so-called disturbing function) can be written in a generic form as

$$\mathcal{R} = \sum_{j_1,j_2,j_3} A_{j_1,j_2,j_3}(a_1, a_2, e_1, e_2) \cos(j_1 Q + j_2 \theta_1 + j_3 \Delta \varpi) \tag{2.1}$$

(e.g. Beaugé and Michtchenko 2003) where the index j_i can take any integer value, A_{j_1,j_2,j_3} are coefficients that depend on the semimajor axes and eccentricities, and $Q = \lambda_1 - \lambda_2$ is the synodic angle. Perturbation terms with $j_1 = j_2 = 0$ are usually called *secular terms*. Terms that only depend on θ_1 and $\Delta \varpi$ (i.e. $j_1 = 0$) are called *resonant perturbations*, while those that contain Q in their arguments are usually known as *short-period terms*. If the resonance relation is dominant, short-period contributions have smaller amplitudes (i.e. small A_{j_1,j_2,j_3}) and their effects cancel out over longer timescales. Thus, if we are interested in the long-term dynamics of the system, we can eliminate the gravitational interactions associated to the angle Q and reduce the disturbing function to resonant and secular terms only. This averaging of $\mathcal{R}$ can be carried either by analytical (e.g. Beaugé and Michtchenko 2003) or semi-analytical methods (e.g. Michtchenko et al. 2006).

The averaged resonant planetary problem (planar case) then depends only on two independent angles θ_1 and $\Delta \varpi$, and constitutes a two degrees-of-freedom dynamical system. If the orbital elements are not astrocentric, but calculated with Jacobi or Poincaré canonical variables, then the problem contains three integrals of motion, given, up to second order in the masses, by:

$$\mathcal{F} = G\frac{m_0 m_1}{2a_1} + G\frac{m_0 m_2}{2a_2} - \frac{1}{2\pi}\int_0^{2\pi} \mathcal{R} dQ$$

$$\mathcal{L} = m_1 n_1 a_1^2 \sqrt{1-e_1^2} + m_2 n_2 a_2^2 \sqrt{1-e_2^2} \tag{2.2}$$

$$\mathcal{K} = (p+q)m_1 n_1 a_1^2 + p m_2 n_2 a_2^2$$

where $\mathcal{F}$ is the Hamiltonian (total orbital energy), $\mathcal{L}$ is the total angular momentum and $\mathcal{K}$ is sometimes referred to as the *spacing parameter* (see Michtchenko and Ferraz-Mello 2001, Michtchenko et al. 2008). Given any initial condition in the immediate vicinity of the MMR, its orbital evolution will be such that all three functions remain constant for all time. It is important to emphasize that expressions (2.2) are not integrals of motion if the orbital elements are astrocentric.

Equilibrium solutions constitute a particular, but very important, type of configurations. They are given by the conditions

$$\frac{da_1}{dt} = \frac{da_2}{dt} = \frac{de_1}{dt} = \frac{de_2}{dt} = \frac{d\theta_1}{dt} = \frac{d\Delta\varpi}{dt} = 0 \tag{2.3}$$

and correspond to periodic orbits in the unaveraged system. If the solution is stable,

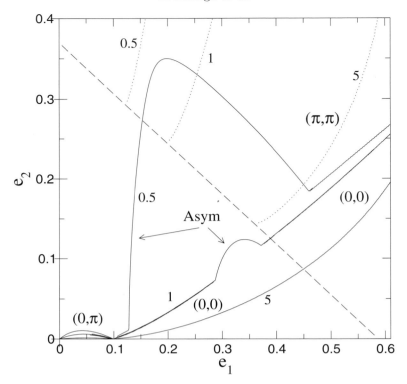

Figure 1. Families of stable zero-amplitude ACR for the 2/1 MMR, in the plane of the eccentricities, and for three different mass ratios $m_2/m_1 = 0.5, 1, 5$. (π,π)-ACR (dotted lines) exist only above the collision curve, shown as a dashed diagonal line. Other ACRs are shown in continuous curves.

initial conditions in its vicinity will display oscillations around the stationary point. In particular, both the resonant angle θ_1 and the secular angle $\Delta\varpi$ will librate around fixed values. Such behavior corresponds to a simultaneous capture into a mean-motion and a secular resonance; we have called such solutions Apsidal Corotation Resonances, or ACR for short (Ferraz-Mello et al. 2005b). Among the numerous families of ACR (See Michtchenko et al. 2006 for a catalog of ACR for several MMR), the most important are those associated to global maxima of the Hamiltonian function $\mathcal{F}$. From (2.2) and (2.3) it can be seen that maxima of $\mathcal{F}$ correspond to minima of the disturbing function $\mathcal{R}$. In other words, these ACR are configurations where the gravitational interactions between the planets are minima, and are thus preferential stable configurations for resonant planets. In the case of high-eccentricity planetary systems, where physical encounters are possible and most orbital configurations invariably lead to unstable motion, ACR is an effective mechanism that protects planets from close approaches and allow the system to survive.

The protective role of ACR makes these solutions natural nesting places for moderate-to-high eccentricity planetary systems, and it is not surprising that many (if not all) resonant planets have stable orbital fits consistent with ACR. Perhaps the most well-known example is GJ876 (e.g. Laughling and Chambers 2001, Lee and Peale 2002), where both outer planets are locked in an ACR inside the 2/1 MMR, displaying small amplitude librations around $\theta_1 = \Delta\varpi = 0$. We refer to such a solution as a $(0,0)$-ACR, where the numbers in the brackets mark the equilibrium value of θ_1 and $\Delta\varpi$, respectively. HD82943 is another example. Up to 2003, the two planets of this system were believed to be locked in a (π,π)-ACR (oscillations around $\theta_1 = \Delta\varpi = \pi$), although later updates in the masses and orbital elements placed the system in an unstable orbit around a $(0,0)$-ACR. Finally, until recently the orbital fit of the planets 55Cnc-c and 55Cnc-d was consistent with an ACR in the 3/1 MMR, where both θ_1 and $\Delta\varpi$ librate around values different from 0 or π. We call this solution an asymmetric-ACR (Beaugé et al. 2003, Ferraz-Mello et al. 2003). With the discovery of the fifth planet in 55Cnc (Fischer et al. 2007), the new orbital solution puts the planets outside the 3/1 commensurability. However, as we will discuss in next section, orbital fits for multi-planetary systems are very imprecise and often show solutions with different dynamical behavior within the 1-σ level of confidence of the best fit. Thus, it is perhaps too early to discard the resonant relation for this system, at least until further observations confirm the more recent fits.

In Beaugé et al. (2003) we showed that, at least in a first approximation, the location of the ACR depends on the planetary masses and semimajor axes only through the ratios m_2/m_1 and a_1/a_2. This permits a global study of ACR, constructing maps or catalogs of these stationary solutions that will be valid for any planetary system. Figure 1 plots the most important families of stable ACR for the 2/1 MMR in the eccentricities plane, and for three mass ratios (values shown near each curve). Except for (π,π)-ACR that only exist above the collision curve, all the other solutions form continuous families (one per mass ratio) that include quasi-circular orbits. Asymmetric solutions only exist for $m_2/m_1 \leqslant 1.015$ (Ferraz-Mello et al. 2003, Lee 2004); for larger mass ratios only symmetric configurations are found.

Notwithstanding their importance, ACR only give limited information on the resonant structure. To obtain a more complete picture and identify other possible configurations, the analysis must be extended beyond the stationary solutions. A study in this direction is presented in Michtchenko et al. (2008). Although the reader is referred to that paper for a more detailed analysis, here we reproduce some of the results.

Figure 2 shows four dynamical maps of the 2/1 MMR, considering a mass ratio slightly larger than unity. Each map was constructed from the analysis of a grid of initial conditions centered around a given ACR. The gray scale shows the spectral number SN (Michtchenko and Ferraz-Mello 2001) resulting from a Fourier analysis of the eccentricity of the inner planet, obtained from a numerical integration over more than 10^5 orbital periods. White strips mark the presence of stable periodic orbits of the averaged problem, while darker tones indicate increasingly chaotic motion. Zones of forbidden motion are shaded in red, and the red circle in each frame is the stable ACR.

As can be seen in the two top plots, the region accessible to the planets is divided into two zones, separated by a chaotic layer. The inner zone, around the ACR, is the region of resonant motion while, outside, the orbits are non-resonant (i.e. secular). This latter region is dominated by the two modes of secular motion (called Modes I and II), corresponding to apsidal alignments around $\Delta\varpi = 0$ and $\Delta\varpi = \pi$, respectively. Inside the resonant region there are two main families of periodic orbits, whose intersection coincides with the ACR. The vertical branch is associated to a fixed value of θ_1 and large amplitude variation of $\Delta\varpi$. In the horizontal branch the opposite occurs: $\Delta\varpi$ is fixed while θ_1 suffers large amplitude oscillations. As can be seen in the plot for $e_1 = 0.1$, other white strips representing additional families of periodic orbits are also present. These are related to secondary resonances where the frequency of libration of θ_1 is commensurable with the circulatory frequency of $\Delta\varpi$.

In the frame for $e_1 = 0.2$ the chaotic layer defining the boundaries of the resonant region has grown significantly, pushing the secular regime practically outside the limits of the graph. The strong chaotic zone implies that the secular and resonant regions are virtually isolated, making any transition difficult. In particular, it helps to explain why

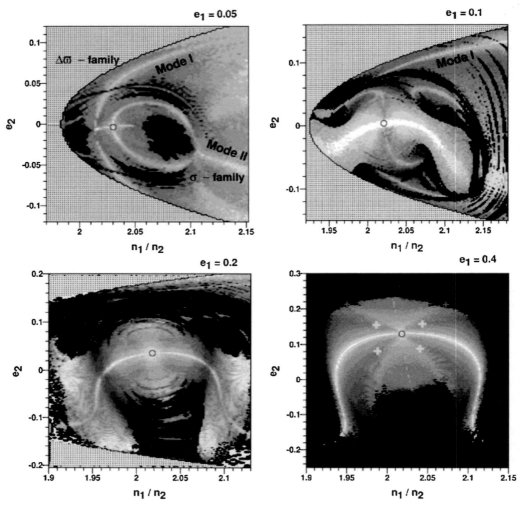

Figure 2. Dynamical maps of the 2/1 MMR for $m_2/m_1 = 1.064$ around four diffrent symmetric ACR. Each is identified by the eccentricity of the inner planet. Each plot shows the spectral number resulting for a numerical simulation for 130000 orbital periods. Initial conditions were chosen with equal values of the angular momentum integral $\mathcal{L}$ and scale parameter $\mathcal{K}$. In each frame, stable ACR is identified by a red circle, periodic orbits by white strips, and chaotic motion by increasingly dark tones of gray. The red dotted region corresponds to forbidden initial conditions. The indicated colors refer to the electronic version of the paper. Figure reproduced from Michtchenko *et al.* (2008).

resonance trapping has a lower probability for initial eccentric orbits. Finally, for $e_1 = 0.4$ the only region with stable motion is associated to resonant motion, and all secular orbits are chaotic with very short lifetimes. For e_1 larger than some limiting value between 0.2 and 0.4, the only stable orbits possible for planetary systems in the near vicinity of the 2/1 MMR are resonant, and correspond to oscillations around the ACR. An example, the red crosses in the lower right-hand frame correspond to initial conditions equal to the orbital fit of the HD82943 planets by Mayor *et al.* (2004), whose dynamical instability appears very clearly from its location at the edge of the resonant region. Conversely, the light blue crosses correspond to the initial conditions given by the stable Fit B

(Ferraz-Mello et al. 2005a) for the same system. This solution is located much closer to the zero-amplitude ACR, and corresponds to a stable ACR with moderate amplitude.

Figure 2 shows a very complex structure for the 2/1 MMR, even restricted to the planar case. Especially for lower eccentricities, several different modes of stable motion are present which, in principle, could represent possible configurations for real planetary systems. Curiously, the phase space becomes simpler for growing eccentricities, until all possible stable motion becomes dominated by the central ACR.

3. Planetary Migration and Resonance Capture

Extrasolar planets usually have smaller semimajor axes and higher eccentricities than expected from classical cosmogonic theories. Either these planets formed by a mechanism different from the one that gave birth to our Solar System, or the exoplanets suffered significant orbital evolution after their formation. This latter proposal is sometimes known as the *Hypothesis of Planetary Migration*. Although several migration scenarios have been proposed, such as planetary scattering or interactions with a remnant planetesimal disk, the most probable process is a byproduct of the interactions between the planets and its surrounding gaseous disk. This interaction causes an exchange of energy and angular momentum between both components (gas and solid bodies) leading to an infall of the orbital distance of the planets (e.g. Ward 1997).

Even though the orbital evolution of the semimajor axis of the planet under planet-disk interactions is well established, there is still ongoing discussion as to the evolution of the orbital eccentricity. Depending on the relative strength of the Lindblad and corotational resonances with the disk (Goldreich and Sari 2003), the eccentricity can either increase or decrease as the planet falls towards the star. This is a fundamental issue in the hypothesis of planetary migration. On one hand, the existence of solitary giant planets with large eccentricities points towards an eccentricity pumping scenario. On the other hand, largescale orbital decay of resonant planets requires a damping of the same orbital element (Lee and Peale 2002, Beaugé et al. 2006). Of course planetary systems with only one known mass could hide additional bodies, and the observed eccentricities could have been excited by other means, such as planetary scattering (e.g. Ford et al. 2005) or perhaps even jet acceleration of the host star (Namouni and Zhou 2006).

Although planetary migration has several unsolved problems, it is probably our best explanation for the current orbital characteristics of exoplanets. However, in order to consider it more than just an interesting hypothesis, we require some evidence that it really occurred. Multiple-planet systems in mean-motion resonances are specially suited for this task. Not only may they serve as evidence of such a past migration, but what type of resonant solution they inhabit at present may also give us valuable information about details of the orbital decay itself.

Figure 3 shows the result of a numerical simulation of the orbital evolution of two fictitious planets ($m_1 = 1 M_{Jup}$ and $m_2 = 0.5 M_{Jup}$) initially placed in circular orbits with $a_1 = 5.4$ AU and $a_2 = 9.5$ AU. Planetary migration was modeled with a Stokes-type dissipative force (Beaugé et al. 2006) which affected only the outer body. The drag coefficients were chosen such that a_2 decayed with an e-folding time of $\tau_a = 10^6$ years, but did not cause any secular change in the eccentricity (i.e. $\tau_e = \infty$). Initially only the outer body suffers an orbital decay, while the semimajor axis of the inner planet is practically constant. However, once resonance capture occurs in the 2/1 MMR ($t \sim 10^5$ years), the ratio of mean motions becomes locked. Since the dissipative force is still acting on m_2, the outer planet still falls, but now with a slower rate since it must push the inner body with it. Although both planets still evolve towards smaller orbits, the resonance relation

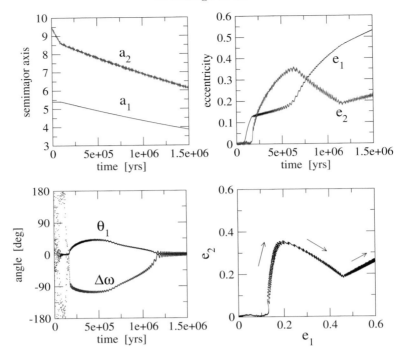

Figure 3. Numerical simulation of planetary migration leading to a capture of two planets ($m_2/m_1 = 0.5$) in the 2/1 MMR. Notice the switch from symmetric to asymmetric ACR at $t \approx 2 \times 10^5$ years, and back to symmetric at $t \approx 1.2 \times 10^6$ years. In the lower right-hand frame, arrows show the direction of orbital variation.

is maintained. As soon as the resonance lock is achieved, the eccentricities begin to grow and the angular variables acquire small oscillations around equilibrium values.

From this point onwards, all orbital evolution occurs within the resonant region of the 2/1 MMR. The lower right-hand frame of Figure 3 shows the relationship between e_2 and e_1 as the bodies evolve inside the commensurability. A comparison with Figure 1 shows that the system follows closely the family of stable ACR for this mass ratio. In other words, the families of ACR that were shown in Section 2 not only give the different equilibrium solutions as function of the angular momentum $\mathcal{L}$, but also correspond to *evolutionary tracks* within the resonance. Thus, catalogs of ACR such as the one depicted in Figure 1 may not only show us the present location of resonant planetary systems, but also how they evolved (see also Zhou *et al.* 2008).

As long as the migration is sufficiently slow (adiabatic limit), the evolution of the system within the resonance will not depend on the particular values of τ_a and τ_e, but only on the ratio

$$K = \frac{\tau_a}{\tau_e} \tag{3.1}$$

(see Lee and Peale 2002). In the particular case of Figure 3, the drag parameters were chosen such that $K = 0$. Of course different values of τ_a will imply faster or slower orbital decay, but the only effect in the plots will be a modification of the timescale. In particular, the evolutionary track in the (e_1, e_2) plane will suffer no change. The adiabatic limit depends on the planetary masses as well as on the resonance (Beaugé *et al.* 2006). Typically, the smaller the mass ratio, the slower must the migration be to guarantee adiabaticity. If this condition is not satisfied, resonance lock may still occur,

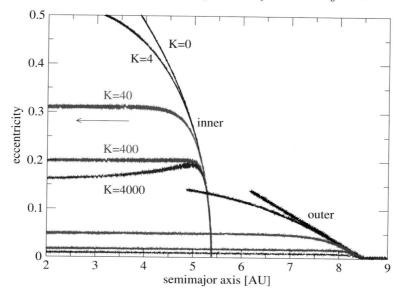

Figure 4. Simulations of planetary migration and capture in the 2/1 MMR for fictitious planets with mass ratios similar to the GJ876 resonant planets. The plot shows the evolutionary curves of the eccentricities as function of the decreasing semimajor axes (arrow indicates direction of evolution). Each line corresponds to a different simulation with varied value of K. Note that for $K \geqslant 10$ the eccentricities reach equilibrium values.

but the orbital evolution within the commensurability will follow a different path from the families of ACR and, typically, with a much larger amplitude of libration.

The simulation shown in Figure 3 continues until the eccentricity of the inner planet reaches values close to unity and the system is disrupted. This occurs when the semimajor axes are still well above 1 AU. However, a different result is obtained if the migration model includes a significant damping in the eccentricity (i.e. finite values of τ_e and $K > 0$). Figure 4 shows the results of a series of simulations with $m_2/m_1 = 3.15$, similar to the latest edge-on fit for the resonant planets of GJ876 (Laughlin *et al.* 2005). The plot shows the evolution of the eccentricities as a function of the decaying semimajor axes. Thus, in the graph the system evolves from right to left. Initial conditions were the same as in the previous figure; curves starting at $a = 9.5$ AU correspond to the outer planet, while the other branches mark the orbit of the inner body. Each line presents an integration with different value of K. For $K = 0$ (no eccentricity damping) the eccentricity of the inner planet is quickly excited, reaching almost parabolic orbits after an infall of less than 2 AU. Larger values of K imply smaller eccentricities at a given semimajor axis until, for $K \geqslant 10$, the eccentricities reach an equilibrium value that is constant with respect to the semimajor axis. Note that for $K = 4000$ there appears to be a transient interval where e_1 decreases before reaching a plateau.

The existence of these equilibrium eccentricities alllow the system to suffer large scale migrations while maintaining orbital stability. More importantly, if the planets were initially formed far from the star and captured from quasi-circular orbits, there is only one value of K compatible with the present configuration. In this way, the present eccentricities of the resonant planets could yield information about the characteristics of the gaseous nebula at the final stages of planetary formation. However, this is not an easy task. For example, the current eccentricities of the GJ876 resonant planets seem to indicate that the migration followed a value of $K \sim 40$, while hydrodynamical simulations usually give values of K close to unity (Kley *et al.* 2005). This discrepancy could be evidence in favor of an inner disk in the system (Sandor *et al.* 2007) or a timely dispersal of the nebula before the eccentricities were excited beyond their present values. A similar study was undertaken for the HD72536 system by Sandor *et al.* (2007), where the authors discuss the ejection of a hypothetical third planet to account for the large amplitude of libration around the stable ACR.

4. Orbital Determination of Resonant Planets

Orbital determination of exoplanets from radial velocity data is a complex process, especially in the case of systems with two planets in mean-motion resonances. The orbital commensurability can give origin to a noticeable periodicity in the radial velocity curve, complicating the separation of both components from the signal. Moreover, two of the most important orbital elements for dynamical studies, the eccentricities e and longitudes of pericenter ϖ are also the most difficult to estimate, since they are given by asymmetries in the quasi-periodic signal. Then, if the individual signals of two planets are mutually affected by their resonant configuration, the precision of the estimation of both e and ϖ can be seriously impaired.

The question then is whether our current knowledge of the planetary parameters are sufficiently reliable to perform detailed dynamical studies, particularly those in which precise values of the orbits are required to reconstruct characteristics of the migration process. Some resonant systems appear well constrained, for example GJ876. Although the orbital elements (in particular the eccentricities) have changed somewhat in the last few years due to an increasing observational database, the overall dynamics of the resonant planets has remained almost invariant. All orbital fits are consistent with a low-amplitude $(0,0)$-ACR. Other resonant systems offer a more complex scenario. A good example of an ill-defined system is HD82943. At present there are two known planets in this system, with orbits consistent with a 2/1 MMR. Both were discovered by the Geneva group (see Ji *et al.* 2003 for an initial orbital solution), and at present the total number of observations is $N = 165$. Of these, 142 were obtained from CORALIE (Mayor *et al.* 2004), while the remaining 23 come from Keck (Lee *et al.* 2006). The orbital fits calculated with the complete data set show significant differences with respect to the solution with CORALIE data alone, although in both cases the best fits lead to unstable configurations (Ferraz-Mello *et al.* 2005a, Lee *et al.* 2006).

Since it is expected that the real planets should lie in stable regions of the phase space, both Ferraz-Mello *et al.* (2005a) and Lee *et al.* (2006) searched for nearby stable orbits with similar weighted r.m.s of the residuals ($wrms$). All solutions were found to be consistent with resonant motion, usually corresponding to a large-amplitude $(0,0)$-type ACR, although some individual solutions with a circulation of $\Delta\varpi$ were also found. More recently, Goździewski and Konacki (2006) showed that two planets in stable co-orbital motion and non-coplanar orbits could yield similar radial velocity curves, albeit with significantly larger residuals. However, the existence of co-orbital giant planets is far from being established, and there is evidence that planetary bodies with masses larger than $\sim 0.7 M_\oplus$ could not be accreted in the equilateral Lagrange points of giant exoplanets (Beaugé *et al.* 2007). Even so, it is very curious how co-orbital bodies can mimic the radial velocity curve of two bodies in a 2/1 MMR.

Dynamically unstable best fits for resonant planetary systems are not uncommon (e.g. HD73526, HD128311, HD202206), and do not seem very surprising considering the existence of large regions of chaotic motion surrounding the stable resonant regions. However, independent of the dynamical stability, we are also interested in the statistical stability of the orbital solution. In other words, how far can we trust a given best fit, even if

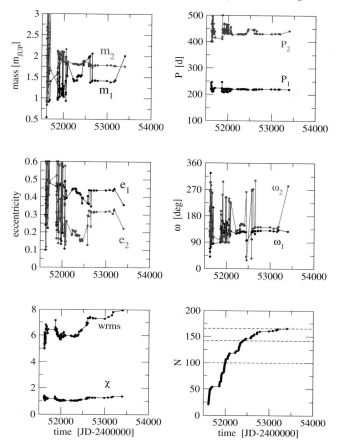

Figure 5. Variation of the parameters of the best multi-Keplerian fit for partial sets of observations, as functions of the times corresponding to the last observation in the set. Data corresponds to complete CORALIE-Keck set.

it leads to stable motion? One possible way is to study the variations of the best fits as function of the number of observations (Beaugé *et al.* 2008). If the current solution is robust, then we should expect only small and smooth changes in the parameters as a function of the data set number. More importantly, we can then expect that future incorporation of additional observations will not significantly change our knowledge of the system.

Results for multi-Keplerian fits of HD82943 are shown in Figure 5. The lower left-hand frame shows the change in *wrms* and $\sqrt{\chi^2}$. Notice the increase in the value of *wrms*, particularly over the last few tens of observations. Thus, more data points does not necessarily imply better orbital fits. The lower right-hand plot shows the number of data points as a function of time. The "times" corresponding to the three published orbital fits are identified by horizontal broken lines. With this we can identify in each of the other frames the values of the parameters for each time interval.

Perhaps the most important conclusion from this figure is an apparent non-convergence of the parameters towards defined values. Thus, there is no confidence that future observations will not change the masses and orbital elements of the planets once again, and the current orbital fit of the HD82943 planets does not seem to be reliable. For instance, the passage from $N = 164$ to $N = 165$ causes very drastic changes in some elements,

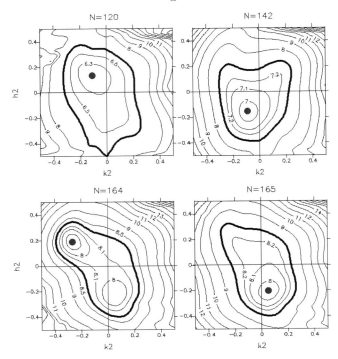

Figure 6. Level curves of constant *wrms* for four multi-Keplerian orbital fits considering fixed values of $(k_2, h_2) = (e_2 \cos\omega_2, e_2 \sin\omega_2)$ in a regular grid. Each plot was drawn using a different number of data points N, simulating the information available at the moment of the publication of each orbital solution. Global minimum is identified with large bullet. Broad level curves represent the 1-σ confidence level.

although practically no change in the *wrms* of the fit (see lower-left hand frame). This seems to imply that both solutions, although very different, are equally consistent with the observational data.

The degree of the statistical instability in the orbital fits is so noticeable that the removal of a single observation can lead to completely different values in some of the orbital parameters. To analyze this sensitivity more globally, we performed a series of multi-Keplerian orbital fits over a grid in the variables $(k_2, h_2) = (e_2 \cos\omega_2, e_2 \sin\omega_2)$. For each point the numerical values of these parameters were fixed, and the fit was done only on the remaining parameters. These were allowed to vary with no restriction, and the resulting *wrms* corresponds to the best fit for those values of (k_2, h_2). We then plotted level curves of *wrms* in this grid, which give information on the shape and relative depth of that fitness function in the plane.

This procedure was applied to four sets of observations, corresponding to $N = 120$, $N = 142$, $N = 164$ and $N = 165$. Results are shown in Figure 6. In each case, the global minimum wrms$_{\min}$ is identified by a large bullet. Although close to the minimum the level curves are approximately elliptical, distorsions appear for larger values, indicating that confidence levels obtained from analysis of the correlation matrix are not valid beyond the immediate vicinity of wrms$_{\min}$. The broad level curve in each frame is the 1-σ confidence level (see Beaugé et al. 2008).

For $N = 164$ the figure shows two local minima with almost the same *wrms*. The global minimum is located in $\omega \sim 140$ degrees, while the other solution ($\omega \sim 270$ degrees) lies near the global minimum obtained for $N = 165$. Thus, not only is the fitness function very shallow around the best-fit solution, but sometimes two local minimums can be observed, and very small changes in the data set can lead to one (or the other) being identified as the best fit. Although the individual best fits do change a lot, the shape and extension of the 1-σ confidence region appears more robust. Even so, many different types of dynamical configurations coexist within these regions. Thus, although at present it is not possible to choose between them, it seems possible that the HD82943 system lies somewhere inside this region.

A possible origin for this sensitivity to the data set is that HD82943 contains an additional planet. This hypothesis was initially proposed by Goździewski and Konacki (2006) and more recently by Beaugé et al. (2008). The periodograms of the residuals after a two-planet fit and the level of *wrms* seem to indicate that an additional planet with an orbital period of approximately 900 days could be present.

5. Conclusions

Studies of mean-motion resonances (MMR) in exoplanets are far from being purely academic. Several known planetary systems (e.g. GJ876, HD82943, HD73526, HD128311) are believed to lie deep inside the 2/1 MMR, and other systems may inhabit other commensurabilities, such as the planets 55Cnc-b and 55Cnc-c in the 3/1 or the HD202206 system in the 5/1 MMR. Consequently, a detailed knowledge of the structure of each resonance can yield important insights towards understanding possible locations of real extrasolar bodies.

The large proportion of resonance relations seems to be related to two distinct characteristics. First, MMR are natural nesting places of planets undergoing convergent migration, for example, due to disk-planet interactions. Second, resonances act as a protective mechanism, generating configurations of stable motion in regions of the phase space dominated by instability and close encounters. Thus, MMR constitute privileged locations where exoplanets could be found with large eccentricities and nearby orbits.

However, the dynamical evolution of resonant systems in very sensitive to the initial conditions, and even small uncertainties in orbital fits (as obtained from radial velocity data) may lead to large uncertainties in the evolution of a given multi-planet system. Although some resonant systems are well known (such as GJ876) and their dynamics well mapped, others are less certain. Possibly the most volatile example is HD82943, where even the substraction of a single RV data point can yield completely different orbital solutions and dynamics.

Compared to dynamical studies of our Solar System, where masses and initial conditions are well known, extrasolar planetary systems constitute completely different problems, requiring new methods and approaches. In some ways, resonant exoplanetary systems can be treated as dynamical systems whose initial conditions are not masses or orbital elements, but radial velocity data. It is therefore not surprising that dynamical studies of these systems have become increasingly tied to the process of orbital determination.

Acknowledgments

This work has been supported by the Argentinean Research Council -CONICET, the Brazilian National Research Council -CNPq, and the São Paulo State Science Foundation -FAPESP. The authors also gratefully acknowledge the support of the CAPES/Secyt programme for scientific collaboration between Argentina and Brazil.

References

Beaugé, C. & Michtchenko, T. A. 2003, *MNRAS*, 341, 760.
Beaugé, C., Ferraz-Mello, S., & Michtchenko, T. A. 2003, *ApJ*, 593, 1124.

Beaugé, C., Michtchenko, T. A., & Ferraz-Mello, S. 2006, *MNRAS*, 365, 1160.
Beaugé, C., Sándor, Zs., Érdi, B., & Süli, À. 2007, *AA*, 463, 359.
Beaugé, C., Giuppone, C. A., Ferraz-Mello, S., & Michtchenko, T.A. 2008, *MNRAS*, submitted.
Ferraz-Mello, S., Beaugé, C., & Michtchenko, T. A. 2003, *CeMDA*, 87, 99.
Ferraz-Mello, S., Michtchenko, T. A., & Beaugé, C. 2005a, *ApJ*, 621, 473.
Ferraz-Mello, S., Michtchenko, T. A., Beaugé, C., & Callegari Jr., N. 2005, *Lect. Notes Phys.*, 683, 219.
Fischer, D. A., Marcy, G. W., Butler, R. P., Vogt, S. S., Laughlin, G., Henry, G. W., Abouav, D., Peek, K. M.G., Wright, J. T., Johnson, J. A., McCarthy, C., & Isaacson, H. 2007, submitted.
Ford, E. B., Lystad, V., & Rasio, F. A. 2005, *Nature*, 434, 873.
Goldreich, P. 1965, *MNRAS* 130, 159.
Goldreich, P. & Sari, R. 2003, *ApJ*, 585, 1024.
Goździewski, K. & Konacki, M. 2006, *ApJ*, 647, 473.
Ji, J., Liu, L., Kinoshita, H., Zhou, J., Nakai, H., & Li, G. 2003, *ApJ*, 591, L57.
Kley, W., Lee, M. H., Murray, N., & Peale, S. J. 2005, *AA*, 437, 727.
Laughlin, G. & Chambers, J. E. 2001, *ApJ*, 551, L109.
Laughlin, G., Butler, R. P., Fischer, D. A., Marcy, G. W., Vogt, S. S., & Wolf, A. S. 2005, *ApJ*, 622, 1182.
Lee, M. H. & Peale, S. J. 2002, *ApJ*, 567, 596.
Lee, M. H. 2004, *ApJ*, 611, 517.
Lee, M. H., Butler, R. P., Fischer, D. A., Marcy, G. W., & Vogt, S. S. 2006, *ApJ*, 641, 1178.
Mayor, M., Udry, S., Naef, D., Pepe, F., Queloz, D., Santos, N. C., & Burnet, M. 2004, *AA*, 415, 391.
Michtchenko, T. A. & Ferraz-Mello, S. 2001, *AJ*, 122, 474.
Michtchenko, T. A. & Malhotra, R. 2004, *Icarus*, 168, 237.
Michtchenko, T. A., Beaugé, C., & Ferraz-Mello, S. 2006, *CeMDA*, 94, 411.
Michtchenko, T. A., Beaugé, C., & Ferraz-Mello, S. 2008, in preparation.
Murray, C. D. & Dermott, S. F. 1999, *Solar System Dynamics*, Cambridge University Press.
Namouni, F. & Zhou, J. L. 2006, *CeMDA*, 95, 245.
Roy, A. E. & Ovenden, M. W. 1954, *MNRAS*, 115, 296.
Sándor, Z., Kley, W., & Klagyivik, P. 2007, *AA*, 472, 981.
Ward, W. R. 1997, *ApJ*, 482, L211.
Zhou, L.-Y, Ferraz-Mello, S., & Sun, Y.-S. 2008, in: Y. -S. Sun, S. Ferraz-Mello, & J.-L. Zhou, (eds.), *Exoplanets: Detection, Formation and Dynamics*, Proc. IAU Symposium No. 249 (Suzhou,China), p. 485.

Dynamics and instabilities in exoplanetary systems

Eric B. Ford[1]

[1]Department of Astronomy, University of Florida, 211 Bryant Space Science Center, Gainesville, FL 32611-2055, USA
email: eford@astro.ufl.edu

Abstract. Extrasolar planet surveys have discovered over two dozen multiple planet systems. As radial velocity searches push towards higher precisions and longer survey durations, they can be expected to discover an even higher fraction of multiple planet systems. Combined with radial velocity data, dynamical studies of these systems can constrain planet masses and inclinations, measure the significance of resonant and secular interactions, and provide insights into the formation and evolution of these systems. Here, we review the dynamical properties of known extrasolar multiple planet systems and their implications for planet formation theory. We conclude by outlining pressing questions to be addressed by a combination of future observations and theoretical research.

Keywords. planetary systems: formation, celestial mechanics, gravitation, instabilities, scattering, techniques: radial velocities

1. Categories of Multiple-Planet Extrasolar Planetary Systems

As is customary for a young science, we begin by observing the properties of individual specimems (i.e., multiple planet systems) and using their properties to identify categories that could be the result of some underling physical processes that we seek to understand. From the perspective of orbital dynamics, the extrasolar planetary systems can be roughly divided into three categories, based on the strength and timescales of planet-planet interactions. We discuss category each below.

Resonant Systems: Some exoplanets are in (or near) mean motion resonances that result in strong planet-planet interactions. For example, GJ 876 hosts two giant planets with a ratio of orbital periods very nearly 2:1. Indeed, n-body integrations reveal that all three resonant angles associated with the 2:1 resonance to librate about zero (Laughlin et al. 2005). The resonant interactions cause the orbits of both planets to precess by 360° once every $\simeq 9$ years. In this case the precession is particularly rapid due to the short orbital period, low stellar mass, and relatively high planet masses (Laughlin & Chambers 2001). In most known multiple planet systems, the interactions are more subtle and/or have somewhat longer timescales than in GJ 876, making it very difficult to observe the interactions directly.

At the time of writing, five pairs of exoplanets also have a ratio of orbital periods near 2:1, one pair has a period ratio near 3:1, and a few pairs of planet have period ratios of nearly n:1 with $n \geqslant 4$ (Butler *et al.* 2006; www.exoplanets.org, www.exoplanet.eu). While there is strong evidence that some of these systems are indeed in resonance, some of these systems may merely be near a mean motion resonance. Unfortunately, observational uncertainties often make it difficult to differentiate the two cases. This distinction can have significant implications for planet formation models (see §2).

Secular Systems: Even in planetary systems without any mean motion resonances, angular momentum can be transfered between planetary orbits, resulting in significant eccentricity/inclination evolution on secular time scales. Among the presently known multiple planet systems, 10-15 appear to be undergoing significant secular eccentricity evolution (e.g., Barnes & Greenberg 2006 and references therein).

It is useful to subdivide the secular systems according to the mode evolution of the periastron angles. In the classical Laplace-Lagrange perturbation theory for a pair of planets, the angle between the two periastron directions ($\Delta\varpi$) can either librate about 0° (aligned), librate about 180° (anti-aligned), or circulate over the range 0-2π (circulate) (e.g., Zhou & Sun 2003; Adams & Laughlin 2006). In the full 3-body problem, short-period terms of the Hamiltonian can cause $\Delta\varpi$ to librate and sometimes circulate for systems near the boundary (e.g., Ford, Lystad & Rasio 2005; hereafter FLR). We label such systems as "borderline" and will make use of them in §2, when we describe how they could be useful for testing planet formation models. In practice, the observational uncertainties can make it difficult to determine the mode of secular evolution even for systems that do not lie particularly close to the boundary between librating and circulating initial conditions.

Hierarchical Systems: Some planetary systems do no contain any known planets near a mean motion resonance and do not appear to undergo significant secular eccentricity evolution. We label these systems hierarchical, since each planet basically orbits the barycenter of the star and interior planets. The planetary perturbations cause each orbit to regress, but there is no apsidal lock or large amplitude eccentricity oscillations. Among the presently known multiple planet systems, $\simeq 6$ are currently categorized as hierarchical systems. Hierarchical systems with large gaps between the known planets may be fertile hunting groups for additional planets (e.g., Barnes & Raymond 2004).

Frequency of Resonant, Secular, and Hierarchical Systems: As exoplanet searches discover more multiple planet systems, it will become possible to ask questions about the relative frequency of different types of systems. At least two factors complicate the interpretation of the apparent frequency of resonant, secular, and hierarchical systems. First, in some cases, the uncertainties in the current orbital elements preclude determining whether the system is in a mean motion resonance and/or has significant secular evolution. Second, the potential discovery of additional planets in or near a mean-motion resonance could cause a system to be reclassified as resonant. Similarly, the discovery of non-resonant planet(s) in a system believed to be hierarchical could result in a tighter dynamical couple between previously known planets and reveal that the system is undergoing more significant secular evolution. Third, in some cases the discovery of additional planet(s) can result in quantitative or even qualitative changes in the orbital elements of the previously known planets. Historically, the masses and/or eccentricities of the known planets has often been revised downward, reducing the strength of dynamical interactions. Thus, the discovery of additional planets can either increase or decrease the significance of secular evolution.

2. Implications for Planet Formation Theory

Resonant and Near-Resonant Systems: Given the wide range of planet semi-major axes, eccentricities, orientations, and orbital phases, it seems unlikely that two planets would form in the relatively narrow range of parameter space that would result in a low-order resonance. However, if the planets formed in orbits with a larger ratio of orbital periods before a slow and smooth migration caused the planets to approach each other, then such systems could naturally become captured into a resonance. Therefore, planets in mean motion resonances have been taken as evidence supporting models that result in the convergent migration of giant planets (Lee & Peale 2002). Even more compelling support for migration leading to resonant capture is provided by the current eccentricities of the two giant planets orbiting GJ 876. Assuming that two planets are captured into the 2:1 resonance while on nearly circular orbits, further migration will result in eccentricity excitation for both planets. The observation that the two eccentricities fall along the one parameter family of solutions predicted by this model provides further support for this model (Lee & Peale 2002).

It is important to note that merely having a ratio of orbital periods close to n:1 does not guarantee that the system is actually participating in a mean motion resonance. To be in a resonance, there must be a resonant angle that liberates about some value. Thus, determining whether or not a planetary system contains a resonance requires dynamical modeling to determine how the system will evolve in time. The results of such integrations will depend on the initial conditions which are chosen based on the available observations. In many cases, the existing observations leave a considerable uncertainty in the choice of initial conditions (Ford 2005, 2006), preventing a definitive determination of whether a system is in resonance (e.g., Gozdziewski *et al.* 2007). Identifying a viable orbital solution that results in resonance would demonstrate that a resonance plausible. However, in order to make the statement that the system is in resonance, one would need to demonstrate that there are no orbital solutions that are: 1) consistent with existing observations, 2) dynamically stable for the age of the system, and 3) not in resonance. Typically, many more observations will be required to satisfy this more demanding criteria than are required to identify a near commensurability of orbital periods. Nevertheless, such observations will be essential for the successful theoretical interpretation of multiple planet systems. For example, a system deep in a resonance could provide evidence for a smooth migration in a dissipative disk (Lee & Peale 2002), while a similar system merely near the same resonance would be more likely to have arisen by chance. As another example, some theoretical models suggest that resonant capture may be followed by additional dynamical interactions that result in the planets evolving out of resonance (Thommes *et al.* 2007; Narayan *et al.* 2002). Thus, resonant systems should be targeted for frequent follow-up observations so as to distinguish models that predict planets near mean motion commensurabilities from models that predict planets actually participating in mean motion resonances.

<u>Secular Systems</u>: Shortly after the discovery of the three planets around υ And, theorists recognized that the system was undergoing significant secular eccentricity variations and proposed two models that might explain the significant eccentricities of the two outer giant planets (c & d). Chiang & Murray (2002) proposed that a protoplanetary disk beyond planet d could *adiabatically* torque planet d. If the system was initially configured so that the longitudes of periastron were circulating, then this torque would drive the system towards solutions where the longitudes of periastron librate about an aligned configuration. Once the system was in the librating regime, the torque would damp the libration amplitude. Thus, this model would predict that the pericenters of the outer two planets would currently be librating with small amplitude about an aligned configuration and that the secular evolution would cause only small variations in the eccentricities.

In an alternative scenario, Malhotra (2002) proposed that the outer planet could have been perturbed *impulsively*. In this model, the periapses of the two planets could be either librating or circulating about an aligned (or anti-aligned) configuration, depending on the state of the system at the time of the impulsive perturbation. If the system were librating, then this model would generally predict that the libration amplitude would be large and that there would be significant eccentricity oscillations. What could cause such an impulsive perturbation? The most natural candidate is a close encounter with another planet (Malhotra 2002). The extra planet could have been ejected from the system or might still remain bound, but in a wide orbit and hence undetected. Other possibilities include the rapid halting of inward migration (Sandor & Kley 2006), perhaps due to an edge or the rapid dispersal of the disk (but see online supplement of FLR).

The viability of these two very different models for eccentricity excitation in the υ And system demonstrated that the mode of evolution could be used to constrain planet formation theory and highlighted the importance of measuring the masses and current orbital elements with enough precision to predict the secular eccentricity evolution. Thanks to an intensive radial velocity campaign, the current angle between the two periapses was measured to be $\Delta\omega \simeq 38° \pm 5°$ (FLR). This implies a large libration amplitude and therefore supports models with an impulsive eccentricity perturbation over models with an adiabatic torque. While the eccentricity of the outer planet undergoes small oscillations, the eccentricity of the middle planet undergoes very large oscillations with e ranging from from 0.34 to very nearly zero. Such behavior is characteristic of "borderline" systems, where the system lies near the boundary separating librating and circulating regimes, and is the natural outcome of a strong impulsive perturbation. FLR used a Bayesian statistical analysis of the radial velocity observations to determine that the eccentricity of the middle planet periodically returns to nearly zero for *all allowed* orbital solutions (see Fig. 2 of FLR). This provides a strong constraint on the timescale for eccentricity excitation in υ And and supports the model of the outer planet being perturbed impulsively, most likely by strong planet-planet scattering.

Variations on a Theme: Other multiple planet systems likely offer additional insights into their orbital histories and planet formation. For example, HD 128311 contains a pair of planets near a 2:1 mean-motion resonance, again suggesting convergent migration leading to resonant capture. However, the outer planet appears to be undergoing large eccentricity oscillations, quite unlike those of GJ 876, suggesting an eccentricity excitation mechanism subsequent to resonant capture. Sandor & Kley (2006) proposed a hybrid scenario that invokes convergent migration, resonant capture, and strong planet-planet scattering to explain the current orbital dynamics of the best-fit orbital solution. Tinney *et al.* (2006) and Sandor *et al.* 2007 have suggested similar scenarios for explaining HD 73526. While the published radial velocity data for HD 128311 and HD 73526 are consistent with an impulsive scenario (Sandor & Kley 2006; Sandor *et al.* 2007), the observations are still consistent with a range of orbital solutions that is too broad to allow a unique interpretation (e.g., Gozdziewski & Konacki 2006).

These and several other multiple planet systems might be examples of borderline secular evolution, as observed or suggested by several authors. For example, Zhou & Sun (2003) found that each of HD 12661 bc, HDH 82943 bc, and 47 UMa bc could be near the borderline between circularization and libration, and the recently announced two planet solution for HD 155358 bc also exhibits the large eccentricity oscillations characteristic of borderline secular evolution (Cochran *et al.* 2007). Barnes & Greenberg (2006, 2007) claimed that a large faction of the known multiple-planet solutions appear to exhibit borderline secular evolution (which they refer to as "near separatrix motion"). They claim that a high abundance of such systems would suggest that the impulsive perturbation is unlikely to have been delivered by a planet on a nearly circular orbit at the time of the first close encounter. While a single close encounter between comparable mass planets on circular orbits does not result in ejection (Katz 1997), numerical simulations show that a rapid succession of close encounters can produce eccentric orbits. In order for the perturbation to remain impulsive, the duration of strong interactions must be less than the secular timescale. FLR presented an example requiring only one additional planet, and also suggested that a second additional planet could have helped to make the strong perturbation impulsive by rapidly raising the periastron distance of the planet that interacted strongly with v And d. Barnes & Greenberg (2007) have proposed an alternative scenario in which the impulsive perturbation comes from a planet that is already on an eccentric orbit at the time of the first close encounter.

Words of Caution: We eagerly look forward to studies of the relative frequency of various types of dynamical behavior in multiple planet systems. However, before jumping to conclusions, it is important to recognize the limitations of existing observations. Unfortunately, for some secularly evolving planetary systems, the observations are not yet able to measure important orbital parameters (e.g., eccentricity and argument of periastron) with sufficient precision to determine whether the systems must be undergoing borderline secular eccentricity evolution. In particular, one should distinguish between the statements: "the observations of a system are consistent with borderline secular evolution" and "the observations imply that the system is near the borderline dividing circulation and libration". The latter statement is much more powerful, but typically requires many more observations accompanied by a detailed dynamical and statistical analysis. We have begun a program to perform such analyses of several additional multiple planet systems. In the mean time, we caution against drawing conclusions based on the dynamical behavior of the best-fit model (as opposed to the dynamical behavior of all models consistent with observations and dynamical stability).

3. Future Directions

In the coming years, it will be particularly important to follow-up discoveries of multiple planet systems with intensive radial velocity campaigns to nail down the secular evolution. In some cases (e.g., systems where the orbital period of the outer planet is comparable or greater than the time span of high-precision observations), even the orbital period and radial velocity amplitude can be highly uncertain (Ford 2005), allowing for qualitatively different orbital solutions and limiting the power of dynamical modeling. Therefore, multiple planet systems containing giant planets with short or modest orbital periods appear the most promising for precision dynamical studies and providing constraints on planet formation theory (Ford *et al.* 2007 and references therein).

We are particularly interested in determining the frequency of multiple planet systems that are in or near low-order mean motion resonances. Accurate determination will require discovering more multiple planet systems, measuring the current orbital elements of known multiple planet systems more precisely, and careful consideration of observational selection effects (Ford 2006). Ongoing radial velocity surveys will soon be complimented by searches based on the transit timing variation method, which is particularly sensitive to planets in or near mean motion resonances (Agol *et al.* 2005; Holman & Murrary 2005; Ford & Holman 2007). As more transiting planets are discovered and subjected to follow-up observations, we expect that transit timing will become a powerful tool for detecting multiple planet systems and constraining planet formation models (e.g., Nelson & Papaloizou 2002; Papaloizou & Szuszkiewicz 2005; Cresswell & Nelson 2006; Fogg & Nelson 2007).

As radial velocity and transit searches continue to increase the sample of exoplanets, it will become increasingly practical to compare the distribution of observed masses and orbital elements to those predicted by various models. Theorists have begun making testable predictions for based on various models planet-planet scattering and planet-disk interactions (e.g., Moorhead & Adams 2005; Veras & Armitage 2006; Chatterjee *et al.* 2007; Ford & Rasio 2007; Juric & Tremaine 2007). For example, radial velocity observations can be used to test the prediction that eccentric planets will be more common when the ratio of escape velocity from the planet's surface exceeds the escape velocity from the host star at the planet (Ford & Rasio 2007). As another example, planet scattering predicts a correlation between eccentricity and inclination (Chatterjee et al. 2007), enabling tests with both astrometric measurements of the relative inclination between orbits (McArthur et al. 2007) and Rossiter measurements of the inclination of orbital angular momentum to the stellar spin axis (Narita et al. 2007; Nagasawa in this volume). We encourage theorists to continue developing models to the point where they make alternative predictions that can be tested by upcoming observations.

References

Adams, F. C. & Laughlin, G. 2006, Astrophysical Journal, 649, 992
Agol, E., Steffen, J., Sari, R., & Clarkson, W. 2005, MNRAS, 359, 567
Barnes, R. & Greenberg, R. 2007, Astrophysical Journal, 659, L53
Barnes, R. & Greenberg, R. 2006, Astrophysical Journal, 652, L53
Barnes, R. & Raymond, S. N. 2004, Astrophysical Journal, 617, 569
Butler, R. P., et al. 2006, Astrophysical Journal, 646, 505
Chatterjee, S., Ford, E. B., & Rasio, F. A. 2007, arXiv:astro-ph/0703166
Chiang, E. I. & Murray, N. 2002, Astrophysical Journal, 576, 473
Cochran, W. D., Endl, M., Wittenmyer, R. A., & Bean, J. L. 2007, ApJ, 665, 1407
Cresswell, P. & Nelson, R. P. 2006, Astronomy and Astrophysics, 450, 833
Fischer, D. A., et al. 2007, arXiv:0712.3917
Fogg, M. J., & Nelson, R. P. 2007, Astronomy and Astrophysics, 472, 1003
Ford, E. B. 2006, Astrophysical Journal, 642, 505
Ford, E. B. 2005, Astronomical Journal, 129, 1706
Ford, E. B. 2006, Astrophysical Journal, 642, 505
Ford, E. B., et al. 2007, arXiv:0705.2781
Ford, E. B. & Holman, M. J. 2007, arXiv:0705.0356
Ford, E. B., Lystad, V., & Rasio, F. A. 2005, Nature, 434, 873
Ford, E. B. & Rasio, F. A. 2007, arXiv:astro-ph/0703163
Gozdziewski, K., Migaszewski, C., & Konacki, M. 2007, arXiv:0705.1858
Goździewski, K. & Konacki, M. 2006, Astrophysical Journal, 647, 573
Holman, M. J. & Murray, N. W. 2005, Science, 307, 1288
Juric, M. & Tremaine, S. 2007, arXiv:astro-ph/0703160
Katz, J. I. 1997, Astrophysical Journal, 484, 862
Laughlin, G., Butler, R. P., Fischer, D. A., Marcy, G. W., Vogt, S. S., & Wolf, A. S. 2005, Astrophysical Journal, 622, 1182
Laughlin, G. & Chambers, J. E. 2001, Astrophysical Journal, 551, L109
Lee, M. H., & Peale, S. J. 2002, Astrophysical Journal, 567, 596
Malhotra, R. 2002, Astrophysical Journal, 575, L33
McArthur, B., Benedict, G. F., Bean, J., & Martioli, E. 2007, AAS 211, Abstract #134.17
Moorhead, A. V. & Adams, F. C. 2005, Icarus, 178, 517
Narayan, R., Cumming, A., & Lin, D. N. C. 2005, Astrophysical Journal, 620, 1002
Narita, N., Sato, B., Ohshima, O., & Winn, J. N. 2007, arXiv:0712.2569
Nelson, R. P. & Papaloizou, J. C. B. 2002, MNRAS, 333, L26
Papaloizou, J. C. B. & Szuszkiewicz, E. 2005, MNRAS, 363, 153
Sándor, Z. & Kley, W. 2006, Astronomy and Astrophysics, 451, L31
Sándor, Z., Kley, W., & Klagyivik, P. 2007, Astronomy and Astrophysics, 472, 981
Thommes, E. W., Bryden, G., Wu, Y., & Rasio, F. A. 2007, arXiv:0706.1235
Tinney, C. G., Butler, R. P., Marcy, G. W., Jones, H. R. A., Laughlin, G., Carter, B. D., Bailey, J. A., & O'Toole, S. 2006, Astrophysical Journal, 647, 594
Veras, D. & Armitage, P. J. 2006, Astrophysical Journal, 645, 1509
Zhou, J.-L. & Sun, Y.-S. 2003, Astrophysical Journal, 598, 1290

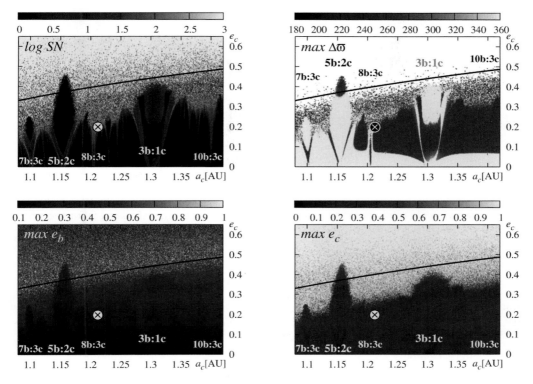

Figure 2. Dynamical maps of putative HD 155358 coplanar configuration of two Jovian planets (~ 0.5–0.9 m$_J$). The osculating elements of the N-body solution at the epoch of the first observation in Cochran et al. (2007) are given in Table 1 (fit II) and marked with crossed circle. The top-left panel is for the Spectral Number. Colors mark the stability regime: black is for regular solutions, yellow is for strongly chaotic solutions. The top-right panel is for the maximal amplitude of apsidal angle $\Delta\varpi = \varpi_c - \varpi_b$. Panels in the bottom row are for the max e indicator (i.e., the maximal eccentricity attained after the integration period $\sim 30,000$ yr). The most prominent mean motion resonances between the planets are labeled.

encounter (see evolution of the critical argument of this resonance in the bottom-right panel in Fig. 4).

Yet the fit parameters are determined within some error ranges that should be interpreted with taking into account the structure of the phase space (see Fig. 2). To illustrate this problem we examined more closely the neighborhood of the best fit II. At this time, we performed two experiments. In the first search, we applied the hybrid code without stability constraints driven by the "usual" N-body model of the RV. The results are illustrated in the bottom-left panel in Fig. 3. The quality of fits within $1\sigma, 2\sigma$ and 3σ confidence intervals of the best fit (marked with crossed circle; see Table 1, fit II) is color coded with blue, light-blue and gray, respectively. The best fits only marginally worse from the best one, are marked in red. Curiously, the plot reveals a subtle structure with three additional local minima of $(\chi^2_\nu)^{1/2}$, in relatively small range of $a_c \in [1.1, 1.3]$ AU. Moreover, these minima are spread over wide range of $e_c \in [0, 0.7]$. Simultaneously, this zone covers many low order resonances, between 5:2 and 3:1 MMR and the $(\chi^2_\nu)^{1/2}$ "valley" is crossed by the collision line. Close to this line, the stability could be preserved only if the planets are protected from close encounter through an MMR.

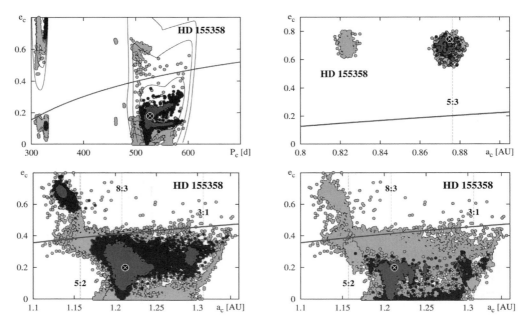

Figure 3. The top-left panel is for the $(\chi_\nu^2)^{1/2}$ of the 2-planet Keplerian solutions to the HD 155358 RV data published in Cochran et al. (2007) derived with the hybrid algorithm. The fit parameters are projected onto the $(P_{\rm C}, e_{\rm C})$-plane. Their quality is color coded: dark blue is for 1σ-, light-blue is for 2σ- and grey is for 3σ-confidence levels of the best fit marked with crossed circle. Contours are for the confidence levels obtained in the systematic scan (see the right panel in Fig. 1). The top-right panel is for stable solutions in the range $P_{\rm C} \sim 300$ days derived with GAMP driven by the N-body model. The stable best-fit solution in this area is marked with crossed circle, and its elements are given in Table 1, fit I. The bottom-left panel is for the ensemble of fits gathered with the hybrid algorithm driven by the N-body model of the RV data around the dominant minimum of $(\chi_\nu^2)^{1/2}$. The bottom-right panel illustrates stable solutions gathered in the GAMP search. The set of solutions within 3σ of the best stable fit (elements given in Table 1, fit II) but obtained without stability constraints, as shown in the bottom-left panel, are again shown as a gray-filled contour. Approximate positions of the most prominent MMRs are labeled.

Now, we could examine the stability of every fit that we found but we choose a new search for stable solutions in a self-consistent manner with GAMP. The results are shown in the bottom-right panel in Fig. 3. In this panel, we overplot the stable solutions within $1\sigma, 2\sigma$ and 3σ confidence interval of the best stable fit II over all solutions within 3σ level which are found in the previous search (i.e., without stability constraints). It now is evident that only a part of the $(\chi_\nu^2)^{1/2}$ valley can consist of dynamically stable solutions, nevertheless the acceptable fits are spread over significant range of $\Delta a_{\rm C} \sim 0.2$ AU. Apparently, this error is quite small but in fact it is large enough to cover a few low-order MMRs. We conclude that the current set of RV data cannot fully characterize the system state and new observations are required to constrain the elements of the outer planet.

Finally, both Keplerian and Newtonian 2-planet solutions lead to apparent excess of the residuals, in particular at the end parts of the RV curve. It may indicate that the 2-planet model does not fully explain the RV variability. In particular, the system may involve more than two planets. A heuristic argument supporting such a claim may be the proximity of the best fits to the collision line. We know similar cases, for instance μ Arae (Goździewski et al. 2007; Pepe et al. 2007), or HD 37124 (Vogt et al. 2005). To

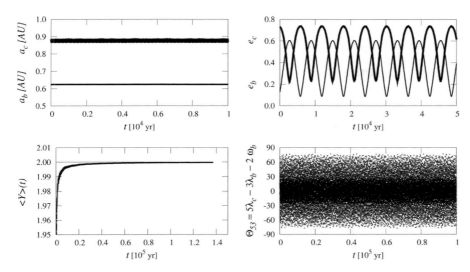

Figure 4. The temporal evolution of orbital elements in the 2-planet, Newtonian solution to the RV data of HD 155358 related to highly eccentric orbits (see the top-left panel in Fig. 3). The osculating elements are given in Table 1 (fit I). The top-left panel is for the semi-major axes, the top-right panel is for the eccentricities. The MEGNO is plotted in the bottom-left panel. The bottom-right panel is for the critical angle of the 5:3 MMR.

check such hypothesis we looked first for 3-planet Keplerian solutions with the hybrid code. The best fit found yields $(\chi^2_\nu)^{1/2} \sim 0.87$ and significantly better rms ~ 4.6 m/s but is unstable. The GAMP search yields stable configurations with *quasi-circular* orbits of the outermost planets, yielding rms ~ 5.5 m/s. An example fit of this type, yielding $(\chi^2_\nu)^{1/2} \sim 1.07$ and an rms ~ 5.6 m/s, given in terms of osculating element at the epoch of the first observation in tuples of $(m \text{ [m}_\text{J}], a \text{ [AU]}, e, \omega \text{ [deg]}, \mathcal{M}(t_0) \text{ [deg]})$ is the following (0.115, 0.383, 0.025, 281.3, 154.4), (0.770, 0.627, 0.040, 108.9, 173.7), (0.490, 1.187, 0.000, 359.1, 65.7), for planets d, b, c respectively, the offset $V_0 = 10.21$ m/s. We note the small mass of the innermost planet. Its RV signal is at the level of noise, so additional observations would be required to confirm of withdrawn such a model.

5. Trojan planets in the τ^1 Gruis system?

Laughlin & Chambers (2002) predict that the reflex signal of a single planet in a quasi-circular orbit may be also interpreted by two Jovian Trojan planets, i.e., two objects sharing similar orbits (involved in 1:1 MMR). That possibility is intriguing because stable Trojan companions to the stars may be quite common. It can be indicated by a number of stable Trojan configurations in the Solar system. Some argue that they can be a frequent by-product of planet formation and and/or dynamical evolution (Laughlin & Chambers 2002). However, the genesis of Trojan planets is not quite clear because on the contrary, there is some evidence that formation of such bodies could be difficult (Beaugé *et al.* 2007). Still, many authors expect that Trojan planets can exist [see, for instance, the work of Dvorak *et al.* in this volume and references therein, also (Ford & Gaudi 2006)].

Recently, we found a similar kind of ambiguity of the RV models concerning 2:1 MMR configurations. At present, we know five extrasolar systems presumably involved in

2:1 MMR, i.e., Gliese 876 (Marcy et al. 2001), HD 82943 (Mayor et al. 2003), HD 128311 (Vogt et al. 2005), HD 73526 (Tinney et al. 2006), and μ Arae (Jones et al. 2003; Goździewski et al. 2007; Pepe et al. 2007). However, the 2:1 MMR model of the radial velocity observations can also be non-unique. The periodogram of the 2:1 MMR RV signal is very similar to that one of the 1:1 MMR. Indeed, the RV variability of HD 128311 and HD 82943 can be explained by highly inclined systems in 1:1 MMR (Goździewski & Konacki 2006). We also found that the RV of HD 73526 can be modeled with two highly inclined Jovian Trojans. The modeling of the 1:1 MMR is a challenging problem because Jupiter-like planets sharing eccentric orbits with similar semi-major axes interact heavily and the collisional configurations are generic. Hence, stability constraints are critical in the search for optimal and stable configurations. This seems to be one of the best applications of GAMP like algorithms.

Among a few cases we analyzed so far, the τ^1 Gruis appears to be a particularly interesting example of the possible "Jupiter on circular orbit"–"two Trojans" ambiguity. A Jovian companion to the G0 dwarf τ^1 Gruis in a wide and almost circular orbit has been announced in the work by Jones et al. (2003). In our analysis, we use updated RV data comprising of 59 precision measurements (Butler et al. 2006). The best-fit single planet model to these data yields $P_b \sim 1300$ days, and $e_b \sim 0.1$. We re-analyse the data to look for possible Trojan planet solutions. Curiously, we quite easily found many *stable, coplanar* configurations involved in 1:1 MMR (Fig. 5) yielding similar or slightly better fit quality (rms $\in [5,6]$ m/s). The reflex signal of the 1:1 MMR (Fig. 6) can hardly be distinguished from that of a single-planet system. The osculating elements of the Trojans are given in Table 1 (fit III). In this case, both planets would move on quasi-circular orbits and these would be *coplanar*. The dynamical maps shown in Fig. 5 (also accompanying other best-fits solutions with acceptable quality which we found in the search) illustrate the extreme variability of the 1:1 MMR islands. The map for the best-fit with elements in Table 1 is shown in the top-left panel of Fig. 5. Another peculiar solution is illustrated in the top-right map in Fig. 5. The initial eccentricities are moderate, and the system would be found in extremely large island of stable motions. It spans whole range of e_b. The possibility of existence of such extended stable zones may strength the hypothesis of stable extrasolar Trojans.

The ambiguity of the RV fits implies interesting issues concerning the models of creation and stability of Earth-like planets interior to the orbits of the putative Jovian Trojans. In the τ^1 Gruis, the space interior to the Jovian planet is "empty" as no smaller planets have been yet detected. So we can try to predict in which regions of the habitable zone (~ 1 AU) smaller planet could survive. For this purpose we computed dynamical maps for putative Earth-like masses with initial conditions varied in the (a_0, e_0) plane, and initial orbital angles set to $0°$. We considered two dynamical environments: the one with the best-fit Jovian companion in close to circular orbit and the second one with Trojans in quasi-circular orbits (their elements are given in Table 1, fit III). The results are shown in Fig.7. For the first configuration, we detect an extended zone of stable motions. Additional experiments regarding creation of Earth-like planets through coagulation of Mars and Moon-size protopolanets (see Raymond, 2008 in this volume) performed with the Mercury code (Chambers 1999) assures us that such planets emerge easily in that zone. In the case of a configuration with Trojans, the stable zone shrinks significantly. Moreover, the creation of Earth-like planets is much more difficult. We found that they could form only in the zones of relatively stable motions, up to ~ 0.9 AU and in the "gap" between the 4:1 MMR and the border of global instability. Yet in that case, the simulations are very difficult to carry out due to frequent close encounters between planetesimals and the Jovian planets.

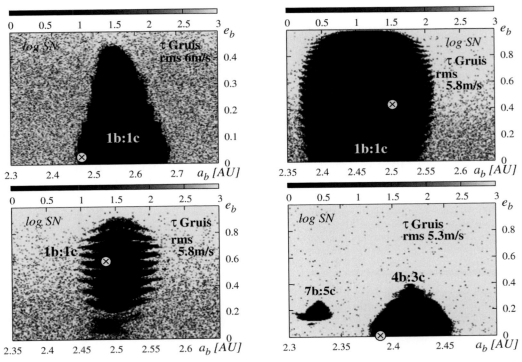

Figure 5. Dynamical maps of putative τ^1 Gruis coplanar, edge-on configuration of two Jovian planets ($\sim$ 0.5–0.9 m$_J$) involved in low-order resonances. The best-fits yield an rms $\sim$ 5–6 m/s and $(\chi^2_\nu)^{1/2} \sim 1$. Their quality is similar to that of the single-planet solution (an rms about of 6 m/s).

6. Conclusions

In this work we consider some problems related to modeling observations of stars hosting multi-planet systems. It is well known that the phase space of such system has a non-continuous and complex structure with respect to any stability criterion. Hence, when searching for initial conditions, one has to take into account the dynamical character of putative planetary configurations. Due to narrow observational windows, significant measurement errors, stellar jitter and other uncertainties, the formal best-fits may appear very unstable. Searching for stable solutions in their neighborhood of the phase space by trial and error, we should not expect that the results could be statistically optimal. Thus, an intuitively natural approach is to eliminate unstable configurations during the fitting process, through penalizing unstable solutions with a suitably large value of the $(\chi^2_\nu)^{1/2}$ function, or of another measure of the fit quality. In that way, the stability plays a role of an additional, implicit observable. That method is suitable for multi-body systems with Jovian planets presumably involved in low-order mean motion resonances (MMRs). In particular, we considered two new examples in which the model of the RV may be non-unique. The RV of HD 155358 by Cochran et al. (2007) permit a few local minima of $(\chi^2_\nu)^{1/2}$ related to different orbital configurations. We also found an example illustrating the ambiguity of Keplerian, close to circular single-planet solutions. The RV data of τ^1 Gruis could be equally well modeled with coplanar configurations of Jovian planets involved in 1:1 MMRs.

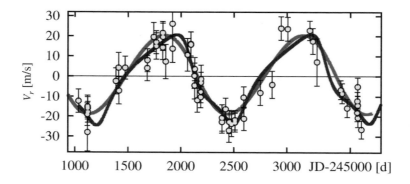

Figure 6. The RV data of τ^1 Gruis and the synthetic signals. The red curve is for the the best, single-planet Keplerian fit yielding $(\chi^2_\nu)^{1/2} \sim 1$ and an rms ~ 6.1 m/s. The blue curve (darker one) is for coplanar, edge-on configuration (an rms ~ 5.9 m/s) involved in 1b:1c MMR (see the top-left panel in Fig. 5 for the dynamical map).

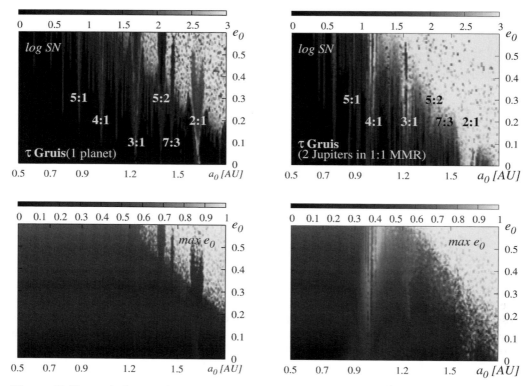

Figure 7. Dynamical maps for Earth-mass planets in the coplanar τ^1 Gruis system. The left column is for the stability map for the best fit configuration with one Jovian planet in quasi–circular orbit ($a \sim 2.5$ AU), the right column is for the systems with Trojans (Table 1, fit III). Panels in the top row are for the stability indicator $\log SN$. Panels in the bottom row are for the max e indicator (the integration period $\sim 50,000$ yr). Some MMRs between the Earth-like planet and the Jovian companions are labeled.

Parameter	HD 155358 (fit I) planet b	HD 155358 (fit I) planet c	HD 155358 (fit II) planet b	HD 155358 (fit II) planet c	τ^1 Gruis (fit III) planet b	τ^1 Gruis (fit III) planet c
$m \sin i$ [m$_J$]	0.827	0.490	0.863	0.497	0.401	0.923
a [AU]	0.623	0.875	0.628	1.212	2.471	2.565
e	0.121	0.743	0.128	0.198	0.027	0.053
ω [deg]	131.6	86.11	161.9	272.0	99.9	163.8
$\mathcal{M}(t_0)$ [deg]	106.32	313.7	130.3	198.8	3.5	3.6
$(\chi_\nu^2)^{1/2}$	1.14		1.08		1.12	
σ_j [m/s]	5		5		4	
rms [m s^{-1}]	6.31		5.98		5.96	
V_0 [m s^{-1}]	12.69		10.69		-0.04	
$M_\star$ [$M_\circ$]	0.87		0.87		1.25	

Table 1. The best-fit astro-centric, osculating Keplerian elements of stable, coplanar and edge-on planetary configurations at the epoch of the respective first observation. Original errors of the data are rescaled by adding the "jitter" σ_j in quadrature.

Moreover, our fitting method, used mainly for RV data, is quite general and may be applied to other types of observations as well. As the stability criterion, one can use the maximal Lyapunov exponent, the most stringent and formal characteristic of stable/unstable motion. Other suitable indicators like the maximal eccentricity, the spectral number, or the diffusion rate of fundamental frequencies may also be applied. These fast indicators help us to search for and find long-term stable solutions, but also make it possible to efficiently explore and to visualize the sophisticated and varying structure of the phase space. We can see the planetary system in its dynamical environment.

Many multi-planet systems are found on the edge of long-term dynamical stability. It is not clear yet whether this is a general property of multi-planet systems, the outcome of poor statistics or just the consequence of a bad choice of the RV model. Large eccentricities in multi-planet systems may "hide" other, unknown planets. Yet in that case, the dynamical modeling of the RV with stability constraints provides valuable information on the dynamical structure of the putative planetary configurations. Finding the best fits on the very edge of stable zones may provide good hints and motivation to look for alternate models of the RV.

Acknowledgments

K. G. thanks the organizers of the IAU 249 symposium for the invitation and great hospitality. We are grateful to the anonymous referee for comments that improved the manuscript. Many thanks to Boud Roukema for corrections of the text. This work is supported by the Polish Ministry of Science, Grant 1P03D 021 29.

References

Arnold, V. I. *Mathematical methods of classical mechanics*. New York: Springer, 1978.
Barnes, R. & Greenberg, R. *ApJL*, 665:L67–L70, 2007.
Baluev, R. V. *arXiv:0712.3862*, 2007.
Beaugé, C., Michtchenko, T. A., & Ferraz-Mello S. *MNRAS*, 365:1160–1170, 2006.
Beaugé S., Sándor, Zs., Érdi, B., & Süli, Á. *A&A*, 463:359-367, 2007.
Benettin, G., Galgani, L., Giorgilli, A., & Strelcyn, J.-M. *Meccanica*, pages 9–30, 1980.

Bevington, P. R. & Robinson, D. K. *Data reduction and error analysis for the physical sciences*. McGraw-Hill, 2003.
Butler, R. P. et al. Catalog of Nearby Exoplanets. *ApJ*, 646:505–522, 2006.
Chambers, J. E. *MNRAS*, 304:793–799, 1999.
Charbonneau, P. *ApJs*, 101:309–+, 1995.
Cincotta, P. M., Giordano, C. M., & Simó, C. *Physica D*, 182:151, 2003.
Cincotta, P. M. & Simó, C. *&s*, 147:205–228, 2000.
Cochran, W. D. Endl, M. R., Wittenmyer, A., & Bean, J. L. *ApJ*, 665:1407–1412, 2007.
Correia, A. C. M. et al. *A&A*, 440:751–758, 2005.
Dvorak, R., Freistetter, F., & Kurths, J. *Lecture Notes in Physics, Springer Verlag*, 683, 2005.
Ferraz-Mello, S., Michtchenko, T. A., & Beaugé, C. Chaotic Worlds: from Order to Disorder in Gravitational N-Body Dynamical Systems, 255–+, 2006.
Ferraz-Mello, S., Michtchenko, T. A., & Beaugé, C. *ApJ*, 621:473–481, 2005.
Ford, E. B. *AJ*, 129:1706–1717, 2005.
Ford, E. B., Lystad, V., & Rasio, F. A. *Nature*, 434:873–876, 2005.
Ford, E. B. & Gaudi, B. S. *ApJL*, 652:L137-L140, 2006.
Froeschlé, C. *Celest. Mech. & Dyn. Astr.*, 34:95–115, December 1984.
Goździewski, K., Breiter, S., & Borczyk, W. MNRAS, 710, 2008 (in press).
Goździewski, K. & Konacki, M. *ApJ*, 647:573–586, 2006.
Goździewski, K., Maciejewski, A. J., & Migaszewski, C. *ApJ*, 657:546–558, 2007.
Goździewski, K. & Migaszewski, C. *A&A*, 449:1219–1232, 2006.
Gregory, P. C. *MNRAS*, 374:1321–1333, 2007.
Holman, M. J. &Murray, N. W. *AJ*, 112:1278+, 1996.
Ji, J. et al. *ApJL*, 591:L57–L60, 2003.
Jones, H. R. A. et al. *MNRAS*, 341:948–952, 2003.
Laskar, J. *Icarus*, 88:266–291, 1990.
Laskar, J. *Celest. Mech. & Dyn. Astr.*, 56:191–196, 1993.
Laughlin, G., Chambers, J. E, & Fischer, D. *ApJ*, 579:455–467, 2002.
Laughlin, G. & Chambers, J. E. *ApJ*, 551:L109–L113, 2001.
Laughlin, G. & Chambers, J. E. *AJ*, 124:592–600, 2002.
Lecar, M. et al. *Annual Rev. of Astron. & Astroph.*, 39:581–631, 2001.
Lee, M. H., et al. *ApJ*, 641:1178–1187, 2006.
Lee, M. H. & Peale, S. J. *ApJ*, 592:1201–1216, 2003.
Lissauer, J. J. *Rev. Mod. Phys.*, 71(3):835–845, 1999.
Lissauer, J. J. & Rivera, E. J. *ApJ*, 554:1141–1150, 2001.
Malhotra, R. In *ASP Conf. Ser. 149: Solar System Formation and Evolution*, pages 37+, 1998.
Marcy, G. W., et al. *ApJ*, 556:296–301, 2001.
Mayor, M., et al. *A&A*, 2003. astro-ph/0310316.
Michtchenko, T. A. & Ferraz-Mello, S. *ApJ*, 122:474–481, 2001.
Morbidelli, A. *Modern celestial mechanics: aspects of Solar system dynamics*. Taylor & Francis, 2002.
Murray, N. & Holman, M. *Nature*, 410:773–779, 2001.
Nesvorný, D. & Morbidelli, A. *Celest. Mech. & Dyn. Astr.*, 71:243–271, 1999.
Pepe, F., et al. *A&A*, 462:769–776, 2007.
Press, W. H., Teukolsky, S. A., Vetterling, W. T., & Flannery, B. P. *Numerical Recipes in C. The Art of Scientific Computing*. Cambridge Univ. Press, 1992.
Rivera, E. J. & Lissauer, J. J. *ApJ*, 402:558–392, 2001.
Robutel, P. & Laskar, J. *Icarus*, 152:4–28, 2001.
Šidlichovský, M. & Nesvorný, D. *Celest. Mech. & Dyn. Astr.*, 65:137–148, 1997.
Smart, W. M. *Text-Book on Spherical Astronomy*. Cambridge Univ. Press, 1949.
Szebehely, V. *Celest. Mech. & Dyn. Astr.*, 34:49–64, December 1984.
Tinney, C. G., et al. *ApJ*, 647:594–599, 2006.
Vogt, S. S., et al. *ApJ*, 632:638–658, 2005.
Wisdom, J. & Holman, M. *AJ*, 102:1528–1538, 1991.

On the dynamics of Trojan planets in extrasolar planetary systems

R. Dvorak[1], R. Schwarz[2] and Ch. Lhotka[1]

[1]Institute for Astronomy, University of Vienna,
Türkenschanzstrasse 17, A-1180, Vienna, Austria
email: dvorak@astro.univie.ac.at, lhotka@astro.univie.ac.at

[2]Department of Astronomy, Eötvös University,
Pázmány Péter sétány 1/A, H-1117 Budapest, Hungary
email: schwarz@astro.univie.ac.at

Abstract. In this article we examine the motion of fictitious Trojan planets close to the equilateral Lagrangean equilibrium points in extrasolar planetary systems. Whether there exist stable motion in this area or not depends on the massratio of the primariy bodies in the restricted three body problem, namely the host star and the gasgiant. Taking into account also the eccentricity of the primaries we show via results of extensive numerical integrations that Trojan planets may survive only for $e < 0.25$. We also show first results of a mapping in the 1:1 resonance with a gas giant on an eccentric orbit which is applied to the extrasolar planetary systems HD 17051. We furthermore study the influence of an additional outer planet which perturbs the motion of the gasgiant as well as the Trojan cloud around its L_4 Lagrangean point.

Keywords. Celestial mechanics, mapping, Trojan planets, HD 17051

1. Introduction

The search for terrestrial planets (=TP) in extrasolar planetary systems (=EPS) is one primary subject of scientists involved in this rapidly growing field of astronomy. Although up to now we have only knowledge of TPs with several Earth masses we hope to find quite soon a second Earth via space-based observations. It is the task of astrodynamical investigations to look for possibly stable orbits of additional planets in EPS where already one, or even more gas giants (=GG) have been detected. Out of the different possible configurations for TPs, namely orbiting inside a GG (like in our Solar System), outside a GG (we know of many hot Jupiters) we also must take into account that a TP may have a GG with a TP as a satellite (e.g. Saturn's Titan). We concentrate in this study on coorbital TPs, which could be realized as Trojan TPs (=TTP).

There are several important studies on the Trojan problem: e.g. Marzari & Scholl 1998, Nauenberg 2002, Laughlin & Chambers 2002, Menou & Tabachnik 2003, Morbidelli et al. 2005.

In chapter 2 we show the results of numerous integrations which have been undertaken to establish the largeness of the stable regions around the equilateral points of a large planet depending on the mass and the eccentricity of the primaries. We also show a mapping in the 1:1 resonance in the elliptic restricted three body problem, which gives a good indication of the dynamical structure of the Trojan region. In chapter 3 we concentrate on the perturbation of the stable region caused by a large planet outside the GG hosting the TP which mimics quite well the Solar system situation (e.g. Jupiter and Saturn). Finally, in the last chapter, we summarize the results which are interesting with

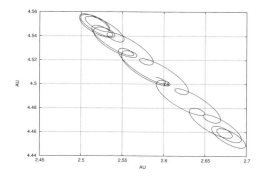 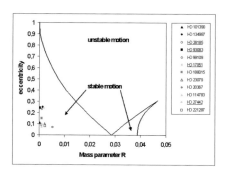

Figure 1. Typical motion of a Trojan asteroid in a rotating frame close to the equilateral equilibrium point L_4 (left graph). The stability of the equilateral equilibrium points depending on the mass ratio P and the eccentricity of the primary bodies after Marchal, 1990 (right graph).

respect to formation mechanism (Beaugé et al. 2007) and the chances to observe such TTPs (e.g. Ford & Gaudi, 2006 and Goździewski & Konacki 2007).

2. The Trojan Regions in Extrasolar Planetary Systems

It is known since the first discovery of an asteroid staying always in the vicinity of the Lagrangean equilibrium point (Fig. 1) that the work of Lagrange and Euler in the restricted three body problem † was not only of theoretical interest. This first asteroid was discovered by Max Wolf in Heidelberg in 1906 who named it after Achilles, the hero of the Trojan War. In the following years many more asteroids were discovered in Jupiter's Lagrangian points and were given names associated with the Iliad‡.

2.1. Numerical determination of the stable region for Trojans

Part of the work was to find out how large is the extension of the stable regions around the equilateral Lagrangian points (Efthymiopoulos & Sándor, 2005). From theory we know that the equilateral points are linearly stable even for large eccentricities of the primaries since the first work of Danby (1964). Fig. 1 shows zones of stability for motions of Trojan bodies in terms of the mass parameter P and the eccentricity. In addition we plotted the positions of all habitable¶ extrasolar systems. The mass parameter P is defined through the equation (see Marchal, 1990):

$$P = \frac{(m_2 + m_3)}{M} + \frac{m_2 \cdot m_3}{m_1^2} + O\left(m_2^3 \cdot \frac{m_3}{m_1^4}\right), \qquad (2.1)$$

† A massless regarded planet is moving under the gravitational influence of Sun and a gasgiant like Jupiter in the same plane, where the two massive bodies, called the primies have circular orbit.

‡ This group of asteroids is called 'Trojans'. The ones close to the L4 point are named after Greek heroes, the ones close to the L5 point are given names of the heroes of Troy. 617 Patroclus, a Greek warrior, is wrong placed as it is in the 'defending' Trojan camp, whereas 624 Hektor, a Trojan warrior, is in the Greek camp.

¶ Which means Extrasolar systems where one gas giant stays partly or fully in the habitable zone.

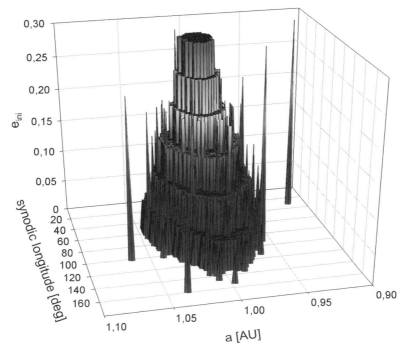

Figure 2. Stable area around the Lagrange point L_4 in the restricted three body problem for a dynamical model Sun–Jupiter depending on the eccentricity of the primaries.

m_1, m_2, m_3 are the masses of the star, the perturbing planet and the TTP respectively and M is the total mass of the system. Several studies have already been dedicated to this problem for extrasolar planetary systems (e.g. Dvorak and Schwarz 2005, Érdi and Sándor 2005, Érdi et al. 2007, Schwarz et al. 2005, 2007a and 2007b). The objectives of these investigations were to examine the dynamics of a terrestrial planet in 1:1 mean motion resonance with a Jovian-like planet. We show in Fig. 2 how the size of the Trojan stability regions for a Jupiter sized planet depends on the eccentricity of its orbit. This stable area was determined making use of the results of the integration of the equations of motion of a fine grid of initial conditions around the L_4 point for 1 million periods of the primaries. Finally a catalogue of hypothetical habitable Trojan planets (http://www.univie.ac.at/adg/) was computed, where the largeness of the stable region depending on the mass ratio of the star to the planet and of the eccentricity of the primaries' orbits is shown.

2.2. Analytical mapping model for terrestrial trojan planets

To obtain a mapping model for the 1:1 resonance of the elliptic restricted three body problem, we expanded the corresponding disturbing function R of the Hamiltonian (where $\mu = m_2/(m_1 + m_2)$ is the mass ratio of the primaries (we assume $m_1 \gg m_2$ and set $x = \sqrt{\frac{a}{a'}} - 1$), see also Hajidemetriou 1993:

$$H = H_0 + \mu H_1 = -\frac{1}{2(1+x)^2} - (1+x) - \mu R, \qquad (2.2)$$

with respect to the eccentricities (e, e') and the variation in the semi-major axis of the ratio between the axis of the asteroid (a) and the disturbing planet (a') to suitable high orders ($'$ indicates the disturbing body).

The set of variable used to describe the systems are the Delaunay variables (Érdi & Sándor, 2005) defined by the quantities:

$$\tau = \lambda - \lambda', \; y = \sqrt{\frac{a}{a'}}\left(\sqrt{1-e^2} - 1\right), \; \omega, \qquad (2.3)$$

where λ is the orbital longitude and ω means the argument of pericenter of the TTP. In the case of HD 17051 we carried out the expansion up to 7th order in the eccentricites to cover the full domain of possible motions in the extra-solar Trojan configuration, such that the disturbing function turns out to be:

$$R(\tau, \omega, x, y; a', e') = \sum_\nu B_\nu(x, y; a', e') \cos \Phi_\nu(\tau, \omega, \lambda), \qquad (2.4)$$

where B_ν are polynomial functions and Φ_ν are the corresponding phase functions holding together terms of equal order ν. A detailed description on the developement of the disturbing function can be found e.g. in (Hajidemetriou, 1993 and Lhotka et al., 2008). Using the method of Hadjidemetriou (1991) we setup the generating function:

$$W = x_{n+1} \cdot \tau_n + y_{n+1} \cdot \omega_n + T \cdot \bar{H}(\tau_n, \omega_n, x_n, y_n, x_{n+1}, y_{n+1}), \qquad (2.5)$$

where $\bar{H} = H_0 + \mu \bar{R}$ and $\bar{R}$ is the averaged disturbing function with respect to the orbital longitude of Jupiter λ' and T is the period of the mapping. The mapping model is derived via the equations

$$J_{i,n} = \frac{\partial W}{\partial \theta_{i,n}}, \theta_{i,n+1} = \frac{\partial W}{\partial J_{i,n+1}}, i = 1, 2, \qquad (2.6)$$

where $J = (x, y)$ and $\theta = (\tau, \omega)$ and the mass parameter in (2.2) is set equal to $\mu = 0.01795$.

The resulting mapping is defined on the 4 dimensional Poincaré surface of section, but given in its implicit form. Former studies (Lhotka et al. 2008) showed that the radius of convergence of the proposed mapping method becomes limited to the librational regime of the asteroid's motion, if we expand the generating function to make the mapping explicit by series reversion. Therefore we introduced a simple root-finding algorithm to iterate the mapping at each iteration step without expanding it into explicit form, so that we can preserve all possible dynamical behaviour of the mapping. For this reason we used the inital values $(\tau_n, \omega_n, x_n, y_n)$ not only as a starting point for the mapping iteration itself, but also as starting values for the root finding procedure to find $(\tau_{n+1}, \omega_{n+1}, x_{n+1}, y_{n+1})$. In the respective figures we show two projections of the phase portrait, which is originally a 4 dimensional manifold in 4D phase space. Nevertheless the Poincaré surface of section $((\tau, \omega, x, y) : \omega = \omega_0, \omega_{n+1} - \omega_n > 0, (\tau, \omega, x, y) : \tau = \tau_0, \tau_{n+1} - \tau_n > 0, x_n > 0)$ reveals the resonant structure of the system (chain of islands and the librational and rotational behaviour of the asteroid (Fig. 3)).

3. Perturbations of a large planet on terrestrial trojan planets

We know the extension of the stable regions in the elliptic restricted problem quite well; even when the third body has a small mass, that means in the general three body

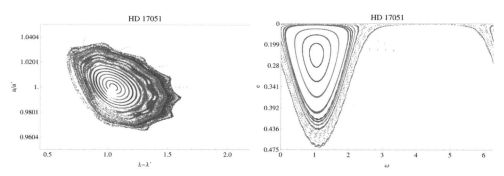

Figure 3. Mapping of the area around the Lagrange point L_4 in the system HD 17051: semimajor axis – libration angle (left graph) and eccentricity – pericenter (right graph).

problem there are stable areas. We now pose the question what happens when another large planet is perturbing these bodies, a situation comparable to a configuration in our Solar System, where Saturn perturbs the Jupiter Trojans. In this case, where the distance of Saturn is only twice the distance of Jupiter with both planets on low eccentric orbits, we have still a large population of Jupiter Trojans. Because nowadays we know already 25 EPS hosting more than one large planet, we study – as first approach to this question – how a second outer planet can disturb the Trojan region of an inner planet.

We put one Jupiter sized planet in a distance of 1 AU and populated the L_4 region with massless Trojan bodies; a second large planet was included in our computations as an outer perturber in different distances ($1.2 AU < a < 3.5 AU$). The orbital eccentricities of both giants were set to equal values: e = 0., 0.05, 0.10, 0.15 and 0.20. To be able to catch the most favorable and the most unfavorable initial configurations we investigated 8 different initial position (IP) for the two giants, which we call aligned (4 positions) and antialigned (4 positions) initial conditions for the inner and outer planet† P(eriastron)P (1,5), PA(poastron (2,6), AP (3,7),4: AA (4,8); the numbers correspond to the y-axis in Fig. 4 and Fig. 5. As integration method we used the *mercury6* symplectic integrator (Chambers, 1999) and the integration time was set to 1 million years for 100 fictitious Trojans around L_4. The stability criterion was a simple check of escape from that region, the largeness was defined as the number (percentage) of remaining Trojans after the end of the integration time. We need to emphasize that especially when we put the perturbing (outer) planet close to the inner planet, which hosts the Trojans, the mean motion resonances in connection with the different initial conditions are important for the stability of an orbit.

The discussion of the results leads to the following conclusions:
• $e = 0$ and $e = 0.05$ (Fig. 4, upper graph): We just show the results of the slightly eccentric orbits of the two planets because they are quite similar to the initially circular case. From $a_2 = 1.8 AU$ on almost 40 orbits survived; there is only a slight decrease of the number of survivors for larger distances. The antialigned IP led to a smaller stable region.
• $e = 0.10$ (Fig. 4, lower graph): For this moderate eccentricities the stable region

† P=Pericenter, A=Apocenter.

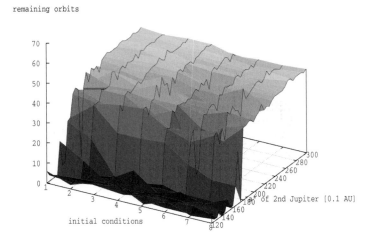

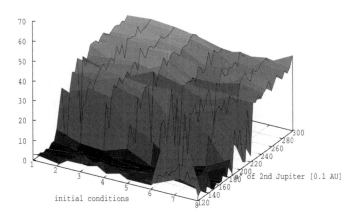

Figure 4. Initial condition diagram of the number of surviving Trojans (z-axes) around L_4 depending on the initial distance of the perturbing Jupiter (x-axis) and the position of the two planets (y-axes): $e = 0.05$ (upper graph) and $e = 0.10$ (lower graph) initial eccentricities of the two planets.

extends to a perturbing Jupiter as far as $a = 1.8 AU$ and the the difference for aligned and antialigned IP starts to be significant.

- $e = 0.15$ (Fig. 5, upper graph): for the aligned IP of the two primaries the difference to the former picture is not very large, but for the antialigned IP the unstable region extends to 2 AU and globally the percentage of the surviving orbits close to L_4 is halved.
- $e = 0.20$ (Fig. 5, lower graph): In both positions the unstable region is quite large (extends to 1.8 AU) for the aligned IP and up to 2.2 AU for the antialigned IP, where almost no Trojans survived.

Already for $e = 0.15$ and 0.20 one can see that even when the perturbing outer planet is more than two AU from the Sun, the MMRs start to cut the stable region into different 'slices' of stable and unstable areas for most of the initial conditions. From $e = 0.25$ of the inner Jupiter almost no stable Trojans survived independent of the perturbing outer Jupiter and thus we needed not to continue our investigations in this sense.

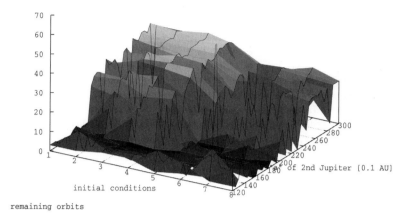

Figure 5. Caption like in Fig. 4 for $e = 0.15$ (upper graph) and $e = 0.20$ (lower graph).

4. Conclusion

We discussed how large the Trojan regions extends for single planet system depending on the eccentricity of the planet's orbit. In an analytical mapping model we also showed the structure of the stable region for the system HD 17051. Because more and more EPS with several large planets are observed we extended our investigations to an additional perturber outside the Trojan hosting large planet. The results show that – with a certain dependence on the eccentricities – the probability for the existence of stable Trojans of the inner large planet in EPS consisting of at least two large planets smaller than a ratio of the semimajor axes of 1:2 is quite large.

Acknowledgements: R. Schwarz thanks the Schrödinger grant of the FWF (J2619-N16); the work of C.Lhotka was fully supported by the FWF project P-18930.

References

Beaugé, C., Sándor, Zs., Érdi, B., & Süli, Á.: 2007, *A&A*, **463**, 359.
Chambers, J. E.: 1999, *MNRAS*, **304**, 793.

Danby, J. M. A.: 1964, *AJ*, **69**, 165.
Dvorak, R. & Schwarz, R.: 2005, *CeMDA*, **92**, 19.
Efthymiopoulos C. & Sándor, Z.,: 2005, *MNRAS*, **364**, 253.
Érdi, B. & Sándor, Z.: 2005, *CeMDA*, **92**, 113.
Érdi, B., Nagy, I., Fróhlich, G., Sándor, Zs. and Súli, Á.: 2007b, *MNRAS*, **381**, 33.
Ford, E. B. & Gaudi, B. S.: 2006, *ApJ*, **652**, 137.
Goździewski, K. & Konacki, M.:2007, *ApJ*,**647**, 573.
Hadjidemetriou, J.: 1991, In A. E. Roy (Ed.), "Predictability, Stability and Chaos in N–Body Dynamical Systems', Plenum Press, 157.
Hajidemetriou, j.: 1993, *CeMDA*, **56**, 563.
Lhotka Ch., Efthymiopolous, C. & Dvorak, R.: 2008, *MNRAS*, **384**, 1165.
Laughlin, G. & Chambers, J. E.: 2002, *AJ*, **124**, 592.
Marchal, C.: 1990, The three-Body Problem, Elsevier, 49.
Marzari, F. & Scholl, H.: 1998, *A&A*,**339**, 278.
Menou, K. & Tabachnik, S.: 2003, *AJ*, **583**, 473.
Morbidelli, A., Levison, H. F., Tsiganis, K., & Gomes, R.: 2005, *Nature*,435, 462.
Nauenberg, M.: 2002, *AJ.*, **124**, 2332.
Schwarz, R., Pilat-Lohinger, E., Dvorak, R., Érdi, B., & Sándor, Zs.: 2005, *AsBio*, **5**, 579.
Schwarz, R., Dvorak, R.,Pilat-Lohinger, E., Süli, Á., & Érdi, B.: 2007a, *A&A*, **462**, 1165.
Schwarz, R., Dvorak, R., Süli, Á., & Érdi, B.: 2007b, *A&A*, **474**, 1023.

Extrasolar planet interactions

Rory Barnes[1] and Richard Greenberg[1]

[1]Lunar and Planetary Lab, University of Arizona, 1629 E. University Blvd., Tucson, AZ, USA
email: rory@lpl.arizona.edu

Abstract. The dynamical interactions of planetary systems may be a clue to their formation histories. Therefore, the distribution of these interactions provides important constraints on models of planet formation. We focus on each system's apsidal motion and proximity to dynamical instability. Although only ∼25 multiple planet systems have been discovered to date, our analyses in these terms have revealed several important features of planetary interactions. 1) Many systems interact such that they are near the boundary between stability and instability. 2) Planets tend to form such that at least one planet's eccentricity periodically drops to near zero. 3) Mean-motion resonant pairs would be unstable if not for the resonance. 4) Scattering of approximately equal mass planets is unlikely to produce the observed distribution of apsidal behavior. 5) Resonant interactions may be identified through calculating a system's proximity to instability, regardless of knowledge of angles such as mean longitude and longitude of periastron (e.g. GJ 317 b and c are probably in a 4:1 resonance). These properties of planetary systems have been identified through calculation of two parameters that describe the interaction. The apsidal interaction can be quantified by determining how close a planet is to an apsidal separatrix (a boundary between qualitatively different types of apsidal oscillations, e.g. libration or circulation of the major axes). This value can be calculated through short numerical integrations. The proximity to instability can be measured by comparing the observed orbital elements to an analytic boundary that describes a type of stability known as Hill stability. We have set up a website dedicated to presenting the most up-to-date information on dynamical interactions: http://www.lpl.arizona.edu/∼rory/research/xsp/dynamics.

1. Introduction

One of the most striking differences between known exoplanets and the giant planets of our Solar System involves the observed orbits: Observed exoplanets tend to have large eccentricities e, and small semi-major axes a, whereas the gas giants of the Solar System have small e and large a. Recently, however, it has been shown that the dynamical interactions in many multiple planet systems (including the Solar System) show certain features in common (Barnes & Quinn 2004; Barnes & Greenberg 2006a [BG06a]; Barnes & Greenberg 2006c [BG06c]; Barnes & Greenberg 2007b [BG07b]). These shared traits suggest that the character of dynamical interactions (over $10^3 - 10^4$ years), rather than the present orbits, may be a more meaningful constraint on the origins of planetary systems (Barnes & Greenberg 2007a [BG07a], BG07b). Considerations of shared dynamical properties have even resulted in the first successful prediction of the mass and orbit of an extrasolar planet, HD 74156 d (predicted by Barnes & Raymond [2004] and Raymond & Barnes [2005]; found by Bean *et al.*2008).

Now with over 200 extra-solar planets known, including ∼ 25 multi-planet systems, we can look at this population as a whole, with increasing confidence regarding which common characteristics may be more than statistical flukes. We have identified commonalities among planetary interactions that may be key constraints on the origins of planetary systems. We have identified parameters that quantify an interaction's proximity to boundaries between qualitatively different types of motion. These two types of boundaries are the "apsidal separatrix" and the dynamical stability boundary. The apsidal separatrix is the boundary between different types of apsidal oscillations, e.g. libration and circulation. The stability boundary separates regions in which all planets are bound to the host star from those in which at least one planet is liable to be ejected. These two parameters are fixed quantities that do not vary over time, but they constrain how the systems evolve. Systems tend to lie close to these two boundaries.

In this chapter we review our derivation of ϵ, which quantifies proximity to an apsidal separatrix, in § 2. Then we describe β, which parameterizes a system's proximity to dynamical instability, in § 3. In § 4 we present the observed distributions of these quantities and use them to constrain formation models. Finally in § 5 we draw our general conclusions.

2. The Apsidal Separatrix

Eccentricities in multiple planet systems oscillate due to secular interactions (see e.g. Laughlin & Adams 1999; Rivera & Lissauer 2000; Stepinski et al.2000; Michtchenko & Malhotra 2002; BG06a). Thus a currently observed eccentricity may not be its average value over the ~10,000 year secular period. Validation of formation models through the eccentricity distribution is therefore inadequate; the eccentricity oscillations are a better description of a multiple-planet system's properties.

We consider analytic and numerical models of planet-planet interactions to describe the oscillations. The analytic approach is secular theory which considers long-term averages of forces between planets. This method was originally developed independently by Laplace and Lagrange, but see BG06a for a presentation that does not involve matrix manipulation. We compare the secular solutions to N-body numerical models that solve the force equations directly, and are therefore arbitrarily accurate, for a sufficiently small timestep, using the MERCURY6 code (Chambers et al.1999).

In general, secular theory predicts that e's and $\Delta\varpi$'s (the difference between two longitudes of periastron, ϖ) oscillate and that a is constant. Therefore, conservation of angular momentum requires that as one planet's eccentricity drops, another must rise (as e increases, orbital angular momentum decreases). The type of oscillation depends on initial conditions. If $\Delta\varpi$ oscillates about 0, the system is experiencing "aligned libration." If $\Delta\varpi$ librates about π, then the system is undergoing "anti-aligned libration." If $\Delta\varpi$ oscillates through 2π then the apsides undergo "circulation". The boundaries between these qualitatively different types of behavior are known as "separatrices".

Recently it has been noted that many systems lie near an "apsidal separatrix" (Ford et al.2005; BG06a, BG06c). For systems of just two planets, the apsidal separatrix can only separate circulation and libration. This type of separatrix is a "libration-circulation separatrix". An example of the libration-circulation separatrix is shown in the left panels of Fig. 1.

In systems of more than two planets, things get more complicated. In addition to the libration-circulation separatrix, the system may interact with different numbers of rotations of $\Delta\varpi$ through 360° during one eccentricity oscillation. The boundary between interactions with different numbers of circulations in one eccentricity cycle is a "circulation-mode separatrix", and an example is shown in the right panels of Fig. 1. Note that the gray curve in the top right panel of this figure circulates once and librates once during one eccentricity cycle, whereas the black curve circulates twice.

For an interaction to lie near a separatrix, the amplitude of eccentricity oscillations is generally two orders of magnitude or more. Since $0 \leqslant e < 1$ for bound planets, this

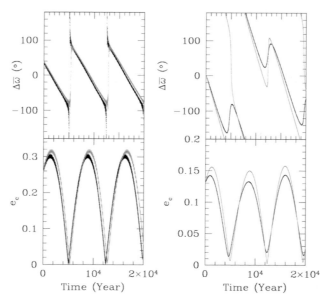

Figure 1. Examples of near-separatrix motion in planetary systems. *Left Panels:* A libration-circulation separatrix. Two possible evolutions of υ And c and d (the middle and outer planet of the system) assuming the current orbits. The black points are the system from Butler *et al.*(2006), the gray from Ford *et al.*(2005). Although the best-fit orbits in these two cases are very similar, they result in qualitatively different types of evolution of $\Delta\varpi$ (top): The older data predict aligned libration, whereas the updated data predict circulation. Note that the evolution of e_c is similar in both cases, and periodically reaches near-zero values (bottom). *Right Panels:* A circulation-mode separatrix. HD 69830 c and d evolve near the circulation-mode separatrix. The black data are from Lovis *et al.*(2006), and the gray data are for a fictitious system in which the inner planet's, b's, eccentricity was changed from 0.1 to 0.15. In the first 2×10^4 years, the actual $\Delta\varpi$ undergoes 1 complete rotation through $360°$, but in the fictitious system, $\Delta\varpi$ undergoes 2 complete circulations (top). We again see that the middle planet's eccentricity periodically drops to near-zero values (bottom).

means that at least one planet in near-separatrix interactions (both libration-circulation and circulation-mode) periodically is on a nearly circular orbit. The proximity to the separatrix can be parameterized as

$$\epsilon \equiv \frac{2[\min(\sqrt{x^2 + y^2})]}{(x_{max} - x_{min}) + (y_{max} - y_{min})}, \quad (2.1)$$

where $x \equiv e_1 e_2 \sin(\Delta\varpi)$; $y \equiv e_1 e_2 \cos(\Delta\varpi)$, and the subscripts min and max refer to the extreme values these variables attain over a complete secular cycle (BG06c). (Note that ϵ is a fixed quantity even as e and $\Delta\varpi$ change periodically.) When $\epsilon = 0$ the pair is on an apsidal separatrix, and one eccentricity periodically reaches zero. Note that in the case of mean motion resonant interactions, ϵ may not necessarily reveal how close an interaction is to an apsidal separatrix (BG06c).

3. The Dynamical Stability Limit

Observational uncertainties in the orbits of exoplanets often include regions of dynamical instability in which one or more planets would be ejected within 1 Myr, *i.e.* the system is dynamically "packed" (e.g. Barnes & Quinn 2001, 2004; Goździewski 2002;

Kiseleva-Eggleton 2002; Érdi et al.2004; BG06b, BG07b). This timescale appears to be long enough to identify nearly all unstable configurations (Barnes & Quinn 2004). Representative cases of these investigations are shown in Fig. 2. The shading in this figure correlates with the likelihood of stability: White regions are stable, black unstable, and grays contain a mix of stable and unstable orbits. In these cases we define stability in the "Lagrange" sense for $\sim 10^6$ orbits: Escapes and exchanges of planets are forbidden.

The preponderance of systems that lie near the stability limit naturally raises the question of where, exactly, in orbital element space, the stability boundary lies. Surprisingly, a general analytical expression for the location of the stability boundary is lacking, even though it has been the goal of considerable research (e.g. Szebehely & McKenzie 1981; Wisdom 1982; Marchal & Bozis 1982; Gladman 1993; Laskar 1990; Holman & Wiegert 1999; Sosnitskii 1999; David et al.2003; Cuntz et al.2007). These investigations all made significant progress toward defining a stability boundary, but none found it.

Given that many planetary systems lie near the dynamical stability limit, we have considered how to quantify the proximity to stability (BG06b, BG07b). We will focus on one formulation by Marchal & Bozis (1982; see also Gladman 1993) that considers "Hill stability". Hill stability only requires the ordering of the planets to remain constant for all time; the outer planet may escape to infinity.

A coplanar, two-planet system, outside of a resonance is Hill stable if

$$-\frac{2M}{G^2 M_*^3}c^2 h > 1 + 3^{4/3}\frac{m_1 m_2}{m_3^{2/3}(m_1+m_2)^{4/3}} - \frac{m_1 m_2(11m_1 + 7m_2)}{3m_3(m_1+m_2)^2} + ..., \qquad (3.1)$$

where M is the total mass of the system, m_1 is the mass of the more massive planet, m_2 is the mass of the less massive planet, m_3 is the mass of the star, G is the gravitational constant, $M_* = m_1 m_2 + m_1 m_3 + m_2 m_3$, c is the total angular momentum of the system, and h is the energy (Marchal & Bozis 1982). Here c and h must be calculated in barycentric coordinates. If a given three-body system satisfies the inequality in Eq. (3.1), then the system is Hill stable. If this inequality is not satisfied, then the system may or may not be Hill stable. In this inequality, the left-hand side is a function of the orbits, but the right-hand side is only a function of the masses. This approach for identifying stable orbits is fundamentally different from other common techniques for determining stability which exploit resonance overlaps (Wisdom 1982; Quillen & Faber 2006), chaotic diffusion (Laskar 1990; Pepe et al.2007), fast Lyapunov indicators (Froeschlé et al.1997; Sándor et al.2007), or periodic orbits (Voyatzis & Hadjidemetriou 2006; Hadjidemetriou 2006).

The ratio of the two sides of the inequality in Eq. (3.1) (left-hand side over right-hand side) allows a quantification of a system's proximity to the Hill stability boundary. We define this ratio as β. If $\beta < 1$ a system's Hill stability is unknown, if $\beta > 1$, Hill stability is guaranteed, if $\beta = 1$, the system is on the boundary. Note that for any system of two planets outside of resonance with fixed mass, the locus of Hill-stable orbits can be easily calculated.

Fig. 2 includes contour lines of constant β for the four systems considered. For the bottom two cases (systems with no mean motion resonance), the $\beta = 1$ lines are near the transition from Lagrange stable orbits (white) to unstable (black). These two cases therefore show that when $\beta \gtrsim 1$, a system is not only Hill stable, it is also Lagrange stable.

However, in the top two panels of Fig. 2, the $\beta = 1$ contour cuts through a swath of stability located at resonance. This qualitatively different behavior is no surprise because the criterion for Hill stability, Eq. (3.1), does not apply to systems in resonance. Note however that in the top left panel, outside of resonance, the β contours do follow the

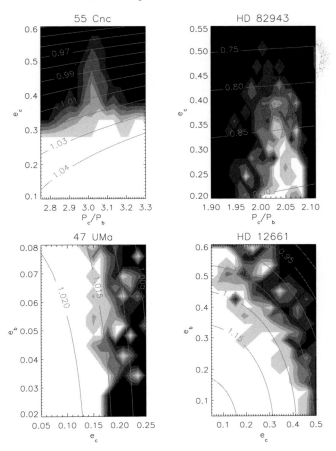

Figure 2. Lagrange stability boundary in relation to the Hill stability boundary for four exoplanetary systems. White regions represent bins in which all configurations were stable, black bins contained no stable configurations, darker shades of gray correspond to regions in which the fraction of stable simulations were smaller (e.g. Barnes & Quinn 2004). The curves represent contour lines of β. Contour lines follow the shape of the Lagrange stability boundary, except in resonance, where the Lagrange stability region is extended to lower values of β.

Lagrange stability boundary, again at values near 1. In the top right, we see that all β values are less than 1, hence there are no non-resonant, Lagrange stable configurations in the region of orbital element space considered.

4. Dynamical Properties of Exoplanets

In this section we will tabulate the observed distributions of ϵ and β and interpret these distributions in the context of models of planet formation. Observed extrasolar planets tend to lie near the apsidal separatrix. In the top panel of Fig. 3, we plot the observed distribution of ϵ values (solid line). These values were calculated through an N-body integration that required energy conservation to be better than 1 part in 10^4 with a symplectic integration technique. Over 40% of known systems have $\epsilon < 0.01$, including two of the three giant-planet pairs in the Solar System.

We tabulate the dynamical properties of multi-planet systems in Table 1†, where we also list the sources of the initial conditions we used. In this table AM stands for "apsidal motion" and the possibilities are circulation (C), aligned libration (A), or anti-aligned libration (AA). The MMR column lists the resonance, if applicable. The proximities to the apsidal separatrix, ϵ, are the values from the literature (BG06c), as are proximities to the Hill stability boundary, β (BG07b). The "Class" distinguishes orbits whose evolution is dominated by tidal (T), resonant (R) or secular (S) interactions. Table 1 includes the dynamical properties of the giant planets in our Solar System for comparison.

Given the strong tendency for systems to be near the apsidal separatrix, it is natural to wonder how systems formed that way. The apsidal behavior is tied to the eccentricities. Considerable work by previous researchers has focused on reproducing the distribution of eccentricity values. The two most studied processes are "planet-planet scattering", *i.e.* gravitational encounters between planets (e.g. Rasio & Ford 1996; Weidenschilling & Marzari 1996; Lin & Ida 1997; Ford *et al.*2001; Marzari & Weidenschilling 2002), and torquing by remnant protoplanetary disks that pumps up eccentricities (e.g. Artymowicz 1992; Boss 2000; Papaloizou *et al.*2001; Chiang & Murray 2002; Goldreich & Sari 2003; D'Angelo *et al.*2006). Simulations of planet-planet scattering can produce eccentricity distributions that are similar to the observed distribution (Ford *et al.*2001), but models of disk torquing tend to produce eccentricities that are too low (D'Angelo *et al.*2006).

Ford *et al.*(2005) were the first to consider how scattering may lead to near-separatrix behavior. They showed that the ejection of a hypothetical additional planet could produce the observed near-separatrix behavior among the planets orbiting υ And. In their model three planets, initially on circular orbits, formed too close together to be stable. Scattering between two planets results in one of them being ejected. The third planet, orbiting at a safe distance from the scattering planets, remained on a circular orbit. The planet that remained bound to the star after the ejection received an impulsive kick and its eccentricity quickly jumped to a relatively large value. This sudden change created a new "initial condition" for the secular interaction. The planet that did not partake in the scattering then began a new secular evolution, but since apsidal behavior is periodic, that planet's eccentricity would return to zero and the motion was near-separatrix.

However, Ford *et al.*only considered one such unstable case, and it was unclear how likely such a scenario was. The observed ϵ statistics (solid line in the top panel of Fig. 3) place a strong constraint on planet-planet scattering models. BG07a considered ~ 400 hypothetical systems in order to provide a statistical test of the planet-planet scattering model's likelihood to reproduce the observed apsidal behavior of extrasolar planetary systems. In the main part of that experiment, they considered systems in which all orbital elements were initially the same except for the mean longitude of the middle planet (of a three-planet system) which was shifted by 1 degree for each successive case (*i.e.* 360 simulations). Stability is most likely independent of this parameter (see Eq. 3.1), therefore there should be no correlation between initial mean longitude and ϵ. The resulting distribution of ϵ is shown in the top panel of Fig. 3 by the dashed line. This scattering (dashed line) produced one-tenth the fraction of near-separatrix ($\epsilon < 0.01$) cases as is observed (solid line).

The planet-planet scattering model had done a reasonable job of reproducing the observed eccentricity distribution (Ford *et al.*2001), but it may be a challenge for it to reproduce the observed apsidal behavior (BG07a). BG07a noted that the path to small ϵ values seemed quite constrained: The scattered planet needed to be ejected immedi-

† see http://www.lpl.arizona.edu/~rory/research/xsp/dynamics for an up-to-date list of these properties

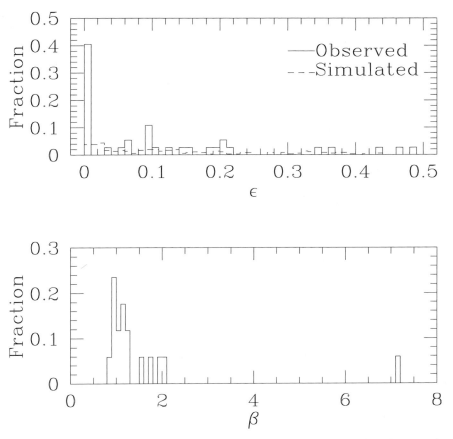

Figure 3. *Top Panel:* Distributions of ϵ. The solid line is the observed distribution, the dashed line is that predicted by a model involving the ejection of an additional Jupiter mass planet (Ford et al.2005; BG07a). *Bottom Panel:* Observed distribution of β in known two-planet systems. Note that we may only calculate β for two-planet systems.

ately. This common scattering route to near-separatrix motion led BG07a to suggest that the probability of near-separatrix motion could be enhanced if the perturber is relatively small and/or on an initially eccentric orbit. Such a "rogue planet" may have a better chance of being removed from a planetary system after one encounter with a giant planet, but this model will require further testing. The actual distribution of ϵ must be reproduced by any model of planet formation.

Next we discuss the implications of the β distribution, that is, how close a system is to instability. We tabulate β (Table 1) values to produce another distribution function that describes the range of dynamical properties of planetary systems. The observed distribution of β is plotted in the bottom panel of Fig. 3. Most systems have $\beta \sim 1$, nearly all have $\beta < 2$, and the exception is HD 217107 ($\beta > 7$). This clustering of systems near the Hill stability limit suggests that many systems form near the edge of dynamical stability.

In Fig. 3 and Table 1 we see that many systems in fact have $\beta < 1$. These systems are all in mean motion resonance; the interaction is stabilized by the resonance. In fact,

HD 108874 is the only known resonance system to have $\beta > 1$. Therefore it seems that resonances tend to form such that $\beta < 1$.

Table 1 Dynamical Properties of Multiple Planet Systems

System	Pair	MMR	AM	ϵ	β	Class	Reference
47 UMa	b-c	-	C[a]	0	1.025	S	Butler et al.(2006)
55 Cnc	e-b	-	C	0.067	-	T	Butler et al.(2006)
	b-c	3:1	C	0.11	-	R	
	c-d	-	C	0.158	-	S	
GJ 317	b-c	4:1?	?[b]	-	0.98	R?	Johnson et al.(2007)
GJ 876	d-c	-	C[a]	0	-	T	Butler et al.(2006)
	c-b	2:1	A	0.34	-	R	
Gl 581	b-c	-	C	0.15	-	T	Udry et al.(2007)
	c-d	-	C	0.20	-	T	
HD 12661	b-c	-	C	0.003	1.199	S	Butler et al.(2006)
HD 37124	b-c	-	C	0.009	-	S	Butler et al.(2006)
	c-d	-	A	0.096	-	S	
HD 38529	b-c	-	C	0.44	2.070	S	Butler et al.(2006)
HD 69830	b-c	-	C	0.095	-	T	Lovis et al.(2006)
	c-d	-	C	0.04	-	S	
HD 74156[c]	b-c	-	C	0.36	1.542	S	Butler et al.(2006)
HD 73526	b-c	2:1	AA	0.006	0.982	R	Butler et al.(2006)
HD 82943	b-c	2:1	C	0.004	0.946	R	Butler et al.(2006)
HD 128311	b-c	2:1	C	0.091	0.968	R	Butler et al.(2006)?
HD 108874	b-c	4:1	C/AA[d]	0.2	1.107	R	Butler et al.(2006)
HD 155358	b-c	-	AA	0.21	1.043	S	Cochran et al.(2007)
HD 168443	b-c	-	C	0.22	1.939	S	Butler et al.(2006)
HD 169830	b-c	-	C	0.33	1.280	S	Butler et al.(2006)
HD 190360	c-b	-	C	0.38	1.701	T	Butler et al.(2006)
HD 202206	b-c	5:1	C	0.096	0.883	R	Butler et al.(2006)
HD 217107	b-c	-	C	0.46	7.191	T	Butler et al.(2006)
Hip 14810	b-c	-	AA	0.05	1.202	T	Wright et al.(2007)
SS	J-S	-	C	0.19	-	S	JPL
	S-U	-	C	0.006	-	S	
	U-N	-	C	0.004	-	S	
μ Ara	c-d	-	C	0.002	-	T	Pepe et al.(2007)
	d-b	2:1	C	0.003	-	R	
	b-e	-	C	0.13	-	S	
υ And	b-c	-	C	1.8×10^{-4}	-	T	Butler et al.(2006)
	c-d	-	C	2.8×10^{-4}	-	S	

[a] The current eccentricity of one planet is 0, placing the pair on an apsidal separatrix.
[b] The values of mean longitude and longitude of periastron are unknown.
[c] These values do not incorporate the new planet Bean et al.(2007)
[d] This pair alternates between circulation and anti-aligned libration.

One noteworthy system in this context is GJ 317 (Johnson et al.2007), with a β value of 0.98 (see Table 1). This value is only permissible if the system is in a mean motion resonance, and, indeed, the ratio of the periods is near a whole number (4.02). Usually a mean motion resonance is identified by calculating the "resonant argument", which depends on mean longitude and longitude of periastron (see, e.g. Murray & Dermott 1999). However, in the case of GJ 317, these two angles are unknown, and, hence, the resonant argument is also unknown. Nonetheless, the system must be in resonance (barring substantial revisions in the masses, semi-major axes and eccentricities of the planets) based on our stability analysis. Therefore consideration of proximity to instability in terms of the Hill stability boundary may provide an alternative method for identifying mean motion resonances. As the orbital parameters of the two planets in GJ 317 are refined, we predict that they will, in fact, be found to be in resonance.

5. Conclusions

We have described here new approaches for parameterizing the dynamical interactions of extrasolar planet systems. Many systems interact such that they are near the stability limit and the apsidal separatrix. Most mean-motion resonance interactions would be unstable if not for the resonance, a feature which may be used to identify resonant interactions (e.g. GJ 317 b and c). The distribution of ϵ shows that the scattering of approximately equal mass planets is unlikely to produce the observed distribution of apsidal behavior. The distribution of β values shows that many systems formed near the limit of dynamical stability.

These results demonstrate the benefits of the consideration of the dynamical interactions of multiple planet systems, which may constrain models of planet formation. For example, as we discussed above, the proximity of so many systems to an apsidal separatrix ($\epsilon = 0$) suggests that "rogue planets" may have played a major role in scattering planets to higher eccentricities. The proximity of so many systems to the stability limit ($\beta = 1$) suggests many systems form in densely packed configurations. Moreover, consideration of planetary systems in these terms has revealed that our Solar System shares dynamical traits with the known multiple planet systems. Perhaps constraining planet formation models through dynamical properties, rather than observed orbital elements, will lead to a universal model of planet formation.

About half of planetary systems are multiple (Wright et al.2007), predictions of additional companions are being borne out (Barnes & Raymond 2004; Raymond & Barnes 2005; Bean et al.2008), and the current distribution of planet masses suggest there will be many planets with a mass equal to that of Saturn or less (Marcy et al.2005), i.e. below current detection limits. These three observations imply many multiple planet systems will be detected in the future. Hence characterizing extrasolar planet interactions will be a critical aspect of the study of planet formation for the foreseeable future.

References

Artymowicz, P. 1992. PASP 104, 679.
Barnes, R. & Greenberg, R. 2006. ApJ 638, 478. (BG06a)
Barnes, R. & Greenberg, R. 2006. ApJ 647, L153. (BG06b)
Barnes, R. & Greenberg, R. 2006. ApJ 652, L53. (BG06c)
Barnes, R. & Greenberg, R. 2007. ApJ 659, L53. (BG07a)
Barnes, R. & Greenberg, R. 2007. ApJ 665, L67. (BG07b)
Barnes, R. & Quinn, T.R. 2001. ApJ 550, 884.
Barnes, R. & Quinn, T.R. 2004. ApJ 611, 494.
Barnes, R. & Raymond, S.N. 2004. ApJ 617, 569.
Bean, J. L. et al.2008. ApJ 672, 1202.

Boss, A. P. 2000. ApJ. 536, L101.
Butler, R. P. et al.2006. ApJ 646, 505.
Chambers, J. E. 1999. MNRAS 304, 793.
Chiang, E. I. & Murray, N. 2002. ApJ 576, 473.
Cochran, W. D. et al.2007. ApJ 665, 1407.
Correia, A. C. M. et al.2005. A&A 440, 751.
Cuntz, M. et al.2007. ApJ 667, L105.
D'Angelo, G., Lubow, S. H. & Bate, M. R. 2006. ApJ 652, 1698.
David, E. -M. et al.2003. PASP 115, 825.
Érdi, B. et al.2004. MNRAS 351, 1043.
Ford, E. B., Havlikova, M., & Rasio, F. A. 2001. Icarus 150, 303.
Ford, E. B., Lystad, V. & Rasio, F. A. 2005. Nature 434, 873.
Gladman, B. 1993. Icarus 106, 247.
Goldreich, P. & Sari, R. 2003. ApJ 585, 1024.
Goździewski, K. 2002. A&A 393, 397.
Hadjidemetriou, J. D. 2006. CeMDA 95, 225.
Holman, M. J. & Wiegert, P. A. 1999. AJ 117, 621.
Kiseleva-Eggleton, L. et al.2002. ApJ 578, L145.
Laskar, J. 1990. Icarus 88, 266.
Laughlin, G. & Adams, F. C. 1999. ApJ 526, 881.
Lin, D. N. C. & Ida, S. 1997. ApJ 477, 781.
Lovis, C. et al.2006. Nature 441, 305.
Malhotra, R. 2002. ApJ 575, L33.
Marchal, C. & Bozis, G. 1982. CeMDA 26, 311.
Marcy, G. W. et al.2005. ApJ 619, 570.
Marzari, F. & Weidenschilling, S. 2002. Icarus 156, 570.
Michtchenko, T. A. & Malhotra, R. 2004. Icarus 168, 237.
Milani, A. & Nobili, A. M. 1983. CeMDA 31, 213.
Murray, C. D. & Dermott, S. F. 1999. *Solar System Dynamics*. Cambridge UP, Cambridge.
Papaloizou, J. C. B., Nelson, R. P. & Masset, F. 2001. A&A 366, 263.
Pepe, F. et al.2007. A&A 462, 769.
Quillen, A. C. & Faber, P. 2006. MNRAS 373, 1245.
Rasio, F. A. & Ford, E. B. 1996. Science 274, 954.
Rasio, F. A. et al.1996. ApJ 470, 1187.
Raymond, S. N. & Barnes, R. 2005. ApJ 619, 549.
Raymond, S. N., Barnes, R. & Kaib, N. A. 2006. ApJ 644, 1223.
Rivera, E. J. & Lissauer, J. J. 2000. ApJ 530, 454.
Sándor, Zs. et al.2007. MNRAS 375, 1495.
Sosnitskii, S. P. 1999. AJ 117, 3154.
Stepinski, T. F., Malhotra, R. & Black, D. C. 2000. ApJ 545, 1044.
Szebehely, V. & McKenzie, R. 1981 CeMDA 23, 3.
Udry, S. et al.2007. A&A 469, L43.
Voyatzis, G. & Hadjidemetriou, J. D. 2006. CeMDA 95, 259.
Weidenschilling, S. & Marzari, F. 1996. Nature 384, 619.
Wisdom, J. 1982 AJ. 87, 577.
Wright, J. T. et al.2007. ApJ 657, 533.
Zhou, J. -L. & Sun, Y. -S. 2003. ApJ 598, 1290.

Secular evolution of exoplanetary systems and close encounters

M. Šidlichovský [1] and E. Gerlach [2]

[1] Astronomical Institute, Academy of Sciences of the Czech Republic,
Boční II 1401, 141 31 Prague
email: sidli@ig.cas.cz

[2] Technical University, Institute for Planetary Geodesy, Lohrmann Observatory,
Dresden, Germany
email: enrico.gerlach@tu-dresden.de

Abstract. We investigate the secular evolution of non-resonant exoplanetary systems consisting of a central star and two co-planar planets using a semi-numerical averaging method of the first order in planetary masses (in this case equivalent to "averaging by scissors" or simply dropping the fast periodic terms). The resulting Hamiltonian level curves for different exoplanetary systems were compared to those obtained by direct numerical integration. Studying the dependence of the reliability of the averaging method (as well as chaoticity of numerically integrated trajectories) upon the initial conditions, we found that the averaging methods fails even for Hill stable systems. Based on the Hill stability criterion we introduced empirically a more restrictive stability condition, that enabled us to give an estimate for the region of validity of the averaging method in the plane of initial conditions.

Keywords. Celestial mechanics, exoplanets, stability

1. Introduction

There are several possibilities to study the secular behaviour of exoplanetary systems consisting of a star and two co-planar planets, which are not in a mean motion resonance. The simplest approach is the classical Laplace-Lagrange secular solution, which takes into account only terms up to the second order in eccentricities. The resulting equations of motion may be written as linear differential equations which can be solved easily. Very interesting insight into topology and artificial singularities of this problem, when reduced to one degree of freedom, was presented by Pauwels (1983), where the representation of motion is depicted on a sphere. But being limited to almost circular orbits this method will fail to describe correctly the secular evolution of most exoplanetary systems which have in general larger eccentricities.

A better possibility is the expansion of the perturbing function in powers of eccentricities as was done by Libert & Henrard (2005) who showed that an expansion to 12th order

Table 1. Elements of exosystems used in our calculations

System	$M[M_\odot]$	$M_P[M_J \sin I]$	$P[d]$	$a[AU]$	e	$\varpi[°]$
HD 12661	1.11	2.34	262.53	0.83	0.361	116.3
		1.83	1679	2.86	0.017	218.0
HD 169830	1.43	2.9	225.62	0.82	0.31	328.0
		4.1	2100	3.62	0.33	72.0
HD 108874	1.0	1.37	395.27	1.05	0.068	70.0
		1.02	1599	2.68	0.253	200

is able to describe correctly most of the exoplanetary systems. The third possibility, the so called averaging method, was used for instance by Michtchenko & Malhotra (2004) and for 3-D problem by Michtchenko et al. (2006). Here the Hamiltonian is numerically averaged over the mean anomalies and is therefore not restricted to low degrees in eccentricities. The dependence of variables may be obtained by means of level curves of the Hamiltonian. On the other hand this method does not give the transformation between osculating and mean elements. Finally numerical integration can be used to check, how well the aforementioned methods solve the problem.

In this study we try to understand how well the averaging method describes the secular evolution in dependence on the initial conditions to come to conclusions on the validity of this method.

2. Canonical variables and the averaged Hamiltonian

Let us consider a system consisting of a star of mass M and two co-planar planets of mass m_1 and m_2. Introducing the relative position vectors $\mathbf{r}_j, (j = 1, 2)$ of the planets to the star and their conjugate momenta $\mathbf{p}_j = m_j \dot{\boldsymbol{\rho}}_j$, where $\boldsymbol{\rho}_j$ are the position vectors relative to the centre of gravity, we get the four degree of freedom Hamiltonian (Michtchenko & Malhotra 2004) $H = H_0 + H_1$, with

$$H_0 = \sum_{j=1}^{2}\left(\frac{\mathbf{p}_j^2}{2\mu_j} - \frac{G(M+m_j)\mu_j}{r_j}\right), \quad (2.1)$$

and

$$H_1 = -G\frac{m_1 m_2}{\Delta} + \frac{\mathbf{p}_1 \mathbf{p}_2}{M}. \quad (2.2)$$

G is the gravitational constant, $\mu_j = Mm_j/(M+m_j)$ is the reduced mass of the j-th body and $\Delta = |\mathbf{r}_2 - \mathbf{r}_1|$.

We may define the formal osculating Keplerian elements $a_j, e_j, \varpi_j, \lambda_j$ as the usual elements of the Keplerian motion with H_0 after the perturbation H_1 is switched off and $\mathbf{r}_j$ and $\mathbf{p}_j$ do not change at this moment. The final set of canonical variables is then:

$$L_j = \mu_j\sqrt{G(M+m_j)a_j}, \qquad \lambda_j = \text{mean longitude},$$
$$S_j = L_j(1 - \sqrt{1-e_j^2}), \quad s_j = -\varpi_j = \text{minus longitude of periastron}.$$

The averaged Hamiltonian of the problem is $H_s = H_0 + \bar{H}_1$, where

$$H_0 = -\sum_{j=1}^{2}\frac{G^2(M+m_j)^2\mu_j^3}{2L_j^2}, \quad (2.3)$$

and

$$\bar{H}_1 = -\frac{1}{4\pi^2}Gm_1m_2\int_0^{2\pi}\int_0^{2\pi}\frac{1}{\Delta}d\lambda_1\,d\lambda_2, \quad (2.4)$$

as the indirect part $\mathbf{p}_1\mathbf{p}_2/M$ does not contribute any secular term.

Being independent of the mean longitudes the conjugate momenta L_j (and thus the semi-major axes) are now constant. Introducing a new canonical set:

$$K_1 = S_1, \qquad k_1 = s_1 - s_2 = \varpi_2 - \varpi_1 = \Delta\varpi \text{ and}$$
$$K_2 = S_1 + S_2, \quad k_2 = s_2$$

it becomes obvious that the resulting Hamiltonian $H_s = H_s(K_1, k_1)$ depends only on

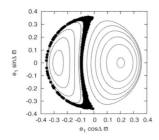 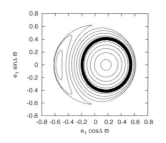 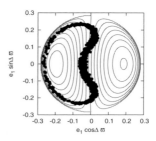

Figure 1. Level curves for the averaged Hamiltonian and numerical integration (dots) for HD 12661 - left, HD 169830 - centre and HD 108874 - right.

one coordinate, representing thus a problem with one degree of freedom. The second constant of motion K_2, introduced as Angular Momentum Deficit (AMD) by Laskar, (Laskar 2000) corresponds to total angular momentum conservation (for constant L_1 and L_2) and couples both eccentricities.

The Hamiltonian $\bar{H}_1$ may be obtained numerically for a fixed value of K_2 from (2.4) and its level curves drawn in the (x_j, y_j) plane, where $x_j = e_j \cos k_j$ and $y_j = e_j \sin k_j$. Since due to the conservation of AMD the eccentricities are not independent the level curves can be converted directly into each other.

Fig. 1 shows exemplarily the level curves of the Hamiltonian for 3 exoplanetary systems: HD 12226, HD 169830 and HD 108874. The calculations were done for a fixed value of K_2 defined by the initial conditions taken from the catalogue by Butler et al. (2006) and given in Table 1. In this figure we included also the nominal trajectories of these exoplanetary systems (shown as dots) obtained by numerical integration using the program *Mercury* by Chambers (1999).

We can see that for the first two systems the agreement between the averaging method and numerical integration is good in comparison to the poor agreement for HD 108874. The reason here is the proximity of the system to the 4:1 mean motion resonance. If one changes here the initial semi-major axis of the outer planet from 2.68 AU towards the nominal value of the resonance at 2.66 AU one can observe a growing disagreement between the averaging method and the real behaviour of the system. The reason for the failure of the averaging method is clear: the double averaging over mean longitudes is of course impossible in case of mean motion resonances.

Therefore in the following we are interested in systems such as HD 12226 and HD 169830, which are not in or close to such mean motion resonances. However it is clear that, if the planets during the system evolution will be close enough for the mean semi-major axes to be changed during such encounters the averaging method must fail as well. In the remaining part we will define the area for which the averaging method is working well to predict the secular evolution.

3. Stability map in the plane of initial conditions

Following Michtchenko & Malhotra (2004) we introduce the plane of initial conditions as follows: from Fig. 1 it is clear that each trajectory, no matter whether librating or circulating, goes through the line defined by $y_1 = 0$. Therefore each set of initial conditions can be represented by a point in the (x_1, e_2) plane, where the initial value of $\Delta\varpi$ is fixed to either $0°$ or $180°$.

Fig. 2 shows the plane of initial conditions for HD 12661 and HD 169830 with energy levels (solid lines) and AMD levels (dashed lines). Let us remark that for the linear

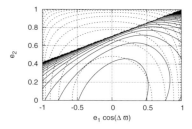 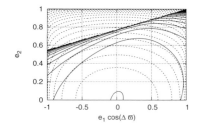

Figure 2. The plane of initial conditions with energy levels (solid lines) and AMD levels (dashed lines) for HD 12661 - left and HD 169830 - right.

approximation both of these level curves are ellipses, where the inclination of the axes for energy level ellipse would be 45°. The orbits which are initially crossing are above the line defined by $a_1(1 - e_1 \cos \Delta \varpi) = a_2(1 - e_2)$.

Using this plane we calculated with the initial condition of the therewith defined planetary systems the maximum Lyapunov Characteristic Exponent (LCE). The masses and semi-major axis were taken from Table 1. As numerical integrator we used the *ODEX* code - an extrapolation method proposed by Hairer & Wanner (1995). Fig. 3 shows the logarithm of the LCE (dark areas refer to stable regions) in the plane of initial conditions. The line of initially crossing orbits is given additionally in each of the figures.

To see how well the averaging method is working in dependence on the stability, three different points were chosen from Fig. 3: for the same e_1 one with $e_2 = 0.1, 0.3$ and 0.4 (shown as light dots in the left part of the figure). The predicted behaviour by the averaging method in comparison to the direct numerical integration for all three points can be found in Fig. 4. While for $e_2 = 0.3$ there is merely a shift to the contour lines, there is complete disagreement for $e_2 = 0.4$. The reason for the latter becomes clear if one looks at the semi-major axis as a function of time, as is shown in Fig. 5. For an initial value of $e_2 = 0.4$ the semi-major axis of the inner planet has small jumps of about 1% as the result of closer approaches between the planets. Numerical experiments show that the averaged Hamiltonian (2.4) gives the correct picture only for log LCE $\leqslant$ -5.5 (or Lyapunov time larger 100 000 years).

For most exoplanetary systems discovered by radial velocity measurement only lower limits for the planetary masses are known. To account and test also for this indetermination we multiplied both planetary masses by κ, keeping the mass ratio of the planets constant and we calculated stability maps as a function of κ. It can be recognized that

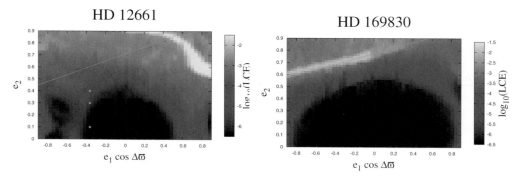

Figure 3. The stability map for HD 12661 and HD 169830.

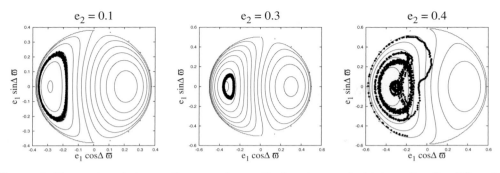

Figure 4. Comparison between the averaging method and numerical integration for different values of e_2. See text for explanation.

the region of applicability of the averaging method slowly shrinks as expected with increasing κ. But the general features remain the same: a stable area for nearly circular orbits enclosed by strongly chaotic orbits and in between some sort of fuzzy layer where the semi-major axis has short periodic variations with growing amplitude but its mean value still does not show any apparent jumps. Here the level curves given by the averaging method are slightly shifted compared to the numerical integration as already mentioned. Our boundary curve will be fitted to the region where the mean semi-major axes starts to jump.

4. Hill stability

The Hill stability condition for the elliptic case of one host star with mass M and two planets with masses m_i is given by Gladman (1993) and based on results of Marchal & Bozis (1982). It reads to lowest order as

$$C\alpha^{-3}\left(\mu_1 + \frac{\mu_2}{\delta^2}\right)(\mu_1\gamma_1 + \mu_2\gamma_2\delta)^2 > 1 + \mu_1\mu_2\left(\frac{3}{\alpha}\right)^{4/3}, \qquad (4.1)$$

where $C = 1$, $\mu_i = m_i/M$, $\gamma_i = \sqrt{1-e_i^2}$ for $i = 1, 2$. $\alpha = \mu_1 + \mu_2$, $\delta = \sqrt{1+\Delta}$ and $\Delta = (a_2 - a_1)/a_1$.

Gladman showed that when this inequality is fulfilled, the planets are forbidden to undergo close approaches for all the time. As a close approach he defines the separation, when one planet is inside the sphere of influence, $2\mu^{2/5}$, of the more massive one. But even for Hill stable systems, one has still a chaotic region close to the boundary defined by (4.1) with short Lyapunov times. Here one finds no crossing of the planetary orbits during several 10^5 conjunctions, which gives the impression of bounded semi-major axes preventing close approaches. But still jumps in the mean value of the semi-major

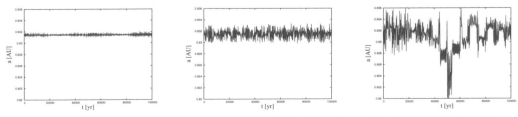

Figure 5. The evolution of the semi-major axis of the inner planet for the same trajectories as in Fig. 4.

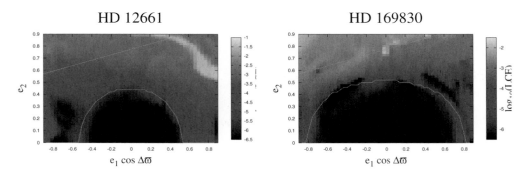

Figure 6. The stability map for HD 12661 for $\kappa = 1$ - left and for HD 169830 $\kappa = 3$ - right. The region of applicability of averaging method is roughly defined by boundary curve given by modified Gladman formula.

axes as shown in the right plot of Figure 5 can occur. The reason for these are called *small* encounters by Gladman. Further away from the boundary curve the amplitudes of short-periodic variations in the semi-major axis decrease and one finds quasi-periodic behaviour.

5. Conclusions

As already shown the Hill stability condition is not sufficient for the averaging method to work due to small encounters. Therefore such condition has to be stronger and so we introduced the constant C into (4.1) to scale the left hand side. We started with finding numerically the value of e_2, for which there is a jump in the semi-major axis for both exoplanetary systems and with different κ. Hereafter we fitted the value of C with least square method. The resulting value $C = 0.89$ leads to the required sharpening of (4.1). Fig. 6 shows the resulting boundary curves for both systems κ. These curves approximate the boundary of applicability of the averaging method.

This condition is useful to get very fast a first impression on the stability of some orbital configurations avoiding expensive (from the point of CPU time) calculations of dynamical maps. But of course for more detailed analysis more sophisticated tools have to be employed, since this method is working only in regions where strong mean-motion resonances are not important.

References

Butler, R. P., Wright, J. T., Marcy, G. W., Fischer, D. A., Vogt, S. S., Tinney, C. G., Jones, H. R. A., Carter, B. D., Johnson, J. A., McCarthy, C. & Penny, A. J.: 2006, *ApJ*, **646**, 505.
Chambers, J. E.: 1999, *MNRAS*, **304**, 793.
Gladman, B.: 1993, *Icarus*, **106**, 247.
Hairer, E. & Wanner, G.: 1995, *http://www.unige.ch/math/folks/hairer/*, in electronic form.
Laskar, J.: 2000, *Phys. Rev. Lett.*, **84**, 3240.
Libert, A. & Henrard J.: 2005, *Celest. Mech. Dyn. Astron.*, **93**, 187.
Marchal , C. & Bozis, G.: 1982, *Celest. Mech.*, **26**, 311.
Michtchenko, T. A. & Malhotra R.: 2004, *Icarus*, **168**, 237.
Michtchenko T. A., Ferraz-Mello S. & Beaugé C.: 2006, *Icarus*, **181**, 555.
Pauwels, T.: 1983, *Celest. Mech.*, **30**, 229.

Formation and transformation of the 3:1 mean-motion resonance in 55 Cancri System

Li-Yong Zhou[1], Sylvio Ferraz-Mello[2] and Yi-Sui Sun[1]

[1] Department of Astronomy, Nanjing University
Nanjing 210093, China
email: zhouly@nju.edu.cn; sunys@nju.edu.cn

[2] Instituto de Astronômico, Geofísica, e Ciências Atmosféricas, Universidade de São Paulo
Rua do Matão 1226, 05508-900 São Paulo, Brazil
email: sylvio@astro.iag.usp.br

Abstract. We report in this paper the numerical simulations of the capture into the 3:1 mean-motion resonance between the planets b and c in the 55 Cancri system. The results show that this resonance can be obtained by a differential planetary migration. The moderate initial eccentricities, relatively slower migration and suitable eccentricity damping rate increase significantly the probability of being trapped in this resonance. Otherwise, the system crosses the 3:1 commensurability avoiding resonance capture, to be eventually captured into a 2:1 resonance or some other higher-order resonances. After capture into resonance, the system can jump from one orbital configuration to another one if the migration continues, making a large region of the configuration space accessible for a resonance system. These investigations help us understand the diversity of resonance configurations and put some constraints on the early dynamical evolution of orbits in the extra-solar planetary systems.

Keywords. Planetary systems, celestial mechanics, methods: numerical, methods: analytical

1. Introduction

Up to date, more than 200 extra-solar planetary systems have been found. Those hosting more than one planet are multiple planet systems. In some of the multiple planet systems, planets are observed to be locked in mean-motion resonance (MMR), for example, the well-known 2:1 MMR in GJ876 system (e.g. Marcy et al. 2001) and the 3:1 MMR in 55 Cancri system (e.g. Zhou et al. 2004). In a 3:1 MMR, at least one of the three resonant angles ($\theta_1 = \lambda_1 - 3\lambda_2 + 2\varpi_1$, $\theta_2 = \lambda_1 - 3\lambda_2 + 2\varpi_2$, $\theta_3 = \lambda_1 - 3\lambda_2 + \varpi_1 + \varpi_2$, where $\lambda_{1,2}$ and $\varpi_{1,2}$ are the mean longitudes and periastron longitudes of the inner and outer planet respectively) librates. The resonant systems are particularly attractive because not only of the complicated dynamics of the resonance but also of interesting information about its origin and early evolution buried in the configuration and dynamics.

The 55 Cancri system is the only example of 3:1 MMR found in the extra-solar planetary systems till now. Four planets have been reported in this system and two of them, 55 Cnc b and 55 Cnc c, seem to be locked in a 3:1 MMR. In our previous work, we have found that they are most likely in one of the three possible configurations (Zhou et al. 2004). G.Marcy declared in his lecture in this Symposium that a fifth planet has been dug out from observing data of this system, enriching the whole story with more connotations. In the dynamical simulations of their 5-planet solution (Fischer et al. 2007), they did not found the resonance between the two planets, although their orbital periods are very close to the 3:1 commensurability. However, the current orbit determinations may not be robust enough to determine the resonant angles, especially for orbits with low eccentricities as be given in this new determination, thus the resonant character of

Table 1. Orbital elements and masses. The mass of the central star is $1.03 M_\odot$ (McArthur et al. 2004). In this paper we adopt the planet masses by assuming $\sin i = 1$.

Parameter	55 Cnc e	55 Cnc b	55 Cnc c	55 Cnc d
$M \sin i\ (M_J)$	0.045	0.784	0.217	3.92
P (days)	2.81	14.67	43.93	4517.4
a (AU)	0.038	0.115	0.240	5.257
e	0.174	0.0197	0.44	0.327

this pair of orbits may change as more observations are included (Beaugé et al. 2008). Furthermore, what we will discuss in this paper has some general significance so that we may still adopt the orbital elements and planetary masses from the previous literature (McArthur et al. 2004), as listed in Table 1

It is widely accepted that a planet may experience (generally inward) migration, and through differential migrations of planets, different commensurabilities among planets can be attained (e.g. Nelson & Papaloizou 2002). For example, the 2:1 MMR in GJ876, whose configuration can be obtained through the inward migration of the outer planet (Lee & Peale 2002).

On the other hand, the resonant capture probability depends on the order of the resonance, the migration rate and the initial planetary eccentricity (Quillen 2006). The formation process of the 3:1 MMR may be very different from the one of the 2:1 MMR. As the only example of 3:1 MMR in extra-solar planetary systems, the 55 Cancri system deserves a detailed investigation. In this paper, we will report our numerical simulations of the capture into resonance, and the evolution of the resonance thereafter, provided the migration continues.

2. Numerical simulations of resonance formation

There are more planets than the two (planet b and c) in our focus of attention in this planetary system. However, besides these two planets, the massive one (55 Cnc d) is too far and the close one (55 Cnc e) is too small. Moreover, even the newly reported planet (55 Cnc f) is neither very massive ($M_f \sin i = 0.144 M_{\rm Jup}$) nor very close to this planet pair ($a_f = 0.781$ AU, $P_f = 260$ days). Therefore as a simplified model, it is reasonable to discuss only the two planets under the influences of the central star and the disc.

2.1. Numerical Model

As usual, the influence of the disc on the planets is simply simulated by an artificial force, which is acting on the outer planet to drive it to migrate inward. The force is defined so that the semi-major axis of the planetary orbit will change following an exponential law:

$$a(t) = a_0 + \Delta a \times e^{-t/\tau}. \qquad (2.1)$$

where a_0 is the initial semi-major axis of the planet, and τ is the timescale of the migration. The two planets are assumed to be located initially on orbits with semi-major axes of 1.25 AU and 0.5 AU, with an initial semi-major axes ratio of 2.5 (periods ratio 3.95), which will evolve downward to the value 2.08 corresponding to the 3:1 MMR. Obviously the value of τ determines the migration speed. The smaller the τ, the faster the migration, and vice versa. Different τ values were adopted in different papers. The standard migration scenario (type II) gives an estimate of migration rate (Ward 1997):

$$\left|\frac{\dot{a}}{a}\right| = 9.4 \times 10^5 \left(\frac{\alpha}{4 \times 10^{-3}}\right) \left(\frac{H/a}{0.05}\right)^2 P^{-1}, \qquad (2.2)$$

where α is a dimensionless coefficient describing the disk viscosity, H is the vertical thickness of the disk and P is the orbital period in year of the planet. Taking the typical parameters $\alpha = 4 \times 10^{-4}$ and $H/a = 0.05$ (see e.g. Kley 2000), the τ for a planet starting from 1.25 AU is $\sim 1.43 \times 10^4$ yrs. Some other rules have been applied in different papers. According to these rules, the τ would be, for example, 1.56×10^4 yrs (Lee & Peale 2002), 1.38×10^4 yrs (Nelson & Papaloizou 2001), and 2.65×10^4 yrs (Kley 2003; Kley et al. 2004). In our simulations, it is set to be $2 \times 10^4, 1 \times 10^5, 2 \times 10^5$ and 5×10^5 yrs. Some large values (meaning slow migration) are adopted here to guarantee the capture into the 3:1 MMR.

Although it is still not very clear now how a certain disk will damp or stimulate the eccentricity of an embedded planet, people hope the disk would circularize an orbit if it was too eccentric. In order to control the unlikely eccentricity increasing during the artificially induced migration, we have included an eccentricity-damping force in our simulations. A parameter K was introduced to describe the damping rate, with which we control the eccentricity e of a planet by:

$$|\dot{e}/e| = K |\dot{a}/a|. \qquad (2.3)$$

In our simulations, the parameter K has the values 0 (no damping), 1, 10 and 100. The damping is always put on planet 55 Cnc c with the migrating force modifying its semimajor axis simultaneously.

In the simulations, the two orbits are nearly coplanar (in the practice of numerical simulations we adopt an arbitrarily defined small inclination between these two orbits, say 0.1 degree), and the initial orbital eccentricities of planet 55 Cnc b (e_1^0) and 55 Cnc c (e_2^0) are set to be 0.001 (a typical value for a nearly circular orbit) as well as $0.01, 0.05, 0.1, 0.2$. For each model (with certain τ, K, e_1^0, e_2^0), we numerically integrate 100 systems with randomly selected orbital angles (longitudes of periastrons, mean longitudes, and ascending nodes). Each testing try is integrated up to a time of 2τ. During an integration, if the two semi-major axes are locked in a definite value and keep this value at least for ten percent of the integration time, that is $\tau/5$, we say the system is in the given commensurability.

2.2. Results

Generally, the migration may drive the two planets into a given commensurability. The inner planet 55 Cnc b will also migrate after it has been captured into a commensurability. The final configuration of a system depends sensitively on the migration speed and the initial conditions. We summarize some remarkable outcomes from our numerical simulations as follow.

(1) The migration with $\tau \sim 2 \times 10^4$ yrs is too fast to form the 3:1 MMR. If we start from nearly circular orbits ($e_1^0 = e_2^0 = 0.001$) and neglect the eccentricity damping ($K = 0$), all the 200 runs for $\tau = 2 \times 10^4$ yrs and $\tau = 1 \times 10^5$ yrs cross the 3:1 MMR without being trapped and are eventually captured into the 2:1 resonance. The 3:1 MMR is observed to occur only when $\tau = 2 \times 10^5$ yrs. When $\tau = 2 \times 10^4$ yrs, the number of simulation runs, in which the 3:1 MMR forms, is very small, if not null, no matter what e_1^0, e_2^0 and K are.

(2) A slow migration favors the formation of the 3:1 MMR. For initially near circular orbits, when $\tau = 2 \times 10^5$ yrs, we have obtained 53 out of the 100 runs trapped in the 3:1 MMR and the number increases to 100 when $\tau = 5 \times 10^5$ yrs.

(3) The initial eccentricities strongly affects the resonance trapping. A nonzero but small initial eccentricity of either orbit (e_1^0 or $e_2^0 \sim 0.01$) will increase significantly the probability of the 3:1 MMR. For the runs with e_1^0 or $e_2^0 = 0.01$, although there is no 3:1 MMR when $\tau = 2 \times 10^4$ yrs, we found 163 out of the totally 200 runs trapped in this resonance when $\tau = 1 \times 10^5$ yrs. Higher initial eccentricities ($e_1^0, e_2^0 \sim 0.05, 0.10$) will bring

in more higher-order resonances, such as the 5:2, 7:3 and 7:2 MMRs. It is interesting to see that a few runs were trapped in the 11:4 resonance. Even higher eccentricities ($e_1^0, e_2^0 \sim 0.20$) generally leads the system to catastrophic planetary scattering.

(4) The eccentricity damping affects the resonance capture too. Without the eccentricity damping ($K = 0$), higher initial eccentricities would likely lead to high order commensurabilities, while low initial eccentricities would cause eccentricities to grow too much after the planets are locked into a low-order resonance. On the other hand, a high K value restrains the eccentricity increasing and therefore causes the system to evolve preferably to a 2:1 MMR rather than the 3:1 MMR. For example, when $K = 100$ all the test systems starting from circular orbits ($e_1^0 = e_2^0 = 0.001$) are driven into the 2:1 MMR if $\tau \leqslant 2 \times 10^5$ yrs. Even in a very slow migration ($\tau = 5 \times 10^5$ yrs), the 2:1 MMR captures 41 out of the 100 runs. So we need a suitable damping rate for the formation of 3:1 MMR. In all our simulations, the most probable modes for the formation of the 3:1 MMR were a slow migration ($\tau \geqslant 1 \times 10^5$ yrs), moderate initial eccentricities ($e_{1,2}^0 = 0.01 \sim 0.05$) and moderate eccentricity damping ($K \sim 10$).

3. Evolution of orbital configuration in resonance

After being captured into the 3:1 MMR, the system will continue to evolve if the migration does not halt. In this section we will discuss the evolution of the system in the resonance.

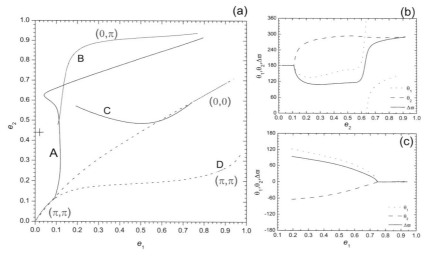

Figure 1. Periodic solutions in the 3:1 MMR. (a) Periodic solution families on the (e_1, e_2) plane. The stable symmetrical apsidal corotation resonance (ACR) solutions are thick solid curves, the unstable symmetrical ACR solutions are short-dashed curves, and the thin solid curves indicate the asymmetrical solutions. The value of ($\theta_1, \Delta\varpi$) of the symmetrical solutions appear as labels along the curves. We also label different solution families with **A**, **B**, **C** and **D**. The eccentricities (e_1, e_2) = (0.0197, 0.44) in Table 1 are indicated by a cross. (b) The variation of the angles θ_1 (dotted), θ_2 (dashed) and $\Delta\varpi$ (solid) along the solution family **A**. Note $\Delta\varpi$ begins from 180° (symmetrical ACR). (c) The same as (b) but for solution family **C**, where $\Delta\varpi$ reaches the symmetrical value 0° at the end.

3.1. Periodic solutions in 3:1 MMR

Fig. 1 shows the periodic solutions, both stable and unstable, symmetrical and asymmetrical, computed using the mass ratio of the planets 55 Cnc b and 55 Cnc c and assuming coplanar orbits. (For other mass ratios and for details of the method used see Michtchenko et al. (2006)).

In the real system, if we assume that it evolved from nearly circular orbits through an adiabatical migration, as suggested in many references, the configuration of two planets captured into a 3:1 resonance should be located on or near the family **A** of periodic solutions shown in Fig. 1(a). In such a migration, the angles $(\theta_1, \Delta\varpi)$ evolve from the symmetrical (π, π) values to asymmetrical values. Since each solution family is not connected by stable periodic solutions to other families (**A** and **B** are not connected because they have different $\theta_1, \Delta\varpi$ values at the two crossings in Fig. 1(a)), there is no reason to expect a system in resonance with orbital configuration far away from the family **A**. In fact, the eccentricities (e_1, e_2) in Table 1 are not on the family **A**. However, we have found that this configuration could be obtained by a resonance capture from initial eccentric orbits. An example of such kind of resonance trapping is shown in Fig. 2(a). Perhaps a perturbation happening to a system in the equilibrium periodic solution may also lead to this observed configuration. We just show here another possibility of the resonance formation.

3.2. Jump between solution families

If the disc does not disappear just after the resonance capture happens, the outer planet will continue to migrate inward accompanied by the inner planet locked in the resonance. During such after-capture migration stage, the system will evolve along the solution family **A** as shown in Fig. 1, provided the migration is "adiabatic". But if the migration is faster (not adiabatic), as some of our numerical simulations, the solution may jump from a family to another. An example of jump from family **A** to **C** is illustrated in Fig. 2(b). Comparing the variations of angles in Fig. 2(c) and Fig. 1(b), (c), it is clear that the jump happens around $t = 1.9 \times 10^4$ yrs. At this moment, if the system goes along family **A**, the angles θ_1 and $\Delta\varpi$ will go up and cross the value 180° one after another with the increasing eccentricity e_2. But they fail and turn around following another evolving route, that is, family **C**. This turnaround is due to some perturbation (in our case it's the migration itself), and it happens around the critical point where the apsidal difference $\Delta\varpi$ crosses the symmetrical value 180° in an asymmetrical ACR. In fact, the stable region of motion around this critical turning point (on family **A** with $(e_1, e_2) = (0.0423, 0.623)$ in Fig. 1(a)), is very small. Therefore a slight disturbance could drive the system out of the family **A**.

The jump from family **A** to family **B** has also been observed in our simulations. The angles $\theta_1 = \theta_2 = 0°$ in family **B**, that is to say, both ϖ_1 and ϖ_2 have to adjust to new values in a jump from **A** to **B**. On the other hand, as we see from Fig. 1(b),(c) and Fig. 2(c), we need only an adjustment of ϖ_1 but not ϖ_2 for a jump from family **A** to **C**, since this jump does not affect the θ_2. When the jump happens, the inner planet is on a near circular orbit ($e_1 = 0.0423$) while the outer one is on a highly eccentric orbit ($e_2 = 0.623$). As a result, it is much easier to adjust ϖ_1 than ϖ_2. This analysis tells why we observe much more **A** to **C** jumps than **A** to **B** jumps.

These jumps between different solution families make it possible for a resonant system to occupy much larger potential volume in the orbital elements space, increasing the diversity of resonant configuration in extra-solar planetary systems.

4. Conclusions

The 3:1 mean-motion resonance in the 55 Cancri planetary system is far from a necessary result of the differential planetary migration. Our numerical simulations showed that

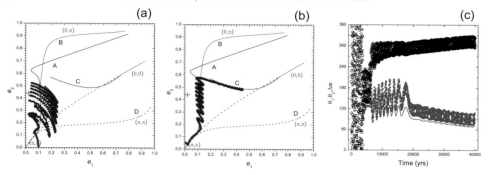

Figure 2. The evolution of orbital configuration in the 3:1 MMR. (a) An example of evolution starting from eccentric initial orbits. The eccentricities in Table 1 can be reached by the solution in this case. Solution families are shown in the same way as in Fig. 1, and the dots indicate the evolution of one run in our numerical simulations. (b) An example of orbital configuration jump from solution family **A** to family **C**. (c) The time variations of angles θ_1 (small dots), θ_2 (open circles) and $\Delta\varpi$ (solid thick curve) for the numerical simulation shown in (b).

the favourable scenario for the formation of the 3:1 resonance is moderate initial eccentricities (0.01 ∼ 0.05), relatively slower migration ($\tau \sim 10^5$ yrs) and suitable eccentricity damping rate ($K \sim 10$). After being captured in the resonance, the system may exhibit some evolutions different from the behaviours in an adiabatic migration, and evolve to some unexpected orbital configurations. All these put some constraints on the planetary migration and early dynamical evolution in the extra-solar planetary systems.

Acknowledgements

We thank T.A. Michtchenko for many helpful discussions and for some of the codes used to find the location of the periodic solutions. This work was supported by the Natural Science Foundation of China (No. 10403004), the National Basic Research Program of China (973 Program, 2007CB814800) and São Paulo State Research Foundation (FAPESP, Brazil).

References

Beaugé, C., Ferraz-Mello, S., Michtchenko, T. & Giuppone, C. 2008, *Proceedings of IAUS249*, this volume
Fischer, D., Marcy, G., Butler, P., Vogt, S., Laughlin, G., Henry, G., Abouav, D., Peek, K., Wright, J., Johnson, J., McCarthy, C. & Isaacson, H. 2007, *preprint*
Kley, W. 2000, *MNRAS* 313, L47
Kley, W. 2003, *Cel. Mech. & Dyn. Astron.* 87, 85
Kley, W., Peitz, J. & Bryden G. 2004, *A&A* 414, 735
Lee, M. & Peale S. 2002, *ApJ*, 567, 596
Marcy, G., Butler, P., Fischer, D., Vogt, S., Lissauer, J., & Rivera, E. 2001, *ApJ*, 556, 296
McArthur, B., Endl, M., Cochran, W., Benedict, F., Fischer, D., Marcy, G., Butler, P., Naef, D., Mayor, M., Queloz, D., Udry, S. & Harrison, T. 2004, *ApJ*, 614, L81
Michtchenko, T., Beaugé, C. & Ferraz-Mello, S. 2006, *Cel. Mech. & Dyn. Astron.*, 94, 411
Nelson, R. & Papaloizou, J. 2002, *MNRAS*, 333, L26
Quillen, A. 2006, *MNRAS*, 365, 1367
Ward, W. 1997, *Icarus*, 126, 261
Zhou, L.-Y., Lehto, H., Sun, Y.-S. & Zheng, J. 2004, *MNRAS*, 350, 1495

Analysis of near-separatrix motion in planetary systems

Su Wang and Ji-Lin Zhou

Department of Astronomy, Nanjing University,
210093, Nanjing, China
email: suwang@nju.edu.cn,zhoujl@nju.edu.cn

Abstract. Near-separatrix motion is a kind of motion of two planets with their relative apsidal longitude near the boundary between libration and circulation. Observed multiple planetary systems seem to favor near-separatrix motions between neighboring planets. In this report, we study the probability that near-separatrix motion occurs with both the linear secular system and full three-body systems. We find that generally the ratio of near-separatrix motion is small unless the eccentricities of the two planets differ from each other by an order of magnitude, or they are in mean motion resonance. To explore the dynamical procedures causing the near-separatrix motion, we suppose a modification to scattering model by adding a mass-accretion process during the protoplanet growth. Statistics on the modified scattering model indicate that the probability of the final planet pairs in near-separatrix motion is high ($\sim 85\%$), which may explain the high occurrence of near-separatrix motions in observed planetary systems.

Keywords. methods: analytical, methods: n-body simulations, (stars:) planetary systems

1. Introduction

To date, around 260 extra-solar planets are detected, among them there are 25 multiple planet systems. Statistical properties show that although planets with $a < 0.05$AU have almost circular orbits, eccentricities in the range 0-0.8 are common(Marcy etal. 2005, Papaloizou & Terquem 2006). Various mechanisms have been proposed to generate large eccentricities after planet formation. e.g., planet-planet scattering (e.g. Ford *et al.*, 2005, Zhou *et al.* 2007), resonance trapping of planets during migration (e.g., Kley *et al.* 2004), etc..

The apsidal alignment is a configuration believed to favor the stability of a planet system, especially for planets on highly eccentric orbits(Zhou & Sun 2003). Based on the new catalog of exoplanets(Butler *et al.* 2006), Barnes & Greenberg (2006) studied the observed multiple-planet systems and found that the relative apsidal longitude ($\Delta\varpi$) of a large fraction of systems lie near the boundary between libration and circulation motions (where the authors called near-separatrix motion), and the systems that exhibit libration were rare. A plausible scenario generating near-separatrix motions is the planet-scattering model. In this model the eccentricities of the two planets are excited by a third planet's close encounters and ejection, resulting in a two-planet system in near-separatrix motion, a configuration similar to that of the upsilon Andromedae system(Ford *et al.* 2005). However, the statistics of the occurrence rate of near-separatrix motions in this model is too low ($\sim 10\%$) as compared to that derived from observed systems($\sim 40\%$, Barnes 2008).

To solve the paradox and to seek the origin of near-separatrix motion, in this report we investigate the probability that the near-separatrix motion occurs in a two-planet system,

http://exoplanet.eu/

with both analytical approach and numerical simulations. The planet-planet scattering model leading to this motion is also revised. In §2 we analyse the two-planet system with the secular perturbation theory. Results from N-body simulations are present in §3. In §4 we modify the classical scattering model by adding a mass-accretion process, i.e., the additional planet grows from 10 Earth mass to Jupiter mass by accreting nearby gas. Statistics for the outcomes of this refined scattering model is studied, with special attentions paid on the cause of near-separatrix motion. We present our conclusions and discussions in §5.

2. Analysis from secular perturbation theory

We set up a model planetary system with a solar-mass star ($m_0 = 1M_\odot$) and two Jupiter-mass planets ($m_1 = m_2 = 1M_J$). Suppose a_i, e_i, ϖ_i, M_i ($i = 1, 2$) are the semi-major axis, eccentricity, longitude of pericenter and mean anomaly of the orbit of m_i, respectively. We fix the initial semi-major axis of the inner planet ($a_{10} = 1$AU) as the unit length, and vary the initial position of the outer planet between 1.3AU and 3AU. The inner boundary 1.3AU is chosen to avoid Hill unstable region of m_1(Gladman 1993).

Based on the definition of near-separatrix motion (Barnes & Greenberg, 2006a,2006b) we choose the minimum value of ($e_1 e_2$) as a criterion of whether a planet-pair is in near-separatrix motion or not. According to the theory of linear secular perturbation(e.g., Zhou & Sun 2003), the criterion during the evolution can be written as

$$(e_1 e_2)_{min} = | \frac{(\rho_1 + \rho_2)FG - | \rho_1 F^2 + \rho_2 G^2 |}{(\rho_1 - \rho_2)^2} |, \tag{2.1}$$

where

$$\begin{aligned} F &= (\rho_2^2 e_{10}^2 - 2\rho_2 e_{10} e_{20} \cos \triangle \varpi_0 + e_{20}^2)^{1/2}, \\ G &= (\rho_1^2 e_{10}^2 - 2\rho_1 e_{10} e_{20} \cos \triangle \varpi_0 + e_{20}^2)^{1/2}, \\ \rho_1 &= \tfrac{1}{2c_0}[1 - \xi - \sqrt{(1-\xi)^2 + 4c_0 \xi}], \\ \rho_2 &= \tfrac{1}{2c_0}[1 - \xi + \sqrt{(1-\xi)^2 + 4c_0 \xi}], \end{aligned} \tag{2.2}$$

and $\xi \approx \frac{m_1}{m_2}\alpha^{1/2}$, $c_0 = b_{3/2}^{(2)}(\alpha)/b_{3/2}^{(1)}(\alpha)$, $\alpha = a_1/a_2 < 1$, where $b_{3/2}^{(i)}(\alpha)(i=1,2)$ are the Laplace coefficients.

To show the situation in the whole plane of (e_{10}, e_{20}), we express the criterion in the forms of $\eta = (e_1 e_2)_{min}/e_{10}^2$ or $\eta' = (e_1 e_2)_{min}/e_{20}^2$. Figure 1a shows the result from equation (2.1), where the boundary lines, defined in the following equations, divide the whole region into two parts (Zhou & Sun 2003 based on Laughlin et al. 2002),

$$\frac{e_{20}}{e_{10}} = \frac{2\rho_1 \rho_2}{\rho_1 + \rho_2} \cos \triangle \varpi_0, \quad \text{or} \quad \frac{e_{20}}{e_{10}} = \frac{\rho_1 + \rho_2}{2} \frac{1}{\cos \triangle \varpi_0}. \tag{2.3}$$

The plot of η' in the case of $e_{10} < e_{20}$ is similar. Since in secular dynamics, $e_1 > e_2$ or $e_1 < e_2$ alternates, we can always choose suitable epoch to let $e_{10} > e_{20}$, thus we neglect the discussion of η'.

We find that the boundary line defined above, equivalent to $\eta = 0$, is in the middle of the $\eta < 0.1$ region. Therefore, around the boundaries, the near-separatrix motion region may be defined by the value of η, for example $\eta < 0.1$ or $\eta < 0.01$. Figure 2 shows the variation of ratio of near-separatrix region in the whole plane of ($\triangle \varpi_0, e_{20}/e_{10}$) in Fig.1a. As we can see, the probability that a planet system fall in the near-separatrix motion

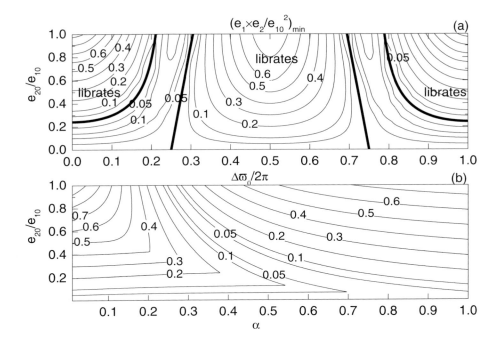

Figure 1. (a) Contours of η changed with e_{20}/e_{10} and $\Delta\varpi_0$ ($\alpha = 0.5$), the wide lines represent the criterion lines for libration and circulation. (b) Contours of η changed with e_{20}/e_{10} and α ($\Delta\varpi_0 = 2\pi/9$).

is very small in secular motion. But as long as $e_{20}/e_{10} \leqslant 0.1$, η is smaller than 0.1 in most cases. With the increase of e_{20}/e_{10}, the area of near-separatrix region decreases. It means that, as the initial ratio of the eccentricity becomes smaller, the opportunities for the system suffering near-separatrix motion becomes larger. Thus, to obtain small values of e_{20}/e_{10} is a key factor for the two-planet system to fall in near-separatrix motion.

3. 3-body simulations

In this section, we study the evolution of two planets in a full three-body model to reveal the occurrence of near-separatrix motion. Considering a model planetary system with a solar-mass star and two Jupiter-mass planets, we carry out a series of simulations with differente_{20}/e_{10}, α and $\triangle\varpi_0$. Initially all the orbits are coplanar and the semi-major axis of the inner planet is fixed at 1AU. The results are summarized in Figure 3. From Figure 3 we can find that the result is quite similar to that from the secular perturbation theory except near the positions of mean motion resonances between the two planets, where the secular dynamics approximation fails. In these resonant locations, the value of η is smaller than that from the linear secular dynamics. In the plot the most obvious location is the 3 : 1 resonance at $a_{20} = 2.1$AU (α=0.48). Therefore, resonance configuration seems to reduce the values of η. In order to study this phenomenon in more details, we compare two systems. The first one is in mean motion resonance initially, with $a_{20} = 2.1$AU, while the second one is in non-resonance region, with $a_{20} = 2.3$AU. Other

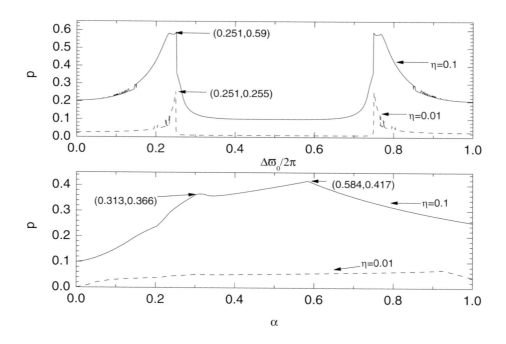

Figure 2. Ratio of near-separatrix region in the whole plane of $(\Delta\varpi_0, e_{20}/e_{10})$ in Fig.1a changed with α and $\Delta\varpi_0$.

initial conditions are the same, i.e., $a_{10} = 1$AU, $e_{10} = 0.2$, $e_{20} = 0.1$. Figure 4a shows the evolution of the eccentricities in the system of mean motion resonance while Figure 4b presents eccentricity evolution of non-resonance system. The visible difference between the two systems is that one of the eccentricities of the resonance system always stays in lower values than the one in non-resonance. As a result, the value of $(e_1 e_2)_{min}$ is 8.67×10^{-4} in resonance system. In comparison, $(e_1 e_2)_{min} = 0.013$ in the non-resonance system. We checked lots of orbits and confirm this scenario. Thus systems initially in a mean motion resonance seems to be easy to fall in near-separatrix motion.

4. Scattering process

Ford et al. (2005) proposed a planet-planet scattering model that could generate near-separatrix motion. In this model, a third planet is added to the previous two-planet system. Due to the gravitational perturbation, the added planet that is initially put on a circular orbit was scattered out, leaving two planets in a near-separatrix motion. This scenario requires that the added planet initially on a circular orbit must be unstable.

To achieve such an initial state, we suppose a reasonable situation as follows. During the formation of gas giants, a protoplanetary embryo will form through the cohesive collisions between planetesimals. When the protoplanet grows to a critical mass ($\sim 10 M_\oplus$), a long period of stage named quasi-hydrostatic accretion sets in. As the gas envelop mass becomes comparable to the core mass, a runaway gas accretion occurs until the gas giant reaches the present mass (Pollack et al. 1996). The final stage of

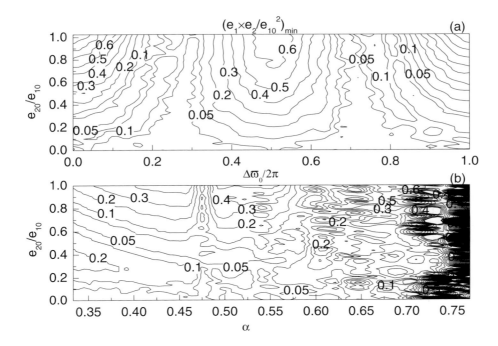

Figure 3. Results of general three-body simulations. (a) Contours of η changed with e_{20}/e_{10} and $\Delta\varpi_0$ ($\alpha = 0.5$). (b) Contours of η changed with e_{20}/e_{10} and α ($\Delta\varpi_0 = 2\pi/9$).

runaway gas-accretion could be very fast, with a mass-doubling time of a Jupiter-mass-planet being few thousand years (Zhou & Lin 2007). Under the fast grow of planet-mass increasing, the previously stable system becomes unstable if the previous distance is small compared with the enlarged Hill radius of growing planet. Thus the newly formed gas giant can perturb the previous formed gas giant, and their mutual interaction may lead to the ejection of one of the gas giants.

Consider a model planetary system similar to upsilon And (Butler et al. 2006) with a $1.3 M_\odot$ star and two planets 1, 2. The mass of planet 1 and 2 is 1.94 and 3.95 Jupiter mass and semi-major axis is 0.83 and 3 AU, respectively. In this model planet 2 has consumed the gas in its feeding zone (1 Hill radius = 0.33AU). Thus when we fix the additional planet 3 at 4.05 AU (1 Hill radius = 0.28AU), there is still enough gas for the accretion process which leads planet 3 grow from an initial mass of 10 Earth-mass to a Jupiter-mass. In the simulations we let its mass accrete according to a formula of Ikoma et al (2002) in a timescale of 10^5 years. For simplicity, we let all the three planets on circular and coplanar orbits initially.

We perform 120 sets of numerical integrations by varying initial longitude of planets in the system. Figure 5 shows a typical evolution. From figure 5 we know that in the beginning the evolution of the system is regular. But after planet 3 reaching a Jupiter mass, it is ejected and planet 2 get a large eccentricity while the eccentricity of planet 1 is oscillating near zero, resulting a system undergo near-separatrix motion.

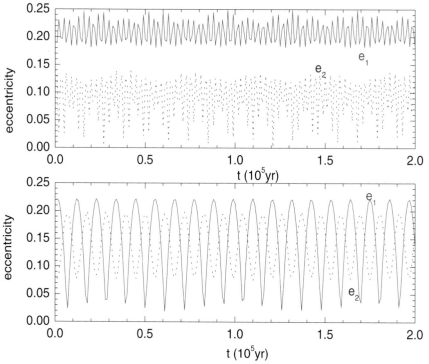

Figure 4. (a) Eccentricity evolution of the system undergo mean motion resonance. (b) Eccentricity evolution of the system $a_{20} = 2.3\,\mathrm{AU}$

For the 120 sets of simulations, 7 cases show strong chaotic motion with $e \sim 0.6$, so that the system will tend to be destroyed after a very long time, leaving one planet finally. For the left 113 cases, 57 cases results in all the three planets survived after the third planet accreting to Jupiter mass, and 56 cases with one planet scattered and two planets left. So there are $56 + 57 \times 2 = 170$ planet pairs for our 113 sets of integration. Our statistics shows, among the 170 planet pairs, 145 ($\sim 85\%$) pairs undergo near-separatrix motion (the eccentricity of at least one planet in the pair oscillate near zero), 21 cases are in circular motion, only 4 systems are in libration motions (2 systems with pericenters of the two planets aligned, and 2 anti-aligned). The ratio of near-separatrix motion is large enough as compared with the observed system (over 40%, Barnes 2008). Most of the near-separatrix motion occur in the cases that the third planet has not been scattered out, and it's perturbation to the inner two planets is too small, thus their eccentricities have changed slightly (with maximum values < 0.1).

5. Conclusions and discussions

Statistics about recent observed multiple-planet systems indicate that near-separatrix motion could be quite common. Based on the linear secular dynamics, a two-planet system will undergo near-separatrix motion provided one of the planet in near circular orbit initially. Due to either dynamical friction or tidal damping of the gas disk, a protoplanet will most possibly be on near circular orbits after they formed. However, two procedures may disrupt the near circular configurations: (1) migration of protoplanet in a gaseous

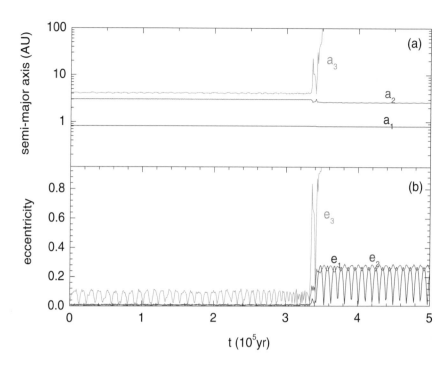

Figure 5. (a) Semi-major axis evolution of the 4-body system. (b) Eccentricity evolution of the 4-body system, e_1 undergoes large amplitude oscillation, will e_2 undergoes small amplitude oscillation.

disk; (2) scattering among planets. As we know, type II migration which leads to the resonance trapping of two gas giants may cause the increase of eccentricity of both orbits (e.g., Kley 2004). Unless in some rare cases, if type-II migration ceases just at the epoch when the two gas giants being captured into resonances, the system will be in a near-separatrix motion (Sándor & Kley 2007). For the scattering model, although the results of Barnes (2008) indicate that the final results can not produce the observed high rate of near-separatrix motion, they neglected those unsuccessfully scattered affairs, which must occurr in the real cases. The window that the third planet can be scattered is small, while most of cases the third planet is not ejected out, resulting in a slight changing of eccentricities of three plants and the near-separatrix motion remains. Our preliminary result in this report indicates this could be a possible explanation of the high probability of near-separatrix motion of multiple planets system. More detailed investigation on this proposition will be done in the future.

We thank Dr. R. Barnes and W. Kley for very helpful discussions. This work is supported by NSFC(10778603), National Basic Research Program of China(2007CB4800).

References

Barnes, R. & Greenberg, R. 2006, *ApJ*, 638, 478
Barnes, R. & Greeberg, R. 2006, *ApJ*, 652, L53
Barnes, R. 2008 , this issue.
Buter, R. P., *et al.* 2006, *ApJ*, 646,505
Ford, E. B., Lystad, V., & Rasio, F. A. 2005, *Nature*, 434, 873

Gladman, B. 1993, *Icarus*, 106, 247
Ikoma, M., Nakazawa, K., & Emori, H. 2000, *ApJ*, 537, 1013
Kley, W., Peita, J., & Bryden, G. 2004, *A & A*, 414, 735
Laughlin, G., Chambers, J., & Fischer, D. 2000, *ApJ*, 579, 455
Marcy, G., *et al.* 2005, *PThPS*, 158, 24
Papaloizou, J. C. B. & Terquem, C. 2006, *RPPh*, 69, 119
Pollack J. B. *et al.* 1996, *Icarus*, 124, 62
Sándor, Zs., Kley, W. 2007, *A & A*, 472, 781
Zhou, J. L. & Sun, Y. S. 2003, *ApJ*, 598, 1290
Zhou, J. L. & Lin, D. N. C 2007, *ApJ*, 666, 447
Zhou, J. L., Lin, D. N. C. & Sun, Y. S. 2007, *ApJ*, 666, 423

Habitable zones for Earth-mass planets in multiple planetary systems

Jianghui JI[1], Lin LIU[2], Hiroshi KINOSHITA[3] and Guangyu LI[1]

[1]Purple Mountain Observatory, Chinese Academy of Sciences, Nanjing 210008, China
email: jijh@pmo.ac.cn

[2]Department of Astronomy, Nanjing University, Nanjing 210093, China
email: xhliao@nju.edu.cn

[3]National Astronomical Observatory, Mitaka, Tokyo 181-8588, Japan

Abstract. We perform numerical simulations to study the Habitable zones (HZs) and dynamical structure for Earth-mass planets in multiple planetary systems. For example, in the HD 69830 system, we extensively explore the planetary configuration of three Neptune-mass companions with one massive terrestrial planet residing in 0.07 AU $\leqslant a \leqslant 1.20$ AU, to examine the asteroid structure in this system. We underline that there are stable zones of at least 10^5 yr for low-mass terrestrial planets locating between 0.3 and 0.5 AU, and 0.8 and 1.2 AU with final eccentricities of $e < 0.20$. Moreover, we also find that the accumulation or depletion of the asteroid belt are also shaped by orbital resonances of the outer planets, for example, the asteroidal gaps at 2:1 and 3:2 mean motion resonances (MMRs) with Planet C, and 5:2 and 1:2 MMRs with Planet D. In a dynamical sense, the proper candidate regions for the existence of the potential terrestrial planets or HZs are 0.35 AU $< a < 0.50$ AU, and 0.80 AU $< a < 1.00$ AU for relatively low eccentricities, which makes sense to have the possible asteroidal structure in this system.

Keywords. methods:n-body simulations-planetary systems-stars:individual (HD69830, 47 UMa)

1. Introduction

To date, over 260 extrasolar planets have been discovered around the nearby stars within 200 pc (Butler *et al.* 2006; The Extrasolar Planets Encyclopaedia†) mostly by the measurements of Doppler surveys and transiting techniques. The increasing numbers of known extrasolar planets are largely attributed to increasing precision in measurement techniques. Observational improvements will likely lead to more substantial discoveries, including: (1)diverse multi-planetary systems, of which more than 20 multiple systems with orbital resonance or secular interactions are already known;(2)low-mass companions around main-sequence stars (so-called super-Earths), e.g., 55 Cancri (McArthur *et al.* 2004), GJ 876 (Rivera *et al.* 2005), HD 160691 (Santos *et al.* 2004; Gozdziewski *et al.* 2007); (3)a true Solar System analog, with several terrestrial planets, asteroidal structure and a dynamical environment consistent with terrestrial planets in the Habitable Zone (HZ) (Kasting *et al.* 1993) that could permit the development of life, e.g., Gl 581 (von Bloh *et al.* 2007); (4)a comprehensive census of a diversity of planetary systems, which will provide abundant clues for theorists to more accurately model planetary formation processes (Ida & Lin 2004; Boss 2006).

Lovis *et al.* (2006) (hereafter Paper I) reported the discovery of an interesting system of three Neptune-mass planets orbiting about HD 69830 through high precision measurements with the HARPS spectrograph at La Silla, Chile. The nearby star HD 69830 is of spectral type K0V with an estimated mass of $0.86 \pm 0.03 M_\odot$ and a total luminosity

† As of Nov. 8, 2007, see http://exoplanet.eu/catalog.php and http://exoplanets.org/

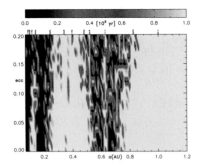 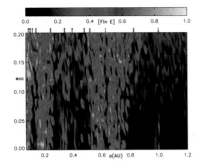

Figure 1. *Left panel*: Contour of the surviving time for Earth-like planets for the integration of 10^5 yr. *Right panel*: Status of their final eccentricities. Horizontal and vertical axes are the initial a and e. Stable zones for the low-mass planets in the region between 0.3 and 0.5 AU, and 0.8 and 1.2 AU with final low eccentricities.

of $0.60 \pm 0.03 L_\odot$ (Paper I), about 12.6 pc away from the Sun. In addition, Beichman et al. (2005) announced the detection of a large infrared excess owing to hot grains of crystalline silicates orbiting the star HD 69830 and inferred that there could be a massive asteroid within 1 AU. Subsequently, Alibert et al. (2006) and Paper I performed lots of calculations to simulate the system and revealed that the innermost planet may possess a rocky core surrounded by a tiny gaseous envelope. This planet probably formed inside the ice line in the beginning, whereas the two outer companions formed outside the ice line from a rocky embryo and then accreted the water and gas onto the envelope in the subsequent formation process. Hence, it is important for one to understand the dynamical structure in the final assemblage of the planetary system (Asghari et al. 2004; Ji et al. 2005), and to investigate suitable HZs for life-bearing terrestrial planets (Jones et al. 2005; Raymond et al. 2006; von Bloh et al. 2007; Gaidos et al. 2007) advancing the space missions (such as CoRot, Kepler and TPF) aiming at detecting them, thus one of our goals is to focus on the issues of the potential Earth-mass planets in the system.

2. Dynamical Structure and Habitable Zones in HD 69830 system

Modern observations by *Spitzer* and *HST* indicate that circumstellar debris disks (e.g., AU Mic and β Pic) are quite common in the early planetary formation. Beichman et al. (2006) used *Spitzer* to show that $13 \pm 3\%$ of mature main sequence stars exhibit Kuiper Belt analogs. They further point out that the existence of debris disks is extremely important for the resulting detection of individual planets, and related to the formation and evolution of planetary systems. As mentioned previously, Beichman et al. (2005) also provide clear evidence of the presence of the disk in HD 69830. Subsequently Paper I's best-fit orbital solutions were for three Neptune-mass planets with well-separated nearly-circular orbits, which may imply that the HD69830 system is similar to our Solar System in that it is dynamically consistent with the possible presence of terrestrial planets and asteroidal and Kuiper belt structures. Hence, it deserves to make a detailed investigation from a numerical perspective.

To investigate the dynamical structure and potential HZs in this system, we performed additional simulations with HD69830's three Neptune-mass companions in coplanar orbits, and one massive Earth-like planet. In the runs, the mass of the assumed terrestrial planet ranges from 0.01 $M_\oplus$ to 1 $M_\oplus$. The initial orbital parameters are as follows: the numerical investigations were carried out in $[a, e]$ parameter space by direct integrations, and for a uniform grid of 0.01 AU in semi-major axis (0.07 AU $\leqslant a \leqslant$ 1.20 AU) and 0.01 in eccentricity (0.0 $\leqslant e \leqslant$ 0.20), the inclinations are $0^0 < I < 5^0$, and the angles of the nodal longitude, the argument of periastron, and the mean anomaly are randomly distributed between 0^0 and 360^0 for each orbit, then each terrestrial mass body was numerically integrated with three Neptune-mass planets in the HD 69830 system. In total, about 2400 simulations were exhaustively run for typical integration time spans from 10^5 to 10^6 yr (about $10^6 - 10^7$ times the orbital period of the innermost planet) (see also Ji et al. 2007 for details).

Figure 1 shows the contours of the survival time for Earth-like planets (*left panel*) and the status of their final eccentricities (*right panel*) for the integration over 10^5 yr, where horizontal and vertical axes are the initial a and e. The *left panel* displays that there are stable zones for a terrestrial planet in the regime between 0.3 and 0.5 AU, and 0.8 and 1.2 AU with final eccentricities of $e < 0.20$. Obviously, unstable zones exist near the orbits of the three Neptune-mass planets where the planetary embryos have short dynamical survival time, and their eccentricities can quickly be pumped up to a high value ~ 0.9 (*right panel*). In these regions the evolution is insensitive to the initial masses. The terrestrial bodies are related to many of the mean motion resonances of the Neptunian planets and the overlapping resonance mechanism (Murray & Dermott 1999) can reveal their chaotic behaviors of being ejected from the system in short dynamical lifetime. Furthermore, most of terrestrial orbits are within $3R_{hill}$ sphere of the Neptune-mass planets, and others are involved in the secular resonance with two inner companions.

Analogous to our Solar system, if we consider the middle planet (HD 69830 c) as the counterpart as Jupiter, we will have the regions of mean motion resonances: 2:1 (0.117 AU), 3:2 (0.142 AU), 3:1 (0.089 AU) and 5:2 (0.101 AU), 2:3 (0.244 AU). In Fig. 1, we notice there indeed exist the apparent asteroidal gaps about or within the above MMRs (e.g., 3:1 and 5:2 MMRs), while in the region between 0.10 AU and 0.14 AU for $e < 0.10$, there are stable islands where the planetary embryos can last at least 10^5 yr. In addition, for Planet D, most of the terrestrial planets in 0.50 AU $< a <$ 0.80 AU are chaotic and their eccentricities are excited to moderate and even high values, the characterized MMRs with respect to the accumulation or depletion of the asteroid belt are 3:2 (0.481 AU), 2:1 (0.397 AU), 5:2 (0.342 AU), 4:3 (0.520 AU), 1:1 (0.630 AU), 2:3 (0.826 AU), 1:2 (1.000 AU), and our results enrich those of Paper I for massless bodies over two consecutive 1000-year intervals, showing a broader stable region beyond 0.80 AU. Note that there exist stable Trojan terrestrial bodies in a narrow stripe about 0.630 AU, involved in 1:1 MMR with Planet D, and they can survive at least 10^6 yr with resulting small eccentricities in the extended integrations. The stable Trojan configurations may possibly appear in the extrasolar planetary systems (see also Dvorak 2008; Psychoyos & Hadjidemetriou 2008 in this issue), e.g., Ji et al. (2005) explored such Trojan planets orbiting about 47 Uma, and Gozdziewski & Konacki (2006) also argued that there may exist Trojan pair configurations in the HD 128311 and HD 82943 systems. Ford & Gaudi (2006) developed a novel method of detecting Trojan companions to transiting close-in extrasolar planets and argue that the terrestrial-mass Trojans may be detectable with present ground-based observatories. Terrestrial Trojan planets with low eccentricity orbits close to 1 AU could potentially be habitable, and are worthy of further investigation in the future.

3. Summary and Discussion

In this work, we investigated the planetary configuration of three Neptune-mass companions similar to those surrounding HD 69830 and added one massive terrestrial planet in the region of 0.07 AU $\leqslant a \leqslant$ 1.20 AU to examine the dynamical stability of terrestrial mass planets and to explore the asteroid structure in this system. We show that there are stable zones of at least 10^5 yr for the low-mass terrestrial planets located between 0.3 and 0.5 AU, and 0.8 and 1.2 AU with final eccentricities of $e < 0.20$. Moreover, we also find that the accumulation or depletion of the asteroid belt is also shaped by orbital resonances of the outer planets, for example, the asteroidal gaps of 2:1 and 3:2 MMRs with Planet C, and 5:2 and 1:2 resonances with Planet D. On the other hand, the stellar luminosity of HD 69830 is lower than that of the Sun, thus the HZ should shift inwards compared to our Solar System. In a dynamical consideration, the proper candidate regions for the existence of the potential terrestrial planets or HZs are 0.35 AU $< a < 0.50$ AU, and 0.80 AU $< a < 1.00$ AU for relatively low eccentricities. Finally, we may summarize that the HD 69830 system can possess an asteroidal architecture resembling the Solar System and both the mean motion resonance (MMR) and secular resonances will work together to influence the distribution of the small bodies in the planetary system. In other simulations, we also show the potential Habitable zones for Earth-mass planets in the 47 UMa planetary system (see Ji et al. 2005), and the results imply that future space-based observations, e.g., CoRot, Kelper and TPF will hopefully produce a handful of samples belonging to the category of the terrestrial bodies.

Acknowledgements

We thank the anonymous referee for informative comments and suggestions that helped to improve the contents. This work is financially supported by the National Natural Science Foundations of China (Grants 10573040, 10673006, 10203005, 10233020) and the Foundation of Minor Planets of Purple Mountain Observatory.

References

Alibert, Y., et al. 2006, A&A, 455, L25
Asghari, N., et al. 2004, A&A, 426, 353
Beichman, C. A., et al. 2005, ApJ, 626, 1061
Beichman, C. A., et al. 2006, ApJ, 652, 1674
Boss, A. P. 2006, ApJ, 644, L79
Butler, R. P., et al. 2006, ApJ, 646, 505
Dvorak, R. 2008, IAU S249, this issue
Ford, E. B., &, Gaudi, B. S. 2006, ApJ, 652, L137
Gaidos, E., et al. 2007, Science, 318, 210
Gozdziewski, K., & Konacki, M. 2006, ApJ, 647, 573
Gozdziewski, K, et al. 2007, ApJ, 657, 546
Ida, S., & Lin, D. N. C. 2004, ApJ, 604, 388
Ji, J., Liu, L., Kinoshita, H., & Li, G.Y. 2005, ApJ, 631, 1191
Ji, J., Kinoshita, H., Liu, L., & Li, G.Y. 2007, ApJ, 657, 1092
Jones, B. W., Underwood, D. R., & Sleep, P. N. 2005, ApJ, 622, 1091
Kasting, J. F., Whitmire, D. P., & Reynolds, R. T. 1993, Icarus, 101, 108
Lovis, C., et al. 2006, Nature, 441, 305 (Paper I)
McArthur, B.E., et al. 2004, ApJ, 614, L81
Murray, C. D., & Dermott, S. F. 1999, Solar System Dynamics (New York: Cambridge Univ. Press)
Psychoyos, D., &, Hadjidemetriou, J. D. 2008, IAU S249, this issue
Raymond, S. N., Mandell, A. M., & Sigurdsson, S. 2006, Science, 313, 1413
Rivera, E. J., et al. 2005, ApJ, 634, 625
Santos, N.C., et al. 2004, A&A, 426, L19
von Bloh, W., Bounama, C., Cuntz, M., & Franck, S. 2007, arXiv:0705.3758

Habitability of super-Earths: Gliese 581c & 581d

W. von Bloh[1], C. Bounama[1], M. Cuntz[2] and S. Franck[1]

[1] Potsdam Institute for Climate Impact Research,
P.O. Box 601203, Potsdam, Germany,
email: bloh@pik-potsdam.de, bounama@pik-potsdam.de, franck@pik-potsdam.de

[2] University of Texas at Arlington,
P.O. Box 19059, Arlington, TX 76019, USA
email: cuntz@uta.edu

Abstract. The unexpected diversity of exoplanets includes a growing number of super-Earth planets, i.e., exoplanets with masses smaller than 10 Earth masses. Unlike the larger exoplanets previously found, these smaller planets are more likely to have similar chemical and mineralogical composition to the Earth. We present a thermal evolution model for super-Earth planets to identify the sources and sinks of atmospheric carbon dioxide. The photosynthesis-sustaining habitable zone (pHZ) is determined by the limits of biological productivity on the planetary surface. We apply our model to calculate the habitability of the two super-Earths in the Gliese 581 system. The super-Earth Gl 581c is clearly outside the pHZ, while Gl 581d is at the outer edge of the pHZ. Therefore, it could at least harbor some primitive forms of life.

Keywords. Astrobiology, planetary systems, stars: individual (Gliese 581)

1. Introduction

Very recently, Udry *et al.* (2007) announced the detection of two super-Earth planets in the Gliese 581 system; namely, Gl 581c with a mass of 5.06 $M_\oplus$ and a semi-major axis of 0.073 AU, and Gl 581d with 8.3 $M_\oplus$ and 0.25 AU. Both mass estimates are minimum masses uncorrected for $\sin i$. The luminosity of Gl 581 can be estimated as $L = 0.013 \pm 0.002 L_\odot$ with a stellar temperature of $T_e = 3480$ K and a stellar age of at least 2 Gyr. The main question is whether any of the two super-Earths around Gl 581 can harbor life, i.e., that any of the planets lie within the habitable zone (HZ). Typically, stellar HZs are defined as regions around the central star, where the physical conditions are favorable for liquid water to exist at the planet's surface for a period of time long enough for biological evolution to occur (Kasting *et al.* 1993).

In the following, we adopt a definition of the HZ previously used by Franck *et al.* (2000). Here habitability does not just depend on the parameters of the central star, but also on the properties of the planet. In particular, habitability is linked to the photosynthetic activity of the planet, which in turn depends on the planetary atmospheric CO_2 concentration, and is thus strongly influenced by the planetary dynamics. We call this definition the photosynthesis-sustaining habitable zone (pHZ). In principle, this leads to additional spatial and temporal limitations of habitability.

2. Estimating the Habitability of a Super-Earth

To assess the habitability of a super-Earth, i.e., a rocky planet smaller than 10 Earth masses (Valencia *et al.* 2006), an Earth-system model is applied to calculate the evolution of the temperature and atmospheric CO_2 concentration. The numerical model couples the stellar luminosity, the silicate-rock weathering rate, and the global energy balance to obtain estimates of the partial pressure of atmospheric carbon dioxide P_{CO_2}, the mean global surface temperature T_{surf}, and the biological productivity Π as a function of time t. The main point is the persistent balance between the CO_2 sink in the atmosphere-ocean system and the metamorphic (plate-tectonic) sources. This is expressed through the dimensionless quantities

$$f_{wr}(t) \cdot f_A(t) = f_{sr}(t), \tag{2.1}$$

where $f_{wr}(t)$ is the weathering rate, $f_A(t)$ is the continental area, and $f_{sr}(t)$ is the spreading rate normalized by their present values of Earth. The connection between the stellar parameters and the planetary climate can be formulated by using a radiation balance equation. The evolution of the surface temperature is derived directly from the stellar luminosity, the distance to the central star and the geophysical forcing ratio (GFR $:= f_{sr}/f_A$). For the investigation of a super-Earth under external forcing, we adopt a model planet with a prescribed continental area. The fraction of continental area to the total planetary surface is varied between 0.1 and 0.9.

The thermal history and future of a super-Earth is determined by calculating the GFR values. Spreading rates can be derived from the mantle temperature. Assuming conservation of energy, the average mantle temperature T_m is obtained as

$$\frac{4}{3}\pi\rho c(R_m^3 - R_c^3)\frac{dT_m}{dt} = -4\pi R_m^2 q_m + \frac{4}{3}\pi E(t)(R_m^3 - R_c^3), \tag{2.2}$$

where ρ is the density, c is the specific heat at constant pressure, q_m is the heat flow from the mantle, E is the energy production rate by decay of radiogenic heat sources in the mantle per unit volume, and R_m and R_c are the outer and inner radii of the mantle, respectively. The photosynthesis-sustaining HZ (pHZ) is defined as the spatial domain encompassing all distances R from the central star where the biological productivity is greater than zero, i.e.,

$$\text{pHZ} := \{R \mid \Pi(P_{CO_2}(R,t), T_{surf}(R,t)) > 0\}. \tag{2.3}$$

In our model, biological productivity is considered to be solely a function of the surface temperature and the CO_2 partial pressure in the atmosphere. Our parameterization yields zero productivity for $T_{surf} \leqslant 0°C$ or $T_{surf} \geqslant 100°C$ or $P_{CO_2} \leqslant 10^{-5}$ bar (Franck et al. 2000). To calculate the spreading rates for a planet with several Earth masses, the planetary parameters have been adjusted following Valencia et al. (2006) as

$$\frac{R_p}{R_\oplus} = \left(\frac{M}{M_\oplus}\right)^{0.27}, \tag{2.4}$$

where R_p is the planetary radius and M is the mass, with subscript $\oplus$ denoting Earth values. See von Bloh et al. (2007) for details and specified parameter values.

3. Results and Discussion

The habitable zone around Gl 581 for planets with five and eight Earth masses has been calculated for $L = 0.013 L_\odot$. The results for Gl 581c and Gl 581d are shown in Fig. 1a,b. The simulations have been carried out for a maximum CO_2 pressure of 5 bar (light colors) and 10 bar (dark colors) neglecting the cooling effect of CO_2 clouds. The super-Earth planet Gl 581c is found to be clearly outside the habitable zone. On the other hand, one might expect that life had a chance to originate on Gl 581d because it is near the outer edge of the pHZ, and for some combination of system parameters even

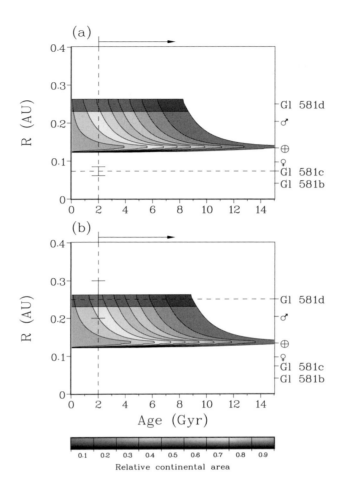

Figure 1. The pHz zone of Gl 581 for super-Earth Gl 581c (a) and Gl 581d (b) with a relative continental area varied from 0.1 to 0.9 and a fixed stellar luminosity of $0.013L_\odot$. The results are given as a function of planetary age. The light colors correspond to a maximum CO_2 pressure of 5 bar, whereas the dark colors correspond to 10 bar. For comparison, the positions of Venus, Earth and Mars are shown scaled to the luminosity of Gl 581.

inside the pHZ (von Bloh et al. 2007). A planet of eight Earth masses has more volatiles than an Earth-size planet and can build up a sufficiently dense atmosphere to prevent it from freezing out due to tidal locking. Planets inside the habitable zone around M stars may be tidally locked, which however does not necessarily thwart habitability (Tarter et al. 2007). The modestly eccentric orbit of Gl 581d ($e = 0.2 \pm 0.1$) further supports habitability, even if the maximum CO_2 pressure is assumed to be as low as 5 bar. The appearance of complex life, however, is unlikely due to the rather adverse environmental conditions. To get an ultimate answer to the profound question of life on Gl 581d, we have to await the TPF/Darwin missions. They will allow for the first time the detection of biomarkers in the atmospheres of the two super-Earths around Gl 581.

References

Franck, S., Block, A., von Bloh, W., Bounama, C., Schellnhuber, H.-J., & Svirezhev, Y. 2000, *Tellus*, 52B, 94
Kasting, J. F., Whitmire, D. P., & Reynolds, R. T. 1993, *Icarus*, 101, 108
Tarter, J. C., *et al.* 2007, *Astrobiology*, 7, 30
Udry, S., *et al.* 2007, *A&A*, 469, L43
Valencia, D., O'Connell, R. J., & Sasselov, D. 2006, *Icarus*, 181, 545
Von Bloh, W., Bounama, C., Cuntz, M., & Franck, S. 2007, *A&A*, 476, 1365

Orbital stability of planets in binary systems: A new look at old results

J. Eberle, M. Cuntz and Z. E. Musielak

Department of Physics, University of Texas at Arlington,
Arlington, TX 76019-0059, USA
email: cuntz@uta.edu, zmusielak@uta.edu

Abstract. About half of all known stellar systems with Sun-like stars consist of two or more stars, significantly affecting the orbital stability of any planet in these systems. This observational evidence has prompted a large array of theoretical research, including the derivation of mathematically stringent criteria for the orbital stability of planets in stellar binary systems, valid for the "coplanar circular restricted three-body problem". In the following, we use these criteria to explore the validity of results from previous theoretical studies.

Keywords. Astrobiology, methods: numerical, binaries, planetary systems

1. Introduction

Observational evidence for the existence of planets in stellar binary (and higher order) systems has been given by Patience *et al.* (2002), Eggenberger *et al.* (2004, 2007), and others. Eggenberger *et al.*, presented data for more than thirty systems, mostly wide binaries, as well as several triple star systems, with separation distances as close as 20 AU (GJ 86). These findings are consistent with previous theoretical results which showed that planets can successfully form in binary (and possibly multiple) stellar systems (e.g., Kley 2001, Quintana *et al.* 2002), known to occur in high frequency in the local Galactic neighborhood (Duquennoy & Mayor 1991, Lada 2006, Raghavan *et al.* 2006). More recently, Bonavita & Desidera (2007) performed a statistical analysis for binaries and multiple systems concerning the frequency of hosting planets, leading to the conclusion that there is no significant statistical difference between binary systems and single stars. The fact that planets in binary systems are now considered to be relatively common is also implied by the recent detection of debris disks in various main-sequence stellar binary systems using the *Spitzer Space Telescope* (Trilling *et al.* 2007). The research team observed 69 main-sequence binary star systems in the spectral range of A3 to F8.

In our previous work, we studied the stability of both S-type and P-type orbits in stellar binary systems, and deduced orbital stability limits for planets (Musielak *et al.* 2005). P-type orbits lie well outside the binary system, where the planet essentially orbits the center of mass of both stars, whereas S-type orbits lie near one of the stars, with the second star acting as a perturbator. The limits of stability were found to depend on the mass ratio between the stellar components. This topic has recently been revisited by Cuntz *et al.* (2007) and Eberle *et al.* (2008a, 2008b), who used the concept of Jacobi's integral and Jacobi's constant (Szebehely 1967, Roy 2005) to deduce stringent criteria for the stability of planetary orbits in binary systems for the special case of the "coplanar circular restricted three-body problem". These criteria are used to contest previous results on planetary orbital stability in binary systems available in the literature.

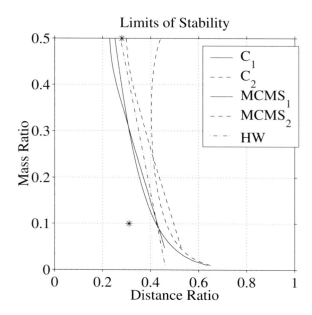

Figure 1. Limits of stability for planetary orbits for different mass ratios μ between the secondary star and the sum of the two stellar masses. We show the result based on Jacobi's constant C_1 (thick solid line) (criterion of stability) and based on Jacobi's constant C_2 (thick dashed line). For comparison, we also show the dividing lines between the regions of stability and marginal stability (thin solid line) [MCMS$_1$] and between marginal stability and instability [MCMS$_2$] (thin dashed line) previously obtained by Musielak et al. (2005). The two asterices ($*$) indicate the stability limit by David et al. (2003) corresponding to 10^6 orbits. The stability limit from the earlier work by Holman & Wiegert (1999) [HW] is depicted as a thin dash-dotted line.

2. Results and Conclusions

By focusing on the coplanar circular restricted three-body problem, we have been able to deduce absolute stability limits for small-mass planets, which only depend on the mass ratio between the two stellar components (see Fig. 1). The analysis by Eberle et al. (2008a) elucidates the physical backround of these limits. They are closely related to the so-called "surfaces of zero velocity" (or zero velocity contours) that form the boundaries of regions where the planet must be found. Thus, planets are unable to escape from these regions, preventing them from leaving the system entirely or from being captured by the other stellar component. This property is tantamount to orbital stability, although it does not necessarily imply stability in the sense of quasi-periodicity. The rationale of the contour is that it limits the allowable region of the planet due to its limited available energy.

In the following, we compare the absolute stability limit of Cuntz et al. (2007) and Eberle et al. (2008b) to results from time-dependent orbital stability simulations obtained by Holman & Wiegert (1999), David et al. (2003), and Musielak et al. (2005) (see Fig. 1).

The comparison of this newly found absolute stability limit with the stability limits previously deduced by Musielak et al. shows that their numerically derived stability limit largely agrees with the C_1-related stability criterion (Cuntz et al. 2007, Eberle et al. 2008b), although for a small range of mass ratios their stability criterion is too strict. In addition, there are discrepancies between our C_2 criterion for unstable orbits and their criterion for marginally stable / unstable orbits. Previous work by Holman & Wiegert (1999) considered a large range of eccentricities for the stellar binary components. For circular orbits, their stability limit is similar to the C_1-based stability limit in our study as well as the stability limit of Musielak et al. (2005), although a careful analysis shows that small, but noticeable differences exist as the Holman & Wiegert criterion is somewhat less strict for mass ratios between $\mu = 0.5$ and 0.1, but too strict for mass ratios of less than 0.1 — at least as viewed based on the polynomial fit given by the authors.

Other recent results have been given by David et al. (2003). They investigated the range of orbital stability for a similar parameter range as previously discussed. Here the orbital stability of an Earth-mass planet around a solar-mass star is studied in the presence of a companion star for a range of companion masses between 0.001 and 0.5 $M_\odot$. David et al. (2003) consider both circular and elliptical orbits, and derive expressions for the expected ejection time of the planet from the system. For fixed companion masses, the ejection time is found to be a steep function of the periastron distance (or orbital radius for circular orbits) to the primary star. David et al. predict that the domain of orbital stability gets progressively smaller over time according to the logarithm of the time of simulation, even though some cases in their study do not seem to indicate this type of behavior. Note that their principle finding of a progressively decreasing stability limit is fundamentally inconsistent with the result from the Cuntz et al. (2007) and Eberle et al. (2008b) studies that yield an analytically derived absolute limit of stability, although those cases solely focus on the restricted three-body problem. For $\mu = 0.3$ and zero eccentricity, David et al. (2003) deduce a stability limit of $\rho_0 = 0.31 \pm 0.04$ for 10^6 orbits, which is well inside the analytically defined stability region obtained in our study.

In conclusion, it is found that the comparison of the results by Cuntz et al. (2007) and Eberle et al. (2008b) to those previously obtained by Holman & Wiegert (1999), David et al. (2003), and Musielak et al. (2005), show that their criteria for orbital stability are in good agreement with our analytical criterion, although some small, but noticeable differences exist. For planets which do not fulfill our stringent criterion of orbital stability, stability may still be possible for a significant period of time. However, in this case the outcome must carefully be studied by employing long-term simulations and may depend on various factors, such as, e.g., the mass and orbital starting position of the planet.

References

Bonavita, M. & Desidera, S. 2007, *A&A*, 468, 721

Cuntz, M., Eberle, J., & Musielak, Z.E. 2007, *ApJ* (Letters), 669, L105

David, E. -M., Quintana, E. V., Fatuzzo, M., & Adams, F. C. 2003, *PASP*, 115, 825

Duquennoy, A., & Mayor, M. 1991, *A&A*, 248, 485

Eberle, J., Cuntz, M., & Musielak, Z. E. 2008a, in: K. Meech et al. (eds.), *Bioastronomy 2007: Molecules, Microbes and Extraterrestrial Life* (San Francisco: ASP), in press

Eberle, J., Cuntz, M., & Musielak, Z. E. 2008b, *A&A*, submitted

Eggenberger, A., & Udry, S. 2007, in: N. Haghighipour (ed.), *Planets in Binary Star Systems* (New York: Springer), in press

Eggenberger, A., Udry, S., & Mayor, M. 2004, *A&A*, 417, 353

Holman, M. J., & Wiegert, P. A. 1999, *AJ*, 117, 621

Kley, W. 2001, in: H. Zinnecker & R.D. Mathieu (eds.), *The Formation of Binary Stars*, IAU Symp. 200 (San Francisco: ASP), p. 511
Lada, C. J. 2006, *ApJ* (Letters), 640, L63
Musielak, Z. E., Cuntz, M., Marshall, E. A., & Stuit, T. D. 2005, *A&A*, 434, 355
Patience, J., *et al.* 2002, *ApJ*, 581, 654
Quintana, E. V., Lissauer, J. J., Chambers, J. E., & Duncan, M. J. 2002, *ApJ*, 576, 982
Raghavan, D., *et al.* 2006, *ApJ*, 646, 523
Roy, A. E. 2005, *Orbital Motion* (Bristol and Philadelphia: Institute of Physics Publ.)
Szebehely, V. 1967, *Theory of Orbits* (New York and London: Academic Press)
Trilling, D. E., *et al.* 2007, *ApJ*, 658, 1289

Retrograde resonances in compact multi-planetary systems: a feasible stabilizing mechanism

Julie Gayon and Eric Bois

Université Nice Sophia-Antipolis, CNRS, Observatoire de la Côte d'Azur,
Laboratoire Cassiopée, B.P. 4229, F-06304 Nice Cedex 4, France
email: julie.gayon@oca.eu - eric.bois@oca.eu

Abstract. Multi-planet systems detected until now are in most cases characterized by hot-Jupiters close to their central star as well as high eccentricities. As a consequence, from a dynamical point of view, compact multi-planetary systems form a variety of the general N-body problem (with $N \geqslant 3$), whose solutions are not necessarily known. Extrasolar planets are up to now found in prograde (i.e. direct) orbital motions about their host star and often in mean-motion resonances (MMR). In the present paper, we investigate a theoretical alternative suitable for the stability of compact multi-planetary systems. When the outer planet moves on a retrograde orbit in MMR with respect to the inner planet, we find that the so-called retrograde resonances present fine and characteristic structures particularly relevant for dynamical stability. We show that retrograde resonances and their resources open a family of stabilizing mechanisms involving specific behaviors of apsidal precessions. We also point up that for particular orbital data, retrograde MMRs may provide more robust stability compared to the corresponding prograde MMRs.

Keywords. celestial mechanics, planetary systems, methods: numerical, statistical

1. Introduction

To identify the dynamical state of multi-planetary systems, we use the MEGNO technique (the acronym of Mean Exponential Growth factor of Nearby Orbits; Cincotta & Simò 2000). This method provides relevant information about the global dynamics and the fine structure of the phase space, and yields simultaneously a good estimate of the Lyapunov Characteristic Numbers with a comparatively small computational effort. From the MEGNO technique, we have built the MIPS package (acronym of Megno Indicator for Planetary Systems) specially devoted to the study of planetary systems in their multi-dimensional space as well as their conditions of dynamical stability.

Particular planetary systems presented in this paper are only used as initial condition sources for theoretical studies of 3-body problems. By convention, the reference system is given by the orbital plane of the inner planet at $t = 0$. Thus, we suppose the orbital inclinations and the longitudes of node of the inner (noted 1) and the outer (noted 2) planets (which are non-determined parameters from observations) as follows : $i_1 = 0°$ and $\Omega_1 = 0°$ in such a way that the relative inclination and the relative longitude of nodes are defined at $t = 0$ as follows : $i_r = i_2 - i_1 = i_2$ and $\Omega_r = \Omega_2 - \Omega_1 = \Omega_2$. The MIPS maps presented in this paper have been confirmed by a second global analysis technique (Marzari et al. 2006) based on the Frequency Map Analysis (FMA; Laskar 1993).

2. Fine structure of retrograde resonance

Studying conditions of dynamical stability in the neighborhood of the HD 73526 two-planet system (period ratio: 2/1, see initial conditions in Table 1), we only find one stable and robust island (noted (2)) for a relative inclination of about 180° (see Fig. 2a). Such a relative inclination (where in fact $i_1 = 0°$ and $i_2 = 180°$) may be considered to a coplanar system where the planet 2 has a retrograde motion with respect to the planet 1. From a kinematic point of view, it amounts to consider a scale change of 180° in relative inclinations. Taking into account initial conditions inside the island (2) of Fig. 1a, we show that the presence of a strong mean-motion resonance (MMR) induces clear stability zones with a nice V-shape structure, as shown in Fig. 1b plotted in the $[a_1, e_1]$ parameter space. Let us note the narrowness of this V-shape, namely only about 0.006 AU wide for the inner planet (it is 5 times larger in the Jupiter-Saturn case). A similar V-shape structure is obtained in $[a_2, e_2]$ with about 0.015 AU wide. Due to the retrograde motion of the outer planet 2, this MMR is a 2:1 retrograde resonance, also noted 2:-1 MMR.

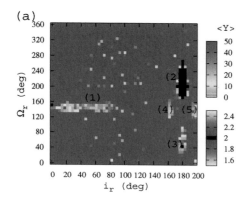 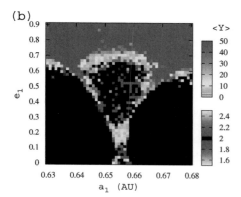

Figure 1. Panel (a): Stability map in the $[i_r, \Omega_r]$ non-determined parameter space of the HD 73526 planetary system (see Table 1). Panel(b): Stability map in the $[a_1, e_1]$ parameter space for initial conditions taken in the stable zone (2) of panel (a). Note that masses remain untouched whatever the mutual inclinations may be; they are equal to their minimal observational values. Black and dark-blue colors indicate stable orbits ($<Y> = 2 \pm 3\%$ and $<Y> = 2 \pm 5\%$ respectively with $<Y>$, the MEGNO indicator value) while warm colors indicate highly unstable orbits.

3. Efficiency of retrograde resonances

Fig. 2 exhibits stability maps in the $[i_r, \Omega_r]$ parameter space considering a scale reduction of the HD 82943 planetary system (see Table 1) according to a factor 7.5 on semi-major axes (masses remaining untouched). The dynamical behavior of the reduced system (Fig. 2b) with respect to the initial one (Fig. 2a) points up the clear robustness of retrograde configurations contrary to prograde ones. The "prograde" stable islands completely disappear while only the "retrograde" stable island resists, persists and even extends more or less. Even for very small semi-major axes and large planetary masses, which should a priori easily make a system unstable or chaotic, stability is possible with counter-revolving orbits.

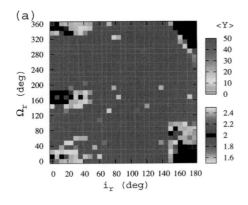 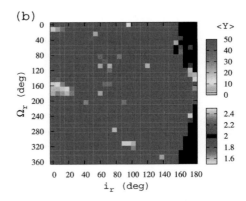

Figure 2. Stability maps in the $[i_r, \Omega_r]$ parameter space. Panel (a): initial HD 82943 planetary system (see Table 1). Panel (b): scale reduction of the HD 82943 planetary system according to a factor 7.5 on semi-major axes. Masses in Panel (a) and Panel (b) are identical. Color scale is the same as in Fig. 1.

In the case of the 2:1 retrograde resonance, although close approaches happen more often (3 for the 2:-1 MMR) compared to the 2:1 prograde resonance, the 2:-1 MMR remains very efficient for stability because of faster close approaches between the planets. A more detailed numerical study of retrograde resonances can be found in Gayon & Bois (2008).

Elements	HD 73526	HD 82943	HD 128311	HD 160691	HD 202206
M_{star} ($M_\odot$)	1.08 ± 0.05	1.15	0.84	1.08 ± 0.05	1.15
$m \sin i_l$ (M_J)	2.9 ± 0.2 2.5 ± 0.3	1.85 1.84	1.56 ± 0.16 3.08 ± 0.11	1.67 ± 0.11 3.10 ± 0.71	17.4 2.44
a (AU)	0.66 ± 0.01 1.05 ± 0.02	0.75 1.18	1.109 ± 0.008 1.735 ± 0.014	1.50 ± 0.02 4.17 ± 0.07	0.83 2.55
e	0.19 ± 0.05 0.14 ± 0.09	0.38 ± 0.01 0.18 ± 0.04	0.38 ± 0.08 0.21 ± 0.21	0.20 ± 0.03 0.57 ± 0.1	0.435 ± 0.001 0.267 ± 0.021
ω (deg)	203 ± 9 13 ± 76	124.0 ± 3 237.0 ± 13	80.1 ± 16 21.6 ± 61	294 ± 9 161 ± 8	161.18 ± 0.30 78.99 ± 6.65
M (deg)	86 ± 13 82 ± 27	0 75.21 ± 1.96	257.6 ± 2.7 166 ± 2	0 12.6 ± 11.2	105.05 ± 0.48 311.6 ± 9.5

Table 1. Orbital parameters of the HD 73526, HD 82943, HD 128311, HD 160691 and HD 202206 planetary systems. Data sources come from Tinney *et al.* (2006), Mayor *et al.* (2004), Vogt *et al.* (2005), McCarthy *et al.* (2004) and Correia *et al.* (2005) respectively. For each system and each orbital element, the first line corresponds to the inner planet and the second one to the outer planet.

Data sources	Period ratio	Prograde MMR	Retrograde MMR
HD 73526	2/1	17	500
HD 82943	2/1	755	1000
HD 128311	2/1	249	137
HD 160691	5/1	ε	320
HD 202206	5/1	ε	631

Table 2. Statistical results. For each type of MMR (prograde or retrograde), 1000 random systems have been integrated in the error bars of each data source. The proportion of stable systems over 1000 is indicated in each case. ε designates a very small value that depends on the size of the random system size. Data sources come from Tinney et al. (2006), Mayor et al. (2004), Vogt et al. (2005), McCarthy et al. (2004) and Correia et al. (2005) respectively (see Table 1).

4. Occurrence of stable counter-revolving configurations

The occurence of stable two-planet systems including counter-revolving orbits appears in the neighborhood of a few systems observed in 2:1 or 5:1 MMR. New observations frequently induce new determinations of orbital elements. It is the case for the HD 160691 planetary system given with 2 planets in McCarthy et al. (2004) then with 4 planets in Pepe et al. (2007). Hence, systems related to initial conditions used here (see Table 1) have to be considered as *academic* systems. Statistical results for stability of these academic systems are presented in Table 2, both in the prograde case ($i_r = 0°$) and in the retrograde case ($i_r = 180°$). For each data source, 1000 random systems taken inside observational error bars have been integrated. Among these random systems, the proportion of stable systems either with prograde orbits or with counter-revolving orbits is given in Table 2. In all cases, a significant number of stable systems is found in retrograde MMR. Moreover, in most data sources, retrograde possibilities predominate.

5. Resources of retrograde resonances

The 2:1 (prograde) MMRs preserved by synchronous precessions of the apsidal lines (ASPs) are from now on well understood (see for instance Lee & Peale 2002, Bois et al. 2003, Ji et al. 2003, Ferraz-Mello et al. 2005). The MMR-ASP combination is often very effective; however, ASPs may also exist alone for stability of planetary systems. Related to subtle relations between the eccentricity of the inner orbit (e_1) and the relative apsidal longitude $\Delta\tilde{\omega}$ (i.e. $\tilde{\omega}_1 - \tilde{\omega}_2$), Fig. 3 permits to observe how the 2:1 retrograde MMR brings out its resources in the $[\Delta\tilde{\omega}, e_1]$ parameter space :

- In the island (1) (i.e. inside the $[a, e]$ V-shape of Fig. 1b), the 2:-1 MMR is combined with a uniformly prograde ASP (both planets precess on average at the *same rate* and in the *same prograde direction*).
- In the island (2) (i.e. outside but close to the $[a, e]$ V-shape of Fig. 1b), the 2:-1 *near*-MMR is combined with a particular apsidal behavior that we have called a *rocking* ASP (see Gayon & Bois 2008): both planets precess at the *same rate* but in *opposite directions*.
- The $[\Delta\tilde{\omega}, e_1]$ map also exposes a third island (3) that proves to be a wholly chaotic zone on long term integrations.

Let us note that the division between islands (1) and (2) is related to the degree of closeness to the 2:-1 MMR.

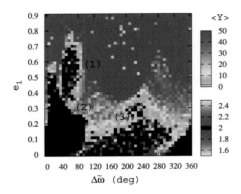

Figure 3. Stability map in the $[\Delta\tilde{\omega}, e_1]$ parameter space. A similar distribution of stable islands is obtained in $[\Delta\tilde{\omega}, e_2]$. Color scale and initial conditions are the same as in Fig. 1 with in addition the i_r and Ω_r values chosen in the island (2) of Fig. 1a.

6. Conclusion

We have found that retrograde resonances present fine and characteristic structures particularly relevant for dynamical stability. We have also shown that in cases of very compact systems obtained by scale reduction, only the "retrograde" stable islands survive. From our statistical approach and the scale reduction experiment, we have expressed the efficiency for stability of retrograde resonances. Such an efficiency can be understood by very fast close approaches between the planets although they are in greater number.

We plan to present an Hamiltonian approach of retrograde MMRs in a forthcoming paper (Gayon, Bois, & Scholl, 2008). Besides, in Gayon & Bois (2008), we propose two mechanisms of formation for systems harboring counter-revolving orbits. Free-floating planets or the Slingshot model might indeed explain the origin of such planetary systems.

In the end, we may conclude that retrograde resonances prove to be a feasible stabilizing mechanism.

Acknowledgements

We thank the anonymous referee for his comments that greatly helped to improve the paper.

References

Bois, E., Kiseleva-Eggleton, L., Rambaux, N., & Pilat-Lohinger, E. 2003, *ApJ*, 598, 1312
Cincotta, P. & Simó, C. 2000, *A&AS*, 147, 205
Correia, A. C. M., Udry, S., Mayor, M., Laskar, J., Naef, D., Pepe, F., Queloz, D., Santos, N. C. 2006, *A&A*, 440, 751
Ferraz-Mello, S., Michtchenko, T. A., Beaugé, C., & Callegari, N. 2005, *Lecture Notes in Physics*, 683, 219
Gayon, J. & Bois, E. 2008, *A&A*, accepted, [arXiv:0801.1089v2]
Gayon, J., Bois, E., & Scholl, H. 2008, *Celestial Mechanics and Dynamical Astronomy*, Special Issue : "Theory and Applications of Dynamical Systems", to be submitted
Ji, J., Kinoshita, H., Liu, L., Li, G., & Nakai, H. 2003, *Celestial Mechanics and Dynamical Astronomy*, 87, 113
Laskar, J. 1993, *Physica D*, 67, 257
Lee, M. H. & Peale, S. J. 2002, *ApJ*, 567, 596
Mayor, M., Udry, S., Naef, D., Pepe, F., Queloz, D., Santos, N. C., Burnet, M. 2004, *A&A*, 415, 391
Marzari, F., Scholl, H., & Tricarico, P. 2006, *A&A*, 453, 341

McCarthy, C., Butler, R. P., Tinney, C. G., Jones, H. R. A., Marcy, G. W. Carter, B., Penny, A. J., & Fischer, D. A. 2004, *ApJ*, 617, 575

Pepe, F., Correia, A. C. M., Mayor, M., Tamuz, O., Couetdic, J., Benz, W., Bertaux, J.-L., Bouchy, F., Laskar, J., Lovis, C., Naef, D., Queloz, D., Santos, N. C., Sivan, J.-P., Sosnowska, D., & Udry, S. 2007, *A&A*, 462, 769

Tinney, C. G., Butler, R. P., Marcy, G. W., Jones, H. R. A., Laughlin, G., Carter, B. D, Bailey, J. A., & O'Toole, S. 2006, *ApJ*, 647, 594

Vogt, S. S., Butler, R. P., Marcy, G. W., Fischer, D. A., Henry, G. W., Laughlin G., Wright, J. T., Johnson, J. A. 2005, *ApJ*, 632, 638

Author Index

Ackerman, A. – 173
Agnor, C. – 301
Aigrain, S. – 89
Alapini, A. – **89**
Allan, A. – 35
Allard, F. – 173
Auvergne, M. – 3

Baglin, A. – 3
Baluev, R.V. – **101**
Barge, P. – **3**
Barnes, R. – 187, **469**
Baruteau, C. – **393**, **397**
Basri, G. – 17
Batalha, N. – 17
Beaugé, C. – **427**
Bessho, T. – 279
Bode, M.F. – 35
Bohlender, D.A. – 151
Bois, E. – 511
Borucki, W. – **17**
Boss, A.P. – 261
Bounama, C. – 503
Bouwman, J. – 359
Brown, T. – 17
Brummelaar, T.T. – 71
Burgdorf, M.J. – 35

Caldwell, D. – 17
Cameron, A.C. – 83, 85, 151
Carmona A., – **359**
Carney, B.W. – 261
Casali, M. – 355
Cassan, A. – bf 31
Charbonneau, D. – 261
Christensen-Dalsgaard, J. – 17
Ciardi, D.R. – 93
CoRoT team – 3
Cochran, W. – 17
Cuntz, M. – 79, **203**, 503, 507

Davis, E. – **313**
Dehn, M. – 173
Delplancke, F. – 119
Dobbs-Dixon, I. – 131, **263**
Döllinger, M.P. – 209
Dominik, M. – **35**
Dong, R.B. – **381**, 389
Dullemond, C.P. – 359
Dunham, E. – 17
Dvorak, R. – **461**
da Silva, L. – 209

de Jong, J. – 119
de Medeiros, J.R. – 209
de Oliveira C.A., – **355**

ESPRI consortium, – 61, 119
Eberle, J. – **507**
Elias II, N.M. – **119**

Fang, X.-S. – 83
Fedele, D. – 359
Ferraz-Mello, S. – 179, 427, 485
Ford, E.B. – 115, **441**
Fouchet, L. – 375, 385
Franck, S. – 503
Fraser, S.N. – 35

Gautier, T.N. – 17
Gayon, J. – **511**
Ge, J. – 115
Geary, J. – 17
Geisler, R. – **61**
Genda, H. – **267**
Gerlach, E. – 479
Gilliland, R. – 17
Girardi, L. – 209
Giuppone, C.A. – 427
Goździewski, K. – **447**
Gonzalez, J.-F. – **375**, 385
Goto, M. – 359
Greenberg, R. – 187, 469
Gu, P.-G. – **145**
Gu, S.-H. – **83**, 85
Guillot, T. – 267
Guinan, E.F. – 203
Gurdemir, L. – 203

Haghighipour, N. – **319**
Hall, J.C. – 79
Han, I. – 53
Hara, T. – **325**
Hatzes, A. – 209
Hauschildt, P. – 173
Helling, Ch. – **167**, **173**
Henning, Th. – 61, 119, 359
Holman, M.J. – 261
Homeier, D. – 173
Hori, Y. – **163**
Horne, K. – 35
Hussmann, H. – 179

Ida, S. – 159, 163, **223**, 267, 279, 305
Ikoma, M. – 163, 267

Inutsuka, S.-I – 401
Izumiura, H. – 53

Jackson, B. – **187**
Jenkins, J. – 17
Ji, J.H. – **407**, **499**
Jones, H. – 111

Kajiura, D. – 325
Kambe, E. – 53
Kane, S.R. – **115**
Kashyap, V.L. – 79
Kerins, E. – 25, 35
Kim, K.-M. – 53
Kinoshita, H. – 499
Kitiashvili, I. – **197**
Kitsionas, P. – **271**
Klessen, R.S. – 271
Kley, W. – **251**
Koch, D. – 17
Kondo, Y. – 17
Kretke, K.A. – **293**, 309
Kubas, D. – 31
Kunitomo, S. – 325
Kurucz, R.L. – 203
Køhler, R. – 119

Laibe, G. – 375, **385**
Laird, J.B. – 261
Latham, D. – 17
Latham, D.W. – 261
Launhardt, R. – 61, 119
Lee, B.-C. – 53
Lhotka, Ch. – 461
Li, G.Y. – 499
Li, S.-L. – 263, **301**
Li, X.-J. – 83
Lin, D.N.C. – **131**, 223, 263, 285, 293, 301, 309, 381, 389
Lissauer, J.J. – 17
Liu, L. – 407, 499
Liu, X.W. – 381, 389
Lodders, K. – 173

Machida, M.N. – 401
Maddison, S.T. – 375, 389
Mao, S.D. – **25**
Marchi, S. C**123**
Marley, M. – 173
Masset, F.S. – **331**, 393, 397
Masuda, S. – 53
Meeus, G. – 369
Michtchenko, T.A. – 427
Migaszewski, C. – 447
Monet, D. – 17
Monnier, J.D. – 71
Moro-Martín, A. – **347**

Mottram, C.J. – 35
Musielak, Z.E. – 507
Musieliński, A. – 447
Muto, T. – **401**
Müller, A. – 61

Nagasawa, M. – **279**
Neuhäuser, R. – 57,65
Niedzielski, A. – **43**, **49**
Nowak, G. – 49

O'Toole, S. – **111**
ODonovan, F.T. – 261
Ogihara, M. – **305**
Ogilvie, G.I. – 145
Omiya, M. – **53**
Ortolani, S. – 123
Ou, S.L. – 407

Pasquini, L. – **209**
Pavlyuchenkov, Ya. – 359
Pedretti, E. – 71
Peeters, E. – 369

Queloz, D. – 61, 119
Quirrenbach, A. – 61, 119

Rattenbury, N.J. – 25, 35
Raymond, S.N. – **233**
Reffert, S. – 61, 119
Rietmeijer, F. – 173
Rodríguez, A. – **179**
Roell, T. – **57**

Saar, S.H. – **79**
Sano, T. – 163
Sato, B. – 53
Schütz, O. – **369**
Schmidt, T. – **65**
Schwarz, R. – 459
Seifahrt, A. – 57
Senshu, H. – 159
Setiawan, J. – 61, 209
Shigeyasu, M. – 325
Shkolnik, E. – **151**
Šidlichovský M. – **477**
Snodgrass, C. – 35
Sozzetti, A. – **261**
Stecklum, B. – 359
Steele, I.A. – 35
Stefanik, R.P. – 261
Sterzik, M.F. – 369
Stilz, I. – 119
Street, R.A. – 35
Sumi, T. – 31
Sun, Y.-S. – 485

Author Index

Tachinami, C. – **159**
Takada-Hidai, M. – 53
Thi, W.F-. – 359
Thureau, N. – 71
Tinney, C. – 111
Torres, G. – 261
Toyota, E. – 53
Tsapras, Y. – 35
Tsuji, T. – 173
Tubbs, R.N. – 119
Turner, N.J. – 293

Urakawa, S. – 53
van den Ancker, M.E. – 359
von Bloh, W. – **503**
von Braun, K. – **93**

Walker, G.A.H. – 151
Wang, S. – **491**
Wang, X.-B. – 83, **85**
Wang, Y. – 381, **389**
Waters, L.B.F.M. – 359

Weise, P. – 61
Weiss, A. – 209
Wheatley, P.J. – 35
Whitworth, A.P. – 271
Winn, J.N. – 261
Woitke, P. – 173
Wolszczan, A. – 43
Wyrzykowski, Ł. – 35

Xie, J.-W. – **419**

Yoon, T.S. – 53
Yoshida, M. – 53

Zhang, H. – **413**
Zhang, L.-Y. – 83, 85
Zhang, X.J. – **309**
Zhao, M. – 71
Zhou, J.-L. – **285**, 413, 419, 491
Zhou, L.-Y. – **485**
Zieliński, P. – 49

Subject Index

Accretion disk(s) – 294, 305, 336, 371, 393, 397, 407, 417, 419
Adiabatic disk – 397, 399
Apsidal corotation(s) – 101, 427, 430
Apsidal precession – 511
Astrobiology – 203, 233, 503, 507
Astrochemistry – 167, 173
Astrometry – 119, 57-59, 61, 63, 66

Baroclinic instability – 407, 408, 410, 411
Bi-fluid – 376, 386
Binary stars:
   β Ceph – 5, 251, 321, 323, 419, 423
   Eclipsing binary(ies) – 5, 10, 12, 13, 15, 20, 22, 68
   Multiple stars – 251, 319

Bondi radii – 263
Borderline – 92, 442, 444, 445

Cassinistate resonance – 139
Celestial mechanics – 187, 285, 319, 427, 441, 461, 479, 485, 511
Chaotic motion – 431, 432, 436, 448, 451, 456
Close encounter(s) – 242, 274, 315, 439, 443-445, 453, 456, 479, 491
Corotation torque – 310, 333, 334, 338, 339, 343, 344, 393, 397-399

Dark matter – 25, 326, 327
Darwins theory – 179, 181, 185
Debris disk – 251, 347-353, 359, 360, 364, 367, 369, 373, 381, 389, 392, 500, 507
Density wave(s) – 236, 309, 378, 395
DNA – 203-205

Einstein radius – 26-28, 36
Energy dissipation – 132, 142, 179, 183, 190, 285, 388
Exposure – 8, 10, 32, 37, 45, 59, 83, 85, 97, 316, 355, 361

Feeding zone – 135-137, 233, 236, 237, 239-241
Free fall – 325-327

Galaxy:
   Galactic centre – 25
   Galactic plane – 10, 18, 20, 33

Galactic bulge – 25, 31-32, 35
Galactic halo – 325
Gas cloud(s) – 325-328
The Galaxy – 10-11, 18, 31, 34

Gaussian distribution(s) – 37, 57, 58, 104, 325-327, 450
Gravitational moment – 163

H & K bands – 170, 171
Habitability – 159, 205, 320, 321, 503, 505
Heavy elements – 134, 164-166, 261, 267-269, 325, 327, 389, 392
Helix nebula – 381, 382
Hierarchical Systems – 442

Impulsive perturbation – 443-445
Inter-penetrating phases – 376

Jean's instability – 326

KAM theorem – 447
Kelvin contraction – 326, 327

Lagrangean equilibrium points(Lagrangean point) – 461, 462
Laplace resonance – 428
Laplace-Lagrange perturbation theory – 442
Light curve(s) – 3-5, 10-15, 22, 23, 25-29, 31, 32, 36-40, 83-86, 89-94, 154
Lindblad resonance – 132, 310, 332-335, 343, 394, 395, 403, 404, 408
Liquid water – 17, 159, 503
Love number – 181, 286
Lyapunov Characteristic Numbers – 511

Mapping – 8, 461, 463-465, 467
Marshak boundary condition – 146
Megno Indicator – 511, 512
Methods:
   Brute-Force approach – 39
   Hamiltonian approach – 515
   Hermit scheme – 382
   Hydrodynamics – 137, 145, 251, 267, 271-273, 276
   Monte Carlo – 57, 59, 85-86, 92, 111, 116, 124, 126, 223-231, 313, 377, 450

Smooth particle hydrodynamics(SPH) – 137, 142, 267-268, 217-277, 301-302, 375-376, 385-387

Mixing length theory – 159, 160
Molecular cloud – 234, 271, 356

Observational biases – 111, 126, 217, 296
Oligarch – 224, 235-236, 313-314, 316
Opacity – 126, 131, 139, 141-143, 146, 171, 176, 226, 295, 301, 343, 373, 375, 395

Photospheric radiation – 203, 205
Planets:
    Asteroid belt – 238, 240, 245, 246
    Atmospheres – 167
    Earth-like planet(s) – 233, 237, 244-246, 286, 287, 289, 291, 313, 320-322, 390, 407, 412, 456, 458, 500, 501
    Habitable zone(s) – 17, 18, 47, 203, 205, 223, 227, 233, 239-245
    Hot Jupiter(s) – 3, 4, 71-73, 76, 126, 127, 139, 145, 148, 149, 151-152, 154-157, 179, 187, 193, 197, 199, 200, 202, 231, 233, 242, 267, 309, 340, 407, 408, 461, 511
    Hot Neptune – 3, 126
    Icy planet(s) – 36
    Kuiper belt – 6, 347-349
    Migration rate – 258, 259, 310, 311, 337-339, 343, 392, 393, 416, 486
    Multiple-planet system – 27, 29, 101, 103, 109, 126, 127, 136, 244-245, 270, 313, 319, 413, 415, 417, 428, 433, 439, 441-445, 447, 457, 459, 469-470, 474, 476-477, 485, 491, 496-497, 501, 511
    Orbital determination – 427, 429, 431, 433, 435-437, 439
    Orbital elements – 60, 103, 183-185, 188-191, 252-253, 255, 258-259, 290, 314-315, 331, 421-422, 429, 431, 436-437, 439, 442, 444-445, 455, 469, 474, 477, 486, 489, 514
    Orbital stability – 435, 447, 507-509
    Protoplanetary disk(s) – 187, 189, 211, 223-225, 233-239, 251-253, 256-259, 267, 270, 271, 285, 293, 296, 299, 306, 309, 315, 331, 340-344, 359-368, 369-373, 375-379, 385-387, 393, 401, 407-408, 411-412, 419, 443, 474
    Rocky planet – 138, 159, 224, 233, 289
    Terrestrial planet(s) – 3, 5, 15, 20-24, 134, 159-162, 167, 223-232, 233-246, 286, 305-308, 309, 313-318, 320-324, 350-351, 365, 370, 419, 461, 563, 499-502
    Trojan – 455-458, 461-468, 501
    Type I migration – 223-232, 305-306, 308, 309-311, 331, 333, 335, 342-345, 393, 395, 397, 400, 408, 415
    Type II migration – 224-225, 228, 269, 283, 286, 293, 312, 331, 335, 337, 340, 344, 408, 413, 415, 496-497
    Type III migration(Runaway migration) – 331, 338-341, 344, 413-417, 408, 413

Poincare surface – 464
Polarization maps – 377
Pop II(Population II stars) – 325,327
Pop III objects – 325
Project:
    ARTEMiS – 35-41
    Kepler mission – 17, 23, 29
    MARVELS – 115
    MOA – 25-28, 31-32, 35, 40
    OGLE – 25-28, 31-33, 36-40, 95, 152, 198-201, 268
    PRIMA – 61, 63, 119-122
    SIGNALMEN – 35-40
    eSTAR – 35, 40
    YSOs – 355, 358
    CoRoT – 3-16, 89-92
    JUNO – 163, 166

Radiative transfer – 139, 167-170, 336, 343, 375, 377, 397
Red noise – 19, 21, 93-100, 318
Relative astrometry – 57-59, 61
Resonance trapping – 429, 432, 487, 489, 491
Resonant capture – 443-444, 486
Restricted three-body problem – 289, 507-509
Rossby wave instability – 407
Runaway migration – 413-417

Self-gravity – 259, 306, 343, 348, 393-396, 412
Snowline – 309, 312
Softening length – 393, 396, 397
Solar System – 15, 28, 45, 111, 113, 134, 155, 163, 179, 181, 197, 223, 227, 233, 237, 244-246, 263, 285-

Subject Index

286, 321, 328, 347-353, 381, 390, 401, 428, 433
Space-based telescope – 17, 34
Spectral energy distribution(s) – 173, 176-177, 351, 364, 371, 377
Spectral type – 5, 18-19, 37, 62, 68, 152, 203-205, 244, 369-370, 499
Spin evolution – 197-202
Spin-orbit resonance – 179, 182, 184-185
Stability criterion – 455, 457, 464, 477, 507
Star (individual):
  14 Her – 105-108
  47 Tuc – 57-60
  47 UMa – 106-109, 319-322, 442, 474, 497, 500
  51 Peg b – 106, 71-75
  55 Cnc – 105-108, 252, 319-322, 474, 481-486
  70 Vir – 105, 108, 321
  α Centauri B – 313
  Boo b – 71-75
  CT Cha – 65-68, 171
  δ Scuti – 3,5
  γ Dor – 5
  GG Tau – 364, 377-379, 386
  GJ 876 – 57, 319-322, 441-444, 476, 499
  HD 149026b – 149, 267, 269
  HD 155358 – 444, 447, 451-455, 457, 459, 476
  HD 17051 – 461, 464, 465, 467
  HD 179949 – 79-81, 151-157
  HD 37124 – 454, 476
  HD 69830 – 313, 316, 351, 352, 471, 476, 499-502
  HD 83443 – 105, 108
  T Tauri – 61, 65, 131-132, 142, 228, 230, 263, 337, 341, 343, 356, 359-368, 369-374, 376, 382-387, 407
  ν And system(ν Andromedae) – 154, 319-320, 322, 444, 491
  ρ Ophiuchi – 355-357

Stars:
  Brown dwarfs – 53, 56, 63, 167, 173, 176, 215
  Circumstellar disc – 274, 276, 355, 356
  Coronae – 79
  Emission line – 360
  Fundamental parameters – 49, 209
  GK-giants – 43-46, 49, 51
  Late-type – 79, 203, 233
  Magnetic field –161
  Metal abundance – 325, 327, 389

Metal-poor star(s) – 212, 261, 262, 325, 328
Planetary systems – 61, 65, 101, 119
Pre-main-sequence – 65, 355
Protostar – 271-274, 285, 366, 419
Pulsating stars – 6
Red giant – 43-47, 50, 52, 220, 384, 392
Stellar activity – 6, 16, 23, 109, 45, 49-52, 61, 79-80, 151-152
Stellar variability – 8, 18-21, 89-92
Variable star – 12, 15, 65, 355
White dwarf – 132, 252, 381-383, 389-392
X-ray activity – 79

Stationary rotation 179-184
Super-Earth(s) – 28-29, 31-32, 159, 161, 230, 232, 499, 503-505
Supernova – 325, 327
Symplectic integrator – 392, 465
Synchronization – 138, 179-185
Synthetic observation – 313

Target-of-opportunity – 37
Techniques:
  Gravitational lensing – 25, 35, 36
  Interferometric – 61, 119
  Interferometry – 71, 89
  IR(near-IR,mid-IR) – 72, 139, 171, 353, 355-358, 359, 360, 362-368, 373
  Photometry – 7-10, 13, 15, 17, 21-22, 45, 49-50, 65-66, 68, 83, 85, 93, 154
  Radial velocity(ies) – 10, 13, 15, 22, 25, 28-29, 43-45, 51, 53-56, 57-59, 61-63, 65, 89-90, 101-109, 115, 190, 194, 210, 215, 228-229, 262, 267, 279, 285, 293, 297, 305, 308, 313-318, 353, 361, 436-439, 441, 444-445, 447, 449, 456, 482
  Spectrograph – 10, 13-15, 22, 43-44, 49, 54, 81, 105, 210, 318, 359-361, 365, 370, 399
  Spectrum(spectra) – 8, 19, 45, 54, 89, 102-103, 112, 152-155, 171, 203, 211, 213, 252, 357, 362, 365, 369, 371, 447-448
  Transit(ing,ion) – 3-16, 17-24, 45, 83-84, 85-87, 102, 89-92, 93-100, 133, 136, 139, 142, 145, 148-149, 152-153, 187-194, 225, 234, 240, 242, 261-262, 267-270, 279, 283, 301, 305, 308, 340, 343, 348, 353, 357, 360, 363, 432, 445, 472, 499, 501
  WFCAM IR imager – 355

The Big Bang – 325-328
Thermal energy – 133, 138, 161, 183
Thermal evolution – 159-161, 503
Thermal tides – 145-149
Tidal barrier – 263-266, 383
Tidal bulge – 79, 180, 197
Tidal friction – 153, 156, 179-185, 327
Tidal heating – 148-149, 187-194
Tidal lag – 179, 181, 286
Tidal perturbation(s) – 197-200

Tidal torque – 132, 136, 179-185, 251, 263, 333, 335, 378
Tidal wave – 179
Tide harmonics – 179-181, 184

Viscosity – 132, 159-161, 197, 230, 234, 252-253, 293-299, 309, 332-341, 386, 408, 414, 487

White noise – 92, 93-100, 316-317

A two-planet system with a disk (Numerical)

This figure comes from the numerical simulation of H. Zhang & J.-L. Zhou in this proceeding(p. 413). It's of the same case of the cover figure. During their hydro-dynamics simulation the two planets first undergo convergent migration and locked in 5:3 MMR for a short while. Then the outer planet's migration reversed and under goes outward runaway migration. Red area denotes high density while blue area denotes low density.